Microsystems and Nanosystems

More information about this series at http://www.springer.com/series/11483

Michael Huff

Process Variations in Microsystems Manufacturing

Springer

Michael Huff
Corporation for National Research Initiatives
MEMS & Nanotechnology Exchange
Reston, VA, USA

ISSN 2198-0063 ISSN 2198-0071 (electronic)
Microsystems and Nanosystems
ISBN 978-3-030-40558-8 ISBN 978-3-030-40560-1 (eBook)
https://doi.org/10.1007/978-3-030-40560-1

This Springer imprint is published by the registered company Springer Nature Switzerland AG
The registered company address is: Gewerbestrasse 11, 6330 Cham, Switzerland

Dedicated to my loving wife, Deanna.

Preface

The main theme of this text is to explain the methods used to estimate and manage the device parameter variations that result from using microsystems manufacturing methods. The broadest definition of microsystems includes devices implemented using any semiconductor-based micro- and nano-fabrication techniques, including silicon integrated circuits (ICs), micro-electro-mechanical (MEMS), photonics, nano-electro-mechanical (NEMS), non-silicon integrated circuits, and others. Nevertheless, the focus of this text is on the class of microsystems devices called MEMS (and NEMS[1]). The words "microsystems manufacturing" were purposefully selected for the title due to the fact that the major topics covered herein are also generally applicable to all of the other types of microsystems technologies.

A companion volume on microsystems manufacturing is directed to the topic of *process integration* techniques for MEMS devices. Managing process variations and process integration are both complex issues, and therefore, it is believed that each of these areas deserves a sufficiently comprehensive coverage in the form of a dedicated volume in order for the reader to fully appreciate and understand these topics. Both of these subjects are critically important for engineers to understand in order to be successful in the development and production of MEMS or any other type of microsystems.

There are several reference books and handbooks that have become available over the past decade that describe the fabrication methods that are commonly used in the implementation of microsystems technologies. Notably, many of these texts cover a number of individual processing steps used in the fabrication of

[1]MEMS and NEMS will be grouped together under the term "microsystems" and used throughout this book to describe the collective of devices and systems that are manufactured using both micro-scale and nano-scale fabrication methods. The reasons are as follows: First, it can be difficult to differentiate these technologies from one another solely based on the dimensional feature sizes. Second, it is awkward to use both the combined term "micro- and nano-systems" as a label for these technologies. Third, and perhaps most importantly, all of these technologies basically use the same set of fabrication techniques regardless of the exact feature sizes involved.

microsystems devices and go into some depth to explain the basic physics behind each of these processing steps. Therefore, a valid question is whether new books are needed or desirable.

The answer to this question is "yes" since the existing references mostly fail to adequately cover two extremely important topics related to manufacturing of microsystems, namely, how to estimate and manage the resultant device parameter variations that are a result of the manufacturing processes used with the purpose of obtaining acceptable production yields (i.e., percentage of number of working devices meeting the performance requirements out of total number of devices made) and how to take a series of individual processing steps and combine them into a process sequence such that the production yields are acceptable.

Process variations are the differences in the dimensions and material properties of the devices being produced from the desired ideal values developed as part of the device design. While every manufacturing operation strives to replicate devices with exacting control of the dimensions and material properties, in reality, each device made will have slight and often measureable differences in the parameters. Surprisingly, this topic is rarely, if ever, mentioned in existing microsystems books despite the fact that it is an exceedingly important issue. One of the reasons for this omission is that there is a widely held misconception that fabricating small devices equates to high-precision manufacturing. This is not true. Instead, when the dimensional variations of the device's critical features are expressed as a ratio of the desired value of the same feature, microsystems fabrication methods are almost always comparatively worse than macroscale production techniques, and they also tend to degrade, as the dimensions get smaller.

With respect to material properties, they also exhibit variations due to slight differences in the processing conditions across the substrate and across the batch of substrates. Moreover, some of the important properties of the materials used in MEMS microsystems are very dependent on the manufacturing process sequence and processing steps used in the implementation. And since most MEMS process sequences are highly customized, there can often be no a priori knowledge of the resultant properties of materials used in the development of MEMS devices.

The consequence of these variations is that if the dimensions and the material properties have variations from their "as-designed" values, the output behavior and performance of the device will also have a variation in its behavior. Furthermore, variations in the device output may result in the device not working or not meeting the required performance specifications. The outcome will likely be wastage of manufactured devices, higher production cost levels, and lowered performance levels.

Part of the reason the topics of process variations and process integration have not been previously been given much attention is that many types of microsystems technologies, such as MEMS, are relatively new and are only recently being successfully commercialized. Therefore, the knowledge of manufacturing issues for the production of MEMS is relatively limited. Additionally, academic researchers have authored many of the previous books on microsystems fabrication methods and rarely focus on the more detailed manufacturing issues.

The major goal of this book on the topic of process variations is to provide the readers with an adequate understanding to allow them to perform the following:

1. Estimate the manufacturing variations of the processing steps and process sequences used in the production of a microsystems device.
2. Recognize what metrology methods are suitable for evaluating whether processing steps performed are acceptable or not.
3. Evaluate the impact of manufacturing variations on the resultant microsystems device output behavior and performance.
4. Estimate the functional and parametric yields of microsystems manufacturing process sequences.
5. Track and analyze important metrics for ensuring that production steps are meeting pre-defined standards.
6. Understand how to modify the design of a microsystems device taking into process variations such that the yields will be acceptable in production.

This book is intended for practicing engineers and scientists working in microsystems development as well as students who are training for employment in the microsystems industry. The prerequisites for understanding this book are a basic knowledge of engineering, math, and physics. Knowledge of fabrication methods is not required but may be useful since these topics are briefly reviewed. While the material is mostly focused on MEMS manufacturing, it is important to state again that the issues of process variations is equally important to other types of microsystems. Therefore, readers working on these other microsystems technologies can also benefit from the material in this volume.

Reston, VA, USA Michael Huff

Acknowledgments

I wish to thank my beloved wife, Deanna, for all of her love and support in this endeavor. Having spent countless nights and weekends working on this book that could have been spent with her, she never complained, always offered her support and encouragement, and was always willing to help.

I sincerely thank the Corporation for National Research Initiatives (CNRI) for all of the support provided while writing this book. I particularly wish to recognize Dr. Bob Kahn and Patrice Lyons. Bob started and manages the CNRI that has been home to the MEMS and Nanotechnology Exchange (MNX) from its inception in 1998. Under Bob's leadership, the CNRI provides an environment where researchers are actively encouraged to help create and support new and impactful technologies. Patrice has always been extremely helpful with recommendations, comments, and questions.

I also thank Dr. Mehmet Ozgur who has helped me on more occasions that can be counted, as well as other members of the MNX team who have helped me over the years, including Christy Short, Dr. Michael Pedersen, Dr. David Binger, Roger Masse, Dr. William Benard, Dr. Paul Sunal, Kip Pettigrew, Andrew Kuching, Neil Schemenauer, Greg Ward, Tommy Short, Lance Oh, Dr. Bill Noonan, and Anton Benard.

My appreciation and respect are given to the editors at Springer, including Merry Stuber, Marta Moldvai, Michael McCabe, Brain Halm, and Nandhakumar Sundar. Merry first encouraged me to think about writing a book after I had contributed a chapter to a *MEMS Materials and Processes Handbook* published by Springer. The staff at Springer has been extremely helpful and accommodating and has shown great patience.

Any errors or omissions are my fault alone, and the opinions expressed herein are mine and do not necessarily reflect the views of CNRI.

Acknowledgments

[illegible]

Contents

Chapter 1
Introduction

1.1 From Vacuum Tubes to Microsystems

From the invention of the first transistor at Bell Labs in 1947 to the present time, semiconductor-based manufacturing technologies have developed at a breakneck pace resulting in enormous economic benefits and countless benefits to society. Initially, rapid scientific discoveries leading to the discovery of the bipolar junction transistor 1948, the first commercial silicon transistor in 1954, the first MOS field-effect transistor in 1959, and the first integrated circuit in 1958–59 quickly propelled the advancements in the field.

Much of the early progress was driven by the realization that semiconductor devices, in the form of discrete electrical components, such as transistors and diodes, could perform the same functions as that of expensive, large in size, heavy in weight, unreliable, and power hungry vacuum tubes. Soon thereafter, discrete semiconductor devices that were wired onto printed circuit boards replaced vacuum tubes. Later, progress was driven by the recognition that printed circuit boards composed of many discrete electronic components could be replaced by integrated circuits (ICs) whereby many active electrical devices such as transistors could be made directly into the semiconductor substrate resulting in much lower cost, much smaller size and weight, better performance, and higher reliability.

The integration of large numbers of small-scale transistors into a small chip was a tremendous improvement over the assembly of circuits using discrete electronic components. By the early 1970s, continued improvements in fabrication methods led to large-scale integrated circuits (LSI), including the first microprocessor by Intel in 1971, namely, the 4004. Thereafter, advancements in the manufacturing of integrated circuits that greatly improved reliability and performance, while simultaneously reducing production costs and device sizes, ensured the rapid adoption of ICs in numerous applications. Since then, the field has followed a growth curve called "Moore's Law," whereby the number of transistors in an integrated circuit approximately doubles every 24 months.

M. Huff, *Process Variations in Microsystems Manufacturing*, Microsystems and Nanosystems, https://doi.org/10.1007/978-3-030-40560-1_1

At present integrated circuit fabrication has evolved into an extremely sophisticated technology with the availability of highly advanced systems composed of hundreds of millions of individual transistors each having gate lengths of 10's of nanometers and working together as a high-performance system integrated onto a single substrate. These devices are being used in almost every type of industry and product that processes and/or stores information.

Much of the progress made in semiconductors and integrated circuits in the early decades was the direct result of focused support from the US Federal Government, specifically from the Department of Defense (DoD) and National Aeronautics and Space Administration (NASA). For example, the US military was a massive early supporter of IC development for application in the Minuteman ICBM program. This and other programs resulted in the US military becoming the dominant consumer of integrated circuits in the 1960s and 1970s. Consequently, the early enormous investments made by the US Government paid back huge dividends to the military and space programs but have also resulted in orders of magnitude with larger dividends to the commercial sector, including radically improved lifestyles for the world's citizens. Indeed, these federal investments have been so successful that IC manufacturing is now almost entirely driven by the commercial sector. Consumers have experienced enormous economic benefits, as well as significant lifestyle improvements as a result of the development of integrated circuit technology.

The technological and economic growth of integrated circuits continues to be hugely successful. Moreover, the same manufacturing methods that have resulted in the tremendous successes of ICs are also enabling enormous technological advances in other areas of microsystems. Specifically, microsystems manufacturing is used to implement a very diverse collection of different technologies, including silicon integrated circuits; non-silicon integrated circuits (e.g., GaAs circuits, GaN devices, MMICs, etc.); micro-electromechanical systems (MEMS); nano-electromechanical systems (NEMS); photonics; vacuum electronics; microfluidics; lab-on-a-chip; and nanotechnology. This volume focuses on MEMS manufacturing issues related to device parameter variations.

1.2 MEMS Microsystems

A general definition of MEMS microsystems has been somewhat elusive even by experts due to the fact that by its very nature the technology is extremely broad and diverse. Nevertheless, MEMS is a technology that in its most general form can be defined as microminiaturized electromechanical devices, mechanical devices, or structures that are implemented using the techniques of micro- and nano-fabrication.

The physical dimensions the components used to implement MEMS devices usually vary from below one micron on the lower end of the dimensional spectrum all the way up to several millimeters for larger components. Likewise, the types of devices encompassed by MEMS can vary from relatively simple structures wherein there are no mechanically moving elements to extremely complex electromechanical systems with multiple moving elements under the control of microelectronics integrated onto the same substrate. Obviously, the one main criterion of MEMS is that there are at least some elements on the substrate having some sort of mechanical functionality whether or not there is any element that is movable.

Common types of MEMS devices are miniaturized sensors and actuators, which are frequently called microsensors and microactuators. Both of these elements are properly categorized as "transducers," which are defined as devices that convert energy from one form to another. In the case of microsensors, the device typically converts a measured mechanical signal into an electrical signal that is representative of the mechanical signal. Importantly, the conversion of a signal from the mechanical to the electrical domain allows the information to be more readily processed and stored for increased system functionality. Over the past several decades, MEMS researchers and developers have demonstrated a wide variety of microsensors for almost every possible sensing modality including temperature, pressure, inertial forces, chemical species, magnetic fields, radiation, etc. [1–11].

Remarkably many of these micromachined sensors have demonstrated performances exceeding those of their macroscale counterparts. For example, the micromachined version of a pressure transducer usually outperforms a pressure sensor made using conventional macroscale manufacturing techniques [12]. Not only is the performance of MEMS devices exceptional, but also their method of production leverages the same batch fabrication techniques employed in the integrated circuit industry – which can translate into low per-device production costs, smaller size and weight, lower power consumption, higher reliability, as well as many other benefits. Consequently, it is possible to not only achieve exceptional device performance but to do so at a relatively low-cost level. Not surprisingly, silicon-based discrete microsensors are being rapidly commercially exploited.

MEMS researchers have also demonstrated a number of microactuators including microvalves for control of gas and liquid flows; optical switches and mirrors to redirect or modulate light beams; independently controlled micromirror arrays for displays; microresonators for a number of different applications; micropumps to develop positive fluid pressures; microflaps to modulate airstreams on airfoils; as well as many others [5, 7, 13].

Surprisingly, even though these microactuators are extremely small, they frequently can cause effects at the macroscale level, that is, these tiny actuators can perform mechanical feats far larger than their size would imply. For example, researchers have located small microactuators on the leading edge of airfoils of an aircraft and have been able to steer this aircraft using only these microminiaturized devices [14].

Advances in MEMS technology promise to revolutionize commercial, medical, and industrial products by bringing together the computational and communication capabilities of integrated circuits with the perception and control capabilities of microsensors and microactuators, thereby enabling *smart systems-on-a-chip* to be mass-produced. The revolutionary potential of microsystems becomes fully realized when these microminiaturized sensors and actuators are combined onto a common substrate along with integrated circuits and even with other types of microsystems technologies, such as photonics and microfluidics (Fig. 1.1).

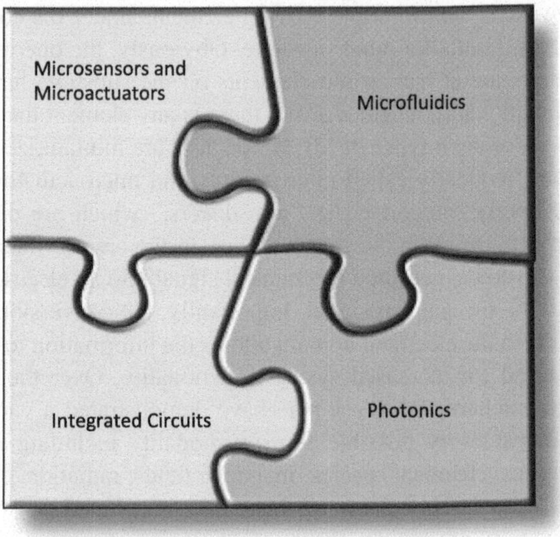

Fig. 1.1 Illustration of the merging of the various microsystems technologies

The microsensors can be used to perceive the surrounding environment (analogous to the eyes, ears, nose, etc., of the microsystem); the integrated circuits process this sensory information as well as make decisions based upon the sensor data (analogous to the brain of the microsystem); and the microactuators take actions based on instructions from the microelectronics to control or manipulate the environment (analogous to the arms and legs of the microsystem). The use of smart systems that can actively and autonomously *sense and control* their environments has far reaching implications for a tremendous number of future commercial and industrial applications and promises significant economic and societal benefits.

Until recently, the advancement of integrated circuits has been measured by the number and size of the transistors that can be packed onto a single substrate, that is, as defined by Moore's Law. However, it is now becoming increasingly clear that in the future the important defining metrics of microsystems will be how capable the microsystem is in terms of a diversity of functions including sensing, actuation, as well as information processing; how well the microsystem can understand and influence the environment; and how the microsystem can assist and/or enlighten the user. Some of the nearer term important applications that underscore these future metrics of microsystems are the use of MEMS microsystems in vehicles and the Internet of Things (IoT) [15].

1.3 Some of the Important Attributes of MEMS Microsystems

There are several unique and advantageous attributes of MEMS microsystems that make this technology distinctively important to a number of applications. First, MEMS is made using integrated circuit-like processes, and this enables the ability

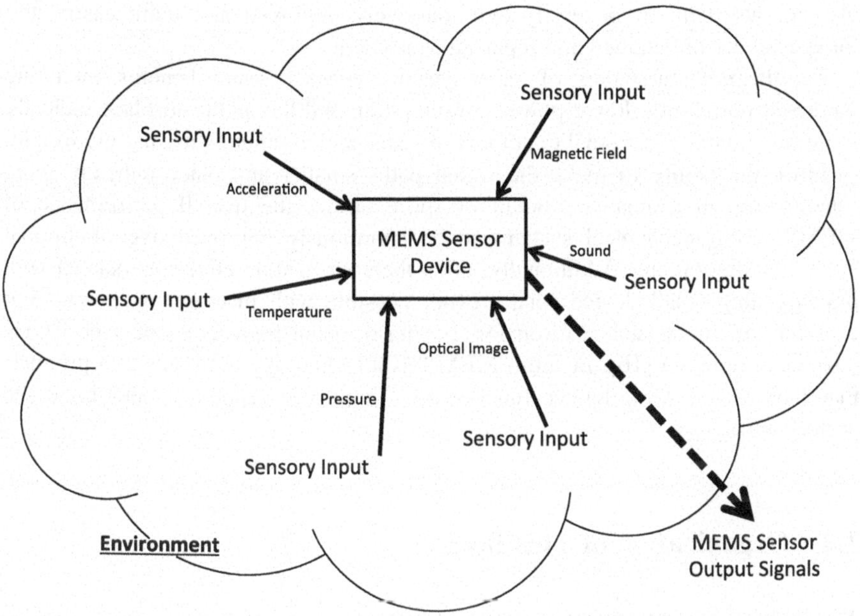

Fig. 1.2 Illustration of MEMS microsystem sensor device immersed in an environment that is receiving sensory input of various types including temperature; pressure; acceleration; optical image; sound; and magnetic field and transducing this sensory information into electrical output signals

to integrate multiple and diverse functionalities onto a single microchip. The ability to integrate microsensors and microactuators along with integrated circuits has sweeping consequences for numerous products, applications, and markets (Fig. 1.2).

Second, MEMS borrows many production techniques of batch fabrication from the integrated circuit (IC) industry, and therefore, the per unit cost of a single die containing complex electromechanical systems can be relatively low compared to alternative technologies. This is similar to the dramatic die cost reductions seen in the IC industry. Although the cost of the development and manufacturing facilities can be relatively high, the fact that this cost can be spread over many dies in batch fabrication production can drastically lower the per die cost.

Third, integrated circuit fabrication techniques coupled with the tremendous advantages of the materials used in semiconducting manufacturing, including silicon and various thin-film materials, allow the reliability of miniaturized electromechanical systems to be radically improved. It is often the case that the most expensive and most unreliable components of conventional macroscale systems are the sensors and actuators. However, if these miniaturized sensors and actuators are integrated onto a single microchip with electronics, enormous improvements in system reliability can be obtained. This is similar to the increased reliability that resulted from the transition from discrete electronics on printed circuit boards to integrated circuits. The lower cost of these miniaturized electromechanical systems also allows them to be easily and massively deployed and more easily and inexpensively maintained and replaced as needed.

Fourth, miniaturization of microsystems enables many benefits including increased portability, lower power consumption, and the ability to place radically more functionality in a smaller amount of space and without increasing the weight.

Fifth, the ability to make the signal paths smaller and place radically more functionality in a smaller amount of space allows the overall performance of MEMS electromechanical systems to be enormously improved over traditional system configurations. Additionally, often there are scaling effects associated with implementing small devices that are not possible with macroscale devices. For example, micromechanical resonators having resonant frequencies of over 1 GHz have been reported [16]. In short, MEMS as a technology translates into products that have lower cost, higher functionality, improved reliability, and increased performance.

1.4 Organization of This Book

The organization of this book is reviewed in this section. In Chap. 2, we will provide a review of many of the various transduction mechanisms used in MEMS-based microsensors and microactuators. The intent here is to give a basic review of the mechanisms of MEMS devices.

Chapter 3 focuses on a review of the various fabrication processing methods commonly used in the IC microsystems manufacturing industry. Most of the process steps are explained relatively briefly with sufficient coverage to provide a basic understanding of how these processes work and the relationships of the important process parameters. Importantly, representative dimensional variations that can be reasonably expected for each of the processing steps covered are provided.

Chapter 4 reviews those processing steps that are not commonly used in the IC microsystems manufacturing industry but instead are specific to MEMS technologies. As in Chap. 3, these processing steps are briefly reviewed, and representative dimensional variations of each processing step are given.

Chapter 5 reviews most of the metrology techniques used in microsystems manufacturing to inspect, measure, and characterize substrates, devices, critical features of devices, and materials throughout the manufacturing processes. A general explanation of each type of metrology capability will be given as well as the measurement resolution, precision, and accuracy that can be expected with actual metrology equipment, if they are known.

Chapter 6 reviews the important subject of microsystems materials. The materials used in microsystems implementation are a very complicated issue since some of the important material properties commonly used in manufacturing are highly dependent on the specifics of the processing step used to deposit the material layers and the process sequence. A consequence of this fact is that often the material properties are not known when the design is performed and may not be able to be determined until they are used in the exact same processing conditions as the complete process sequence and measured. Representative values of some of the more important material properties for different processing steps and the associated variations in these values of material properties are reviewed.

Chapter 7 discusses the topic of process integration for microsystems manufacturing. What is involved in process integration as well as representative examples of process sequences and process technologies used in MEMS microsystems manufacturing is reviewed. Given that the device parameter (i.e. both device dimensions and material properties) variations are largely determined by the process sequence used to implement the device as well as the individual processing steps, information about the types of variations seen in a manufacturing process sequences are reviewed in an illustrative example. The material covered in this chapter will not provide detailed instruction on how to perform process integration since this topic is very complicated and a more complete coverage would easily constitute a complete volume on its own. Therefore, the reader will be directed to a second volume on microsystems manufacturing dedicated to the topic of process integration.

Chapter 8 explains the most important topic of this volume, specifically device parameter variations in microsystems manufacturing and how these variations impact device output behavior and performance. It will be shown that the expected relative variations in device parameters, such as the device dimensions, are commonly larger when using microsystems manufacturing compared to macroscale manufacturing methods. After reviewing some of the definitions and statistics used in analyzing variations, techniques to estimate and analyze the variations and their impact on the device behavior will be explained. To the extent possible, this chapter will use real-world examples of MEMS devices.

Chapter 9 focuses on manufacturing yield and statistical process control (SPC). Manufacturing processes often desire to optimize the yields since this has a direct impact on the device costs. This will be followed by covering the various techniques that are used to estimate yields along with the methods available for determining whether each process step in a manufacturing process is acceptable or not. Various concepts about how yields are impacted will also be reviewed.

Chapter 10 discusses the topic of methods that can be used in design to manage the impact of the device parameter variations so that the output behavior and performance is able to meet the required specifications. This chapter explains each of the methods using real MEMS device examples. Methods to optimize the cost function for MEMS manufacturing are also covered.

If the reader compares the structure of this volume to other reference sources on MEMS or other microsystems for that matter, they will immediately notice that the material appears to be presented in the inverse order. That is, design methods of microsystems are covered last. This is very purposeful and is due to the belief that in order to be able to perform microsystems design, the developer must first have knowledge and understanding of the fabrication issues and in particular the manufacturing device parameter variations.

References

1. C.Y. Change, S.M. Sze, *VLSI Technology, ULSI Technology* (McGraw Hill, New York, 1996)
2. S. Wolf, R.N. Tauber, *Silicon Processing for the VLSI Era, Volume 1- Process Technology* (Lattice Press, Sunset Beach, 1986)
3. S.A. Cambell, *The Science and Engineering of Microelectronic Fabrication* (Oxford Press, New York, 1996)
4. P. van Zant, *Microchip Fabrication* (McGraw Hill, New York, 2000)
5. M. Madou, *Fundamentals of Microfabrication*, 2nd edn. (CRC Press, Boca Raton, 2002)
6. S.D. Senturia, *Microsystem Design*, 2nd edn. (Springer, New York, 2004)
7. G.T.A. Kovacs, *Micromachined Transducers Sourcebook* (McGraw-Hill, New York, 1998)
8. N. Maluf, *An Introduction to Microelectromechanical Systems Engineering*, 2nd edn. (Artech House, Boston/Norwood, 2004)
9. L. Vivoen, L. Pavesi, *Handbook of Silicon Photonics* (CRC Press, Boca Raton, 2013)
10. C. Ashby, A. Baca, *Fabrication of GaAs Devices* (IEE, London, 2005)
11. M.A. Huff, S.F. Bart, P. Lin, Chapter 14: MEMS process integration, in *MEMS Materials and Processing Handbook*, ed. by R. Ghodssi, P. Lin, (Springer Press, New York, 2012)
12. S. Sze, *Semiconductor Sensors* (Wiley, New York, 1995)
13. M.A. Huff, M.S. Mettner, T.A. Lober, M.A. Schmidt, A wafer-bonded electrostatically-actuation microvalve, Solid-State Sensor and Actuator Workshop, 4th Technical Digest IEEE (1990)
14. P.H. Huang, C. Folk, C. Silva, B. Christensen, Y.F. Chen, G.B. Lee, Chen, Minjdar, S. Newbern, F. Jiang, C. Grosjean, C.M. Ho, Y.C. Tai, Applications of MEMS devices to delta wing aircraft: From concept development to transonic flight test, AIAA Paper No. 2001-0124, Reno, Nevada, January 8–11, 2001
15. M. Huff, MEMS: An enabling technology in the Internet of Things (IoT), in *IoT & Data Analytics Handbook, 1e*, (Wiley, New York, 2017)
16. L.C. Popa, D. Weinstein, Switchable piezoelectric transduction in Al-GaN/GaN MEMS resonators, 17th Transducers, p. 2461, June 2013

Other Information

The reader is referred to three very popular additional sources of information concerning MEMS technology. The first source is a Web site called the MEMS and Nanotechnology Clearinghouse which is located at http://www.memsnet.org and is a general informational portal about MEMS technology and includes events, news announcements, directories of MEMS organizations, and a MEMS material database. The second source is the MEMS and Nanotechnology Exchange (MNX) located at http://www.mems-exchange.org. This web site represents a large MEMS foundry network and offers MEMS design, fabrication, packaging, product development, and related services as well as considerable information about MEMS and Nanotechnologies. Lastly, the reader is referred to several electronic discussion groups concerning MEMS technology that have very active participation from several thousand MEMS developers and researchers from around the world. These groups can be accessed through the following URL: http://www.memsnet.org/lists/

Chapter 2
An Overview of MEMS Microsystems

2.1 Introduction

This chapter provides an explanation of some of the physical phenomena employed to implement the electromechanical transduction mechanisms for MEMS devices, specifically microsensors and microactuators. Additionally, this chapter will discuss some of the methods used to manufacture MEMS and explain that MEMS manufacturing typically uses fabrication methods from the integrated circuit (IC) industry and combines them with various specialized fabrication techniques commonly called micromachining. The important concept of batch fabrication that is used in both IC and MEMS manufacturing will then be explained including the significant economic benefits derived from this method of manufacturing. Next, this chapter provides a review of some of the important definitions regarding microsystems manufacturing processes. Lastly, this chapter reviews some of the important material properties of silicon that are relevant to MEMS microsystems manufacturing.

2.2 Microsensors and Microactuators

Microsensors and microactuators are important types of MEMS microsystem devices. Both microsensors and microactuators are classified as transducers, that is, devices that convert energy from one form to another. These are various physical mechanisms used in microsystems devices to implement transduction functionality that are reviewed in this section.

M. Huff, *Process Variations in Microsystems Manufacturing*, Microsystems and Nanosystems, https://doi.org/10.1007/978-3-030-40560-1_2

2.2.1 *MEMS Microsensors*

A very large and diverse number of MEMS microsensors have been demonstrated using micro- and nano-fabrication techniques including pressure sensors; acoustic sensors; temperature sensors; infrared sensors and focal plane imaging arrays; inertial sensors for acceleration; inertial sensors for rate of rotation; magnetic field sensors (Hall, magnetoresistive, and magnetotransistors); force sensors; tactile sensors; strain sensors, optical radiation sensors; radiation sensors; chemical sensors; biological species sensors; and many others [1–5].

There are several basic physical transduction principles used in microsensors including resistive; piezoresistive; capacitive; piezoelectric; tunneling; magnetic; photoconductive; thermoelectric; and diodes. All of these sensing principles and more have been successfully demonstrated in MEMS microsystems for many applications. There are far too many variations of MEMS microsensors to review them all, and therefore we will review only a few of the more important examples. The reader is referred to references [1–5] for more information.

2.2.1.1 Resistive

A resistor is one of the most simple but useful elements used in MEMS microsensors. The basic structure is shown in Fig. 2.1 wherein an element of material having height, h, width, w, and length, L (each having units of meters) has a current I (units of amps) passing through the element from left to right. This current can be established by applying a voltage potential across the length of the element. The resistance, R (units of ohms or Ω), of the element is given by:

$$R = \left(\frac{\rho}{H}\right)\left(\frac{L}{W}\right). \tag{2.1}$$

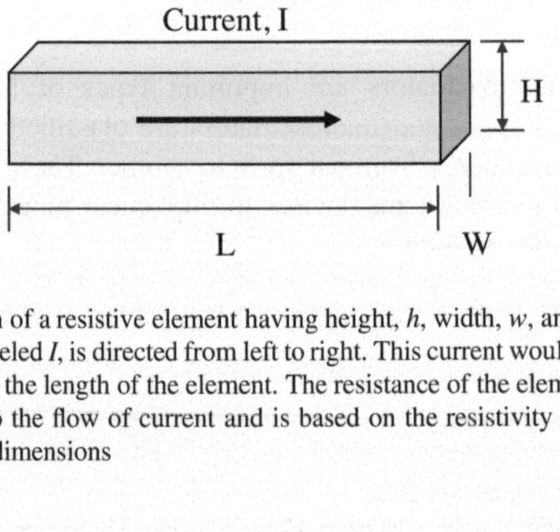

Fig. 2.1 An illustration of a resistive element having height, h, width, w, and length, L. As shown, an electrical current, labeled I, is directed from left to right. This current would be based on a voltage potential applied across the length of the element. The resistance of the element is a measure of the material's opposition to the flow of current and is based on the resistivity of the material and the element's geometrical dimensions

where ρ (units of $\Omega{\cdot}$m) is the resistivity of the material and is an intrinsic property of the material that relates to how much the material resists the flow of current through it. Equation (2.1) can be rewritten as follows:

$$R = \left(\frac{\rho}{H}\right)(\text{Number of squares}) = (\text{sheet resistance}) \times (\text{number of squares}), \quad (2.2)$$

where the (Number of Squares) is a dimensionless quantity equal to (L/W) and the quantity (ρ/H) is a quantity termed the sheet resistance. The benefit of Eq. (2.2) is that it separates the sheet resistance, which is material and process dependent from the number of squares, which is layout dependent. Additionally, once the sheet resistance is known, it is a simple matter to calculate the number of squares on the layout design and thereby the resistance value.

The resistivity of a material has a temperature dependence that can be used for the implementation of thermal sensing capability. Therefore, the transduction mechanism is for the resistor to absorb thermal energy that results in a change in the resistivity value and therefore the resistance of the sensing element. This change in resistance can be measured electrically with a simple circuit. For example, a thin-film metal layer can be deposited and patterned into a resistive element. Most metal materials have a near linear resistance versus temperature relationship over a wide range of temperatures and that is a desirable trait for temperature sensors.

In order to reduce the thermal mass, the resistive sensor element can be thermally isolated. Thermal isolation can be achieved by suspending the resistive sensor over the substrate or by removing some or all of the substrates under the sensor element. The effect of decreasing the thermal mass will be to decrease the thermal capacitance of the device, thereby resulting in a faster thermal response time, and make the sensor more responsive to smaller thermal fluctuations.

An illustration of one method to reduce the thermal mass of a temperature sensor is illustrated in Fig. 2.2, which shows a plan and cross-sectional view of a silicon substrate with a metal line on the surface of the substrate that has been patterned into a meandering resistor element that acts as a temperature sensor. A portion of the silicon substrate under the sensor has been removed using micromachining techniques to significantly lower thermal mass. Micromachined structures like that shown in Fig. 2.2 are common in MEMS microsystems. This type of structure can be useful as a temperature sensor, a hot-wire flow sensor, or an infrared sensor, and others.

Resistive sensing elements can also be made using a conductive material such as metal or polycrystalline silicon that can be deposited and patterned. Resistive elements can also be made from bulk silicon.

The advantages of the resistive effect for sensing transducers are that this principle is very simple and it can be implemented without exotic materials. One issue of concern with resistive elements made from semiconductors, such as single-crystal silicon or polycrystalline silicon, is that the resistor will be very sensitive to any mechanical strains. In comparison, metal temperature resistors tend to have a resistivity that is not as sensitive to strain.

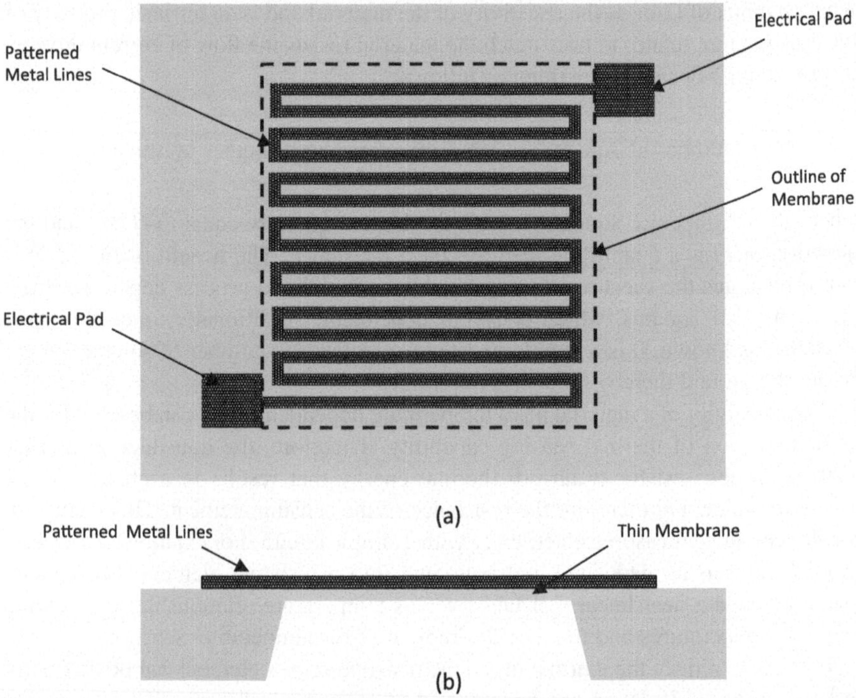

Fig. 2.2 An illustration of a resistive element fabricated onto a thin silicon membrane made by using micromachining methods to remove the bulk of the silicon substrate from underneath the sensing element. This structure can be used for temperature sensing, flow sensing, infrared sensing, and more. (**a**) Plan view. (**b**) Cross section

2.2.1.2 Piezoresistive

A piezoresistive material is one in which the resistance is influenced by applied mechanical strains. This phenomenon is prominent in semiconductors where the strain induces changes in the electronic band structure of the material, thereby making the electrical carrier scattering rates dependent on the direction of transport. This effect is useful for a variety of sensors by placing the piezoresistive elements at a position where a strain is exhibited. With a piezoresistive sensing element, the transduction mechanism is a mechanical strain that results in a change in resistance that can be measured electrically.

A useful quality factor of piezoresistive materials is the gauge factor (GF), which is a dimensionless quantity given by the normalized change in resistance divided by the strain as follows:

$$\text{Gauge factor (GF)} = \left(\frac{(\Delta R)}{R\varepsilon}\right), \tag{2.3}$$

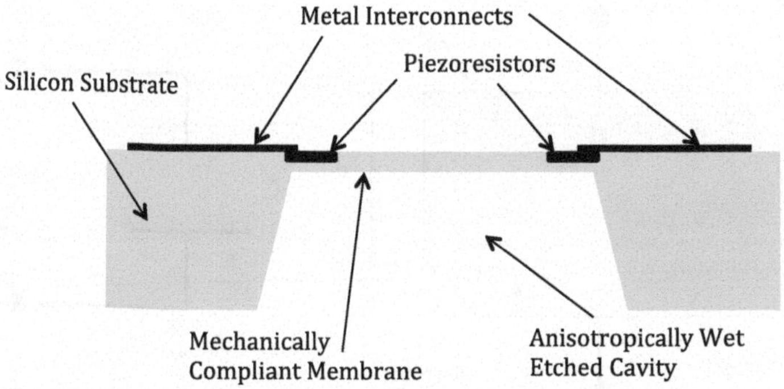

Fig. 2.3 Cross-sectional illustration of a silicon micromachined piezoresistive pressure sensor

where ΔR is the change in resistance (in units of ohms), R is the nominal value of resistance (in units of ohms), and ε is the strain. The strain for an element of length, L, with a tensile force applied is a dimensionless quantity and is given by $\Delta L/L$. Bulk silicon, depending on crystal orientation and doping type and level, has a very high gauge factor approaching 200 in some special configurations [5]. In comparison, metal resistors typically have gauge factors of only 2–5.

Piezoresistors are commonly employed as strain measurement sensors wherein the resistor (strain sensing element) is fabricated on a mechanically compliant surface or structure as such illustrated in Fig. 2.3, which is an illustration of the cross section of silicon micromachined piezoresistive pressure sensor. As can be seen, the silicon substrate has been micromachined so as to remove a section of the substrate to form a thin silicon membrane that is mechanically compliant. Under a uniform pressure loading (not shown in figure), the thin membrane will deflect, thereby resulting in a strain in the membrane. The piezoresistive elements are located on the membrane where the strain is maximized so as to optimize the sensor sensitivity to applied pressures.

Micromachining manufacturing techniques enable the piezoresistive sensing elements to be fabricated in the silicon substrate, which has many important benefits such as lower manufacturing cost, smaller size and weight, and better performance [6]. Additionally, piezoresistors do not require any special materials.

2.2.1.3 Capacitive

The capacitive effect is often used as a transduction method in microsensors due to its inherent simplicity and high performance. The basic equation for a parallel-plate capacitor (ignoring fringing fields) is given by:

$$C = \frac{\varepsilon_0 \varepsilon_r A}{d}, \tag{2.4}$$

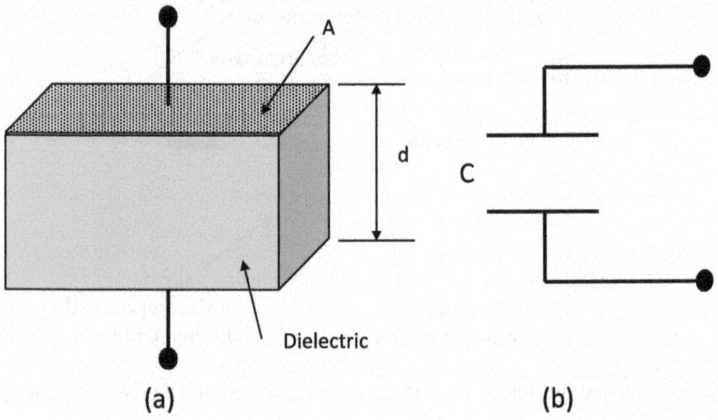

Fig. 2.4 Illustration of a capacitor. In (**a**) is shown a 3-D pictorial of a capacitor having a dielectric between the two electrodes, each having area of *A* and separated by a distance *d*. In (**b**) is shown the electrical circuit diagram of a capacitor

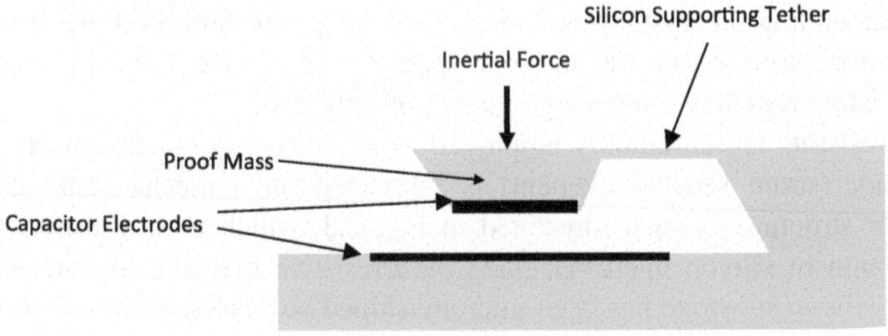

Fig. 2.5 Illustration an inertial sensor using the capacitive effect

where C is the capacitance (in units of farads), ε_o is permittivity of free space (units of farads per meter), ε_r is the relative permittivity of the material between the capacitor electrodes and is dimensionless, d is the separation between the capacitor electrodes (in units of meters), and A is the area of the electrodes (in units of meters^2). An illustration of a capacitive structure is shown in Fig. 2.4. Figure 2.4a shows a 3-D view of a capacitor with a dielectric material between the two electrodes, each of area A. Figure 2.4b shows an electrical circuit equivalent of the capacitor. Capacitors can be used in several different ways as a transducer element by designing a structure that allows the capacitance to be altered using a mechanical movement of some part of the device that will vary its capacitance.

Figure 2.5 shows how the capacitive effect can be used in a MEMS microsensor for measuring inertial forces. A proof mass is micromachined from silicon with a supporting tether that is mechanically compliant. A sense capacitor is formed by a top electrode fabricated on the bottom of the proof mass and bottom electrode made on the substrate surface. Applied inertial forces to the proof mass result in it moving downward, thereby decreasing the gap spacing between the electrodes of the capacitor and varying the capacitance.

Among the advantages of capacitive transduction include: it is relatively simple to implement; no exotic materials are required; and can be easily integrated with electronics. Capacitive-based microsensors also have a high degree of mechanical movement measurement precision. Capacitive sensors also have a very low amount of temperature sensitivity and very low noise levels.

2.2.1.4 Piezoelectric

Another widely used physical phenomena for making MEMS microsensors is the piezoelectric property that is exhibited in some materials. Piezoelectricity is when a mechanical strain produces an electrical polarization (i.e., voltage potential) in the material. Therefore, piezoelectric materials can be used as the transduction mechanism in microsensors to convert a mechanical strain into a voltage potential across the material that can be measured. Materials that lack a center of symmetry such as quartz, lead zirconate titanate (PZT), or zinc oxide (ZnO) are piezoelectric and are commonly used in microsensors.

To calculate the amount of charge generated by an applied strain, we can write:

$$\Delta Q_i = d_{ij}\Delta F_j = d_{ij}\Delta\sigma A \tag{2.5}$$

where ΔQ is the change in charge (in units of C); d_{ij} are constants that are referred to as the piezoelectric charge coefficients (units C/N) and are related to the amount of charge at the surface of the material generated on the i axis by an applied force on the j axis; ΔF is the change in force (in units of N); $\Delta\sigma$ is the change in applied stress (units of Pascals); and A is the area (units of meters^2). Since the electrical potential is given by charge divided by the capacitance or $V = Q/C$, Eq. (2.5) can be rewritten as:

$$\Delta V_i = \frac{d_{ij}\Delta F_j x}{\varepsilon_o \varepsilon_r A}, \tag{2.6}$$

where ε_o and ε_r are the free space and relative permittivity of the piezoelectric material and V is the voltage (in units of volts). Equation (2.6) allows the voltage potential generated by the applied force to be calculated.

Figure 2.6 illustrates the transduction mechanism of the piezoelectric effect wherein at the top is shown a piezoelectric material layer that is unstrained having an electrode fabricated on both sides. The piezoelectric layer and electrodes are fabricated on a mechanically elastic layer.

At the bottom of Fig. 2.6, an applied stress (or force) loading is applied normally to the piezoelectric material layer. This will result in some small amount of strain in the piezoelectric material and a resultant charge generated on the electrodes of opposite polarity. This charge results in a voltage potential across the electrodes that can be measured electrically.

To implement in a microsensor, the piezoelectric material layer could be fabricated on the top surface of a mechanically compliant membrane or flexure, similar to

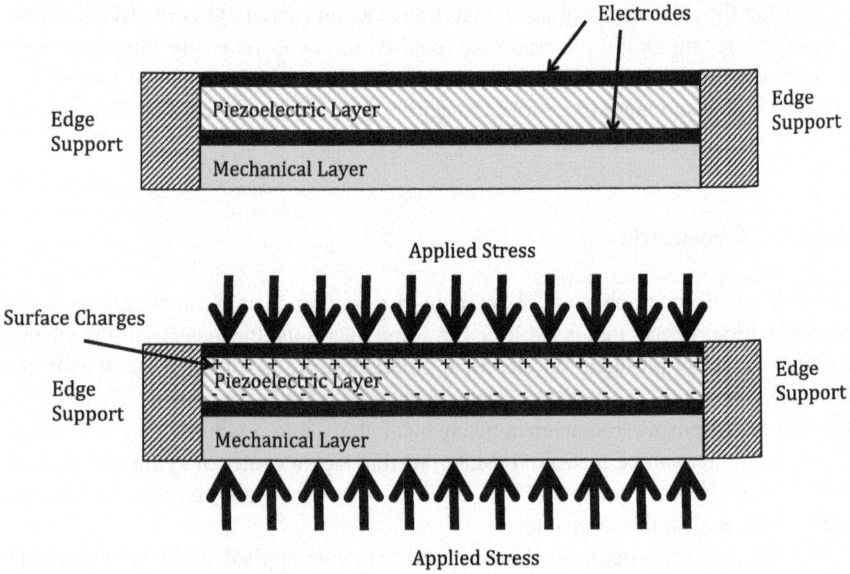

Fig. 2.6 Illustration showing a piezoelectric material layer fabricated between two electrodes positioned on an elastic layer on top. Bottom shows an applied stress to the piezoelectric stack with a resultant developed charge

Fig 2.3 or 2.5, to measure the strain of the mechanical element due to a loading effect (e.g., applied pressure, inertial forces, etc.).

The advantages of piezoelectric sensing mechanisms are that they can be very sensitive and exhibit high electrical and mechanical bandwidths depending on the device design. There are also several disadvantages for the use of piezoelectric materials for MEMS microsensors. First, use of these materials greatly increases the complexity of fabrication processes. Second, piezoelectrics are best characterized as an exotic material choice and will require access to or development of a suitable piezoelectric deposition process. Third, many piezoelectric materials are composed of complex chemical stoichiometry that results in difficult to control deposition processes. Fourth, microsensors made with piezoelectric materials can exhibit a very high level of temperature sensitivity, and therefore care must be exercised in the design and readout electronics to compensate for this effect.

2.2.1.5 Tunneling

The tunneling effect can be used to make extremely precise instruments to scan and image the surfaces of materials at the atomic scale [7]. The working principle is based on a very small radius needle-shaped tip (as small as about one atom at the apex of the tip) that is comprised of a conductive material such as a metal (e.g., tungsten, gold, platinum–iridium, etc.) that is positioned in very close proximity to the material surface to be scanned, so close that electrons can tunnel over the free space gap between the tip and surface (Fig. 2.7). The sharp tip is commonly made using micromachining fabrication techniques.

Using piezoelectric actuation (not shown in figure), the tip is located in very close proximity to a material surface that is to be scanned. If a biasing voltage (labeled "V_{Bias}" in Fig. 2.7) is placed across the gap (labeled "z" in Fig. 2.7) between the tip and material surface and the tip is sufficiently close to the material surface (e.g., within about 5–10 Å), a quantum mechanical tunneling current (labeled $I_{tunneling}$ in Fig. 2.7) passes across the gap between the tip and the material surface. A circuit can be used to measure the tunneling current through the tip and thereby the distance between the tip and surface as it scans across the material surface. This spatially varying tunneling tip current can be used to create an image of the material surface at the atomic dimensional size scale. Alternatively, the tip current is maintained at a constant value, and the actuator voltage to maintain a constant gap distance is measured.

The typical tip tunneling current is on the order of a few hundred pico-amps at a bias voltage of about 1 V. This instrument is extremely sensitive based on the fact that the tunneling current is exponentially dependent on the gap between the tip and material surface; specifically the tunneling current is proportional to e^{-kz}, where k has a value of about 2.2 1/Angstrom and z is the gap between the tip and surface. Consequently, the tunneling current increases by a factor of 10 for each Angstrom the closer the tip gets to the surface. The resolution of the scanning tunneling microscope is approximately 0.1 nm in the lateral (i.e., x and y coordinate) directions and about 0.01 nm in the orthogonal (i.e., z coordinate) direction [7].

Tunneling can be used as a transduction mechanism by using the tip to measure the displacement of a mechanical element. For example, the tip could be located in the middle of the thin membrane as shown in Fig. 2.3 to measure the upward movement of the membrane under pressure loading. Given the extreme sensitivity of the tunneling current to tip displacement relationship, the resolution of the deflection measurement can be very high.

The advantages of tunneling tip microsensors are that they provide for extreme levels of sensitivity compared to almost any other transduction mechanism. However, the disadvantages of this mechanism must also be considered, including a complicated process to make tunneling tips; the tips wear quickly over time; the tunneling effect is highly nonlinear; a complicated closed-loop feedback control is

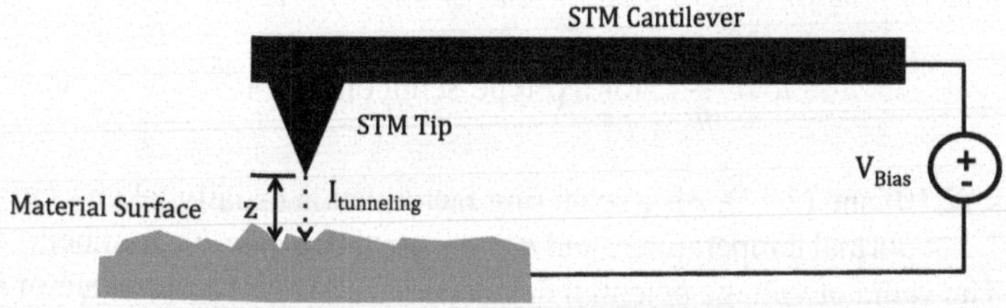

Fig. 2.7 Illustration of a scanning tunneling microscope (STM)

required; and while the sensitivity of this method is extremely good, the selectivity, that is, the ability of the sensor to differentiate an actual signal from a parameter of interest from another parameter, can be challenging.

2.2.1.6 Magnetic

There are a number of magnetic effects that can be used as transduction mechanisms, including Hall effect, magnetoresistive, magnetotransistors, and others. The Hall effect has been the most often used magnetic phenomena for sensing in microsensors, and therefore this method will be described.

A Hall effect sensor can be composed of a semiconductor material such as silicon and is based on the Lorentz force, $\overline{F}$ (units N), placed on moving charges according to the following relationship:

$$\overline{F} = q\overline{E} + q\overline{v} \otimes \overline{B}, \tag{2.7}$$

where q is the charge (units C), $\overline{E}$ is the electric field (units V/m), $\overline{v}$ is the charge carrier velocity (units m/s), and $\overline{B}$ is the magnetic field (units of T). The force, electric field, velocity, and magnetic field are all vector quantities. If there is no electric field, the first term drops out giving:

$$\overline{F} = q\overline{v} \otimes \overline{B}. \tag{2.8}$$

Since the cross product of the velocity and magnetic field is taken to calculate the force, the force will be orthogonal to both of these vectors.

In the case of a semiconductor with one dominant charge carrier type, a voltage will develop across the material laterally that is related to the longitudinal current density, J (units of amps/meter^2), and applied magnetic field, B, as follows:

$$V = RJBW, \tag{2.9}$$

where W is the semiconductor sample width (in units of meters) and R is the Hall coefficient (in units of meters^3/C) that depends on carrier type and is given by:

$$R = -\frac{r}{qn}, \quad \text{for an n-type semiconductor, and} \tag{2.10}$$

$$R = \frac{r}{qp}, \quad \text{for a p-type semiconductor.} \tag{2.11}$$

In Eqs. (2.10) and (2.11), r is a scattering factor that is usually taken as given by a value of 1.2 at normal temperatures, and n and p are the respective numbers of charge carriers. The value of voltage potential developed across the semiconductor is called the Hall voltage, and an illustration of the relationship between the fields and this voltage on the sample is shown in Fig. 2.8.

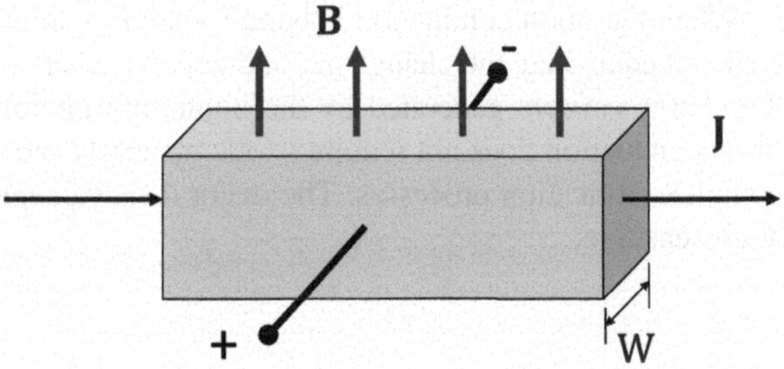

Fig. 2.8 Illustration of the Hall voltage developed on the sides of a semiconductor based on a current density passing longitudinally along the sample and an applied magnetic field directed vertically

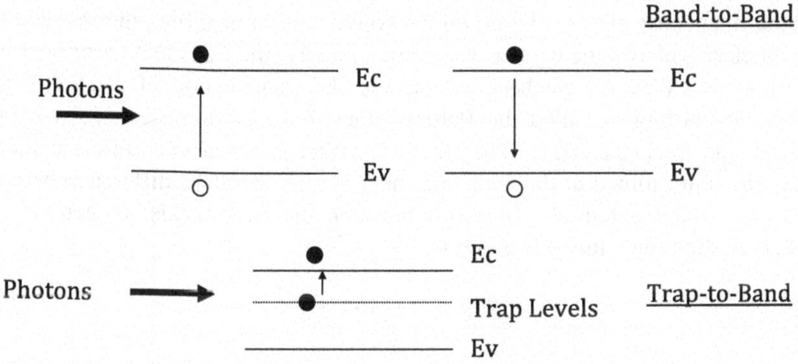

Fig. 2.9 Illustration of the photoconductor effect in a semiconductor

A microsensor using the Hall effect can be used to measure the magnetic field strength. The sensor is usually configured in a bridge or four-point measurement arrangement to negate any contact resistance issues. The Hall effect is relatively simple to implement in a MEMS microsensor and does not require exotic materials or fabrication processes. The major disadvantage of this method of sensing is that it is temperature sensitive.

2.2.1.7 Photoconduction

The photoconduction effect involves the generation of charge carriers by illuminating a material, usually a semiconductor, with photons. The photons must have an energy level that is sufficient to cause electrons to be able to jump from the valence band into the conduction band for a band-to-band photoconductor or from an energy state of a trap within the band gap into the conduction band.

The creation of additional charge carriers due to the incident photon impingement increases the conductivity of the material, and this can be detected using electrical circuitry. The photoconduction effect is illustrated in Fig. 2.9 for both band-to-band and trap-to-band conductors.

When the photoconduction is band-to-band, both electrons and holes are generated and contribute the change in conductivity, whereas for deep level trap-to-band, the charge carriers generated by the impinging photons will only be of one type. Photoconduction does not require exotic materials and can be implemented without complex fabrication processes. The major disadvantage is that the effect is temperature sensitive.

2.2.1.8 Thermoelectric

The thermoelectric effect is based on the conversion of heat (i.e., thermal energy) into an electrical voltage or vice versa. Specifically, the transduction of heat into electricity is called the Seebeck effect, and the transduction of electricity into heating or cooling is called the Peltier effect. For our purposes, we will only consider the Seebeck effect. The Seeback effect is when two different metals (e.g., wires) are joined at the ends and there is a temperature difference between joints; a voltage potential will result between the two metals. In general, the voltage gradient developed is given by:

$$\nabla V = -S(T)\nabla T, \tag{2.12}$$

where ∇T is the temperature gradient between the two metals (in units of degree Kelvin) and $S(T)$ is called the Seebeck coefficient (in units of V per degree Kelvin) and is dependent on the materials involved.

The configuration for using the Seebeck effect to measure a temperature is shown in Fig. 2.10. T_{Sense} is the temperature to be measured, $T_{\mathrm{Reference}}$ is a point that is held at a known reference temperature, metal A is one type of metal wire, metal B is a different type of metal wire, and metal C is a third type of metal wire. Since metal C is on both sides of the junction, its influence cancels out, and the voltage for the temperature to be measured is given by:

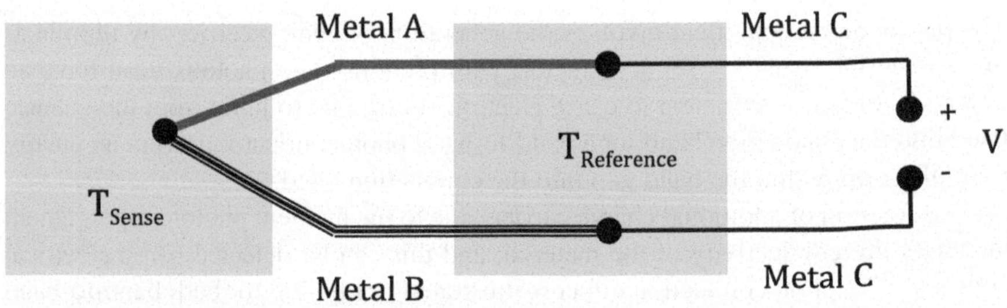

Fig. 2.10 Illustration of a thermocouple with output voltage V created by electromotive force due to temperature differential between T_{Sense} and $T_{\mathrm{Reference}}$

$$V = \int_{T_{\text{Reference}}}^{T_{\text{Sense}}} (S_{+}(T) - S_{-}(T))dT, \tag{2.13}$$

where $S_{+}(T)$ and $S_{-}(T)$ are the Seebeck coefficients of metal A and metal B connected to the positive and negative terminals of the voltage potential at the output.

Instead of solving Eq. (2.13) for each temperature measurement, a characteristic function, $E(T)$, is solved to give the following:

$$V = E(T_{\text{Sense}}) - E(T_{\text{Reference}}), \tag{2.14}$$

where $E(T)$ is given in terms of the Seeback coefficients by:

$$E(T) = \int^{T} S_{+}(T') - S_{-}(T')dT' + \text{constant}. \tag{2.15}$$

Macroscale temperature sensors using the Seebeck effect are called thermocouples, and a typical commercially available device will have an output of 40 mV/°C. Some materials used in thermocouples can span the temperature range from −270 °C to 2500 °C. The calibrated accuracy is on the order of 0.5 °C.

The Seebeck effect is used in MEMS microsensors mostly as temperature and flow rate sensors. Micromachining can be performed in order to reduce the thermal mass similar to what was described in the section above on resistive sensors. A microscale version of a thermocouple can exhibit high sensitivity and fast thermal response times (using micromachining). Multiple thermocouples can be fabricated into a series or parallel arrangement to form thermopiles wherein the output voltage is proportional to the temperature difference as shown in Fig. 2.11.

2.2.1.9 Diodes

Simple diode devices can also be used as transducers. In general the current through a diode is given by:

$$I = I_o e^{\left(\frac{qV}{kT}\right)}, \tag{2.16}$$

where I_o is the saturation current of a diode (units of amps), V is the voltage (units of volts), k is Boltzmann's constant (equal to 1.38×10^{-23} J/K), and T is the temperature (units of degree Kelvin). Shown in Fig. 2.18 is a curve of the relationship between current, I, and voltage, V. Equation (2.16) can be rearranged to the following:

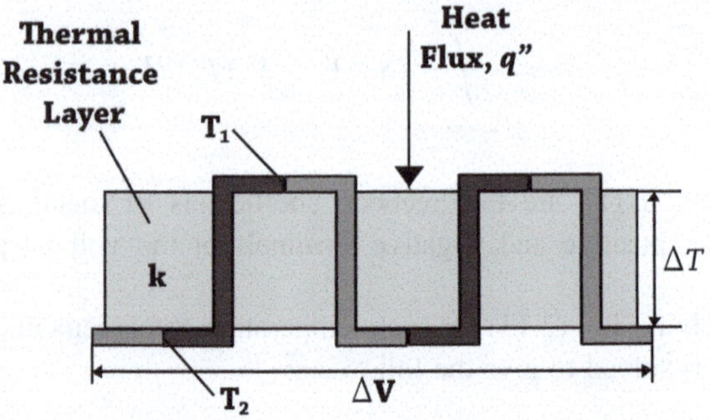

Fig. 2.11 Illustration of a thermopile with two thermocouples in series. The voltage output is proportional to temperature differential across the thermal resistance layer and number of thermocouples in series. The output voltage is also proportional to the heat flux, q''. (Copyright FluxTeq. Used under Creative Commons License)

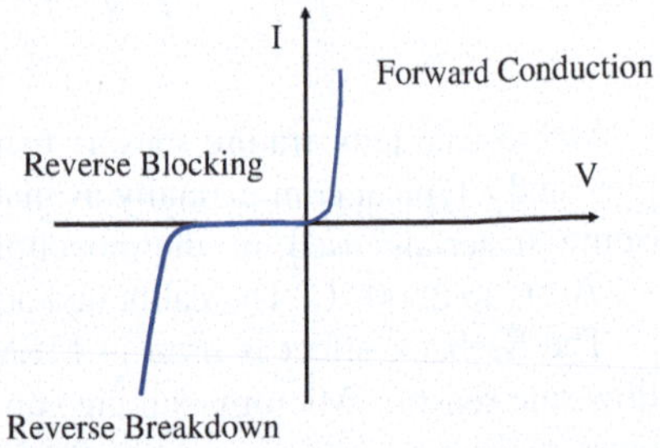

Fig. 2.12 Curve of the current plotted as a function of voltage for a Schottky diode

$$V = \left(\frac{kT}{q}\right) \ln \left(\frac{I}{I_0}\right). \tag{2.17}$$

The temperature dependence of the voltage is appreciable and for silicon is given by:

$$\frac{dV}{dT} = -2.3\,\text{mV}/^{\circ}\text{C}. \tag{2.18}$$

Therefore, strong temperature dependence allows the diode to be used as a temperature sensor. Diodes can also be used as strain sensors as well as photodetectors (Fig. 2.12).

The use of diodes for sensing is attractive since it does not require complex fabrication or exotic materials. The main problem with the use of a diode as a microsensor is to reduce either the temperature or strain sensitivity depending on the parameter to be sensed. This can often be achieved by using a differential pair configuration.

2.2.2 MEMS Microactuators

A variety of different transduction mechanisms are used to implement MEMS microactuators including electrostatic; piezoelectric; thermal; shape-memory alloy; and magnetic. Each of these has its respective advantages and disadvantages, and therefore careful consideration to the specific application requirements should be part of the selection process. We shall review a few of the more popular methods for MEMS microactuators here, and the reader is referred to for a more comprehensive review elsewhere [1–3].

2.2.2.1 Electrostatic

Electrostatic actuation is based on mutual attraction of two oppositely charged electrodes. The force F (units of N) generated between plates under the application of a voltage potential V is given by:

$$F = \frac{1}{2}(\varepsilon_o \varepsilon_r A)\left(\frac{V}{d}\right)^2, \tag{2.19}$$

where ε_o is the permittivity of free space (units of F/m), ε_r is the relative permittivity of a dielectric, A is the area of the electrodes (units of meters^2 and it is assumed that both electrodes are equal in size), V is the applied voltage potential on the plates (units of volts), and d is the separation or gap between the electrodes (units of meters). The transduction mechanism for electrostatic actuation is to convert a voltage potential into a mechanical displacement.

An illustration of one configuration for electrostatic actuation is shown in Fig. 2.13. A fixed electrode has been fabricated onto a substrate surface with an insulating layer made on top of the electrode. The opposing electrode is made and positioned at a distance, d, from the fixed electrode. The opposing electrode is supported by a spring having a spring constant given by k. An applied voltage potential, V, is placed across the electrodes, and this creates an electrostatic attraction force, F, as given by Eq. (2.19) above that will cause the moveable electrode to displace toward the fixed electrode.

Assuming the mechanical restoring force of the spring is sufficiently large, it will prevent the moveable electrode from collapsing into the fixed electrode, which is important since the force, F, is inversely proportional to the distance, d, between the electrodes to the square power. That is, as the moveable electrode displaces toward the fixed electrode, the force between the electrodes increases to the inverse square power of the electrode gap distance. If the mechanical restoring force of the spring is not sufficiently high, the moveable electrode can collapse into the fixed electrode. This could short out the electrodes if not for the presence of an insulating layer on the bottom electrode.

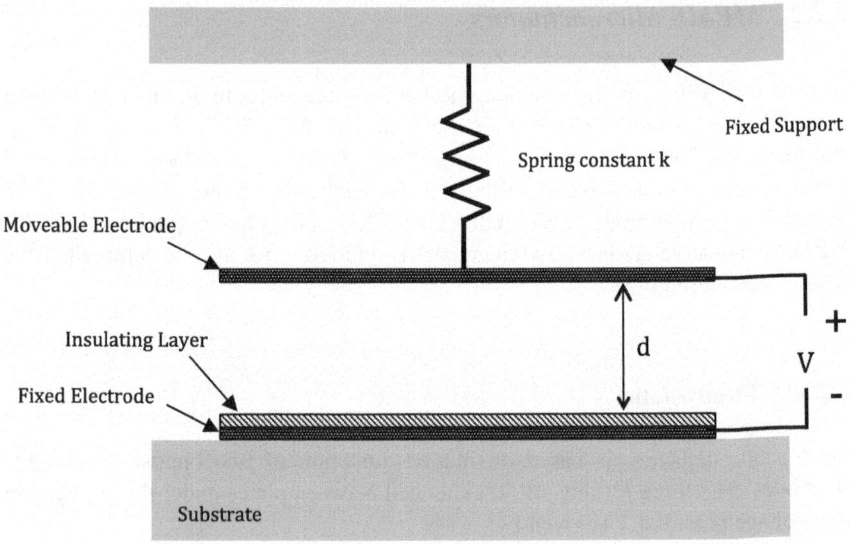

Fig. 2.13 Diagram to show the electrostatic actuation mechanism. A voltage is applied between two electrodes, one moveable and one fixed, that causes an electrostatic attractive force to be generated between two electrodes. The moveable electrode, which is constrained by a spring with spring constant k, moves toward the fixed electrode, and the distance between the electrodes decreases in magnitude

Electrostatic actuation is by far the widely used method of actuation in MEMS microsystems mostly due to the inherent advantages of this method including it has a simple design; it is relatively easy to fabricate; it does not require special materials; it can be easily integrated with electronics; it exhibits a very low power consumption for actuation; and it can provide extremely high mechanical bandwidths. It also is not as temperature sensitive as most other actuation methods.

There are also some disadvantages of electrostatic actuators, including the electrostatic attraction force is nonlinear with displacement and applied voltage; the resultant electrostatic actuation force is relatively small; the initial gap between the electrodes must be relatively small; and the operating voltages can be relatively high.

2.2.2.2 Piezoelectric

Piezoelectric transducers were discussed above for use in MEMS microsensors where it was explained how a voltage potential results across these materials when a strain is applied, and therefore the amount of voltage is indicative of the amount of strain present. Piezoelectric materials can also be used as microactuators as well. Specifically, a voltage potential across the material will result in the material

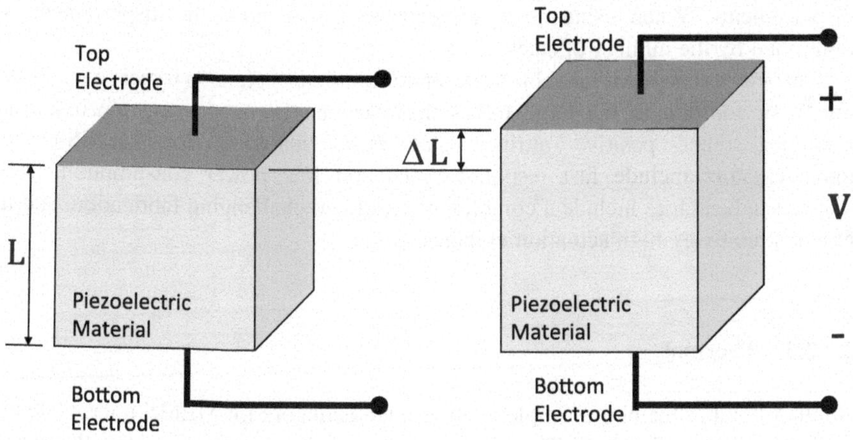

Fig. 2.14 Illustration of a piezoelectric material as a linear displacement actuator. On the left, no voltage is applied to the electrodes on either side of the piezoelectric material block. On the right, a voltage, V, is applied resulting in a change in length of the material from L and $L + \Delta L$

exhibiting a strain, and this strain can be employed to cause a mechanical displacement.

A simple longitudinal piezoelectric actuator is illustrated in Fig. 2.14. As shown on the left, the piezoelectric actuator has a top and bottom electrode that will be a metal conductive layer deposited on these two surfaces. No voltage potential is applied across the piezoelectric material layer in the left-hand side of this figure. With no voltage applied, the material is static.

In the right-hand side, a voltage potential, V, is applied across the electrodes. The electric field in the piezoelectric material is parallel to the direction of polarization, thereby inducing a strain, or displacement, in the direction of polarization in the material, which is given by ΔL. An estimation of the value of displacement for a given applied voltage is given by:

$$\Delta L = d_{33} V, \tag{2.20}$$

where d_{33} is the piezoelectric coefficient (units of meters per volts), which is material-specific parameter. Equation (2.20) is not strictly valid since the displacement will depend on the mechanical stiffness that the actuator must overcome, but it serves as a useful and simple equation to estimate the displacement if the actuator does not have to overcome a large mechanical restoring force.

A typical strain level for a piezoelectric is about 0.1%. Therefore, in general, piezoelectric materials do not deliver a significant amount of mechanical displacement. This can be partly overcome by forming stacks of piezoelectrics, which is a common practice in macroscale piezoelectric actuators to develop larger displacements. When using stacked piezoelectric actuators, the displacement is multiplied by the number of stacks, n.

Piezoelectric actuators are able to generate significant actuation forces (e.g., 30 N/ mm^2). In addition to the large forces that can be generated using piezoelectric actuators, other positive attributes of these materials for MEMS-based microactuators include fast response times and low power consumption. The major disadvantages include a complex material and challenging fabrication, and it requires relatively high actuation voltages.

2.2.2.3 Thermal

Another popular method of implementing microactuators for MEMS microsystems is based on thermal effects. There are a number of different mechanisms that have been reported using thermal effects including bimetallic; thermo-pneumatic; buckle-beam; and other variations. The bimetallic effect is the most commonly used of these and will be described here.

This technique uses the differing thermal expansion coefficients of two different materials to realize a microactuator. When these two materials are made into a composite structure and heated, a thermally induced stress is generated that can result in the structure exhibiting a mechanical displacement, assuming the structure is sufficiently mechanically compliant. The thermal strain is given by:

$$\varepsilon = (\alpha_{\text{film}_A} - \alpha_{\text{film}_B})(T_{\text{elevated}} - T_{\text{ambient}}), \tag{2.21}$$

where α_{filmA} and α_{filmB} are the thermal expansion coefficients of the top and bottom films (in units of 1/degree Kelvin), respectively, and T_{elevated} is the elevated temperature of the bimetallic element, and T_{ambient} is the ambient environmental temperature (both in units of degree Kelvin), respectively.

A simple bimetallic microactuator is illustrated in Fig. 2.15. A free-standing cantilever beam is made from silicon using micromachining fabrication methods. This silicon cantilever is built-in at the edge of the substrate. A thin-film material layer of aluminum has been deposited on the top surface of the silicon cantilever. If an electrical current is passed through the aluminum, it will experience some amount of Joule heating due to the electrical resistance of this layer, thereby resulting in the heating of the structure. Since aluminum has a higher-temperature coefficient of expansion compared to silicon, it will elongate more than the silicon. This will create

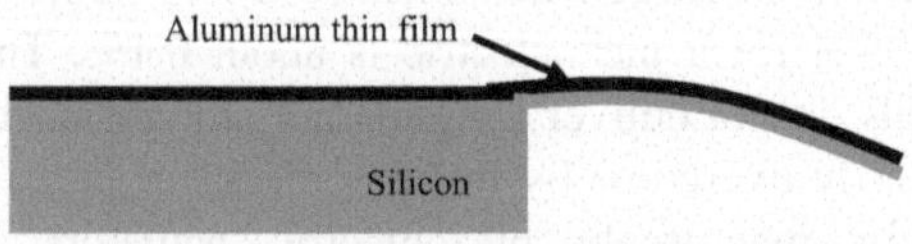

Fig. 2.15 Illustration of a bimetallic microactuator made from a thin-film aluminum layer on a silicon cantilever. When heated, the aluminum has a higher thermal expansion coefficient than silicon and causes the cantilever to deflect downward

a thermal stress in the structure that results in a downward directed force as shown in Fig. 2.15.

If the current in the aluminum layer is turned off, the heating will dissipate, the thermal stresses will eventually reduce back to zero, and the cantilever will return to its unactuated position. This on–off heating and actuation cycle can be repeated.

Some of the positive attributes of a bimetallic microactuator include relatively large displacements which are possible depending on the design, a linear deflection versus electrical power relationship; depending on materials used, the implementation of this actuation approach can be compatible with integrated circuits; and depending on the design, this method of actuation can generate reasonable force levels with moderately complex fabrication. The negative attributes of bimetallic actuators are high power consumption compared to many other actuation methods; requiring continual power to maintain the actuated state; and thermal actuation that usually has a relatively low mechanical bandwidth. Also, since bimetallic actuators are actuated using heating, they are sensitive to the environmental conditions.

2.2.2.4 Shape-Memory Alloys (SMA)

Yet another popular method for implementing microactuators in MEMS microsystems is shape-memory alloys (SMAs). A shape-memory alloy is a material that undergoes a martensite to austenite phase change upon heating. Shape-memory alloys are a special class of metal alloys that have a temperature-dependent crystal structure.

Figure 2.16 is an illustration demonstrating the shape-memory effect. In the lower-temperature phase, called martensite, the material has a low yield strength

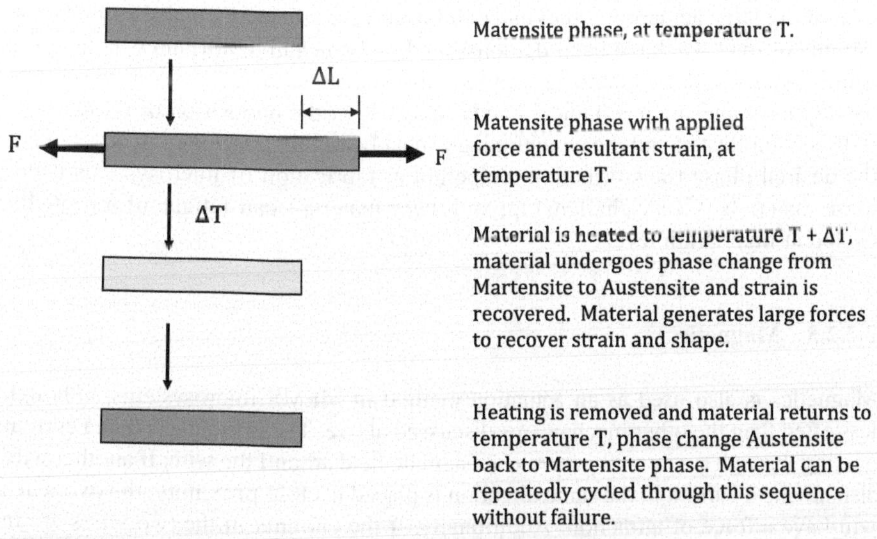

Fig. 2.16 Illustration of the phase changes of a shape-memory alloy material and how the strains can be recovered

making it relatively easy to strain (i.e., deform) into a new shape. The material retains the strained or deformed shape until the temperature of the material is raised to above the phase transition temperature, whereupon the material undergoes a phase state change from martensite to austenite. Once it makes the transition to the austenite phase, the material returns to its previous unstrained or undeformed state, thereby exhibiting the ability to have "memory of" its previous shape. Once the heating is removed, the material will re-transition from the austenite back to the martensite state and can then again be easily strained to form a new shape. This cycle can be repeated many times.

There are several different types of shape-memory alloy materials that are available. The most widely used type is an alloy of titanium and nickel that is abbreviated TiNi and also sometimes called Nitinol. Titanium–nickel shape-memory alloys have good electrical and mechanical material properties, and the martensite to austenite phase transition temperature can be varied from about −100 °C to +100 °C by tailoring the stoichiometry of the material. One of the most unusual properties of these materials is the very high level of recoverable strains that can be obtained. For example, titanium–nickel can exhibit a recoverable strain on excess of 5%. Another important attribute of shape-memory alloys is that these materials are capable of generating the highest-energy density compared to any other method of actuation. Therefore, shape-memory alloys are able to generate very large actuation forces.

Shape-memory alloys can be sputter deposited in thin-film form on silicon wafers. Heating of the film is usually achieved by Joule heating. The shape-memory effect is a reversible effect and can be repeated many times as long as the strain levels are below a certain maximum threshold (usually a few percent strain maximum). The positive attributes of shape-memory alloys for use in MEMS microactuators include exhibiting very high-energy densities, thereby generating large actuation forces and exhibiting very large recoverable strain levels (strains of over 8% have been demonstrated). The negative attributes include the following: since these materials are thermally activated, the power consumption is usually high and the mechanical bandwidth is low; the deposition of these materials is complicated and the stoichiometry must be tightly controlled in order to get the desired phase transformation temperature; fabrication of microsystems using these materials is very challenging; and these materials can fatigue of repeatedly cycled at high strain levels.

2.2.2.5 Magnetic

Magnetics is also used as an actuation method in MEMS microsystems, although less often than the other mechanisms discussed above. The principle is that a current passing through a wire will generate a magnetic field around the wire. If another wire also having a current in the same direction is placed in close proximity, the two wires will have a force of attraction. Additionally, if the currents in the two wires are in opposite directions, the wires will develop a force of repulsion. The magnetic force

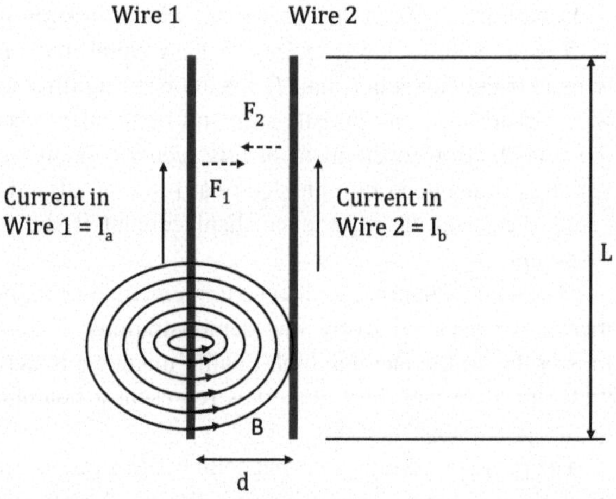

Fig. 2.17 Illustration of two current-conducting wires that generate a magnetic field that results in a force of attraction between the wires

generated between two wires is shown in Fig. 2.17. On the left is wire 1 carrying current I_1, and on the right is wire 2 carrying current I_2 (both in units of amps). As shown, I_1 and I_2 are both in the same direction and of equal magnitude. There is a separate distance between the wires of value d (units of meters), and each wire has a length of L (also in units of meters). As shown, a magnetic field of magnitude B develops around wire 1 as given by Ampere's law. An identical magnetic field will also develop around wire 2 (not shown in Fig. 2.17).

According to the Lorentz equation, a force of attractive will be generated of value F (units of N) between the two wires. The force on wire 2 due to the magnetic field generated around wire 1 is given by[1]:

$$F = I_2 L B. \tag{2.22}$$

The magnetic field around wire 2 from the current in wire 1 is given as:

$$B = \frac{\mu_o I_1}{2\pi d}, \tag{2.23}$$

where B is the magnetic field strength (in units of T) and μ_o is the permeability of free space (units of T-m/A). Combined Eqs. (2.22) and (2.23) result in the following:

$$F = \frac{\mu_o L I_1 I_2}{2\pi d}. \tag{2.24}$$

[1]Equation (2.22) is shown in scalar form. In actuality, some of the variables in this equation, such as the magnetic field, B, and force, F, are best represented by vector quantities. However, for our purposes, the simpler, nonvector form of this relationship is adequate to explain the concept.

Equation (2.24) is the force on wire 2 resulting from the current in wire 1. Newton's third law requires that an equal and opposite force be present on wire 1. If the currents I_1 and I_2 are in opposing directions, then the forces on each wire would have an opposite sign and be repulsive between the wires. Therefore, the transduction mechanism in this scenario is an electrical current in the wires which is transduced into an electrical force of attraction (or repulsion) that would cause the wires to experience displacement if the mechanical compliance was sufficient.

Magnetic actuation can also be implemented using magnetic materials. However, there have been relatively few demonstrations of this type of magnetic actuation mostly due to the fact that high-quality magnetic materials are difficult to deposit in thin-film form and these materials represent a potential contaminant for semiconductor processing facilities.

The positive attributes of magnetic MEMS microactuators include the following: possible to generate reasonable force levels; the force is proportional to current; the force can be both attraction and repulsion with a switching of current direction in one of the wires; fast response times; and can exhibit large actuation displacements. The negative attributes include the following: large currents may result in high power consumption levels. and it requires very complicated fabrication. In practice, the wires are most often configured into multiple windings so as to generate larger force levels. However, this greatly increases the fabrication complexity using microfabrication techniques.

2.3 Microsystems Manufacturing Processes

It is important to establish some basic definitions in order to help clarify the terminology related to microsystems processes. The reason is that certain words used in micro- and nano-fabrication are at best ambiguous and often confusing, particularly to non-experts since the meaning of the word can only be understood by its contextual usage. Specifically, one of the most often used words in microsystems manufacturing is "process." In order to avoid confusion, some definitions are established and will be used throughout this volume (see Fig. 2.18).

> A processing step is defined *as a single process operation that is performed on a substrate or a set of substrates in order to advance the fabrication in the implementation of microsystem device* [1].

A processing step is usually, but not always, performed on a single item of equipment. Chapters 3 and 4 will provide the reader with an overview of a number of examples of processing steps including thin-film depositions, etching, photolithography, planarization, and so on. An example of a processing step under this definition would be the deposition of polycrystalline silicon (polysilicon) on a batch of wafers using low-pressure chemical vapor deposition (LPCVD). The processing step would normally be limited to only the actual deposition of the polysilicon thin-

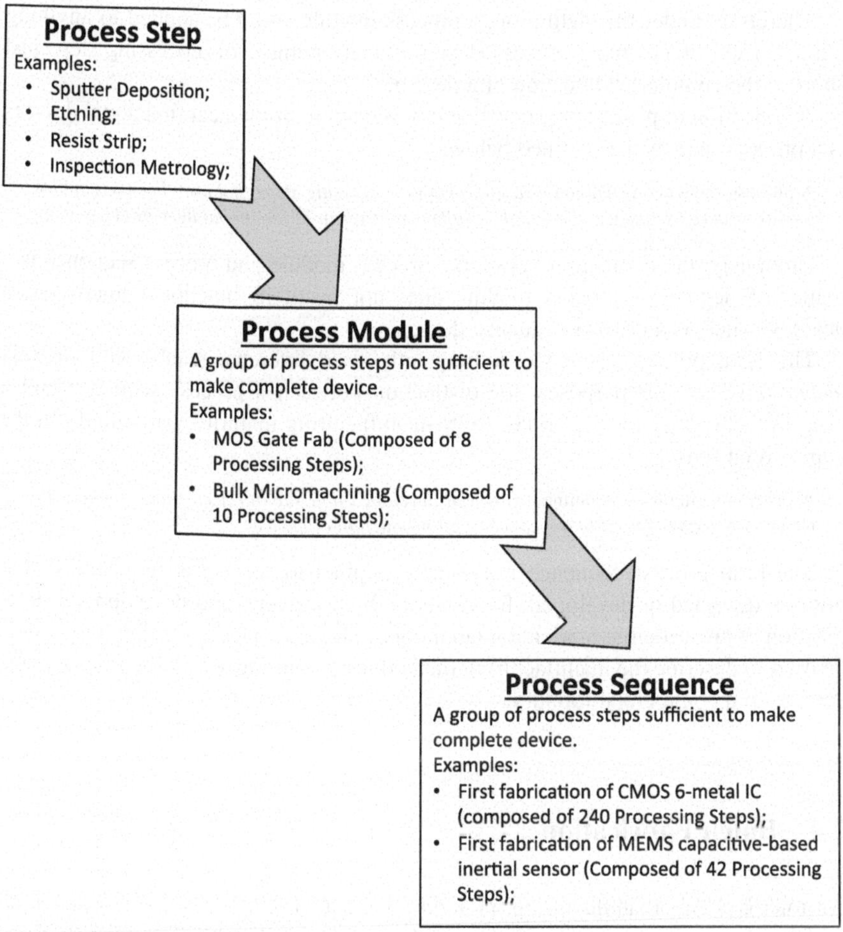

Fig. 2.18 Illustration of relationship between processing steps, process modules, and process sequences

film material on the substrates and not to the cleaning and preparation steps performed on the substrates prior to the deposition or the metrology that would be performed after the deposition was completed.

The substrate cleaning and preparation steps as well as the metrology steps would be considered processing steps on their own. Having said that, one must be careful since these distinctions are often made somewhat arbitrarily by many manufacturers, and it is not uncommon to see several steps grouped together and labeled as a process step.

A definition of a collection of processing steps is a process module that is defined as follows:

> A process module is defined *as a grouping of processing steps that have been assembled into an ordered sequence, but which is not sufficient in and of itself to result in the complete implementation of functional devices* [1].

Therefore, under this definition, a process module could be applied to anything ranging from two or more process steps up to any number of processing steps just short of the complete fabrication of a device.

A collection of processing steps that is sufficient to implement functional devices is a process sequence as defined below:

A process sequence is defined as *a grouping of processing steps that have been assembled into an ordered sequence and which is sufficient to result in functional devices* [1].

Obviously, the distinction between a process module and process sequence is a matter of degree. A process module does not result in functional microsystem devices, whereas a process sequence does.

The term process sequence is frequently used interchangeably with *process technology*. For our purposes, the distinction between a process sequence and a process technology mostly relates to its manufacturing maturity, particularly in the commercial sense.

A process technology is defined as *a well-developed and standardized process sequence, which is repeatedly used in the commercial production of devices.*

The term process sequence is typically applied to a process sequence that is initially designed or developed, for example, by a research and development organization or an academic fabrication laboratory, whereas a process technology would be used to describe the manufacturing method used to produce specific microsystem devices in a commercial foundry.

2.4 Batch Fabrication

All microsystems manufacturing, including both IC and MEMS microsystems, are based on concept of batch fabrication techniques whereby the manufacturing processes involve a number of sequentially performed individual fabrication steps on a batch (also often called a "lot") or grouping of the substrates (see Fig. 2.19). Often the individual fabrication steps are performed at various types of workstations and/or process tools. At any given time, multiple batches of the substrates can be

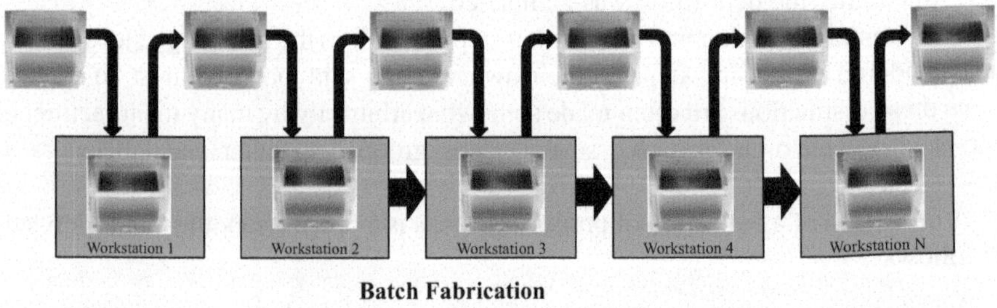

Fig. 2.19 Illustration of the batch fabrication production method

simultaneously involved in the manufacturing process, usually with different batches located at different stages of the individual fabrication steps in the manufacturing process.

The substrates (also called "wafers") that these batch fabrication processes performed on typically have standardized diameters and thicknesses. This standardization is important since much of the equipment and tools used in fabrication are sized according to these standards.

Multiple wafers are usually processed together, as a batch, in a cassette that is carried from workstation to workstation throughout the manufacturing process (see Fig. 2.20). Depending on the equipment used for the processing step, the entire cassette of wafers may be simultaneously processed, such as in a low-pressure chemical vapor deposition (LLPCVD) furnace tube, or sequentially processed one wafer at time through each wafer in the cassette, such as in a cassette-to-cassette single-wafer plasma-enhanced chemical vapor deposition (PECVD) system.

Once all of the individual processing steps have been completed, the substrate is then be diced using a die-saw machine into the individual die (Fig. 2.21) for subsequent packaging. Therefore, each wafer can be thought of as a batch of dies, and each cassette filled with wafers is also a batch.

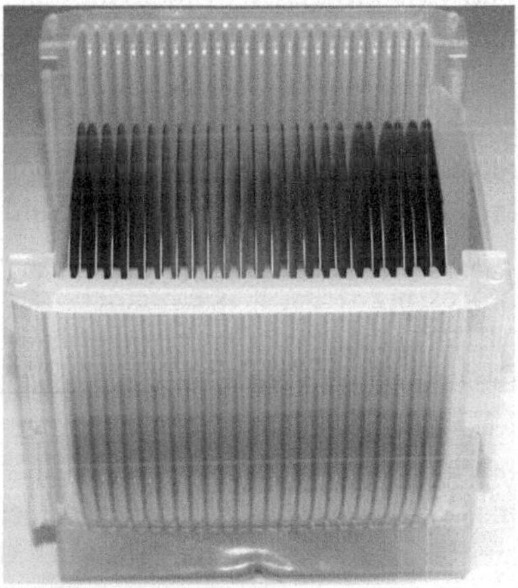

Fig. 2.20 A cassette of a quantity of 25 silicon wafers having diameters of 100-mm

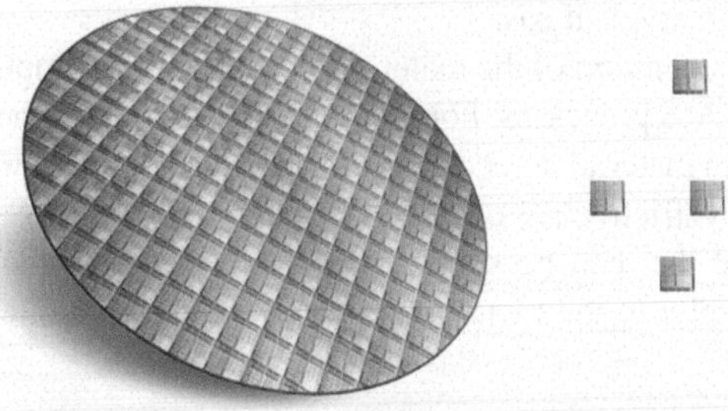

Fig. 2.21 A silicon wafer (left) and some of the resultant integrated circuit dies from wafer (right) after wafer has been diced

Batch fabrication is an extremely important manufacturing advantage since it allows a large number of microsystems die to be made in each lot, thereby resulting in relatively low per die production costs. The production costs of microsystems manufacturing are high, but this high cost can be spread over a large number of die.

2.5 Some Basics of Microsystems Manufacturing

The spectrum of microsystems manufacturing technologies spans from the commonly employed processing steps used to implement conventional integrated circuits (ICs) to a broad and diverse set of specialized MEMS processing steps and process modules, commonly known as micromachining. This section provides a general overview of microsystems manufacturing and discusses some of the high-level differences between IC and MEMS manufacturing methods. Since MEMS microsystems manufacturing employs methods from both of these domains, a more comprehensive review of each will be given in later chapters.

As shown in Fig. 2.22, microsystems manufacturing methods can be categorized into two groups: IC processing steps and MEMS processing steps and process modules.[2] This distinction deserves further explanation.

The IC manufacturing industry uses a standard set of processing steps, including chemical vapor deposition (CVD); photolithography, sputtering; ion implantation; reactive ion etching (RIE); etc. In comparison, the MEMS manufacturing employs a broader and diverse array of specialized processes.

Importantly, while a typical IC manufacturer will employ some or all of the IC processing steps, it is extremely rare for IC manufacturers to ever use any of the MEMS processes. On the other hand, most MEMS microsystem manufacturers do use a number of the IC processing steps. Therefore, it may appear that the overlap of IC and MEMS manufacturing is one-sided wherein IC manufacturing only uses IC fabrication methods, whereas MEMS manufacturing uses both. In general, this is mostly true. However, there are sometimes exceptions. One example would be when an IC microsystems manufacturer discovers a critical need for using a MEMS processing technology, such as the use of high-aspect ratio silicon plasma reactive ion etch technology (e.g., DRIE etching) to form through-wafer vias. Nevertheless, this is an atypical case.

Consideration of the materials used is another important differentiator between IC and MEMS processes. For example, while sputtering may be routinely used as an IC thin-film material layer deposition process as shown in Fig. 2.22, this pertains only to a very restrictive set of materials that are used in IC manufacturing and not to any material that can be sputter deposited. For example, sputtering in the IC domain would refer to the metal material layers used for interconnecting various electronic

[2] As will be explained later, some MEMS processes are better described as process modules since they actually entail a number of processing steps as opposed to a single processing step.

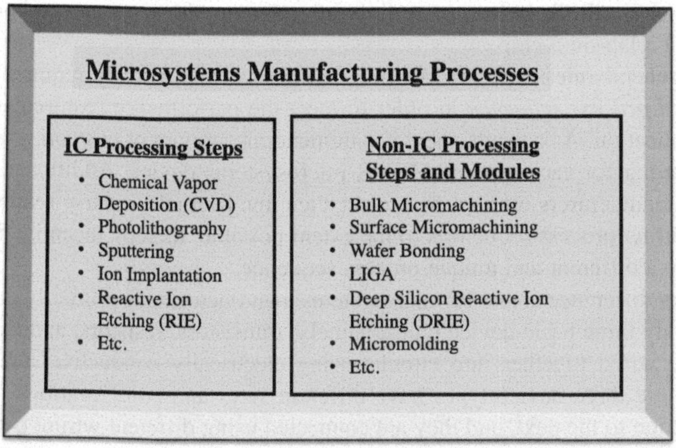

Fig. 2.22 Diagram of microsystems methods

devices. However, it would not refer to the deposition of other material layers or types, such as non-IC compatible metals (e.g., indium, gold, etc.) as well as materials that have specific functionality such as piezoelectric or shape-memory alloys. Nevertheless, these other materials may be readily used in MEMS microsystems fabrication.

Even though a material is commonly used in IC microsystems manufacturing, it still may not be properly categorized as an IC process based on specific parameters of the deposition, such as the material layer thickness. For example, if a material layer is commonly used in IC manufacturing, it has a typical nominal thickness of 100 nm; the deposition of a few microns thickness of this same material may not be considered an IC processing step. Consequently, the differentiation of whether a process is an IC or a MEMS method requires careful consideration of the materials, details of the processing step(s) (e.g., thickness of thin-film layer deposition, depth of etch, etc.), and even equipment that is employed.

2.5.1 Differences Between IC and MEMS Manufacturing

In addition to differences in the processing steps, there are also differences in the process sequences and microsystems device made. The first major difference is that the IC manufacturing domain has converged to a relatively small number of "standardized" or "fixed" *process technologies*, such as CMOS (complementary metal-oxide semiconductor), DRAM (dynamic random access memory), etc. In contrast, MEMS microsystems manufacturing domain tends to use a totally different sequence of processing steps for implementation of each type of device. For example, the sequence of processing steps to implement a MEMS-based accelerometer will likely have very little in common with a process technology to implement a lab-on-a-chip.

As a general rule, each MEMS microsystems device usually requires its *own specialized process sequence* in order to meet the performance requirements for a given application. As a result, there is a tremendous amount of *customization* in the manufacturing for each type of MEMS microsystems device. Additionally, many MEMS manufacturers attempt to protect their intellectual property related to the manufacturing process sequences to the extent possible, thereby incentivizing them to develop a different and unique process sequence.

Another difference is that IC microsystems manufacturing is primarily directed at making only three basic device types, namely, transistors, resistors, and capacitors, which are wired together into circuits using electrically conductive interconnect layers. These three device types have different sizes and configurations from one IC generation to the next, and they are connected using different wiring patterns for each circuit design, but in several decades of development and maturity of IC microsystems, mostly only these three device types are manufactured.

In comparison, MEMS microsystems technology has a vast array of different device types including pressure sensors, light switches, light-emitting diodes, accelerometers, waveguides, lasers, gyroscopes, vibration sensors, optical modulators, magnetic sensors, microvalves, micropumps, optical switches and modulators, microfluidic systems, and blood and serum analysis systems, each with its own unique device design and manufacturing process. Importantly, the design, the manufacturing processing steps, how the individual processing steps are ordered into a process sequence, functional characteristics, and materials are often very different not only for each of these different application areas but also for each device type and designer.

The customized nature of the devices made and the manufacturing process sequences used to implement the devices, combined with the vast array of processing capabilities and materials used to make MEMS microsystem devices, have a very important consequence with regard to the dimensional and material property variations in the devices fabricated. This is because both the dimensional and material property variations will be highly dependent on the processing steps and the process sequence used in the fabrication of a device, and since the process sequence probably has not been previously developed or performed, there may be little or no data on these resultant parameter variations. As a result, the MEMS designer often does not have access to data on the parameter variations for the devices the MEMS designer intends to fabricate. Even if there is some parameter variation data on some or all of the processing steps, the parameter variations will probably still not be known since they will be dependent on the process sequence as well.

While MEMS microsystems developers have a much richer range of choices regarding materials and fabrication techniques than is used or available in the IC microsystems domain, which has many advantages, it also poses some challenges as well. A major advantage is that MEMS microsystem designers have considerably more design and processing freedom to obtain outstanding device performance. On the other hand, the MEMS microsystem developer has comparatively little in the way of pre-existing standardized process technologies from which to leverage and also little to no knowledge of the variations in the device parameters resulting from performing the fabrication. This means that the designer developing a new MEMS microsystem device must choose from an enormous range of process options and parameters of which processing steps will work best for the fabrication of a given device, know how to *integrate* these processing steps into a viable process sequence for production, know how to reasonably estimate the parameter variations that are likely the result in the fabrication, and know how to manage these parameter variations so as to get acceptable yield levels.

2.5.2 Microsystems Feature Sizes

The IC industry has been reducing the size of transistors as well as increasing the number of transistors on each die steadily over several decades (i.e., "Moore's Law"). The critical feature of advanced IC devices that is often compared to one another is the gate lengths made in complementary metal-oxide-semiconductor (CMOS) process technologies. Current state-of-the-art IC manufacturing processes are now able to make gate lengths on the order of approximately 10 nm.

While reduction in minimum feature sizes is a key driver in most IC manufacturing environments, the situation can be more complex for MEMS microsystems. On one hand, the physics of MEMS devices often dictates larger device sizes. For example, a MEMS inertial sensor may need a specific amount of mass in order to be able to be sufficiently sensitive to acceleration changes, and this amount of mass may require that the proof mask have dimensions that are considerably large compared to what is state-of-the-art in IC manufacturing. Therefore, based on this fact alone, it may be concluded that the use of the most advanced (and costly) photolithography technologies from the IC domain is not needed since the patterning of this type of MEMS microsystems device does not require extreme miniaturization of the device dimensions. However, this is not the whole story.

As it turns out, the amount of variation of the dimensions made using photolithography tends to stay about the same for a given tool regardless of the dimensions of the transferred features. As a general rule, the dimensional variation of most photolithography systems is about 1/3rd of the ultimate resolution. This means that a photolithography tool having higher resolution, such as a deep-ultraviolet (DUV) stepper, would have less variation in the patterned dimensions compared to a photolithography system having lower resolution, such as a contact aligner system. Consequently, even though the ultimate resolution of a more advanced photolithography system may not be required for making a MEMS device, the ability to reduce the variations using the more advanced lithography capability may be needed. In short, it may be beneficial to use a more advanced and higher-resolution tool for MEMS despite the higher manufacturing costs. This topic will be examined in more detail in a later chapter.

2.6 Some Material Basics Regarding Semiconductors and Silicon

Most solid-state electronic devices and MEMS are made from a class of materials called "semiconductors," and silicon is the most widely used semiconductor material choice for high-volume integrated circuit manufacturing. Silicon is by far the most widely used semiconductor material in the integrated circuit industry due to a variety of reasons, including its excellent semiconductor properties; ability to easily form a native oxide; ability to introduce impurities to modify the material's electrical behavior; its excellent mechanical properties; ability to make large diameter single-crystal boules that enable larger diameter wafers; and relatively low cost [6].

Silicon is a tetravalent metalloid having a diamond cubic crystal structure (Fig. 2.23) with a lattice spacing of 5.4307 Å. Because of its crystal structure, silicon's material properties are anisotropic, meaning that the material properties vary depending on the crystallographic orientation. Silicon is the most common element in the universe by mass and usually exists in nature as silicon dioxide (i.e., sand). Silicon has an atomic number of 14 and is located in the IVth column of the periodic table. Its electronic structure consists of ten core electrons that are tightly bound and four additional valence electrons that are loosely bound around the atom's nucleus. It is these four loosely held valence electrons that result in this material's chemical properties, including its semiconductor behavior [2–4].

While silicon is the most widely used semiconductor material in MEMS manufacturing, other semiconductor materials are sometimes used as well. Specifically, the other elements in the IVth column of the periodic table, such as germanium, also have four valence electrons and exhibit semiconductor material properties. Germanium is a well-known semiconductor material used in the industry, although to a much lesser extent than silicon. The diamond form of carbon is semiconducting as well although the energy band gap is very large, thereby making it useful for very limited applications. Silicon is also sometimes combined with

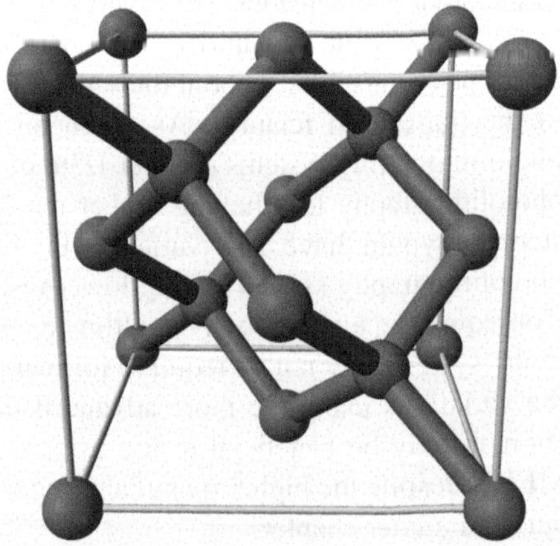

Fig. 2.23 A unit cell of silicon exhibiting the diamond cubic crystal lattice structure

germanium to form silicon–germanium (SiGe) that is used in RF applications. Similarly, silicon is also combined with carbon to form silicon carbide (SiC) that is used in applications involving harsh environments. Combining elements from Columns III and V of the periodic table will also result in some useful binary semiconductors. For example, gallium arsenide (GaAs) is used in microwave electronics, light-emitting diodes, photonics, and laser diodes; indium phosphide (InP) in high-power and high-frequency electronics, photonics, and laser diodes; and gallium nitride (GaN) is used in laser diodes, light-emitting diodes, and high-electron mobility transistors (HEMTs). Table 2.1 lists most of the more common types of semiconductors and their bulk material properties at room temperature.

Silicon wafers are fabricated by growing a large single-crystal boule using a special process called the Czochralski growth method (Fig. 2.24). This process consists of forming a crystal ingot by first melting a batch of ultra-pure silicon material (typically 99.9999999% pure) and then placing a much smaller seed crystal in contact with the surface of a molten liquid of high-purity silicon and pulling the seed crystal slowing from the molten silicon whereby a large continuous crystal is grown starting from the seed crystal. The size of the crystal grown is dependent on the speed at which the seed crystal is pulled, and current silicon crystal growing processes can grow boules having diameters sufficiently large for 450-millimeter diameter wafers.

This process results in a large single crystal of silicon material that is nearly free of defects. This is sometimes called "electronic-grade silicon (EGS)." Small amounts of selected impurity elements such as boron or phosphorus may be added to the molten silicon to change the polarity of the majority of charge carriers in the material to result in what is termed either a p- or n-type semiconductor. These impurities are called dopants, and the process of introducing them into a semiconductor material is called doping.

After the crystal growing process, the large boule is cut into individual slices using a special saw. Semiconductor wafers, specifically silicon wafers, are categorized according to a number of important parameters, including crystallographic orientation, dopant types (i.e., n- or p-type), and grade of material (e.g., prime or test). Special markings, called flats, are typically made in the wafers that provide information about these parameters. The slices are then polished to form the individual wafers and usually packaged in individual cassette carriers each containing 25 wafers.

As noted above, silicon is an anisotropic material wherein the material properties vary depending on how the crystalline structure is oriented. Therefore it is important to know the crystallographic orientation of the substrates in order to be able to accurately predict the resultant material and device behavior. Crystallographic orientations are defined according to the Miller indices. From Fig. 2.25, it can be shown that the silicon unit cell is composed of two interpenetrating face-centered cubic lattice structures of silicon atoms [2–4].

Figure 2.25a illustrates one of these simple cubic cells of silicon wherein the silicon atoms are located at the corners and represented by black dots. The Miller indices allow the labeling of important crystallographic directions wherein directions

Table 2.1 Material properties for silicon, germanium, silicon carbide, gallium arsenide, indium phosphide, gallium nitride at 300 °K

Material property	Silicon	Germanium	Silicon carbide	Gallium arsenide	Indium phosphide	Gallium nitride
Crystal structure	Face-centered diamond cubic	Face-centered diamond cubic	Polymorphic 3C zincblende 4H wurtzite 6H wurtzite	Zincblende	Zincblende	Wurtzite
Lattice constant (Å)	5.43	5.66	3C, 4.4 4H, $a = 3.07$, $b = 10.1$ 6H, $a = 3.07$, $b = 10.1$	5.64	5.87	$a = 3.186$ $c = 5.186$
Atomic or molecular weight (g/mol)	28.09	72.6	40.11	144.63	145.8	83.73
Density (g/cm^3)	2.328	5.3	3.21	5.32	4.81	6.1
Dielectric constant	11.9	16	3C, 9.72 4H, 9.66 6H, 9.66	13.1	12.5	9.7
Melting point (°C)	1412	1211	3103	1238	1062	>2500
Thermal expansion coefficient (/°C)	2.6×10^{-6}	6×10^{-6}	2.8×10^{-6}	6.9×10^{-6}	4.6×10^{-6}	5.7×10^{-6}
Thermal conductivity (W/cm °C)	1.5	0.6	3C, 0.36 4H, 0.37 6H, 0.49	0.46	0.68	1.3
Specific heat (J/g °C)	0.7	0.321	0.75	0.35	0.31	0.33
Thermal diffusivity (cm^2/s)	0.9	0.36	3C, 1.6 4H, 1.7 6H, 2.2	0.44	0.37	0.31
Energy band gap (eV)	1.11	0.67	3C, 2.36 4H, 3.23 6H, 3.05	1.435	1.344	3.4
Poisson's ratio	0.28	0.26	0.14	0.31	0.36	0.31
Young's modulus (GPa)	130.9	103	3C, 250 4H, 220 6H, 220	82.7	61	85

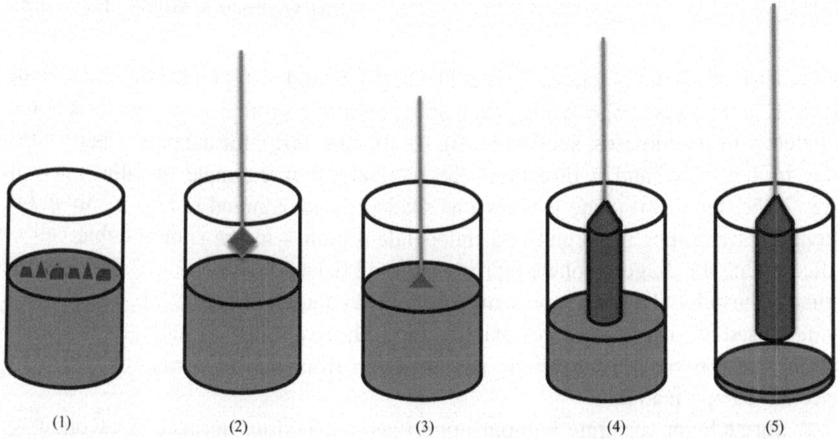

Fig. 2.24 Illustration of the Czochralski crystal growing method used to make single-crystal silicon boules that will be cut into slices to form silicon wafers. (1) Preparing the molten silicon melt. (2) Lowering of a seed crystal into molten silicon bath. (3) Touching of seed crystal onto molten silicon bath. (4) Pulling seed crystal from molten silicon bath to form larger crystal. (5) Completely formed large boule made from melt

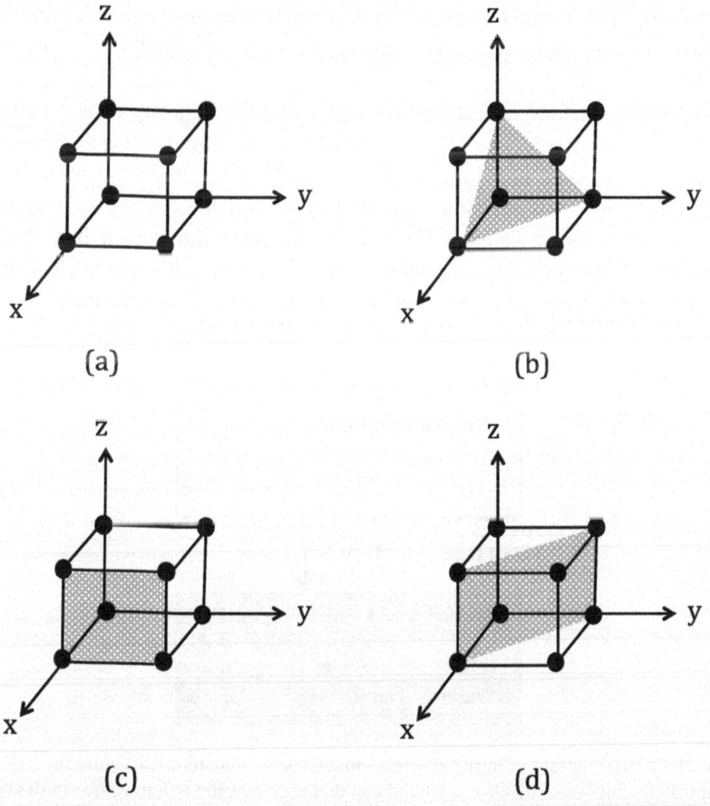

Fig. 2.25 Illustration of the different crystallographic directions and planes of the simple cubic cell that represents silicon. Figures (**b**) through (**d**) show the important and independent different planes of silicon as shaded rectangles [2–4]. (**a**) Silicon simple cubic unit cell. (**b**) Silicon (111) plane. (**c**) Silicon (100) plane. (**d**) Silicon (110) plane

are labeled in square brackets, such as [100], [010], and [001] for the direction along the *x*, *y*, and *z*-axes, respectively. The corresponding crystal plane of this direction is labeled with parentheses, such as (100), (010), and (001), for the planes perpendicular to the *x*, *y*, and *z* directions, respectively. For example, as illustrated in Fig. 2.25c, the (100) plane is shown as shaded. Also as noted in Fig. 2.25b and d, there are two other important and independent planes in the simple cubic cell of silicon. One the diagonal plane represented as (110) as shown in Fig. 2.25d, and the other is the cubic diagonal represented as (111) as shown in Fig. 2.25b. Most silicon wafers used in microelectronics manufacturing have a (100) crystallographic orientation, but the other orientations are available from substrate manufacturers for specialized applications.

A pure silicon substrate without impurities is called an intrinsic semiconductor [2, 3]. A semiconductor has a band gap between the conduction and valence energies bands (Fig. 2.26). This means that an intrinsic semiconductor will behave as an insulator at absolute zero (i.e., 0 °K), since all of the electrons will be confined to the valence band and all energy states in that energy band are occupied, thereby prohibiting electron transport even under the application of an electrical field. However, with the addition of sufficient energy (e.g., thermal energy), some of the electrons from the valence band can to jump over the energy band gap and into the conduction energy band, thereby making the silicon sample able to conduct electricity.

An important attribute of semiconductors is that the majority charge carrier type can be determined by selective introduction of suitable impurities (i.e., dopants) into the material. In silicon, elements from column V of the periodic table including phosphorus, arsenic, and antimony can be used as suitable dopants of silicon. Each of these elements has 5 valence electronics and once introduced into the silicon lattice will substitute for a silicon atom but have an additional valence electron that is very weakly bound to the dopant atom. Because the 5th valence electron is so weakly

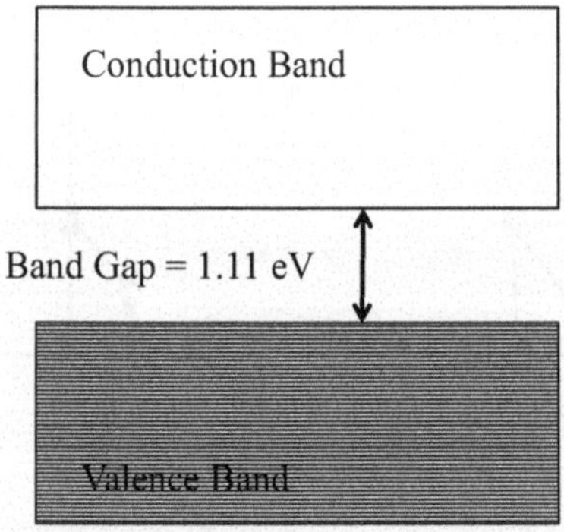

Fig. 2.26 Band gap diagram of intrinsic semiconductor silicon material showing the valence band that is populated with bound electrons, a band gap that separates the valence and conduction bands that is about 1.11 electron-volts for silicon material and the conduction band which is composed of numerous energy states that are unfilled with charge carriers

bound to the column V impurity atom, a very small amount of thermal energy can cause it to jump into the conduction band and thereby increase the electrical conductivity of the silicon material. Silicon material with doping from elements with 5 valence electrons is called "n-type."

Conversely, the introduction of an impurity from column III of the periodic table, such as boron, gallium, indium, and aluminum, can also be used as suitable dopants of silicon. Each of these elements has 3 valence electronics and once introduced into the silicon lattice will substitute for a silicon atom but have a hole (the absence of a valence electron) that is very weakly bound to the dopant atom. Because the hole is so weakly bound to the column III impurity atom, a very small amount of thermal energy can cause it to jump into the valence band and thereby increase the electrical conductivity of the silicon material. Silicon material with doping from elements with 3 valence electrons is called "p-type."

The introduction of charge carriers due to doping with impurities into semiconductors like silicon is an important element in the implementation of active electronic devices such as diodes and transistors and passive components such as resistors, and therefore fabrication methods for introduction of dopants will be discussed in the next chapter.

2.7 Summary

This chapter has reviewed the transduction mechanisms most commonly used in the implementation of MEMS-based microsystems, specifically microsensors and microactuators. Despite the fact that these technologies are relatively recent developments, the breadth and diversity of the different transduction methods used is quite impressive. Perhaps even more impressive is that often these microminiaturized devices often have exceptional performance for use in numerous applications in the macroscale world. Additionally, this chapter discussed some of the important elements that make MEMS manufacturing both more interesting and more challenging compared to IC manufacturing. MEMS design is very interesting since the number of device types and potential application areas is enormous. Nevertheless, MEMS implementation usually involves significant customization of both the design and the process sequence, and therefore the device designer often does not have much prior knowledge of the magnitude of the variations in device parameters. Lastly, this chapter reviewed some basics about semiconductor materials, which are heavily used in MEMS manufacturing.

References

1. M. Huff, S.F. Bart, P. Lin, Process integration, in *MEMS Materials and Processing Handbook*, ed. by R. Ghodssi, P. Lin, (Springer Press, New York, 2012)

2. M. Madou, *Fundamentals of Microfabrication* (CRC Press, Boca Raton, 1997)
3. G.T.A. Kovacs, *Micromachined Transducers Sourcebook* (McGraw-Hill, New York, 1998)
4. M. Elwenspoek, R. Wiegerink, *Mechanical Microsensors* (Springer, Berlin, Germany, 2001)
5. S. Sze, *Semicondcutor Sensors* (Wiley, New York, 1995)
6. K.E. Peterson, Silicon as a mechanical material. Proc. IEEE **70**(5), 420–457 (1982)
7. C. Bai, *Scanning Tunneling Microscopy and its Applications* (Springer Verlag, New York, 2000)

Chapter 3
Microsystems Manufacturing Methods: Integrated Circuit Processing Steps

3.1 Introduction

In this chapter a general overview of the processing steps commonly used in integrated circuit (IC) microsystems manufacturing is reviewed. The specialized processing steps used in MEMS fabrication will be covered in the next chapter.

Integrated circuit manufacturing consists of performing an ordered set of processing steps in a sequential manner. If the ordered arrangement of processing steps results in the complete fabrication of a device, it is called a "process sequence" or a "process technology." The major categories of processing step types used in process sequences include (see Fig. 3.1) thin-film depositions or growths; lithography; etching; impurity doping; and metrology. Additionally, depending on the IC process sequence involved, there may be other types of processes in the sequence as well such as planarization, rapid thermal anneals, cleans, and others.

Metrology steps (labeled as Ⓜ in Fig. 3.1) are typically performed after various important process steps and involve inspections and/or measurements of the thicknesses and uniformity of thin films deposited, measurements of the dimensions of the features made in photoresist after a lithography has been performed, or measurement of the dimensions of features made in a material layer after etching. An overview of metrology technologies regularly used in microsystems manufacturing is given in a later chapter.

The starting wafer used in the process illustrated in Fig. 3.1 is purposefully selected to have the desired crystallographic orientation, dopant type, and dopant levels. Some processing steps are repeated multiple times in the process sequence. For example, the photolithography processing step (an optical form of lithography) may be performed many times in a modern state-of-the-art multilayered metal CMOS IC manufacturing process. Additionally, there will likely be multiple thin-film depositions, etches, and impurity implantations during a process sequence, although the specifics of the processing steps, such as material types and thicknesses, will often vary at various stages of the manufacturing process.

M. Huff, *Process Variations in Microsystems Manufacturing*, Microsystems and Nanosystems, https://doi.org/10.1007/978-3-030-40560-1_3

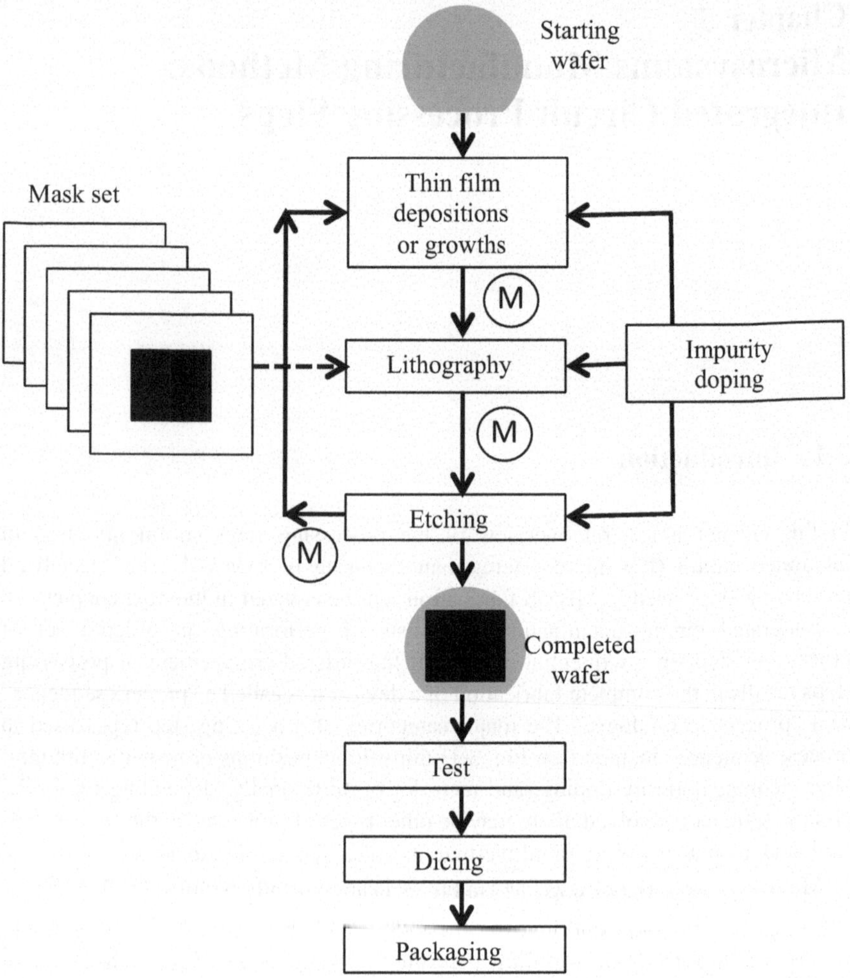

Fig. 3.1 A flowchart to illustrate the manufacturing of integrated circuit microsystems and is composed to major processing step group types such as thin-film depositions, lithography, etching, and impurity doping

A cycle as shown in Fig. 3.1 wherein a deposition of a thin-film layer followed by a photolithography and then followed by an etch may be repeated a number of times, and usually the number of photolithography steps is equal the number of mask plates in the mask set. A mask set is shown at the far left, and typically there may be more than 30 individual mask plates used in an advanced IC manufacturing process. Often the number of photolithography steps that are performed is used as a measure of the complexity of a process sequence.

Once the fabrication of the devices is completed, the completed wafers will usually go through a series of wafer-level tests (sometimes called wafer probings)

to determine the good die from the bad or differentiate the die according to some pre-defined categories of performance specifications. The wafer is then diced into individual die and then packaged. A final test may be performed after packaging to ensure that the packaged device works as designed.

3.2 Basic IC Processing Steps

The next several subsections provide an overview of basic processing steps that are commonly used in integrated circuit manufacturing. We have categorized these basic processing steps into major types, as follows: thin-film growth and deposition; impurity dopant introduction and diffusion; lithography; planarization; etching; and cleaning. Readers with further interest in this subject are directed to other sources that go into greater detail about these processing steps [1–4].

As previously noted, one topic that is often overlooked in the existing references on microsystems fabrication technologies is the resultant variations of the dimensional parameters resulting from these processing steps. Therefore, this chapter will provide guidance on microsystems device dimensional parameter variations. Variations in materials properties will be covered in a later chapter. We will attempt to provide reasonable estimates of the random variations that are typical in a IC microsystems foundry in our review of the processing steps. However, it is noted that these variations should always be measured for each individual processing step and foundry.

3.2.1 *Thin-Film Growth and Deposition Techniques*

The growth and deposition of thin-film material layers play an essential role in the implementation of devices and systems, including providing electrical insulation between adjoining electrical elements in devices; enabling electrical interconnections between electrical devices for implementation of circuits; allowing for the implementation of passive devices such as capacitors and resistors; and providing a means for making electrical connections to devices and circuits from off the die. The review of thin-film growth and deposition begins with thermal oxidation that is categorized as a growth processing step followed by a review of various thin-film deposition technologies including chemical vapor deposition (CVD) processes; atomic layer deposition (ALD); and physical vapor deposition (PVD) processes.

3.2.1.1 Thermal Oxidation

Silicon has a distinct advantage as a semiconductor material choice due to the fact that it readily forms an oxide layer on its surface, specifically silicon dioxide or SiO_2

(also sometimes called oxide, silica, as well as glass) when it is exposed to oxidizing agents such as O_2 or H_2O. The oxidation process can be rapidly accelerated if the silicon substrate is exposed to an oxidizing agent at elevated temperatures.

Thermally grown silicon dioxides have been extensively used in microsystems manufacturing due to the excellent electrical properties of these layers, particularly for use as electrical insulating layers [5–8]. In IC manufacturing, thermally grown silicon dioxide is used for several purposes including to serve as an important element in MOS transistors (specifically as a gate dielectric); to serve as an isolation barrier between devices (specifically as a field oxide); to act as a mask from dopant introduction during implantation or diffusion; and to serve as a passivation layer.

There are several methods that are available to grow a silicon dioxide layer on a silicon substrate including thermal oxidation; wet electrochemical anodization; and plasma anodization. However, thermal oxidation of silicon to grow silicon dioxide is by far the mostly widely used technique.

There are two generally accepted chemical reaction formulas for the thermal oxidation of silicon. The first is silicon reacting with the oxidizing agent of oxygen gas in the form of O_2 and is given by [2]:

$$\text{Silicon} + O_2 \rightarrow SiO_2(\text{dry oxidation}), \tag{3.1}$$

and the second is where the oxidizing agent is water vapor (i.e., steam) and is given by:

$$\text{Silicon} + 2H_2O \rightarrow SiO_2 + 2H_2(\text{wet oxidation}). \tag{3.2}$$

Oxidation of silicon proceeds by a moving interface between the top surface silicon substrate and the silicon dioxide layer that is growing on the surface of the silicon substrate (Fig. 3.2) that advances into the silicon material. The silicon dioxide formed on the surface has a thickness that is about 2.27 times that of the thickness of the consumed silicon material. Or stated another way, for a thermally grown silicon dioxide layer thickness of x, the thickness of the silicon substrate consumed will be $0.44\ x$.

The model widely used for describing the thermal oxidation of silicon under the most common processing conditions is the Deal and Grove Model [9], which is considered accurate for calculating the thicknesses of oxides when the oxidation growth temperatures are in the range from 700 °C to over 1200 °C, the silicon dioxide is grown in thicknesses from 300 Angstroms to 2 microns, and the partial pressures of the oxidants during the oxidation process are between 0.2 and slightly above 1.0 Atm.

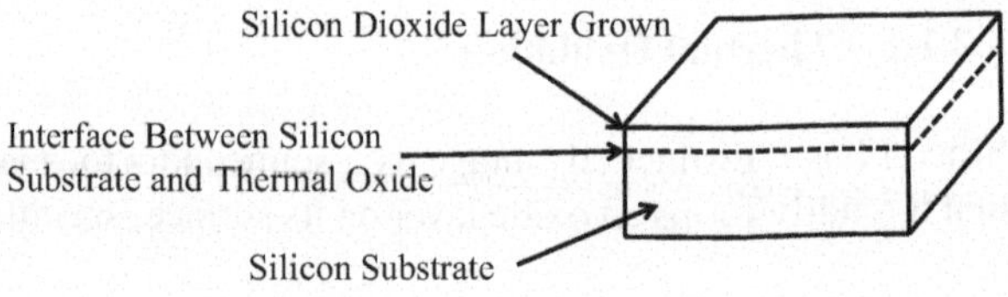

Fig. 3.2 Illustration of the grown of a thermal silicon dioxide layer on the silicon substrate surface

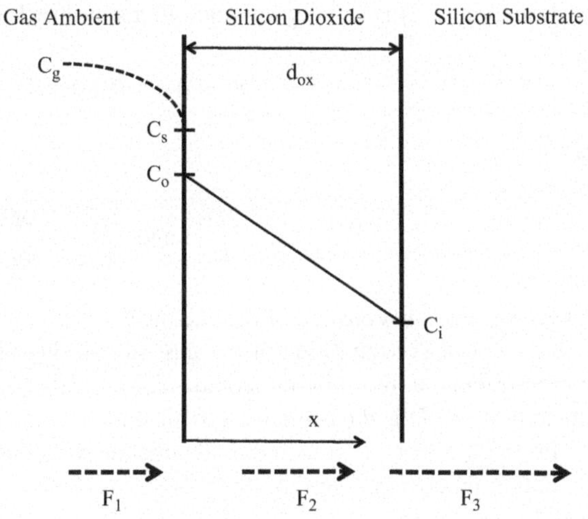

Fig. 3.3 Illustration of the model of thermal oxidation of silicon [12]

The Deal and Grove model is based on evaluating three fluxes in the various media involved, as shown in Fig. 3.3. Each of these fluxes are given by the number of atoms or molecules passing through a unit area in a given unit of time, t. The first flux, F_1, is the flux of oxidizing agents that diffuse from the gas phase to the interface between the gas and the silicon dioxide's growth surface. The second flux, F_2, is the number of oxidizing atoms or molecules that diffuse through the silicon dioxide layer toward the silicon dioxide – silicon interface. And, the third flux, F_3, is number of atoms or molecules that chemically react with the silicon at the silicon dioxide – silicon interface. The flux having the lowest value will determine the overall oxidation rate.

Thermal oxidation systems are operated in steady-state conditions whereby the temperature and flow rate of oxidizing agents are held at constant values once the oxidation process begins. Steady-state conditions would reasonably represent the vast amount of the time for a typical oxidation process.

The derivation of the oxide growth rates is based on solving Fick's laws based on the boundary conditions of an oxidation process. The interested reader can find more information on the deviation in various sources [1–4], and the result is the following equation:

$$x^2 + \frac{2D}{k}x = \frac{2DC_o}{C_1}(t + \tau), \tag{3.3}$$

where τ represents a time shift due to the presence of the initial oxide of thickness d_o present and is given by:

$$\tau = \frac{C_1\left(d_o^2 + \frac{2Dd_o}{k}\right)}{2DC_o} \tag{3.4}$$

Equation (3.3) is usually rewritten in a simplified form as follows:

$$x^2 + Ax = B(t + \tau), \tag{3.5}$$

with:

$$A = \frac{2D}{k} \quad \text{and} \quad B = \frac{2DC_o}{C_1}, \tag{3.6}$$

where k is the rate constant of the chemical reaction of oxidation at the interface, C_o is the oxidant concentration at the gas–silicon dioxide interface, C_i is the oxidant concentration at the silicon dioxide–silicon interface, d_{ox} is the existing oxide thickness, and D is the diffusion coefficient.

From Eq. (3.3), the silicon dioxide thickness for duration of time given by t in the oxidation process is given by:

$$x = \frac{D}{k}\left(\sqrt{1 + \frac{2C_o k^2 (t + \tau)}{DC_1}} - 1\right). \tag{3.7}$$

Equation (3.7) can be approximated for small values of time, t, by:

$$x \cong \frac{C_o k}{C_1}(t + \tau), \tag{3.8}$$

and can also be approximated for large values of time, t, by:

$$x \cong \sqrt{\frac{2DC_o}{C_1}(t + \tau)}. \tag{3.9}$$

Using Eq. (3.5) above, Eqs. (3.8) and (3.9) can be written in the simplified form as:

$$x \cong \frac{B}{A}(t + \tau) \text{ for small values of time, } t, \text{ and} \tag{3.10}$$

$$x \cong \sqrt{B(t + \tau)} \text{ for large values of time, } t. \tag{3.11}$$

The ratio B/A is called the linear rate constant, and B is called the parabolic rate constant. These rate constants are experimentally determined and vary depending on whether the oxidation is performed in oxygen (O_2) or water vapor (H_2). Usually oxidations performed in oxygen are called "dry," whereas those performed in water vapor are called "wet."

Importantly, these rate constants vary with temperature and also with the crystallographic orientation of the silicon substrate. Table 3.1 provides some of the

Table 3.1 Rate constants *A* and *B* for dry oxidations at various temperatures performed in oxygen (O_2) [9]

Temperature (°C)	*A* (microns)	*B* (microns^2/hour)	*t* (hours)
700	–	–	81.0
800	0.37	0.001	9.0
920	0.24	0.005	1.4
1000	0.17	0.012	0.4
1100	0.09	0.027	0.08
1200	0.04	0.045	0.027

Table 3.2 Rate constants *A* and *B* for wet oxidations at various temperatures performed in water vapor (H_2O) [12]

Temperature (°C)	*A* (microns)	*B* (microns^2/hour)	*t* (hours)
920	0.5	0.203	0
1000	0.23	0.287	0
1100	0.11	0.51	0
1200	0.05	0.72	0

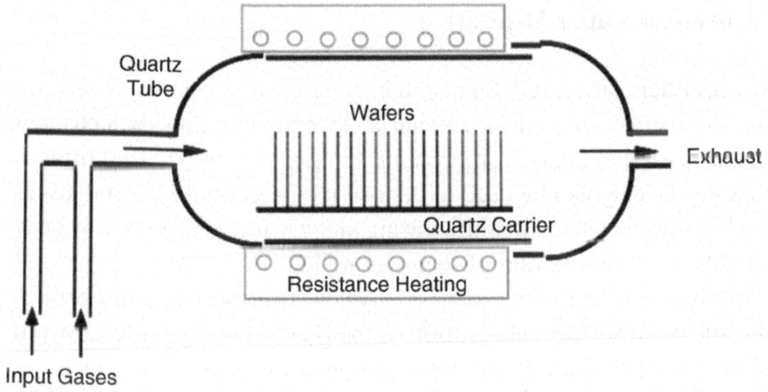

Fig. 3.4 Illustration of an oxidation system

experimentally determined rate constants for dry oxidation at various temperatures, while Table 3.2 does the same for wet oxidations. These are from Deal and Grove [9] and were for (111) oriented silicon substrates. The rate constants for other orientations of silicon can be found in Deal [10]. As noted in comparing the rate constants in Tables 3.1 and 3.2, wet oxidation rates are higher than those for dry oxidations. Nevertheless, the electrical properties of dry oxides are superior to wet oxides.

An illustration of a typical oxidation system is shown in Fig. 3.4. An oxidation system consists of a cylindrical fused quartz glass tube that has adequate length and breadth to process a number of substrates (e.g., 50–100 wafers or more) in a single batch. The substrates are located onto a quartz carrier, with equal distant spacing so as to ensure that adequate gas flows between each substrate. A resistance heater surrounds the fused silica furnace tube to heat the tube to the oxidation temperature, usually in the range between 900 °C and 1150 °C. When not in use, the furnace is kept at a lower idling temperature usually around 350–500 °C, and an inert gas, such as nitrogen, is flowed through the tube when it is idling. The tube has gas inlets to allow the introduction of gases into the tube. The amount of each gas feed into the system is controlled using in-line mass flow controllers (not shown) connected to a programmable process controller. The commonly used process gases for thermal oxidations include oxygen, hydrogen, and water vapor.

A more recent development in oxidation furnace systems is vertically oriented furnaces that provide some advantages, such as reduced contamination; smaller footprints; and improved uniformity.

The uniformity of an oxidation process is highly dependent on the type of equipment being used; the type of oxide being grown; the size of the batches being processed; the maintenance of the equipment; the expertise of the process engineers setting up and running the equipment; and the technology vintage of the oxidation system. Older oxidation furnace systems typically have across wafer and substrate-to-substrate non-uniformity values of ±2.5% or less. More modern vertical furnaces can provide uniformity values of ±1% or better [11].

3.2.1.2 Chemical Vapor Deposition

Chemical vapor deposition (CVD) is extensively used in microsystems manufacturing for the deposition of thin-film material layers including dielectrics and semiconductors. It is used to deposit amorphous and polycrystalline thin films, and under special conditions, can also be used to deposit single-crystal material layers. CVD is widely used in microsystems manufacturing since it is a relatively low cost, reliable, and reasonably uniform method of deposition [12].

CVD involves the following stages: physical transport of the chemical precursors[1] to the substrate surface; absorption of the precursors onto the substrate surface; dissociation of the precursors into their chemically reactive components; migration of the chemical reactive components to chemical reaction sites; chemical reactions; and desorption of reaction byproducts from the substrate surface. It is important to note that the material layer's chemical composition is formed on the surface of the substrate and not in the vapor phase above the substrate. In order to adequately control the complex chemical reactions required for the CVD process, the process parameters including temperature, gas flow rates, and pressure are accurately controlled. Most often the substrate is heated during deposition. Some CVD processes allow for the incorporation of dopants into the deposited material layers. This is commonly called "in situ doped CVD." This is achieved by introducing into the process chamber specific gases that contain the desired dopant species. Other

[1] A chemical precursor is a chemical compound that participates in a chemical reaction that produces another chemical compound.

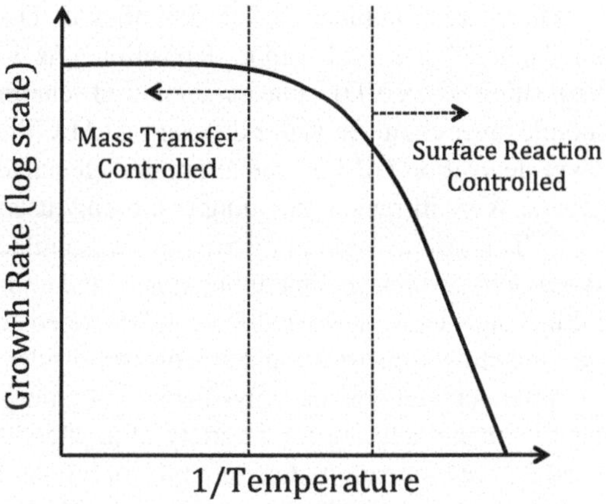

Fig. 3.5 An illustration of the relationship between the growth rate of the thin film being deposited and the temperature (plotted as 1/temperature)

nonreactive gases, such as nitrogen or argon, may also be introduced into the process chamber as carrier gases or purging gases.

A model of CVD growth consists of two fluxes, the first of which is composed of the flux of reactants through a vapor boundary layer at the surface of the substrate, and the second flux is the reactants that are involved in the chemical reactions that result in the thin-film material layer growth [13]. Under steady-state conditions, these two fluxes are equal with the overall process deposition rate that is determined by the lower flux. A typical CVD process will exhibit a growth rate that is dependent on temperature as illustrated in Fig. 3.5.

There are two identifiable regions on the curve. The first is labeled the "mass transfer controlled region" that occurs at relatively high process temperatures. In this region, the process is controlled by the arrival rate of chemical reactive species at the surface of the substrate, and the growth rate has relatively little temperature dependence. However, a process operating in this region requires that the process gases be very uniformly distributed in the chamber over the substrates in order to get a deposition that is uniform across the substrate. Typically these systems are single substrate process chambers. Films deposited in this region can exhibit very high levels of crystallographic order, and it is possible to grow single-crystal material layer using a process called epitaxy.

The second is labeled the "surface reaction controlled" region where the growth rate is dependent on temperature. CVD systems operating in this region must have excellent control of the temperature as well as a uniform temperature distribution. CVD systems operating in this region tend to be large batch systems, which assist in creating a large thermal mass for better temperature stability. Deposition rates are not high in this region, but large number of wafers can be processed simultaneously. The films deposited will also tend to have less crystallographic order.

There is an intermediate region between these two regions wherein there is both mass transfer and surface reaction dependencies. This growth region is usually avoided since control and repeatability in the process results are very difficult to obtain.

There are a number of subcategories of chemical vapor deposition including atmospheric chemical vapor deposition (ACVD); low-pressure chemical vapor deposition (LPCVD); plasma-enhanced chemical vapor deposition (PECVD); atomic layer chemical vapor deposition (ALCVD) [more often referred to as atomic layer deposition (ALD)]; metal organic chemical vapor deposition (MOCVD); and others. We will review the more commonly used CVD techniques.

CVD is widely used in microsystems manufacturing for the deposition of several types of material layers including doped and undoped silicon (amorphous; polycrystalline; and single crystal); doped and undoped silicon dioxides; silicon nitride (with varying stoichiometries); and silicon oxynitrides.

Single-crystal silicon, polysilicon, and amorphous silicon are used in integrated circuits manufacturing for a variety of applications. Single-crystal silicon is mostly used for the fabrication of certain type of bipolar transistors wherein layers of silicon of between 0.5 and 20 microns are used [14]. Typically, these single-crystal silicon layers employ in-situ doping.

Polysilicon is used in older metal–oxide semiconductor (MOS) transistor technologies as a gate metal material, but other materials have mostly displaced polysilicon for this application in more advanced IC process technologies. Polysilicon is also used as an electrical interconnection layer material in integrated circuits; as a material layer for making resistors, particularly high-value load resistors in static memories; and as trench refill material.

Primary uses of CVD SiO_2 include as a dielectric electrical insulation layer between the gate electrode of MOS devices and other electrically conductive layers; as an electrical insulation layer between metal interconnect layers; as a masking layer; as an electrical isolation capping layer; as a passivation layer; and as a gettering material layer. Phosphorus-doped silicon dioxide, also called phosphosilicate glass (PSG), can be deposited using CVD. PSG has the property that it can reflow at lower temps. This is useful in integrated circuits since it can allow the abrupt step heights of patterned oxide insulation layers to be smoothed out, thereby making coverage with a subsequent metal interconnect layer much easier and more reliable [15]. In some cases, the reflow temperature of PSG may be too high for certain types of advanced microelectronics, and in those cases, borophosphosilicate glass, or BPSG, may be used as an alternative [16]. At medium deposition temperatures, between 650 °C and 800 °C, SiO_2 can be deposited using tetraethyl orthosilicate, $Si(OC_2H_5)_4$, also known as TEOS [17]. TEOS can also be doped to form PSG and BPSG [18].

Silicon nitride (SiN) is an amorphous material that is widely used as a passivation, electrical isolation, and barrier material layer in microelectronics manufacturing [19, 20]. SiN has excellent properties as a passivation layer as it is impervious to moisture and impurity diffusion. Additionally, it can be deposited in a very conformal manner and with low defect density. Silicon nitride can be deposited using LPCVD in what is called the stoichiometric form of silicon nitride. While this type of silicon nitride is very conformal, has a good uniformity, and has low defect density, it does suffer from one major problem; the stoichiometric form of silicon nitride has a very high level of residual stress [21]. A solution to this problem is to increase the ratio of silicon in the layer that can be tailored to have a near zero stress state [22].

Table 3.3 List of the most widely used materials deposited using CVD, the associated reactant gases, deposition methods, and deposition temperatures

Material type	Reactant gases	Deposition method	Deposition temperature (°C)	Notes
Polysilicon	SiH_4	LPCVD	580–650	May be in situ doped
Silicon nitride	$SiH_4 + N_2$	PECVD	200–350	
Silicon nitride	$SiH_4 + NH_3$	PECVD	200–350	
Silicon nitride	$SiCl_2H_2 + NH_3$	LPCVD	650–750	
Silicon nitride	$SiH_4 + NH_3$	LPCVD	700–900	
Silicon dioxide	$SiH_4 + O_2$	PECVD	200–350	
Silicon dioxide	$SiH_4 + N_2O$	PECVD	200–350	
Silicon dioxide	$SiH_4 + O_2$	ACVD	200–350	Poor step coverage and uniformity
Silicon dioxide	$Si(OC_2H_5)_4$	LPCVD	650–750	TEOS, very conformal
Silicon dioxide	$SiCl_2H_2 + N_2O$	LPCVD	850–900	Conformal
Phosphosilicate glass	$SiH_4 + O_2 + PH_3$	ACVD	300–500	PSG
Phosphosilicate glass	$SiH_4 + O_2 + PH_3$	PECVD	300–500	PSG
Borophosphosilicate glass	$SiH_4 + O_2 + PH_3 + B_2H_6$	ACVD	300–500	BPSG
Borophosphosilicate glass	$SiH_4 + O_2 + PH_3 + B_2H_6$	PECVD	300–500	BPSG

Table 3.3 provides a summary of the most popular material types deposited using CVD processes as well as their associated reactant chemistries, deposition methods, and temperatures. This table is by no means inclusive of all known materials or processes but instead provides information on some of the most widely used ones. In the next sections, the most widely used CVD deposition methods are reviewed.

3.2.1.2.1 Atmospheric Chemical Vapor Deposition (ACVD)

Atmospheric chemical vapor deposition (ACVD) is a CVD process performed at atmospheric pressure. Since ACVD does not involve the control of the process pressure during deposition, this method is most useful for depositions performed in the mass transfer controlled region of Fig. 3.5. ACVD has been mostly used for the epitaxial deposition material layers such as silicon that is performed at temperature of 1000 °C or higher [23]. The majority of epitaxial material layers are grown using a process called vapor phase epitaxy (VPE), which is a special type of CVD.

ACVD is performed using a relatively simple equipment setup (see Fig. 3.6). The deposition takes place inside a process chamber that is heated by the use of heating elements that are located externally of the process chamber. The substrates are

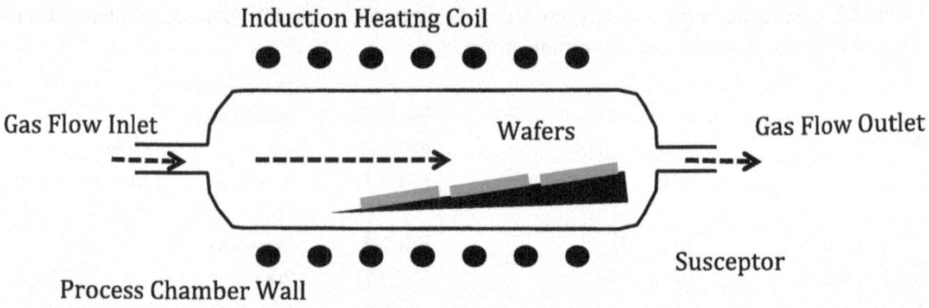

Fig. 3.6 An illustration of an atmospheric chemical vapor deposition (ACVD) system. This configuration is sometimes called a "cold wall reactor" system

located on a susceptor chuck that may also be heated. The susceptor chuck may be wedge-shaped along the reactor chamber gas flow path in order to improve the uniformity of the deposited thin-film layers. Process gases include precursor gases, carrier gases, dopant gases, and inert purging gases. Precursor process and carrier gases at predetermined flow rates under computer control.

The deposition rates of high-temperature ACVD can be relatively high, typically around 1 micron per minute or more. The limitation on deposition rate is mainly due to reactant gas depletion effects, and therefore there is a trade-off between growth rates and the resultant uniformity of the thin films. Some reactor designs employ the ability to create a temperature profile along the gas flow path through the reactor for the purpose of helping to compensate for gas depletion effects.

The uniformity of atmospheric chemical vapor deposition systems varies depending on whether they are high-temperature or low-temperature systems. Low temperature ACVD systems typically have non-uniformities in the range of 10% or higher. Additionally, the particulate formation in this process for the deposition of dielectric such as silicon dioxide has been shown to be very high. The non-uniformity for high-temperature epitaxial deposition using commercial-grade systems is typically in the range of 4% or better [24]. Additionally, if the deposited epitaxial films are in situ doped, there will also be a uniformity of the resultant resistivity of the deposited film. For modern commercial systems, the non-uniformity of resistivity is typically 10% or better.

3.2.1.2.2 Low-Pressure Chemical Vapor Deposition (LPCVD)

Low-pressure chemical vapor deposition (LPCVD) is a commonly used type of CVD process for the deposition of various thin-film material layers in IC microsystems manufacturing. Of the more common materials deposited using LPCVD include various doped and undoped forms of silicon dioxide (e.g., phosphosilicate glass [PSG], borophosphosilicate glass [BPSG], low-temperature oxide [LTO], etc.); doped and undoped forms of polycrystalline silicon (polysilicon); silicon nitride (e.g., Si_3N_4); and oxynitride. These thin-film material layers are important in microelectronic manufacturing for use as MOS gate metal layers, electrical interconnect layers, insulating layers between electrical conducting lines, and passivation layers to protect devices from the environment. LPCVD has found widespread use in the microsystems manufacturing industry due to the fact that it offers several advantages over alternative methods of thin-film deposition including relatively good control of the chemical stoichiometry and crystal structure of the materials deposited; comparatively good uniformity; reasonably good step coverage for modest aspect ratio features (i.e., less than 2 to 1, depth-to-width aspect ratios); reasonably good deposition rates; the ability to deposit material onto large wafer batches (i.e., single deposition batches of 50 or more wafers); and low costs (in terms of both deposition costs and equipment maintenance costs). The ability to process large batches and the resultant low processing costs are the most important advantages of LPCVD.

LPCVD depositions are usually performed in the surface reaction controlled region of Fig. 3.5. Typical LPCVD deposition pressures range from about 10 mTorr to 100 mTorr and temperatures in the range from about 400 to 950 °C. LPCVD is usually performed in a hot-walled reactor system composed of a long tube of fused silica or quartz that is horizontally oriented (Fig. 3.7). One or more inlet gas lines are connected to the tube usually on the substrate loading side of the reactor for introduction of process gases into the reactor. The flow rates of gases are controlled to the desired levels using in-line mass flow controllers. To obtain lower pressures

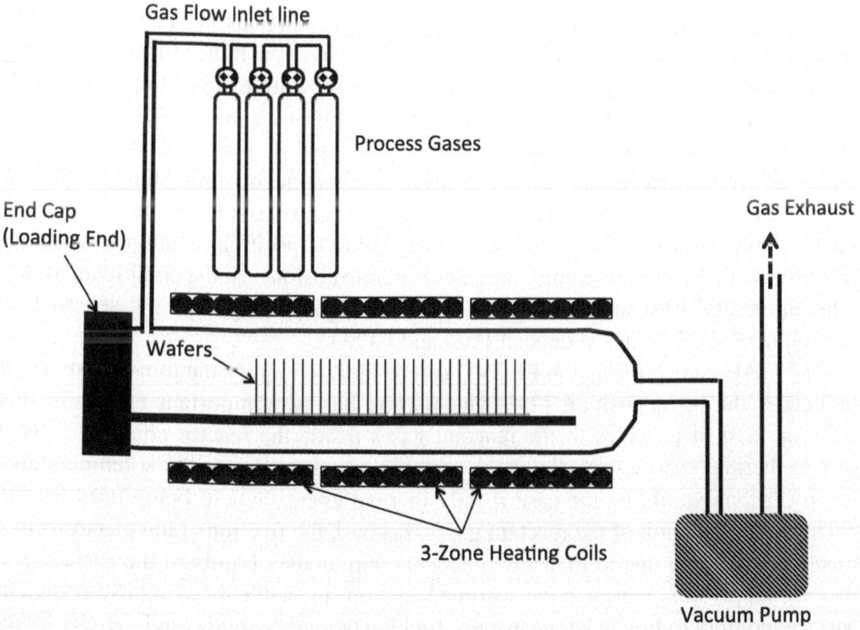

Fig. 3.7 An illustration of a horizontally oriented low-pressure chemical vapor deposition (LPCVD) system

inside the reactor tube, a vacuum pump is used, usually a high-throughput rotary vane type of vacuum pump that may be assisted by a roots blower pump. The reactor tube is heated to the desired deposition temperature using heating elements positioned outside of the reactor tube walls. Typically these are configured with three hearing zones.

There are a variety of material types that are deposited using LPCVD. The most common for use in IC manufacturing are polysilicon (doped and undoped), silicon nitride, and silicon dioxide (doped and undoped).

The uniformity of LPCVD deposition systems varies depending on many factors including the temperature of the deposition; the design of the deposition system; the type of material layers being deposited; etc. In general, the non-uniformity of horizontal hot-walled LPCVD systems is generally less than ±2.5% across a load of substrates. More recently, vertical LPCVD deposition systems have become available, but at higher cost than horizontal systems, and these typically have batch non-uniformities less than 2% [25–27].

3.2.1.2.3 Plasma-Enhanced Chemical Vapor Deposition (PECVD)

LPCVD can provide thin films having good material properties, good step coverage, and uniformities, but for many situations, the deposition temperatures are too high. Plasma-enhanced chemical vapor deposition (PECVD) is an alternative CVD process that allows high-quality thin films to be deposited at lower temperatures. In PECVD energy from the plasma is used to enable the reactions to take place at lower process temperatures. Using PECVD it is possible to deposit thin films at temperatures at 200 °C or lower [28]. Because of the low temperature deposition of PECVD, it is routinely used for depositions of insulating thin-film material layers used in integrated circuit manufacturing. It can safely be used on substrates having previously fabricated electronic devices, metal interconnects, and shallow dopant implants and diffusions. PECVD is widely used to deposit oxides on metal electrical interconnects for use as interlayer dielectrics and to deposit passivation and protective layers of silicon nitride onto integrated circuit substrates at the completion of the fabrication. PECVD can be configured for in situ doping using doping gases, such as phosphine and diborane, to deposit PSG and BPSG.

The presence of a plasma in PECVD can be used to modify the thin-film material properties during deposition [29]. The plasma has two important effects in this process. First, it provides to the reactant gases inside the reactor chamber a form of non-thermal energy that allows the precursor gases to dissociate at temperatures far lower than would be the case if only thermal energy where being used for the chemical dissociation of the reactant gases. Second, the free ions (and electrons to a much lesser extent due to their low masses) continuously bombard the surface and thereby supply non-energy to the absorbed adatoms to enable them to have sufficient surface mobility to find reactant species, find lower energy states (and thereby lower residual stress), as well as improve conformality and uniformity of the deposited thin film.

There are several different PECVD deposition reactor system configurations including cold-walled parallel plate; hot-walled parallel plate; and electron cyclotron resonance (ECR). Most PECVD deposition systems used in industry are either the cold-walled parallel plate or the hot-walled parallel plate. The cold-walled parallel plate PECVD system can have improved uniformity compared to the hot-walled parallel plate configuration, mostly because it is configured as a single-wafer system. The hot-walled system can process a large batch of substrates in a single batch, but the uniformity of the hot-walled system is usually not as good. Shown in Fig. 3.8 is an illustration of the cold-walled parallel plate PECVD configuration. The pressure inside the reactor chamber is lowered using a vacuum pump connected to the chamber. One of more wafers is placed onto a heated platform. Inside the chamber are located two electrodes. One is positioned above the wafers, and the wafers sit on the other electrode. These electrodes are electrically connected to a radio-frequency (RF) generator that operates at 13.56 MHz to create a plasma.

The hot-walled PECVD deposition system is illustrated in Fig. 3.9. This system has a configuration that is very similar to that of a LPCVD reactor. It employs a

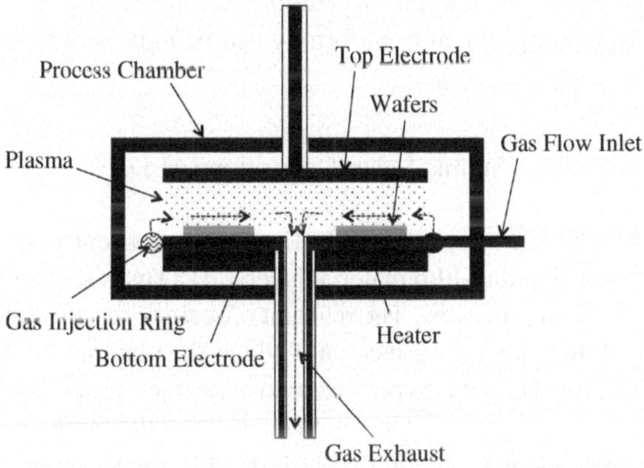

Fig. 3.8 An illustration of a cold-walled parallel plate plasma-enhanced chemical vapor deposition (PECVD) system

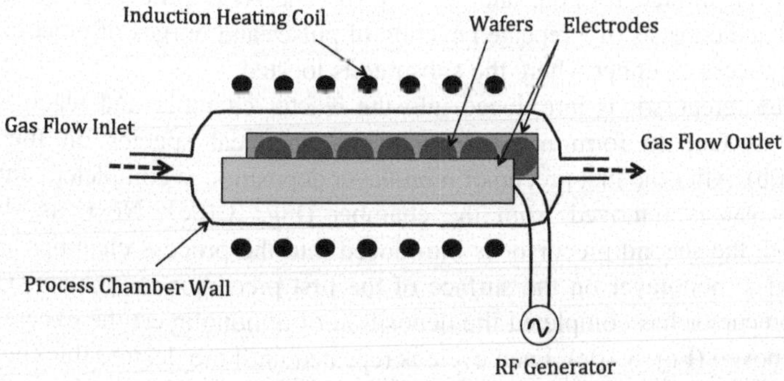

Fig. 3.9 An illustration of a hot-walled plasma-enhanced chemical vapor deposition (PECVD) system

process chamber that is usually made of fused silica that is connected to a vacuum pumping system. The wafers inside the process tube are heated using an induction-heating coil. The wafers are positioned onto a graphite electrode with a counter electrode positioned directly parallel with the wafer electrode. The two sides of the electrodes are connected to a radio-frequency (RF) generator that operates at 13.65 MHz. Process gases are introduced into one end of the process tube, and the reactant by-products are removed through the exhaust at the other end of the process tube. Similar to the LPCVD system described above, the reactant gases can be depleted along the length of the reactor tube, thereby limiting the uniformity of the film thicknesses deposited using this method.

The uniformity of PECVD depends on the equipment configuration, manufacturer, material layers being deposited, deposition temperatures, and operator expertise. As noted above, hot-walled parallel plate systems typically have reduced uniformity compared to single-wafer cold-walled parallel plate PECVD reactors.

The non-uniformity of hot-walled parallel plate systems is typically less than ±2.5% across a process batch. The single-wafer cold-walled parallel plate PECVD reactors typically have a non-uniformity of less than ±2% across each wafer. However, if multiple depositions are performed with rotating the wafer between depositions, the non-uniformity can be improved to around ±1% [28].

3.2.1.2.4 Atomic Layer Deposition (ALD)

Atomic layer deposition (ALD) is a more recently developed CVD process step for depositing thin-film material layers. ALD uses sequential and self-limiting gas phase chemical processes. The reactant process gases are called precursors, and each reacts with the existing surface onto which the material layer is being depositing in a self-limiting fashion. Typically, two precursor gases are used in ALD, and neither is introduced into the deposition chamber at the same time. One precursor is introduced and reacts in a self-limited fashion with the substrate, and once completed, the next phase is started wherein the second precursor is introduced to react with the substrate also in a self-limited fashion, and so on [30]. The ALD process is illustrated in Fig. 3.10 and consists of a repeated cycling of pulses and purges of precursor gases into the process chamber where the substrate is located.

The first precursor is introduced into the reactor chamber and reacts with the substrate surface to form a monolayer of a chemical species on the surface (Fig. 3.10b). After the first precursor monolayer deposition is completed, any excess precursor gas is removed from the chamber (Fig. 3.10c). Next, as shown in Fig. 3.10d, the second precursor is introduced into the process chamber and starts to deposit a monolayer on the surface of the first precursor monolayer. Once the second precursor has completed the deposition of a monolayer, the excess process gas is removed (Fig. 3.10e). Each cycle is repeated until the desired thickness of the material layer has been deposited. The nature of ALD enables very precise control of the deposited film thickness. Typical deposition thickness per cycle is approximately

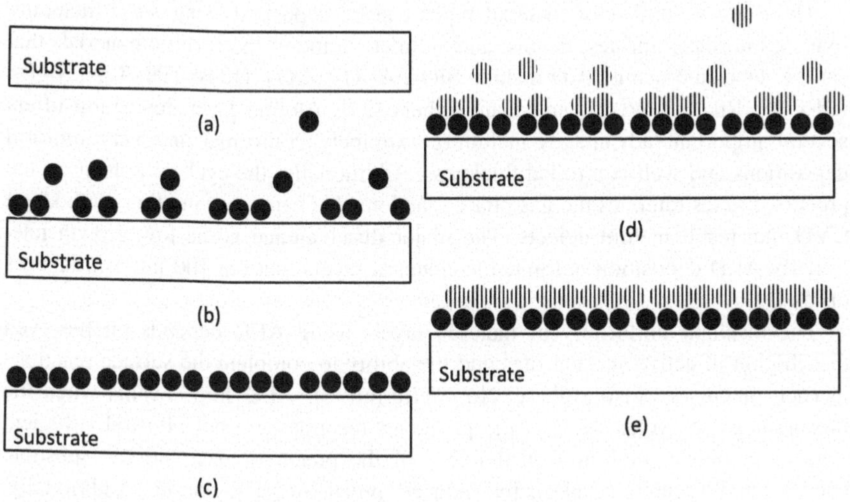

Fig. 3.10 Illustration of the process in atomic layer deposition (ALD)

Fig. 3.11 Illustration of the process reactor design for performing atomic layer deposition (ALD)

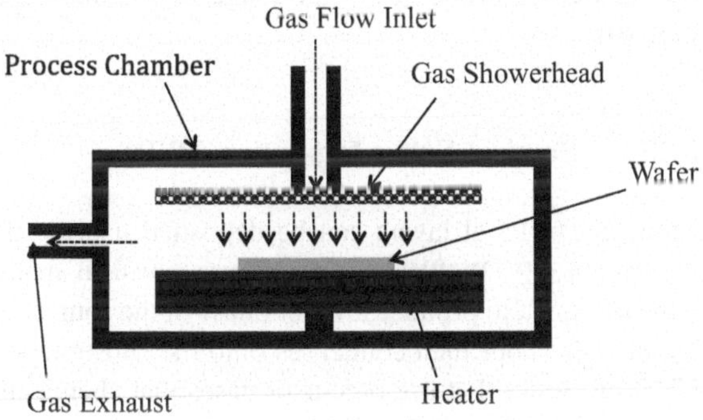

1 Å/cycle, depending on the specific process chemistries involved. ALD deposition may require an elevated temperature depending on the chemical precursors.

An illustration of an ALD deposition system is shown in Fig. 3.11. The design consists of a process chamber that is airtight, a substrate holder that is heated, a gas flow inlet conduit for the introduction of the precursor and purge gases, a gas showerhead to assist in distributing the gas flows uniformly over the substrate, and a gas exhaust where the reactant by-products and unused precursor gases are removed from the process chamber during the cycling using a vacuum pump. The deposition pressures are typically in the range of mTorr. The deposition temperatures are typically in the range from 25 °C to over 400 °C depending on the chemistry being used [30–32]. ALD can be performed at lower temperatures when performed in a plasma environment and is called plasma enhanced atomic layer deposition (PEALD) [31, 33–36].

There are a number of material types can be deposited using ALD including oxides; fluorides; nitrides; metals; and sulfides. Some of the specific materials that can be deposited using ALD include SiO_2; Al_2O_3; ZrO_2; HfO_2; TiN; TaN; MgF_2; AlF_3; Pt; Ru; Ni; W; Ti; and many others [37]. Atomic layer deposition offers several important advantages including extremely conformal and very uniform depositions and well-controlled thickness. Additionally, the cyclical pulsing of the precursor gases removes the gas-phase reactions that can occur in other methods of CVD that result in film defects. The major disadvantage is the low growth rate. Usually ALD deposition is limited to material layers that are 100 nm in thickness or less.

The resultant uniformity of material layers using ALD depends on the even distribution of active surface sites and the ability to complete the surface reactions in each precursor pulsing phase. Non-uniformity of ALD films results when the precursor pulse cycles overlap, the precursor reactions are not allowed sufficient time to complete, non-uniform delivery of the precursor gases to the substrate surface, and premature thermally induced precursor dissociation. Additionally, PEALD requires a uniform plasma over the substrate in order to obtain a uniform material layer deposition. In general, the non-uniformity of a typical ALD process that is well controlled is in the range of less than ± 1 to 2% across the substrate [38, 39].

3.2.1.3 Physical Vapor Deposition (PVD)

Thin-film material layers can be deposited using a process called physical vapor deposition (PVD). PVD processes use physical methods, such as heating or ion bombardment, to produce a vapor phase of the source material to be deposited from a target. The vapor then condenses onto the substrate surface. PVD is different from CVD processes that use precursor gases that chemically react in the chamber or on the substrate. One of the major advantages of PVD deposited materials is that they can be deposited at relatively low temperatures relative to CVD processes. Moreover, it is possible to vary the processing conditions to influence the material properties of PVD deposited materials. There are two main types of PVD, evaporation and sputtering. PVD processes can be used to deposit a wide range of materials and material types including metals, metal–oxides, semiconductors, dielectrics, and polymers. Binary and ternary compounds can also be deposited using PVD methods.

3.2.1.3.1 Evaporation

The deposition of thin films by evaporation is a thermal process whereby a target material is heated in a high vacuum environment and the material from the target changes its phase state from a solid to a vapor (a process called sublimation), the vapor transports to the substrate surface(s), and then the vapor precipitates onto the surface of the substrate, thereby changing its phase state back from a vapor to a solid resulting in a coating of the target material onto the substrate [40]. There are several different means to provide thermal energy to the material to be evaporated with the most common being an electron beam, also called e-beam evaporation. The target material is usually placed in a crucible that acts as an anode that is bombarded with an electron beam from a filament that is usually composed of tungsten.

The electron beam can be generated using several different means including thermionic emission and field emission. The electrons in the beam are accelerated to a very high kinetic energy (i.e., 5–30 keV) and steered to the target material crucible using magnetic and electrical fields [41]. The electron beams striking the target material lose their kinetic energy very rapidly resulting in the target material heating and sublimation. The electron beam powers can be very high, up to over 1200 kW for some evaporator systems, thereby allowing high deposition rates to be obtained (around 1 micron/minute). Evaporation must be performed inside a sealed process chamber at high vacuum levels in order for the electron beam not to be disrupted from the filament to the target material; typically a vacuum level of 10^{-7} torr or lower is obtained prior to initiating the deposition.

Figure 3.12 is an illustration of an evaporative electron-beam physical vapor deposition (EBPVD) system. The system consists of a process chamber connected to a high-vacuum pumping system, typically either a diffusion or cyro-pump.

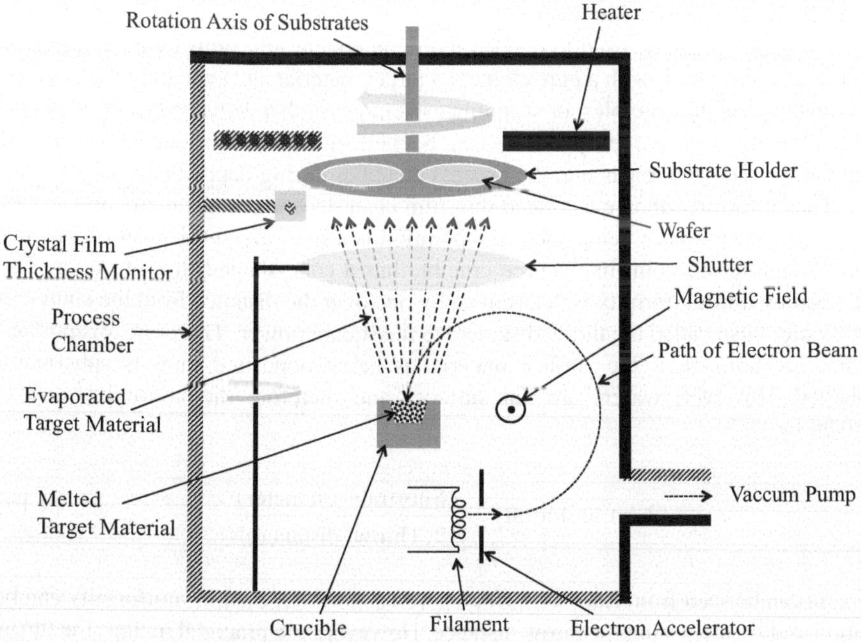

Fig. 3.12 Illustration of the deposition system for electron-beam physical vapor deposition. This is also called evaporation

A filament is connected within a closed-loop circuit to maintain the desired power level to the electron-beam source. An electron beam from the source is accelerated by an electrical field. A magnetic field steers the electron beam toward the target material located in a crucible. Electrical fields may be used to scan the electron beam back and forth over the target material to ensure uniform heating and improve deposition uniformity. The target material usually consists of pure pellets of the material to be deposited. The crucible is held at a positive voltage potential relative to the filament so that it acts as a sink for the electron beam.

The substrates are usually rotated in order to improve the uniformity. Heaters may be used inside the process chamber to provide heating to the substrates during deposition to modify the deposited thin-film material properties and improve uniformity. A crystal sensor for in situ monitoring the thickness of the deposited material layer is typically employed [40].

The advantages of evaporation are that it is simple and the equipment is inexpensive. Additionally, the deposition rate can be well controlled, even at very low deposition rates, thereby allowing very thin layers to be deposited with relatively good thickness control. There are also several disadvantages including it is line-of-sight deposition and therefore the deposition is not conformal to substrate surfaces having any pre-existing topography; it can be performed without heating the substrate; at electron-beam voltages of more than 10 kV, some target materials such as aluminum emit X-rays that can cause damage and trapped charges; and metal droplets are occasionally ejected from the source and can land on the substrates, especially at higher power levels.

Evaporation can be used to deposit multicomponent thin films by having two or more crucibles, each with a pure elemental target material and switching the electron beam between the crucibles, or scanning each target with a dedicated electron beam. Additionally, reactive evaporation can be performed by introducing controlled amounts of a reactive gas into the process chamber during deposition.

The uniformity of an evaporated thin-film layer depends on geometry and size of the deposition system being used. Evaporation is a line-of-sight deposition process and therefore the point-like source crucible has a cone-shaped flux of vapor. The deposition non-uniformity is determined by one over the distance from the source to the wafer (also called the throw distance) to the square power. Therefore, evaporated films are uniform if the surface material is being deposited onto is spherically shaped. However, wafers are flat surfaces and therefore the non-uniformity is given by:

$$\text{Non-uniformity} = \frac{(\text{Substrate diameter})^2}{2(\text{Throw distance})^2}. \tag{3.12}$$

As can be seen from Eq. (3.12), for a given size wafer, the non-uniformity can be improved by increasing the throw distance. However, as a practical matter, the throw distance is not adjustable in an evaporation system and is set by the manufacturer. Additionally, larger throw distances result in lower deposition rates, lower efficiency in target material usage, increased system cost and size, and longer cycle times (due to longer vacuum pump down times since the chamber will be larger). For example, for a 200 mm diameter wafer with a throw distance of 2 feet, the non-uniformity across the wafer would be about 5.4%, while if the throw distance were increased to 3 feet, the non-uniformity improves to about 2.4%. As this example demonstrates, the non-uniformity can vary considerably depending on the dimensions of the system used. In general, about ± 1 to 2% is at the lower end of obtainable non-uniformity for commercial systems used in production [42].

3.2.1.3.2 Sputtering

Sputtering is used extensively in IC manufacturing. A primary application of sputtering is for depositing the metals used as electrical interconnects in integrated circuits. Sputtering uses the bombardment of a target with ions from a plasma to dislocate atoms from the target to create a vapor of the target material [43]. Sputtering is a physical process driven by a momentum exchange between impinging ions from the plasma having high kinetic energies and atoms in the target material. The ions collide with the target atoms and cause some of them to be ejected from the surface. The plasma is generated using an inert gas such as argon. The target may be electrically biased (with either DC or AC signals or both), and this can be used to provide acceleration (i.e., additional energy) to the ions impinging onto the target. If the target material is composed of an insulating material, then a RF bias can be used to reduce charge buildup on the target surface.

An important parameter of sputtering is what is called the "sputtering yield." The sputtering yield is the number of atoms from the target ejected per incident ion and depends on the target material type; the mass of the ions; the angle of incidence of the ions with the target; and the energy of the ions. To increase the efficiency of the process, the target and substrate are located as close to one another as possible. A distance of a few centimeters (<10 cm) is a typical value of separation.

The pressure that sputtering is performed, in the mTorr range, translates into a mean free path of around 1 cm, and therefore the target atoms will collide with other particles in the plasma before reaching the substrate surface. These collisions result in the atoms losing some of their energy before reaching the substrate and also more pronounced scattering. This assists in sputter deposition having more conformal coverage of the topology of the substrate surfaces compared to evaporation, for example. Moreover, modifying the gas pressures during sputtering enables the ability to alter the deposited material properties. Modification of other process parameters, such as substrate temperature and biasing power, can also be used to effect changes in the material properties of thin-film layers deposited using this technique. Residual gases may be incorporated into the film during deposition, particularly if the gases are chemically active. Similar to reactive evaporation, reactive sputtering can be performed by introducing controlled amounts of a reactive gas into the process chamber during deposition. Step coverage of sputter deposition can be improved by either heating the substrate during deposition, increasing the pressure inside the chamber during deposition, applying a RF bias to the substrates, or all of the above.

There are essentially three variations of the configuration of sputtering systems: DC bias sputtering; radio-frequency (RF) bias sputtering; and magnetron sputtering. DC bias sputtering is limited to electrically conductive materials and therefore is used for the deposition or metals. RF and magnetron sputtering can be used for a wider variety of materials types and therefore are used more extensively in integrated circuit manufacturing.

Figure 3.13 is an illustration of a simple DC bias sputtering system configuration. The wafer is mounted onto an electrode that is connected to a negative voltage potential and acts as the anode. The material target is mounted onto another electrode separate from the first that is connected to a positive voltage potential and acts as the cathode. The cathode and the target may be water cooled to better control their temperature during sputtering. The sputtering is performed in a closed and sealed process chamber that is connected to a vacuum pumping system.

An RF bias sputtering system (not shown) replaces the DC electrical connections to the anode and cathode with an AC source operating at RF frequencies, typically 13.56 MHz. A matching network is inserted into the circuit to allow the power efficiency to be increased. A magnetron sputtering system (also not shown) employs a powerful magnetic field (generated by using either a solenoid or magnetics) positioned behind the target in a manner that their magnetic field lines are parallel to the target surface. This results in significantly higher ion efficiency, sputter yield, deposition rates, and uniformity than either DC or RF sputtering, at the expense of a more complicated design configuration. The so-called "S-gun" magnetron sputtering system is very popular since it enables the substrates to be mounted onto a planetary stage, thereby allowing higher throughputs and better uniformity.

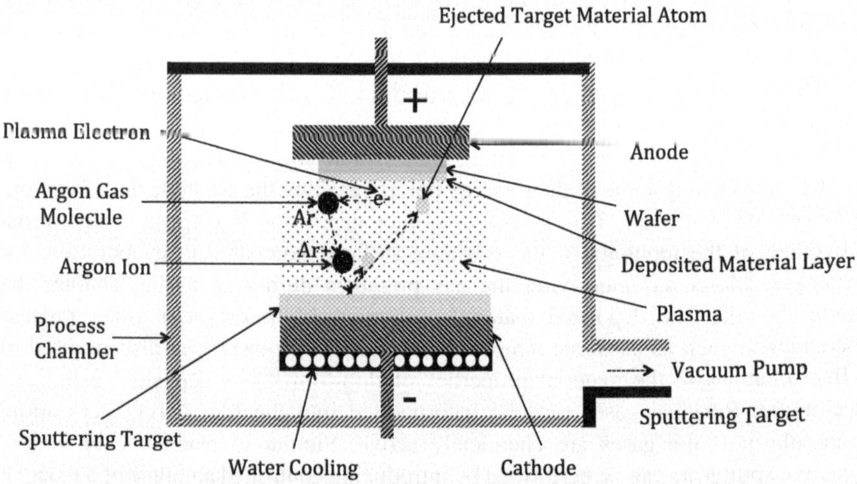

Fig. 3.13 Illustration of the DC bias sputtering deposition system for physical vapor deposition (note that a conventional voltage potential is used to define anode and cathode in figure rather than the electrochemical potentials)

Sputtering can be used to deposit many different types of materials including most metals, dielectrics, and semiconductors. Additionally, it is possible to perform reactive sputtering by introducing a reactive gas into the process chamber during sputtering.

Multiple component alloys can be deposited using sputtering as well. In fact, for most alloys, sputtering is a preferred method of deposition compared to evaporation since it allows much better control of the resultant stoichiometry.

There are some notable advantages and disadvantages of sputtering. The advantages include the following: it can be a reasonably uniform method of deposition; the control of thicknesses is good; the film properties, such as step coverage and crystal structure, can be tailored by varying the bias and heat on the substrates during deposition; the substrates can be sputter cleaned prior to deposition to ensure high-quality interfaces; the target depletion rate is low; and there is no X-rays generated and therefore no X-ray damage to pre-existing devices on the substrates.

The disadvantages of sputtering are as follows: high capital equipment costs; deposition rates that can be comparatively slow for some materials; possible incorporation of impurities into the deposited thin films since the depositions are carried out at moderate vacuum levels; and some pre-existing material types, such as polymers, can be damaged or eroded by impinging ions on the substrate surfaces.

The uniformity of thin-film layers deposited using sputtering depends on the type of sputtering system used, the equipment manufacturer, the number of wafers onto which deposition is performed simultaneously, the materials being deposited, how well the system is maintained, etc. The more uniform deposition systems are usually single-wafer chamber tools configured with a load-lock. Additionally, almost all sputtering systems used in integrated circuit production are magnetron-type sputtering systems in order to have higher deposition rates and throughputs. Most commercial sputtering systems will have less than $\pm 2.5\%$ non-uniformity across the substrate and from wafer-to-wafer$\pm$. Non-uniformities exhibited in the stoichiometry of alloys and compounds deposited using sputter deposition will be $\pm 1\%$ or more [44].

3.2.2 *Impurity Doping*

The introduction of specific elements, often called impurities or dopants, into semiconductor materials is important for fabrication of microsystems devices. Specifically, the introduction of controlled amounts of these elements into defined regions of the semiconductor enables that region to have a defined type of majority charge carrier, either electrons or holes, as well as a defined level of electrical conductivity. The introduction of these elements is also important to create depletion regions wherein the semiconductor has a defined boundary with opposite types of charge carriers on either side the boundary. This is called a p–n junction and signifies that one side of the junction is n-type doped, with electrons as the majority charge carriers, and the other side of the junction is p-type doped, with holes as the majority charge carriers. The formation of p–n junctions is an essential and basic component of making different types of transistors and diodes used in integrated circuits. For each type of semiconductor, there are certain elements that can act as dopants. For example, silicon is a group IV element, and the n-type dopants for silicon are from group V and include phosphorus and arsenic, while the p-type dopants are from group III and include boron.

In order to be useful for integrated circuit devices, the introduction of these elements into semiconductor materials must possess the following attributes: the dopants must be able to be introduced in controlled amounts, usually defined as number of dopants per cubic centimeter; the dimensions of the regions where the dopants are introduced into the semiconductor, including length, width, and depth, must be able to be controlled; the locations on the semiconductor where the dopants are introduced must be controlled; and the dopants introduced into the semiconductor must be electrically active.

There are two basic methods for the introduction of dopants into semiconductors. The first is diffusion whereby there is a net migration of dopants from beyond the semiconductor's surface, where the concentration of dopants is high, into the semiconductor material, where the concentration of dopants is low. Fundamentally, diffusion is based on a concentration gradient. The second, called ion implantation, involves injecting the dopant ions into the semiconductor by accelerating them to sufficient energies to be able to penetrate some distance into the surface of the semiconductor. After implantation, an anneal is subsequently performed to diffuse the dopants to the desired depth in the substrate and also electrically activate the dopants and reduce crystal damage caused by the implantation process. Both of these methods are used in integrated circuit manufacturing; however, the more advanced technologies now mostly rely on ion implantation.

3.2.2.1 Thermal Diffusion

Thermal diffusion was extensively used in the early of ICs, but it is rarely used anymore in more advanced technologies. This is due to the very shallow diffusions that are required in advanced electronic devices. Nevertheless it is still used as a processing method in some older IC processes and MEMS. Various models to predict the diffusion process have been developed. Often these models are less than accurate, but experimental data can be used to improve these models. Additionally, computer simulation software for diffusion processes are available that are very commonly used by engineers for process modeling purposes.

Diffusion is described mathematically using Fick's first and second laws. Fick's first law in one-dimensional states that the flux, J, of a substance is proportional to the negative of the concentration gradient, or:

$$J = -D\frac{\partial C(x,t)}{\partial x}, \tag{3.13}$$

where C is the concentration, $\partial C(x, t)/\partial x$ is the concentration gradient at point x and time t, and D is the diffusion constant (units of length^2/time). Fick's second law examines a volume element wherein the concentration gradient is decreasing with time. Without going through the derivation, if the diffusion constant, D, is independent of location, Fick's second law can be rewritten as:

$$\frac{\partial C(x,t)}{\partial t} = D\frac{\partial^2 C(x,t)}{\partial x^2}. \tag{3.14}$$

The solution to this equation requires two independent boundary conditions. We consider two solutions. The first solution relates to a scenario with a constant and non-time varying flux of dopants at the surface of the substrate. This would be applicable to a substrate immersed into a furnace having an ambient vapor of gas dopants that is not changing with time and is sometimes called a *pre-deposition* since it is used to introduce a uniform and repeatable amount of dopants in the substrate. The boundary conditions for this case at $t = 0$ is given as follows:

$$C(0,t) = C_s \quad \text{and} \quad C(\infty,t) = 0, \tag{3.15}$$

where C_s is the concentration of the dopants at the surface. The initial condition is:

$$C(x,0) = 0, \tag{3.16}$$

The solution for Eq. (3.14) using these boundary and initial conditions is given by:

$$C(x,t) = C_s \mathrm{erfc}\left[\frac{x}{2\sqrt{Dt}}\right], \tag{3.17}$$

where the term "erfc" is called the complementary error function and is tabulated for various values of $x/2\sqrt{Dt}$. The total amount of a dopant diffused into the semiconductor as a function of time, t, is given by $Q(t)$ and is found by integrating Eq. (3.17) from the surface of the substrate to infinity and is given by:

$$Q(t) = \frac{2}{\sqrt{\pi}} C_s \sqrt{Dt}. \tag{3.18}$$

Using Eq. (3.17) the depth of the junction, x_j, wherein the concentration of diffused dopants in the semiconductor is equal to the background concentration of the dopants in the substrate, $C_{\mathrm{substrate}}$, can be found from:

$$x_j = 2\sqrt{Dt}\,\mathrm{erfc}^{-1}\left(\frac{C_{\mathrm{Substrate}}}{C_S}\right). \tag{3.19}$$

The second important situation relates to the circumstance where a relatively thin layer of dopant has been introduced at the surface of the substrate. The dopants may have been introduced using either a pre-deposition or ion implantation. The boundary conditions are as follows:

$$\int_0^\infty C(x,0)dx = Q_o \text{ and } C(x,\infty) = 0, \tag{3.20}$$

where Q_o is the total amount of dopant that was introduced into the surface of the substrate. The initial condition is:

$$C(x,0) = 0. \tag{3.21}$$

The solution using these boundary and initial conditions is given by:

$$C(x,t) = \frac{Q_o}{\sqrt{\pi Dt}} e^{-\frac{x^2}{4Dt}} \tag{3.22}$$

The surface concentration of dopant, C_s, is found from substitution $x = 0$ into Eq. (3.22) which gives:

$$C_s = C(0,t) = \frac{Q_o}{\sqrt{\pi Dt}}, \tag{3.23}$$

and the junction depth where the dopant concentration is equal to the background dopant concentration in the substrate, $C_{\text{substrate}}$, is given by:

$$x_j = \left[(4Dt)\ln\left(\frac{Q_o}{C_{\text{substrate}}\sqrt{\pi Dt}}\right)\right]^{1/2}. \tag{3.24}$$

Equation (3.24) is used for the so-called drive-in of the dopants whereby an initial surface concentration is driven to the desired junction depth. For example, if the dopants were introduced using ion implantation that was followed by an anneal, Eq. (3.24) could be used to calculate the junction depth.

While the equations reviewed above can be used to calculate the diffusion processes for many circumstances, there are factors that may result in inaccuracies. First, it was assumed that the diffusion coefficients, D, are constant in value. The assumption of a constant diffusion coefficient is accurate when the dopant concentration levels are low, but for higher dopant levels, the diffusion coefficient is a function of the dopant concentration level [45]. Secondly, the impurities used in semiconductors are mostly substitutional and ionized, thereby resulting in an electrical field that can affect the movement of dopants. Thirdly, diffusion in an oxidizing ambient can result in the diffusion rates being retarded or enhanced from the normally expected rates depending on the dopant type and substrate crystal orientation [46]. Many of these complicating factors can be modeled using more sophisticated methods such as modeling software called SUPREM for this task and augmented with experimental results to improve the predictive accuracy.[2]

Diffusion processes are also used to dope polycrystalline semiconductor materials, such as polysilicon layers. The diffusion process for polycrystalline semiconductors is different than that of single-crystal materials due to the presence of a large number of grains and grain boundaries [47–49].

The equipment used for diffusion processes are very similar to those used for thermal oxidation that were described above. Like the requirements for oxidation, diffusion furnaces require good control over the temperature levels, temperature uniformity, as well as the gas flow rates. A diffusion furnace consists of a quartz furnace tube surrounded by one or more heating elements, a wafer carrier, and a dopant delivery system. The dopant delivery system for diffusion processes can take several forms including diffusion from a dopant source that is usually in the form of a vapor at the substrate surface; diffusion from a deposited oxide source that contains dopants; and diffusion after introduction of the dopants into the semiconductor after ion implantation. Most diffusion processes use temperatures in the range from about 800 °C to 1200 °C.

The uniformity of diffusion of dopants depends the type of equipment used, the number of wafers onto which deposition is performed simultaneously, the dopants being diffused, the length of the diffusion time, how well the system is maintained, etc. A large determinate in the uniformity of a diffusion process is the uniformity of the dopant sources and the temperature uniformity in the diffusion furnace. Since diffusion systems are not considered state of the art any longer, there has not been a drive to improve the equipment technology as compared to other processing steps. In general, a typical dopant diffusion non-uniformity across a substrate is no better than ±5% across the substrate and from wafer-to-wafer.

3.2.2.2 Ion Implantation

Ion implantation is a process for introducing dopants into a semiconductor material based on creating a beam of ions composed of the dopant species, accelerating these ions under a high electrical field potential to strike the semiconductor, whereupon the ions penetrate some distance into the surface of the semiconductor substrate. Ion implantation provides a higher level of control compared to diffusion, specifically in the amount and depth of the dopants placed into the semiconductor. As a result of this, ion implantation is used more often than diffusion for advanced IC manufacturing.

An undesired attribute of implantation is that the ions implanted into a substrate surface collide with the atoms in the semiconductor material causing damage to the

[2]For interested readers, more information about SUPREM can be obtained at http://www-tcad.stanford.edu/tcad/programs/suprem-IV.GS/Book.html.

crystal lattice structure. Further, the implanted ions are not immediately electrically active since they have not become part of the silicon lattice structure, which is required for them to behave as dopants. Therefore, ion implantation must be followed by a thermal anneal in order to repair the lattice damage and enable the implanted dopants to chemically bond within the silicon lattice to become electrically activated.

Models for ion implantation have been developed to allow the prediction of dopant profiles based on the amount of ions, energies, angles of impingement, and other factors. Since ion implantation involves the introduction of the dopant species followed by a thermal anneal, these models have multiple parts including the first part based on the physics of ions colliding with the semiconductor to derive a resultant implanted species concentration profile and a second part involving the diffusion of the dopants based on their migration during the anneal. The model of the second part, that is the diffusion, was covered in the section above. In this section, a model of predicting the penetration profile is discussed.

The implanted species profile, also called the penetration profile, is the amount of dopant as a function of depth into the semiconductor surface. The model is based on physical interaction of the ions having high energies colliding with the semiconductor atoms. These collisions may be between the ions and the nuclei of the atoms in the substrate or between the ions and the electron clouds surrounding the nucleus [50–52].

The depths that the ions penetrate into the semiconductor surface depend on the mass and energies of the ions as well as other factors including the type of semiconductor that the ion implantation is performed on, the angle of the incident ions, and other factors. As the ions impinge onto the semiconductor surface, they collide with the atoms and electrons in the target material through a series of scattering events until the ion loses sufficient energy to eventually come to rest. This is illustrated in Fig. 3.14 where an ion that is implanted has a path that changes direction as it collides with atoms, thereby taking a scattered pathway through the material. The ion finally comes to rest some distance into the semiconductor material. The ion causes damage to the semiconductor material as it travels through it due to the multiple collisions that occur. This damage is of the type that breaks chemical bonds of the semiconductor and places lattice atoms into interstitial[3] locations.

The orthogonal distance that the ion penetrates into the semiconductor as measured from the material surface is called the projected range and is given by R_p. Importantly, the projected range usually has a different value from that of the total path length of the ion since the ion takes a more convoluted path, each having a slightly different angle into the semiconductor due to scattering events. The total distance of the ions into the semiconductor substrate is given by R [52].

[3]Interstitial atoms are atoms that are located in the crystal but are not part of the crystal lattice of the material.

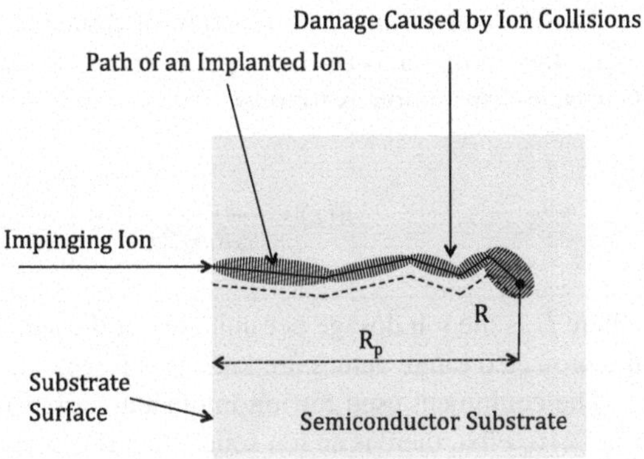

Fig. 3.14 An illustration of an implanted ion's path in a semiconductor. The total distance traveled is given by *R*, and the distance traveled normal to the surface is given by R_p. As shown, the ion undergoes a series of scattering events as it travels through the material and a region of damage is caused in the material surrounding the ion path

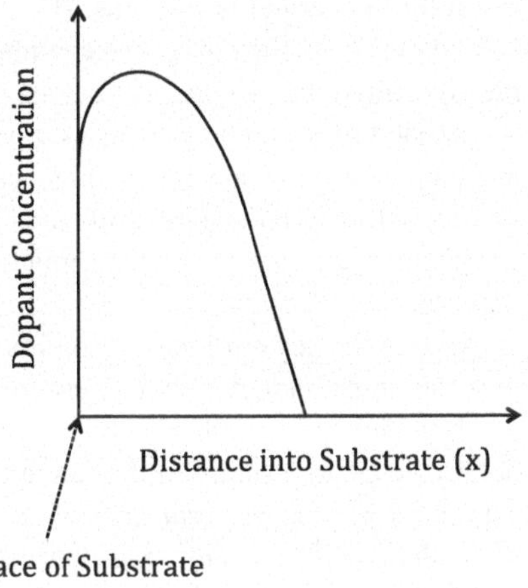

Fig. 3.15 An illustration of a typical implanted ion concentration profile showing that the maximum dopant concentration is located some distance into the substrate surface

A typical profile of the implanted ions as a function of depth is shown in Fig. 3.15. It is noted from this figure that the maximum dopant concentration is not at the semiconductor surface but is located some small distance into the substrate from the surface. The projected range is an average depth of normal penetration of the ions. That is, some ions will undergo a lesser number of collisions per unit distance traveled into the semiconductor and thereby penetrate further into the material than the projected range. And, other ions will have more collisions per unit distance traveled and thereby penetrate a lesser distance into the material than the projected range. Therefore, the projected range will be defined by a distribution function given by the amount of dopant ions per unit depth. The amount of statistical variance in the projected range along the normal direction of penetration into the target is given by the projected straggle, or ΔR_p. As noted above, some of the ions when they collide with atoms in the material scatter in off-normal directions and the amount of variance in the projected range in the direction perpendicular to the angle of ion incidence is given by the projected lateral straggle, or $\Delta R_{\perp}$.

The number of ions as a function of distance, x, into a semiconductor material along the normal axis of the target is given by $n(x)$ and can be approximated by a Gaussian distribution as follows:

$$n(x) = \frac{D}{\Delta R_{\mathrm{p}}\sqrt{2\pi}} \exp\left[-\frac{\left(x - R_{\mathrm{p}}\right)^2}{2\Delta R_{\mathrm{p}}^2}\right], \tag{3.25}$$

where D is the ion dosage per unit area of the surface. Equation (3.25) assumes that the projected range values are known beforehand.

The equipment used for ion implantation consists of multiple parts as shown in Fig. 3.16. First, there is an ion source that is usually composed of a source gas that is broken into ions using a heated filament. For example, a source gas may be BF_3, resulting in charged boron ions, B^+. A voltage potential is employed to extract the source ions from the ion source chamber into a mass analyzer that is used to filter out the specific ions desired. The ions that are selected by the analyzer are then accelerated to high energies using a voltage potential. Various apertures along the ion path make the ion beam more collimated, and the chamber pressure is kept relatively low to reduce ion collisions as they travel toward the target. The resultant ion beam is scanned across the semiconductor substrate using electrical field deflection plates.

Among the key attributes of ion implantation systems is the ion dosage implanted into the semiconductor that is controlled by the ion current and the average ion depth, which is controlled by the acceleration energies. Typically, the ion implantation dosages are usually in the range of 10^{12}–10^{16} ions/cm^2, wherein the dosage is expressed in ions per unit area of the substrate surface, although higher dosages, more than 10^{18} ions/cm^2, may be used for special applications, such as creating

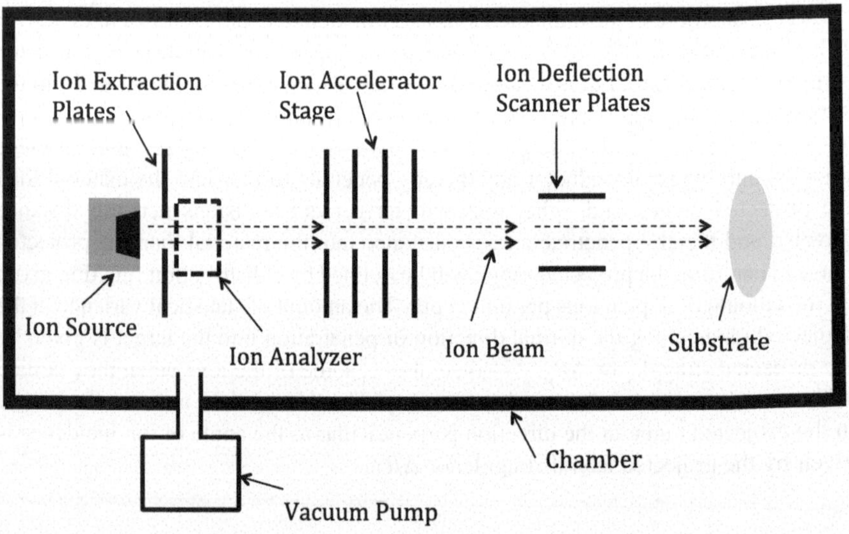

Fig. 3.16 An illustration of an ion implantation system

oxide layers buried in the semiconductor surface. The acceleration energies for implantation usually are in the range from below 1 keV to over 1 MeV and result in average ion depths of 10 nm to about 10 microns.

Among some of the specific advantages of ion implantation compared to diffusion include the following: it has higher levels of control in total dosage and uniformity across the substrate; it has less lateral distribution across the substrate, thereby enabling devices to be made with smaller footprints; the dopants are pure; a single piece of equipment can be used to introduce multiple dopant species into the semiconductors; dopants can be interjected through thin surface layers such as silicon dioxide for threshold adjustment of MOS devices; the dopants can be selectively introduced into the semiconductor using patterned masking layers; and different dopant profiles can be made using multiple implantation. Nevertheless, ion implantation also has some issues and limitations including the following: it causes damage to the semiconductor that must be repaired; the obtainable depths are limited compared to diffusion; and implantation equipment is relatively expensive to both purchase as well as maintain.

With respect to the uniformity of the dopant profiles using ion implantation, there are two principal issues of interest. The first is the uniformity of the implanted dopants across the substrate surface, and the second is the variation in dopant uniformity from wafer-to-wafer. Note that the wafer-to-wafer variation could be the use of the same implanter as well as the use of multiple implanter systems. As a general rule, the uniformity of the dopants implanted across the substrate is quite good and can be better than 1.5%. This is considerably better than can be obtained with diffusion. However, pre-existing mask patterns on the substrate can result in some shadowing effects at the edges of the mask for implanters that use electrostatic beam rastering since the ion beam appears as a point source. Therefore, some newer implanters use a mechanical scanning to reduce this effect.

The wafer-to-wafer uniformity variation depends on several factors. The first is whether it is measured on a single implanter or across multiple implantation systems, and second, how well the implantation system is maintained. Studies have shown wide variations in total dosage levels from implanter to implanter, as much as 14% depending on the dosage levels [53]. The wafer-to-wafer variations in dosage of a single implanter, assuming the system is well maintained, however can be better than 1–2%. The dose uniformity across wafer is usually better than 1.5%. The absolute variation also depends on the atomic weights of the implanted species with lighter ions having smaller variations in general. While the non-uniformity of the ion implantation is of concern to the resultant dopant non-uniformity, it should be noted that implantation will be followed by an anneal to activate and drive in the dopants. Therefore, the thermal process used after the implantation is an important consideration in estimating the overall activated dopant non-uniformity. If diffusion processes are used for the activation of the implanted dopants, then the non-uniformity estimates given previously for diffusion processes are applicable here and may result in an increase of the non-uniformity over and above that of the implantation alone.

3.2.3 Photolithography

Photolithography is perhaps the most important of the basic fabrication methods involved in microsystems manufacturing. The term lithography is a generic term that is used to describe the method of transferring patterns onto the substrate surfaces. These patterns and the means of the transfer process essentially define the shapes and sizes of the devices being manufactured and have a significant impact on the resultant dimensional variations.

There are different forms of lithography including optical-based lithography, also called photolithography; e-beam lithography; X-ray lithography; ion beam lithography; print-based lithography; and extreme ultraviolet lithography, also called extreme-UV or EUV. Each of these lithographic methods have advantages and disadvantages, but in the production domain, photolithography is the almost exclusively used current method in microsystems manufacturing due to its ability to reliably transfer quality high-fidelity replications of the desired device features onto silicon at reasonable cost levels and high-wafer throughputs. Therefore, we will focus only on a discussion of photolithography in this section. Progress in the work to develop more advanced lithographic systems is available in the literature [54–56].

Photolithography uses optical wavelength radiation to transfer the patterns onto the substrate surface. The transfer of the pattern is accomplished using a photosensitive layer that is deposited onto the substrate that can be selectively exposed using optical radiation. Once the photosensitive layer has been exposed, it is developed using a special chemical solution, thereby removing resist from unwanted areas. After the resist has been patterned, it is used to transfer its pattern onto some layer on the substrate surface, usually through a means of etching. Therefore, the photosensitive layer acts as an intermediary in the transfer process.

Performance of photolithography is probably better labeled as a process module since it typically involves multiple steps including the preparation of the surface of the substrate so that a photosensitive polymer can be deposited; the deposition of a thin photosensitive layer onto the substrate surface; the heating of the photosensitive layer to drive off excess solvents (i.e., pre-bake); exposure of this photosensitive layer through a masking layer (that may be aligned to previous layers on the substrate surface); development of the photosensitive layer; a heating process to stabilize the photosensitive layer (i.e., post-bake); and an inspection to determine if the pattern was transfer correctly and with proper alignment accuracy.

A photolithography process is illustrated in Fig. 3.17. As can be seen, there is a pre-existing thin-film layer of material on the surface of the substrate that is to be patterned (Fig. 3.17a). The photoresist layer is deposited on the substrate surface (Fig. 3.17b). A mask with defined patterned is aligned to the substrate, and the photoresist is exposed by projecting optical radiation through the mask (Fig. 3.17c). The photoresist is developed to reveal the pattern that has been successfully transferred from the mask into the photoresist on the substrate (Fig. 3.17d). The substrate can then be etched, thereby resulting in the transfer of the mask pattern from the mask into the thin-film layer on the substrate surface.

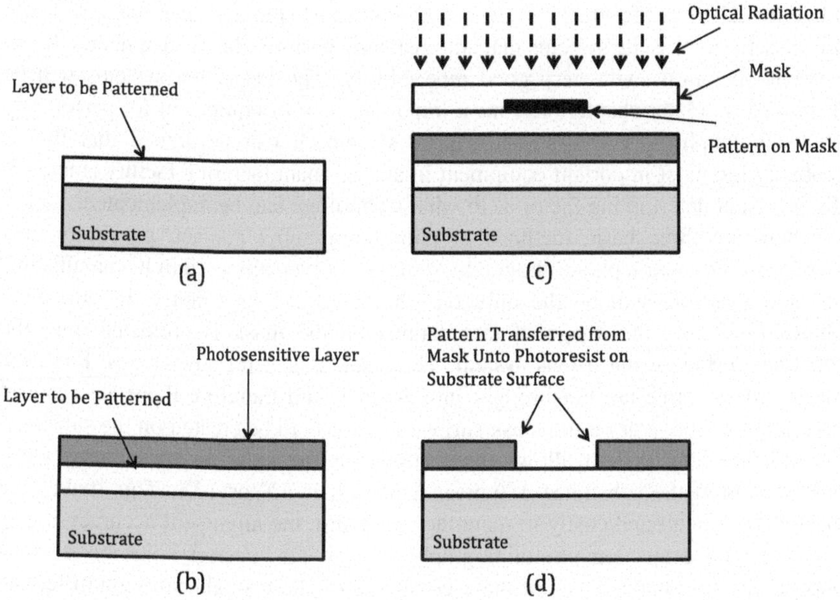

Fig. 3.17 Cross section of substrate as photolithography process is being performed

Photoresists are formulated to be sensitive to specific optical wavelength(s) that the exposure tool operates at. Moreover, the wavelength of the optical radiation has a direct impact on the resolution that can be obtained. Since the minimum feature sizes that can be transferred have been made smaller with the advancement of IC manufacturing, the wavelengths used in photolithography have by necessity also gotten shorter. For example, the 436 nm wavelength, also called the "g-line," was replaced by the use of the 405 nm wavelength, also called "h-line," that was then replaced by the 365 nm wavelength, also called "i-line." These wavelengths were all created using lamp sources. Subsequently, lamp sources were replaced by lasers operating at the desired wavelengths, such as 193 nm, which is referred to as deep-ultraviolet, or deep-UV. Work on developing shorter and shorter light sources continues in order to satisfy the need to obtain smaller feature resolutions.

The photoresist used in photolithography can be either positive or negative in nature. A negative resist transfers the opposite or inverse of the pattern on the mask plate to the photoresist, while a positive resist transfers the same pattern on the mask to the photoresist. Both positive and negative resists require a certain radiation dosage threshold in order to clearly define the patterns, and this is found by multiplying the exposure intensity (mW/cm^2) by the exposure time (seconds) to calculate the exposure energy (mJ/cm^2) radiated onto the resist.

There are a number of separate items that are used in photolithography process steps including a dehydration wafer preparation station; a HMDS deposition system (adhesion promoter on wafer surfaces prior to deposition of photoresist); a photoresist spin station; an align and expose system; a development station; a hard bake station; and a wafer inspection. Most advanced IC manufacturing operations use what are called “wafer-track” systems for photoresist spin and development. These are cassette-to-cassette systems that automatically perform the resist spin and development operations with very good repeatability. The inspection systems will be discussed in a later chapter. The most important item of equipment for performing photolithography is the align-and-expose system. It can be argued that this is probably the most important equipment in any IC manufacturing facility since it is the principal determining factor as to what technology can be implemented.

There are three basic methods of photolithography: contact; proximity; and projection. Projection photolithography is almost universally used in IC manufacturing and therefore will be the only one discussed in this chapter. In projection photolithography, the image of the features on the mask is projected onto the substrate surface using a lens system. This enables several advantages. First, the mask and substrate are not brought into contact, and therefore there are reduced opportunities for wear on the mask surface or defects to be created on the substrate. Second, the lens system allows the features on the mask to be reduced when projected onto the substrate. A typical reduction is 5X or 10X. This makes the mask less difficult and costly to manufacture. Third, the alignment accuracy that is possible with projection photolithography can be far superior since there is no opportunity for slippage. All of these combined to allow smaller minimum feature sizes than possible with other methods. As a result, projection photolithography is almost universally used in IC manufacturing.

An illustration of a typical projection photolithography setup is shown in Fig. 3.18. As with all photolithography systems, there is an exposure light source.

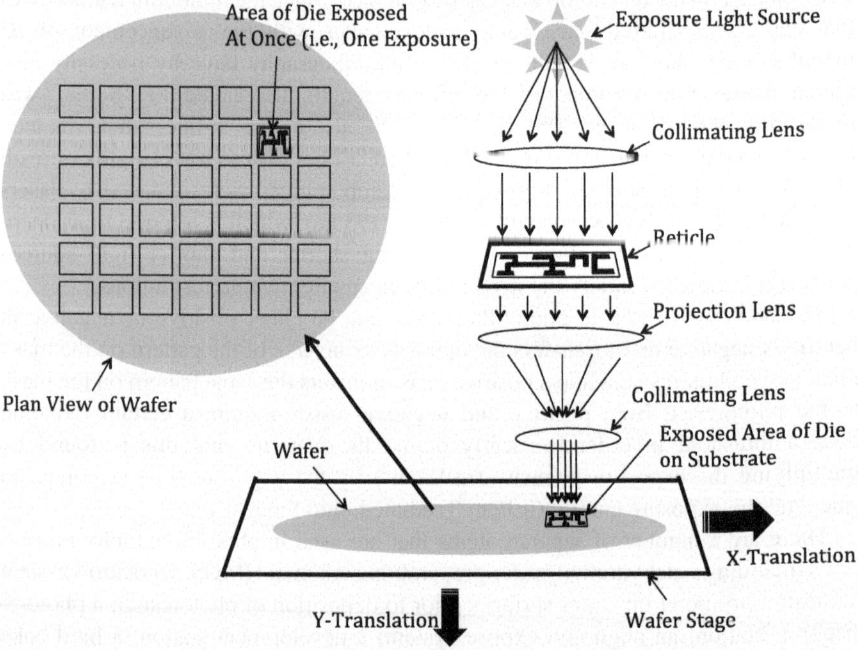

Fig. 3.18 An illustration of projection photolithography using a step and repeat process

For more advanced projection systems, the light source would usually be a laser operating in the deep-ultraviolet spectrum. At the 193-nm-deep, ultraviolet wavelength an argon fluoride (ArF) excimer laser is commonly used. A lens system may be placed between the light source and the reticle if the light source is non-collimated. The collimated light passes through the mask plate, which is often called a reticle. The reticle usually has the pattern of a single die that encompasses one IC die pattern when projected onto the substrate. The light that passes through the reticle is then reduced using a projection lens element. A collimating lens element is placed between the projection lens and substrate in the light path. The collimated light then impinges onto the substrate surface to expose a location that represents a single die.

The substrate is located on a precision mechanical stage that can move in the *x*- and *y*-directions with submicron precision. A limited amount of angular rotation of the substrate is also under precise movement control. Movement of the substrate in the *z*-direction, also with submicron precision, is either by movement of the substrate stage or the optical system. After each exposure for a die has been performed, the precision mechanical stage moves the substrate to the next die position and performs another exposure. Consequently, the layout design on the reticle is stepped out over the substrate surface to create an array of identical die patterns using a precision mechanical stage movement and a light source that is turned "on" and "off" (or shuttered "on" and "off") as the reticle is stepper over the substrate. Therefore, projection photolithography tools are often called "step and repeaters" or "steppers."

In modern projection photolithographic tools, the optics are in diffraction limited regime wherein the resolution, or minimum feature size that can be transferred, is approximately given by:

$$\text{Resolution} = k\left(\frac{\lambda}{\text{NA}}\right), \tag{3.26}$$

where λ is the wavelength of the exposure radiation, NA is the numerical aperture of the optical system, and k is a numerical value that is dependent on the processing conditions and approximately equal to 0.4 for a typical production photolithography process. The numerical aperture, NA, of a lens system is dimensionless number that characterizes the range of angles that the lens can accept or emit light and is given by:

$$\text{NA} = n \sin \theta, \tag{3.27}$$

where n is the index of refraction in which the lens is immersed (e.g., equal to 1 for air) and $\sin \theta$ is the maximal half-angle of the light that can be accepted or emitted by the lens.

The resolution that is possible is directly proportional to the wavelength wherein a shorter wavelength allows smaller features to be printed. The resolution is also inversely proportional to the numerical aperture wherein larger NA values allow smaller features to be printed. However, the numerical aperture cannot be increased without a trade-off in the depth of focus of the optics used in the system. The depth of focus, D_{of}, is the range of distances in the normal direction over which the projected image remains within an acceptably sharp focus and is approximately given by:

$$D_{of} \approx \pm \frac{\lambda}{2(\mathrm{NA})^2}. \qquad (3.28)$$

As can be seen, decreasing λ and increasing NA in Eq. (3.26) result in better resolution, but according to Eq. (3.28), it also results in a degradation of the depth of focus. For example, at the current 193 nm wavelength light sources most commonly used in the production and a NA value of 0.7 would result in a resolution of 110 nm and a depth of focus of 197 nm.

Nevertheless, the resolution of photolithography has been extended downward to about 10 nm in the last decade while mostly still using 193 nm wavelength sources. This remarkable achievement is attained by a number of technical developments including immersion photolithography [57]; the use of high index materials for the lens elements [58]; and the use of multiple pattern lithography (e.g., double, triple, etc.) wherein it is possible to reduce the minimum feature sizes by a factor of the number of repeats of the lithography [59].

There are a number of important issues related to the obtainable dimensional variations using photolithography processes including the flatness of the substrates; the dimensional variations present on the mask features; the thickness and thickness uniformity of the photoresist; the optical resolution of the photolithography tool; the across wafer uniformity of the exposure source; the alignment accuracy of photolithography tool; the type of underlying material to be patterned; whether or not there is a stack of layers to be patterned; the amount of topography on the surface of the substrates; as well as a number of other processing related issues.

The collective of these issues are commonly lumped together to determine a minimum resolution and the variations of these features for each mask layer, as well as the alignment accuracy for different mask layers that includes the specifics as to what is being manufactured, the processing details, manufacturing tools being employed, etc. to develop what are called "design rules" for a given process technology. Design rules will be discussed in more detail in a later chapter.

The resolution, the variance on the feature size, and the layer-to-layer registration accuracy, each depends on a number of factors. Using immersion photolithography and a 193 nm light source, minimum feature sizes of about 32 nm has been reported without resorting to multiple exposures. Multiple exposures can further reduce the resolutions down by a factor of the number of multiple exposures performed. For example, using the self-aligned double patterning (SADP) method would result in minimum feature sizes of about 16 nm [59].

The variation (or tolerance) of the minimum feature size (usually refers to as the critical dimension of "CD") is statistical in nature with a Gaussian distribution and is commonly expressed as a variance in terms of 3σ of about 10% of the minimum feature size[4] [60–62]. For example, if the resolution of a process were found to be 40 nm, 99.7% of the features would fall between 40 nm ± 4 nm. The 10% variation of the CD has propagated over several generations of IC manufacturing technologies.

With regard to layer-to-layer registration alignment accuracy, this is also based on a statistical distribution that is Gaussian. Most processes specify a 3σ layer-to-layer registration that is about 20% of the minimum feature size. Therefore, if the resolution of the process is 40 nm, then it would be expected that 99.7% of the layer-to-layer tolerances are within about 40 nm ± 8 nm [60].

Importantly, these resolutions assume that everything is nearly perfect in terms of processing controls. That is, they are performed on nearly perfectly flat substrates under ideal conditions. Any changes from the ideal, such as variation in the thickness of the layers, types of material layers onto which the lithography is performed, as well as unusual design features will likely increase the minimum feature sizes and variations.

It is also important to note that IC specifications for minimum feature sizes and pitches (line-width dimension plus spacer dimension) are the smallest for the device gate levels and tend to increase in later processing steps. Most IC manufacturing processes use what is called a "self-aligned" process whereby the source and drain regions of the MOS gates are aligned with the gate due to the nature of the process flow. This greatly lessens the challenges of aligning the sources and drains to the gates of the individual transistors. Additionally, the MOS gates are patterned on flat substrates without pre-existing topography, thereby enabling smaller features to be made. In later stages of the manufacturing, such as the processing steps to make electrical interconnections between the various transistor devices, the dimensions tend to increase, and the alignment accuracy tends to degrade compared to the gate level dimensions.

As a general rule, the minimum feature size, variance of the minimum feature size, and layer-to-layer registration alignment accuracy will need to be determined based on the exact process being performed and the tool set being used. It should also be noted that if the most advanced minimum feature sizes are not required for a given application, the use of an advanced photolithographic process for larger features can result in better dimensional control, that is, smaller resolution capability and less variations in the printed dimensions. However, this needs to weighed against the additional cost of more advanced photolithographic processes.

3.2.4 *Rapid Thermal Anneal (RTA)*

In the discussion of diffusion processes above, the migration of impurities in semiconductor materials at elevated processing temperatures was reviewed. While

[4] Gaussian statistical distributions will be covered in a later chapter.

diffusion is an effective method for introducing and driving dopants into a semiconductor, the level of control of the dopant profile using this approach is not suitable for advanced IC manufacturing due to the requirement that the dopant depth profiles be kept very shallow and there be minimal lateral diffusion of the dopants. Therefore, ion implantation is almost exclusively used for introducing dopants into semiconductors in state-of-the-art IC manufacturing. Nevertheless, ion implantation still requires a subsequent high-temperature anneal in order to electrically activate the dopants and repair the damage caused in the semiconductor lattice structure during the implantation process. While a drive-in diffusion could be used for these purposes, the large thermal budgets (i.e., the amount of time spent at elevated temperatures) of batch furnaces with their associated high thermal masses result in far too much migration of the dopants. A solution to this problem is to expose the wafers to the required temperatures to activate dopants and repair crystal damages while also drastically reducing the time that the wafers are exposed to the elevated temperatures. A method called rapid thermal annealing (RTA) has been developed to address this need.

The basic principal of rapid thermal annealing is to obtain the required activation temperatures, but only for as long a time as needed to activate the dopant species and repair the lattice damage. As it turns out, the wafers can be exposed to the required temperature for electrical activation for only a few seconds to activate and recrystallize the material. Therefore, since the time at the high temperature is so short, the amount of dopant diffusion that can occur is severely limited. By using ion implantation combined with RTA, the positioning of the dopants into a semiconductor can be far more accurately controlled than other means such as diffusion. Therefore, RTA is almost always used for advanced IC microsystems manufacturing wherein the critical dimensions of devices are below a few hundred nanometers.

A typical RTA configuration is illustrated in Fig. 3.19. These systems commonly use tungsten halogen bulbs for heating, and the process is performed in a chamber with an inert gas environment. The wafer is positioned on a quartz holder so as to increase the thermal isolation of the wafer, thereby assisting in obtaining fast thermal response times. The inside of the chamber walls has diffuse reflecting surfaces in order to distribute the thermal radiation as uniformly as possible. An important part

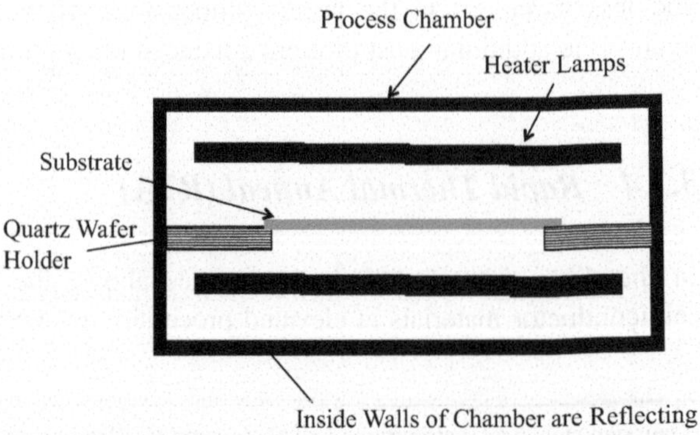

Fig. 3.19 An illustration of a rapid thermal annealing (RTA) system

of the RTA (not shown in figure) is the temperature control system that uses an accurate temperature measurement sensor (e.g., pyrometer) and a closed-loop feedback control.

Ideally, the annealing process should take place over the shortest time period possible. However, as a practical matter, care needs to be exercised to avoid thermal shock and potential for creation of dislocations in the semiconductor materials, and this places restrictions on how quickly the temperature can be ramped up and down. The temperatures and times of rapid thermal annealing processes depend on the type of dopant and semiconductor. Temperatures up to 1100 °C may be used. Interestingly, due to the high-temperature ramp rates involved in RTA, the substrate may not reach thermal equilibrium, and therefore the electrically active dopants can exceed the solid solubility of the dopants in the semiconductor. This effect allows the introduction of dopants into the semiconductor exceeding the solid solubility. Both of these effects are useful for advanced ICs where high dopant gradients are the norm.

The low thermal budgets of RTA has made this technique attractive for other base processing steps such as for the growth of very thin dielectric layers and the nitridization of thin oxide layers. However, other competing methods for thin dielectrics, such as ALD, are mostly used in most IC manufacturing.

Another use of RTA is for the formation of silicides and electrical contacts. Silicides are made by depositing a thin metal layer, such as Ti, W, Mo, Ta, etc., on the top surface of a polysilicon layer, followed by an anneal that alloys the metal with the silicon to form a silicide. Silicides are used for electrical interconnects since they have reduced resistivity compared to even highly doped polysilicon layers. Electrical contacts often must be annealed in order to obtain the lowest value of resistance through the contact. RTA is useful for both of these applications since it provides for a lower overall thermal budget compared to traditional furnace-based anneals.

The uniformity of the temperature using RTA depends on the system being used, but most of the presently available systems are able to obtain a temperature control uniformity of better than 1.5 °C across the substrate [63].

3.2.5 *Planarization*

Planarization is a process capability that removes some or all of the topography on the surface of the substrate that is created due to fabrication steps that have been previously performed. Planarization is particularly important in advanced manufacturing processes for making the metal electrical interconnection layers on the IC. Usually, these are fabrication steps that occur later in the manufacturing process and involve the deposition and patterning of various metal and insulating dielectric material layers to make electrical connections to the various devices located in the semiconductor material, route power, route clocking functionality, etc. For advanced IC processes, there may be five to ten or more metal layers

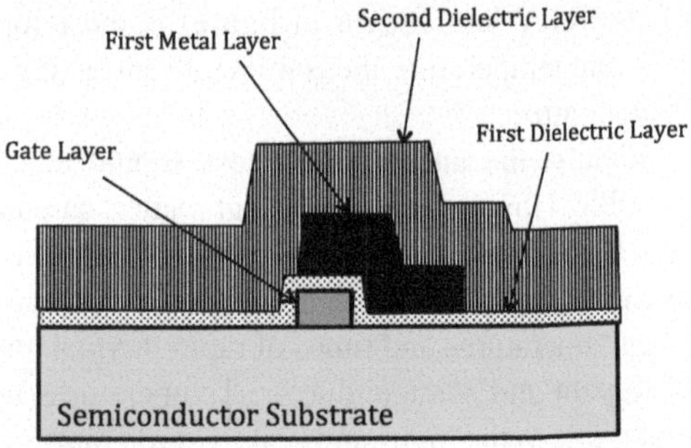

Fig. 3.20 An illustration topography on an IC substrate with only one metal layer

involved. As the number of metal layers increases in the later stages of the IC manufacturing process, the metal interconnections and the insulating layers between these metal interconnections tend to be wider and thicker. This is to ensure that the resistance-capacitance (RC) time delays of these metal interconnection layers are kept manageable.

How multiple metal layers results in increased topography is illustrated in Fig. 3.20 where there is a gate layer, a first layer dielectric, a first metal layer, and a second layer dielectric. If the gate layer has a thickness of 100 nm and the first metal layer has a thickness of 0.5 microns, then the maximum topography will be 0.6 microns for only a one-layer metal process. As the number of metal layers increases, the maximum topography similarly increases.

There have been a number of planarization processes that have been developed over several generations of IC manufacturing processes including the reflow of glasses such as BPSG; intentional slopping of metal sidewalls using RIE etching techniques; thick dielectric layer depositions followed by etch backs; oxide spacers; spun on polyimide dielectrics layers; sacrificial layer etch backs; spin-on glasses (SOG); chemical-mechanical polishing (CMP); and others. Some of these result in partial planarization, whereas others result in a full planarization of the topography. Of these techniques, CMP is by far the most widely used and will be discussed here.

Most current planarization processes involve chemical-mechanical polishing (CMP) since it is the only method that provides for global planarization whereby the entire substrate surface is made more flat. Nevertheless, the adoption of CMP by industry was slowed by concerns of the cleanliness of this process. CMP processes simultaneously employ both chemical and mechanical effects to planarize the substrate surface. Typically, the CMP processes use a chemically reactive slurry that is a colloid mixture of a liquid mild etchant specific to the material to be removed and solids suspended in the liquid that act as abrasives.

The mechanically abrasive action occurs as the substrate surface to be planarized is forced into contract with a polishing material surface that is also called a polishing pad (see Fig. 3.21). The slurry is spread across the polishing pad and between the substrate and pad. In order to obtain even removal rates across the substrate, both the substrate and polishing surfaces each rotate around their respective central axes. The polishing surface has a much larger diameter than the substrate. In order to keep the damage to the substrate surface to a minimum level, the chemical action of the CMP

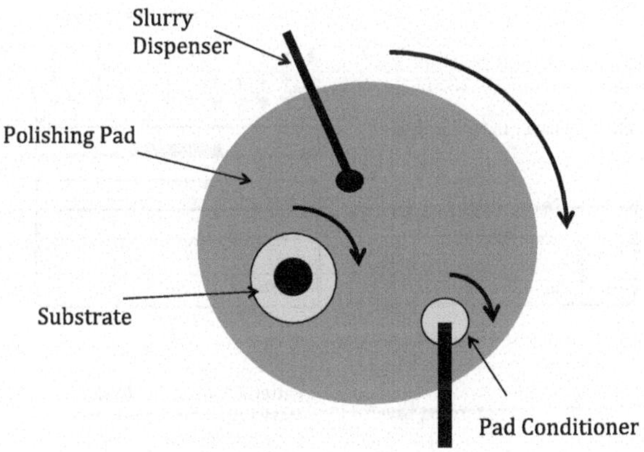

Fig. 3.21 A plan view illustration of the chemical-mechanical polishing (CMP) process whereby a substrate surface is planarized

should dominate over the mechanical action. While CMP is a valuable process for advanced ICs, it is expensive since it has a relatively low-wafer throughput and requires significant post-CMP cleaning time.

The uniformity of CMP processes is a current topic of research and there are a large number of process, equipment, and substrate parameters that have an impact of the resultant uniformity that can be obtained. For example, if there are hard material stops that are positioned across the substrate such that the material removal will be limited once these stops are reached, then the uniformity can be greatly improved compared to attempting to planarize a substrate without any stops. However, even with the use of stops, a softer material may exhibit dishing whereby the soft material layer is thinner at the center of the regions between the stops than immediately adjacent to the stops.

What constitutes a uniform CMP process and how to accurately measure it are not straightforward since the desired result will be a substrate having non-uniformity upper material layer(s) across the substrate surface so as to fill in the topography. Additionally, since the substrate acts as an offset from the materials on the surface being planarized and the chuck onto which the substrate is located, any thickness variations across the substrate will result in those variations being propagated into the layers on the substrate surface. Therefore, substrates having minimal thickness variation are advised if CMP will be used during the manufacturing. Although the uniformity for any specific CMP process must be measured and statistically analyzed, as a general rule, CMP processes will not have better than 5% uniformity across the substrate surface [64].

3.2.6 *Etching*

Photolithography allows the transfer of the mask image into a photoresist layer on the substrate surface. However, the photolithography process alone does not implement that pattern directly into the underlying material. Instead, photolithography is

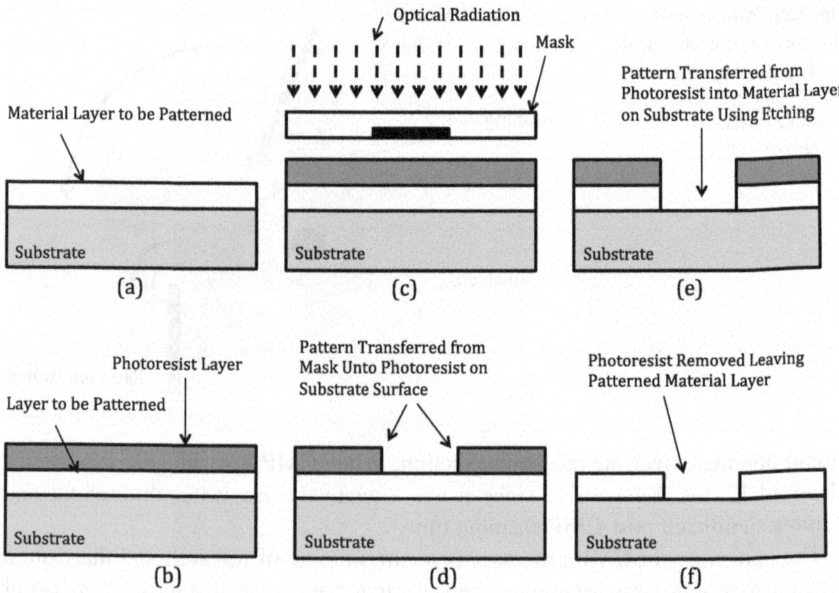

Fig. 3.22 Process flow illustrating the patterning of a material layer on a substrate. (**a**) shows the material layer that has been deposited on the substrate; (**b**) is the substrate after a photoresist layer has been deposited onto the material layer; (**c**) shows photolithography being performed; (**d**) the developed pattern in the photoresist after photolithography has been performed; (**e**) after the etching of the material layer whereby the material layer area not protected by the resist is exposed to the etchant and is removed; and (**f**) after the photoresist has been removed thereby resulting in the patterned material layer

commonly followed by an etching step to transfer the pattern into the material. The patterned photoresist layer protects regions of the substrate surface from etching where the photoresist is left behind and allows other regions of the substrate surface that are exposed to be etched (see Fig. 3.22).

There are different types of etching processes, but the one that is most commonly used in IC manufacturing is called reactive ion etching (RIE). RIE uses a plasma that participates in the etching process. A plasma is a partially ionized gas that consists of ions, electrons, and neutral species that is formed by applying a strong electromagnetic field to a gas at low operating pressures (i.e., less than about 10^2 torr). The applied electromagnetic field strips some electrons from the gas molecules resulting in charged electrons (having relatively low masses and negative charges) and ions (having relatively high masses and positive charges). The free electrons in the plasma are accelerated by the applied electromagnetic field and thereby gain high kinetic energies that collide with neutral gas molecules, thereby resulting in additional electrons, ions, and radicals (neutral atom or molecule that has at least one unpaired electron, but has no charge). Simultaneously, some number of electrons, ions, and radicals also collide to recombine into neutral gas atoms or molecules.

A steady-state condition is reached when the rate of ionization is equal to the rate of recombination.

The plasma discharge is very identifiable and has a color central glow that is called the bulk plasma region. In this region there are nearly equal numbers of ions and electrons. Near the electrodes are dark areas that are the sheath regions. It is in this area that there is a large voltage potential drop that causes the positively charged ions to be accelerated toward the substrate. A plasma can be initiated with many gases, but certain gases that are well suited as reactive process gases can be introduced into the etch chamber and form free radicals. Free radicals are neutral atoms or molecules having an unpaired valence electrons, thereby resulting in a species that is highly chemically reactive. Examples are fluorine and chlorine radicals that are commonly used in RIE plasma etching.

RIE is performed at process pressures between 10^{-1} and 10^{-3} torr. The impacting plasma species impart energy into the exposed atoms of the substrate surface to initiate the etching process and the chemically reacting species react with the substrate atoms, which then desorb from the surface as reaction by-products. As a result, RIE is somewhat material specific while at the same time provides good anisotropy of the etched features.

The advantages of RIE etching include relatively good depth uniformity control; good mask selectivity (the ratio of the etch rate of the material being etched divided by the etch rate of the mask); potentially good selectivity with materials underlying the material layer being etched; less chemical waste handling issues (compared to wet etching); is a relatively clean process; can provide high fidelity and dimensional control of the etched features; and amenable to automation for cassette-to-cassette high-wafer-throughput production.

RIE plasma etch processes involve six individual steps, all of which must occur for etching to proceed (see Fig. 3.23). First, the process gas(es) introduced into the etch chamber must be broken into chemically reactive species (i.e., radicals) by the plasma. Second, these chemically reactive species must diffuse to the substrate surface. Third, the reactive species must be absorbed onto the substrate surface. Fourth, the chemically reactive species must have the ability to move about the surface and react with a material layer. Fifth, the reaction products must desorb from the surface. Sixth, the reaction products must be carried away by the gas stream out of the etch chamber. The overall etch rate will be determined by that step that has the lowest rate.

A typical RIE system uses a process chamber that is connected to a vacuum pump in order to achieve the low base operating pressures (see Fig. 3.24). Process gases are introduced into the etch chamber in accurately controlled amount using in-line mass-flow controllers (not shown in Fig. 3.24). Most advanced RIE configurations are single-wafer etch tools since this results in an improved uniformity and most production etch tools will utilize cassette-to-cassette capability combined with robotic automation and load-lock chambers in order to speed up the wafer throughput (also not shown in Fig. 3.24). The substrate is placed on a chuck that also acts as an electrode. This electrode is electrically connected to an RF generator operating at a frequency of 13.56 MHz with typically a few hundred watts of power. The

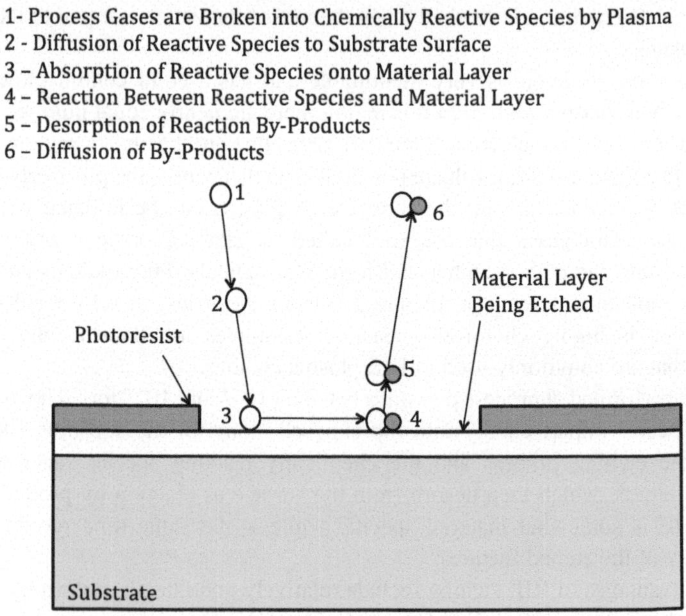

Fig. 3.23 Illustration of the six steps involved in plasma etching

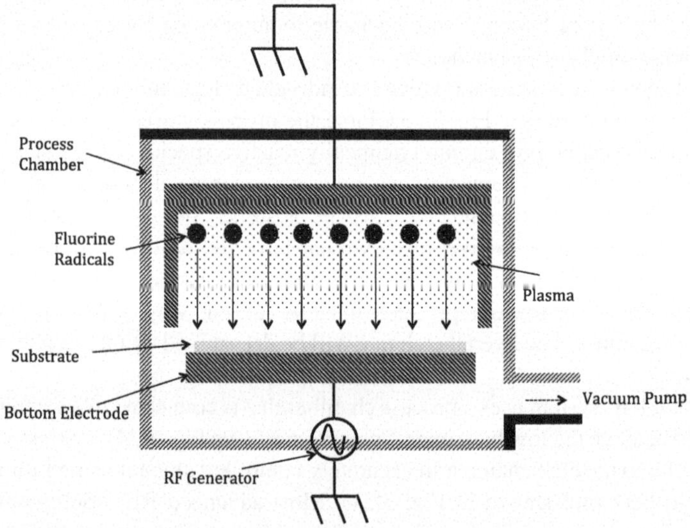

Fig. 3.24 Illustration of a reactive ion etch (RIE) plasma etching system. The substrate is positioned on the bottom electrode that is electrically powered, while the top electrode and chamber walls are electrically grounded. The grounded electrode has a much larger area than the bottom electrode

chamber walls and top electrode are electrically grounded. The electromagnetic field applied to the process gases into the etch chamber result in the electrons being stripped from the gas molecules, thereby initiating a plasma.

The configuration of grounding the chamber walls and top electrode that have large areas, and applying the electromagnetic field to the bottom electrode having a much smaller area, results in an increase in the ion energies that impinge onto the substrate. As the electromagnetic field oscillates, electrons in the plasma can strike both the chamber walls and substrate. The plasma ions, on the other hand, are much more massive and therefore are not able to travel distances equivalent to that of the electrons. The electrons impacting the chamber walls are absorbed by the grounded electrode. The electrons that strike the substrate electrode result in a large negative charge to be created since this electrode is DC isolated. Simultaneously, the plasma sheath builds up a positive charge due to the higher concentration of positive ions compared to free electrons. As a result, a large voltage potential develops on the substrate electrode relative to the top electrode and chamber walls. The potential difference between the substrate electrode and the plasma sheath is usually a few hundred volts and results in the ions in the plasma to be attracted to and impact the substrate with high kinetic energies, thereby imparting energy into the substrate material.

RIE etching proccsses use gases that contain halogens, which are group VII elements, including fluorine, chlorine, bromine, and iodine, due to the fact that these elements are highly electronegative and reactive. When these halogens react with the material being etched, they form chemical compounds known as halides. The volatility of the resulting halide is an important parameter, and for an etching process, it is desirable that the halides have a high vapor pressure.

Specifically, the different process gases that are employed in RIE for etching different material types commonly used in IC manufacturing include the following: fluorine-based gases such as SF_6, CF_4, etc. for the etching of silicon; fluorine-based gases such as C_3F_8, C_2F_6, etc. for the etching of silicon dioxide and silicon nitride; chlorine-based gases such as BCl_3, $CCl_4/Cl_2/BCl_3$, etc. for the etching of aluminum and other metals; and oxygen-based gases such as O_2, CO_2, etc. for the etching of organics such as photoresist; as well as other chemistries for other material types such as silicides and refractory metals.

There are several important parameters of interest for RIE etch processes. The first is the etch rate. A higher etch rate is desirable since it allows the process to be performed quicker, thereby increasing wafer throughput and obtaining lower production costs. However, the etch rate must also be controllable. Therefore, most etch processes have etch rates in the range of hundred to thousands of Angstroms per minute in order to provide a sufficient level of process control.

Another important parameter is the uniformity of the depths of the etched features. Most etch processes have some level of inherent non-uniformity, and deposited thin films also have some non-uniformity across the substrates. As a result of the combined effect of these two non-uniformities, each substrate must be over etched for a sufficient time in order to ensure that all of the features have been completely etched through. This commonly results in some amount of over etch of those features that cleared first, and this can result in increased lateral etch and some amount of etch into the underlying material in those locations.

The mask selectivity, which is defined as the ratio of the etch rate of the material being etched to the etch rate of the masking layer, is also an important parameter. Ideally, the masking layer would have a zero etch rate, but this is never the case. Knowledge of the mask selectivity, the etch rate of the material being etched, and the depth of the features to be etched allow the determination of the mask layer thickness required. A level of higher mask selectivity is more desirable since this allows the mask to be made from a thinner material layer, and a thinner mask improves the dimensional control of the features being etched in the material layer.

The etch selectivity to the underlying material is also important. It is given by the ratio of the etch rate of the material being etched to the etch rate of the underlying material (that presumably is not to be etched). Ideally, once the etch process has proceeded through the material layer being etched, it will not etch into the underlying material. However, there is always some amount of etching in the underlying material.

Another important parameter for etch processes is the amount of lateral etch. This is the amount that the material that is etched sideways resulting in undercutting of the masking layer. While it is desired that there is no lateral etching, some amount of lateral etch is inevitable. The anisotropy of an etch process is defined as:

$$\text{Etch anisotropy} = 1 - \frac{\text{Lateral etch rate}}{\text{Vertical etch rate}}. \tag{3.29}$$

As can be seen, the anisotropy includes both the amount of lateral and vertical etch. Ideally, the lateral etch would be zero, and the anisotropy would be 1. However, in practice this is never the case.

The aspect ratio of an etch is another important parameter and is given by the depth of the feature etched divided by the width of the feature. A higher aspect ratio is more difficult to obtain since as the etched feature gets deeper and more narrow, the diffusion of reactive species and by-products of the etch from the bottom of the features becomes increasingly difficult due to transport limitation effects.

Two other issues related to RIE etching are the loading effect and etch lag. The loading effect in etching is when the etch rate depends on the amount of the substrate area being etched. Specifically, it is observed that when larger amounts of the substrate surface area is exposed to the etchant, the etch rate decreases. Loading effects can be at the macroscale (i.e., across the entire substrate) or at the microscale (i.e., at the die or feature scale). Loading effects are thought to be due to the depletion of reactants and can reduced to some extent by decreasing the etch rate and/or increasing the reactants available for etching. Etch loading can also have a detrimental impact on the resultant etch uniformity. It has been shown that if a large percentage of the substrate surface is exposed to etching, from about 8% exposed area to about 100% exposed area, then the etch rate can be reduced by more than 50%, and the non-uniformity can be increased from 2% to 35% [65]. Therefore, as a general rule, it is prudent to keep the amount of substrate area exposed to etching to below 10%.

Etch lag refers to an etch rate that is lower in smaller feature sizes than in the features with larger dimensions. If the pattern being etched has a large dynamic range of feature sizes, the larger features will be completely etched sooner than the smaller features. In order to complete the etch of the smaller features, the larger features will need to be over etched for a considerable length of time before the smaller features are completely cleared. The severe over etch in the larger features can result in the etching of the underlying material layer (which can affect device performance) as well as some lateral etching of the features and undercutting of the masking layer (which can mean a loss in dimensional control). The amount of etch lag present depends on the dimensions of the smallest features as well as the differences in the sizes of the features and the amount of area exposed to the etchant. The typical remedy for etch lag is to ensure that all of the etched features have the same smallest dimension across the substrate. However, this solution may not always be possible depending on the device design. It has also been shown that etch lag can be reduced by adjusting certain processing parameters [66].

Another parameter in RIE etching is called the aspect ratio dependent etching (ARDE). This refers to the fact that the etch rate may depend on the aspect ratio of the features being etched. As the aspect ratio increases, wherein the ratio of the depth to the width of the feature increases, it becomes increasingly difficult for the reactants to diffuse to the bottom of the feature as well as reaction by-products to diffuse out into the gas stream to be pumped out of the chamber. The impact is to slow rate of the etching.

With regard to the dimensional uniformity that can be obtained using RIE plasma etching, it should be first noted that the accurate control of the dimensions of etched features depends on a number of parameters and therefore this is a complex issue to analyze and estimate compared to many of the other processing steps we have reviewed so far.

To the first order, the dimensional control that can be obtained relates to the uniformity of the etch process across the wafer, the selectivity of the etch process used with respect to the underlying material, and the lateral etch rate (resulting in the undercutting of the masking layer). Further, many of these first-order effects are dependent on the material layer type being etched; the material type underneath the material layer being etched; the masking material being used; the size of the features; the percentage of substrate area exposed to the etchant; the aspect ratio of the etched features; the resultant profile of the sidewalls of the etched features; the equipment that the etch is performed on; the process recipe being used; and others. Additionally, it is important to also note that the dimensional control of the resultant etched features is also dependent on the dimensional control of the masking layer used that, in turn, is dependent on the photolithography process used.

Under the best-case scenario, wherein the material layer is a commonly etched material type (e.g., silicon, Si_3N_4), having a modest thickness (e.g., 0.5 micron or less), with an extremely good selectivity with the etch rate of the underlying material layer, a loading factor below 10%, an aspect ratio 1 to 1 or less, and a near perfect photolithography process, it is possible for single-wafer production-oriented RIE tools to exhibit across wafer etch uniformities of 1–2% or better.

However, while a given RIE etch process may exhibit a relatively good uniformity, this obviously is not the complete picture since other factors have a significance impact on the resultant control of the dimensions of the etched features. For example, if the material layer being etched is non-uniform, then considerable over etch would need to be performed in order to ensure that all of the features were cleared. This over etch may result in some additional etching into the material layer underneath the layer being etched as well as some lateral etching since no etch is perfectly anisotropic.

As probably apparent given the number of parameters involved, not to mention the complex interrelationships between these parameters, dimensional control using etching is very complicated with a number of coupled and complicated factors that will impact the result. Consequently, the uniformity of an etch process must be determined on the specific process (process here meaning both photolithography and etching) and equipment in order to have an accurate estimation of the values of the dimensional variations.

3.2.7 *Clean and Strip*

The flowchart shown in Fig. 3.1 illustrating the manufacturing process of ICs includes a number of cycles of deposition, photolithography, and etching steps, each of which typically involves a different tool and chemistry. Some of these processing steps require that the material be deposited or grown onto the surface of the substrate that is pristine. That is, the substrate surface should be completely devoid of any chemical contaminates, residues, and particulates.

Historically, the most widely used cleaning method in integrated circuit manufacturing was the so-called RCA clean process. The RCA clean involves a sequential process involving (1) removal of organic contaminants, both an organic and particulate clean; (2) removal of a thin native oxide layer (which is optional); and (3) removal of ionic contamination. Each step requires separate chemical solutions that the substrates are immersed into. Some of the solutions are heated. De-ionized water rinses are performed on the substrates between each step and at the end of the clean process to remove any residual cleaning chemicals from the substrates. The resistivity of the DI water may be monitored in order to know that the substrates are adequately cleansed of any residual chemicals. Each cleaning step is performed in a dedicated set of quartz vessels that are also routinely cleaned.

While the RCA cleaning process has been used for a number of years, it has several disadvantages, including the following: it requires a large amount of costly chemicals; there is a large amount of chemical waste that is toxic and expensive to dispose of; and the wafer throughput is relatively low, based on the time it takes to perform each step. Consequently, most advanced IC manufacturing facilities now use single-wafer cleaning tools that are configurable to the process being performed as well as the previous step(s) that were performed. These tools use a number of cleaning chemicals and high-pressure sprays and high-velocity spins to clean and remove contamination and particulates at high-wafer throughput rates [67].

Additionally, the photoresist left on wafers after the photolithography and etch step has been performed must also be stripped from the substrates before any subsequent process steps are performed. In advanced IC manufacturing processes, the resist strips are performed using plasma operation consisting of hydrogen and a weak oxidizing agent, such as carbon dioxide [68].

3.3 Summary

This chapter has reviewed a number of individual processing steps commonly used in the manufacturing of integrated circuits. In the review of each processing step, we have provided an overview to describe how the process works, described the equipment that is commonly used, and provided some general guidance as the best expected dimensional variations that can be obtained in performing the processing step. It was advised that for each process step that will have an impact on the resultant dimensions, it is always best to directly measure these variations. Guidance on how these variations can be measured, as well as how to determine their impact on the device performance, will be covered in a later chapter. Table 3.4 provides a compilation of the expected "best-case" dimensional variations for each of the processing steps reviewed in this chapter for quick reference.

Table 3.4 Best-case dimensional variations for various IC processing steps

Basic processing steps	Typical best-case variation
Thin-film growth and depositions	
Thermal oxidation	Horizontal systems: Across wafer and wafer-to-wafer thickness non-uniformity <2.5%
	Vertical systems: Across wafer and wafer-to-wafer thickness non-uniformity <1%
Chemical vapor deposition	
Atmospheric chemical vapor Deposition	Epitaxy: Across wafer and wafer-to-wafer non-uniformity <4%: Resistivity <10%
	Other ACVD systems: Across wafer and wafer-to-water non-uniformity >10%
Low-pressure chemical Vapor deposition	Horizontal systems: Across wafer and wafer-to-wafer non-uniformity <2.5%
Plasma-enhanced chemical Vapor deposition	Hot-walled systems: Across wafer and wafer-to-wafer non-Uniformity <2.5%
	Cold-walled systems (non-rotating): Across wafer and wafer-to-wafer non-uniformity <2%
	Cold-walled systems (rotating): Across wafer and wafer-to-wafer non-uniformity 1%
Atomic layer deposition	Across wafer and wafer-to-wafer non-uniformity <1 to 2%
Physical vapor deposition	
Evaporation	Across wafer and wafer-to-wafer non-uniformity <1 to 2%
Sputtering	Across wafer and wafer-to-wafer non-uniformity <2.5%
Impurity doping	
Diffusion	Across water and wafer-to-wafer non-uniformity <5%
Ion implantation	Across wafer non-uniformity <1.5%; wafer-to-wafer non-uniformity <2%
Photolithography	Critical Dimension Uniformity ("CDU") 10% of CD; overlay inaccuracy 20% of minimum feature size
Rapid thermal anneal	<1.5 °C across wafer and wafer-to-wafer
Planarization	Across wafer and wafer-to-wafer non-uniformity 5%
Etching	Across wafer and wafer-to-wafer non-uniformity <1 to 2%
Clean and strip	N/A

References

1. C.Y. Change, S.M. Sze, *VLSI Technology, ULSI Technology* (McGraw Hill, New York, 1996)
2. S. Wolf, R.N. Tauber, *Silicon Processing for the VLSI Era, Volume 1- Process Technology* (Lattice Press, Sunset Beach, 1986)

3. S.A. Cambell, *The Science and Engineering of Microelectronic Fabrication* (Oxford Press, New York, 1996)
4. P. van Zant, *Microchip Fabrication* (McGraw Hill, New York, 2000)
5. S.M. Sze, *Physics of Semiconductor Devices* (Wiley Inter-science, New York, 1981)
6. B.E. Deal, Standardized terminology for oxide charges associated with thermally oxidized silicon. IEEE Trans. Electron Devices **ED-27**, 606 (1980)
7. E.H. Nicollian, A. Geotzberger, Bell Syst. Tech. J. **46**, 1055 (1979)
8. R.R. Razouk, B.E. Deal, Extended abstracts of the May 1979 Electrochemical Soc. Meeting, Abstract 135. J. Electrochem. Soc. **126**, 1573 (1979)
9. B.E. Deal, A.S. Grove, General relationship for the thermal oxidation in steam and oxygen. J. Appl. Phys. **36**, 3770 (1965)
10. B.E. Deal, J. Electrochem. Soc. **125**, 576 (1978)
11. J. Spiegelman, Improved oxide growth rate and uniformity through new steam delivery method, RASIRC Publication Tech Notes. See: http://www.rasirc.com/resources/technotes/technote_oxidation.pdf
12. A. Sherman, *Chemical Vapor Deposition for Microelectronics: Principles, Technology, and Applications* (Noyes, Park Ridge, 1987)
13. A.S. Grove, Mass transfer in semiconductor technology. Ind. Eng. Chem. **58**, 48 (1966)
14. B. J. Baliga (ed.), *Epitaxial Silicon Technology* (Academic Press, Orlando, 1986)
15. R.A. Levy, K. Nassau, Viscous behavior of phosphosilicate and borophosphosilicate glasses in VLSI processing. Solid State Technol. **29**, 123 (1980)
16. W. Kern, G.L. Schnable, CVD BPSG for Si devices. RCA Rev. **43**, 423 (1982)
17. D.A.P. Bulla, N.I. Morimoto, Deposition of thick TEOS PECVD silicon oxide layers for integrated optical waveguide applications. Thin Solid Films **334**, 60 (1998)
18. K. Seshan (ed.), *Thin Film Deposition: Processes and Technologies*, 2nd edn. (Noyes Publications, William Andrew Publishing, Norwich, 2002)
19. R.S. Rosler, Low pressure CVD production processes for poly, nitride, and oxide. Solid State Technol. **20**(4), 63 (1977)
20. H.O. Pierson, O. Hugh, *Handbook of Chemical Vapor Deposition (CVD)* (William Andrew Publishing, Norwich, 1992)
21. P. Temple-Boyer, C. Rossi, E. Saint-Etienne, E. Scheid, Residual stress in low pressure vapor deposition SiNx films deposited from silane and ammonia. J. Vac. Sci. Technol. A **16**(4), 2003 (1998)
22. P.J. Frencha, P.M. Sarrob, R. Malleeb, E.J.M. Fakkelddijc, R.F. Wolffenbuttela, Sensors Actuators A Phys. **58**(2), 149 (1997)
23. D.V. Morgan, K. Board, *An Introduction to Semiconductor Microtechnology*, 2nd edn. (Wiley, West Sussex, England, 1991)
24. Gemini-2 Product Brochure, Gemini Research, Inc., Fremont, CA
25. R. Iscoff, Hotwall LPCVD reactors: Considering the choices. Semicond. Int., 60 (1991)
26. A.C. Adams, Dielectric and polysilicon film deposition, in *VLSI Technology*, ed. by S. M. Sze, (McGraw Hill, New York, 1988)
27. T.I. Kamins, Structure and properties of LPCVD silicon films. J. Electrochem. Soc. **127**, 686 (1980)
28. R.D. Compton, PECVD: A versatile technology. Semicond. Int., 60 (1992)
29. S. Nguyen, S. Burton, J.D. Mackenzie, The variation of physical properties of plasma-deposited silicon nitride and oxynitride and their compositions. J. Electrochem. Soc. **131**, 2348 (1984)
30. R.L. Purunen, Surface chemistry of atomic layer deposition: A case study for the trimethylaluminum/water process. J. Appl. Phys. **97**(12) (2005)
31. S.M. George, Atomic layer deposition: An overview. Chem. Rev. **110**, 111–131 (2010)
32. H. Kim, Atomic layer deposition of metal and nitride thin films: Current research efforts and applications for semiconductor device processing. J. Vac. Sci. Technol. A **21**(6), 2231 (2003)
33. A.J.M. Mackus, D. Garcia-Alonso, H.C.M. Knoops, A.A. Bol, W.M.M. Kessels, Room-temperature atomic layer deposition of platinum. Chem. Mater. **25**(9), 1769–1774 (2013)

34. S.E. Potts, W.M.M. Kessels, Energy-enhanced atomic layer deposition for more process and precursor versatility. Coord. Chem. Rev. **257**(23–24), 3254–3270 (2013)
35. H.C.M. Knoops, E.M.J. Braeken, K. de Peuter, S.E. Potts, S. Haukka, V. Pore, W.M.M. Kessels, Atomic layer deposition of silicon nitride from bis(-butylamino)silane and N plasma. ACS Appl. Mater. Interfaces **7**(35), 19857 (2015)
36. E. Langereis, H.C.M. Knoops, A.J.M. Mackus, F. Roozeboom, M.C.M. van de Sanden, W.M.M. Kessels, Synthesis and characterization of low-resistivity TaNx films by remote plasma atomic layer deposition. J. Appl. Phys. **102**(8), 083517 (2007)
37. G.N. Parsons, J.W. Elam, S.M. George, S. Haukka, H. Jeon, W.M.M. Kessels, M. Leskelä, P. Poodt, M. Ritala, History of atomic layer deposition and its relationship with the American Vacuum Society. J. Vac. Sci. Technol. A **31**(5), 050818 (2013)
38. D. Turakhia, Atomic Layer Deposition-1, growth rate and uniformity of Al_2O_3, Tool Data. Paper 5. See: http://repository.upenn.edu/cgi/viewcontent.cgi?article=1004&context=scn_tooldata
39. E. Jia, C. Zhou, W. Wang, Uniformity and passivation research of Al_2O_3 film on silicon substrate prepared by plasma-enhanced atom layer deposition. Nanoscale Res. Lett. **10**, 129 (2015)
40. R. Giang, Vacuum evaporation, in *Handbook of Thin Film Technology*, ed. by L. Maissel, R. Giang, (McGraw Hill, New York, 1970)
41. R.J. Hill, *Physical Vapor Deposition* (Temescal, Berkeley, 1976)
42. M. Ohring, *The Materials Science of Thin Films* (Academic Press, London, 1992)
43. G.K. Werner, G.S. Anderson, The nature of physical sputtering, in *Handbook of Thin Films*, ed. by L. I. Maissel, R. Glang, (McGraw Hill, New York, 1970)
44. J.L. Kurt, System for sputtering uniform optical coatings on flat and curved surfaces without masks, Nov 2016, see: https://www.lesker.com/newweb/blog/post.cfm/system-for-sputtering-uniform-optical-coatings-on-flat-and-curved-surfaces-without-masks
45. P.G. Shewmon, *Diffusion in Solids* (McGraw Hill, New York, 1963)
46. K. Taniguchi, K. Kurosawa, M. Kashiwagi, Oxidation enhanced diffusion of boron and phosphorus in <100> silicon. J. Electrochem. Soc. **127**, 2243 (1980)
47. C.J. Coe, The lateral diffusion of boron in polycrystalline silicon and its influence on fabrication of sub-micron MOS. Solid State Electron. **20**, 985 (1977)
48. B. Swaminathan, K.C. Saraswat, R.W. Dulton, T.I. Kamins, Diffusion of arsenic in polycrystalline silicon. Appl. Phys. Lett. **40**, 795 (1980)
49. S.M. Hu, S. Schmidt, Interaction in sequential diffusion process in a semiconductor. Phys. Rev. **107**, 2 (1957)
50. G. Deanaley, J.H. Freeman, R.S. Nelson, J. Stephen, *Ion Implantation* (New Holland, Amsterdam, 1973)
51. J.W. Mayer, L. Ericksone, J.A. Davies, *Ion Implantation in Semiconductors, Silicon and Germanium* (Academic Press, New York, 1970)
52. J.F. Gibbons, Ion implantation in semiconductors – Part 1, range distribution theory and experiments. Proc. IEEE **56**, 295 (1968)
53. M.I. Current, W.A. Keenan, A performance survey of production ion implanters. Solid State Technol., 139 (1985)
54. V. Bakshi, EUV Lithography, SPIE Press Book, vol. PM178, (2008)
55. Y. Vladimirsky, Chapter 10: Lithography, in *Vacuum Ultraviolet Spectroscopy II*, ed. by J. A. Samson, D. L. Ederer, (Academic Press, Cambridge, 1998), pp. 205–223
56. S.Y. Chou, P.R. Krauss, P.J. Renstrom, Imprint lithography with 25-Nanometer resolution. Science **272**(5258), 85–87 (1996)
57. B.W. Smith, Y. Fan, M. Slocum, L. Zavyalova, 25nm immersion lithography at a 193 nm wavelength, Proceedings, SPIE, (2005)
58. J. Burnett, High-index materials for UV lithography optics, Frontiers in Optics, OSA Technical Digest, Optical Society of America, (2006)
59. X. Hua et al., J. Vac. Sci. Technol. B **24**, 1850–1858 (2006)

60. W.H. Arnold, Towards 3nm overlay and critical dimension uniformity: An integrated error budget for double patterning lithography. Proc. SPIE **6924** (2008)
61. ITRS Lithography 2012 http://www.itrs.net/Links/2012ITRS/Home2012.htm
62. Y.K. Kim, et al., Lithography focus/exposure control and corrections to improve CDU at post etch step, Proc. of SPIE Vol. 6924, (2014)
63. W.S. Yoo, T. Fukada, T. Setokubo, K. Aizawa, J. Yamamoto, R. Komatsubra, Rapid thermal implant annealing using Cold Wall and Hot Wall systems. Proc. Electrochem. Soc. **2002-11**, 21 (2002)
64. C. Mau, Control of wafer-scale non-uniformity in chemical-mechanical planarization by face-up polishing, S.M Thesis, MIT, See: http://hdl.handle.net/1721.1/45202, (2008)
65. J. Karttunen, J. Kiihamaki, S. Franssila, Loading effects in deep silicon etching. Proc. SPIE **4174**, 90 (2000)
66. S.L. Lai, D. Johnson, R. Westerman, Aspect ratio dependent etching lag reduction in deep silicon etch processes. J.. Vac. Sci. Technol. A **24**(4), 1283 (2006)
67. See: http://www.lamresearch.com/products/strip-clean-products#Tab2
68. H.H. Gato, I. Kalinovski, K. Mohamed, Photoresist strip method for low-k dielectrics, USPTO patent, patent no. 8058178 B1, Nov 15, 2011

Chapter 4
Microsystems Manufacturing Methods: MEMS Processes

4.1 Introduction

This chapter provides a general overview of the processing steps and process modules commonly used in MEMS microsystems manufacturing. Like IC manufacturing, MEMS manufacturing consists of performing an ordered set of processing steps in a sequential manner; and, if the ordered arrangement of processing steps results in the complete fabrication of a MEMS device, it is called a "process sequence" or a "process technology." Also similar to IC manufacturing, MEMS manufacturing can be lumped into major categories of processing step types including (see Fig. 3.1) thin-film depositions or growths; lithography; etching; impurity doping; and metrology. Depending on the MEMS process sequence involved, there may be other types of processes in the sequence such as wafer bonding, etc.

While MEMS processing steps share a number of common attributes with the IC processing steps reviewed in the last chapter, there are a number of very important differences. In some cases, MEMS processing steps use the same equipment as IC fabrication, such as LPCVD furnaces, PECVD reactors, PVD systems, etc. However, in these cases a distinguishing feature of the MEMS processing step is that either the material processed is not something used in IC manufacturing or some other material attribute, such as the thickness of the film deposited, is unique to MEMS. Additionally, there are some MEMS processing steps that are unique (i.e., are not performed in IC fabrication) and may use specialized equipment.

Therefore, in order to avoid reviewing the same material covered in Chap. 3, the overview of MEMS processing steps and modules will be structured differently. For the MEMS processing steps that use the same equipment as an IC processing step, only the important differences are described, usually the difference will pertain to some attribute of the material involved in the process. Since the equipment used for the MEMS processing step in these cases is the same as used in IC fabrication, the expected best-case uniformities will be identical. For those MEMS processing steps

M. Huff, *Process Variations in Microsystems Manufacturing*, Microsystems and Nanosystems, https://doi.org/10.1007/978-3-030-40560-1_4

that are unique (and not performed in IC fabrication), the overview will attempt to provide some description of both the processing step and equipment. Other MEMS fabrication methods, such as bulk micromachining, are better labeled as process modules rather than processing steps, and the processing steps and equipment for these will also be described. The substrates used in MEMS manufacturing are also far more diverse than those used in IC manufacturing and will also be reviewed.

4.2 MEMS Substrate Material Types

While silicon is by far the most commonly used substrate in MEMS manufacturing, other types of substrate materials are also used such as non-silicon semiconductors, including germanium; silicon carbide; gallium arsenide and others; glasses of different types, such as fused silica; Pyrex^{TM}, etc.; ceramics, such as LTCC; plastics; and metals. Often the application is the primary driver for the use of non-silicon substrates. For example, plastic substrates may provide for a less expensive manufacturing for some microfluidic devices compared to silicon. And, glass substrates are often used when an optical signal must be transmitted into fluid containing cavities enclosed within substrates for biochemical applications. When non-semiconductor material substrates are used in MEMS manufacturing, they are most often made so as to have the same size and form factor as a silicon wafer in order that they can be processed on the same fabrication equipment using batch fabrication techniques. As a general rule, silicon wafers have tighter tolerances than wafers made from other materials. For example, the total thickness variation (TTV) specification for a commercial-grade silicon substrate will usually be better than for glass, metal, or ceramic substrates. Silicon wafers with TTV of around 1 micron or better are obtainable at a premium price. A tight TTV specification is useful for certain types of micromachining processes such as backside anisotropic etching to make membranes across the substrate that have uniform thicknesses.

One particular type of substrate that is often used in MEMS fabrication is silicon-on-insulator (SOI). Figure 4.1 is an illustration of a SOI wafer. The bottom wafer is called the handle wafer and is usually a standard wafer thickness. The buried oxide layer is usually a high-quality thermal silicon dioxide and can range in thicknesses. The device layer is the top layer and is a single-crystal silicon layer also with thickness that can range over a large number of values.

There are a number of methods for making SOI wafers including SIMOX and wafer bonding.[1] One of the important differentiators in all of these methods is how

Fig. 4.1 Silicon-on-insulator wafer (cross-sectional view)

[1]Wafer bonding is described later in this chapter.

to control the device layer thickness. For most MEMS applications, the SOI wafer are made using a wafer-bonding process followed by a thinning of the device wafer to a desired thickness, usually several to tens of microns using grinding and polishing techniques. The device layers thickness variations has a standard of about ±0.5 microns, but special order and higher costs SOI wafers with a device layer thickness variation of ±0.25 microns are obtainable.

4.3 MEMS Materials Deposition Processing Steps

This section reviews many of the fabrication methods that are commonly employed for the deposition of materials used in the manufacturing of MEMS microsystems. These include deposition processing steps that use the same equipment and similar processing techniques employed in IC manufacturing, except the material layers may be considerably thicker or the material properties may be modified for specific values and characteristics, and highly specialized material types and processing steps that are only performed in MEMS fabrication.

4.3.1 MEMS Thin-Film Materials Deposited on IC Equipment

There is a huge variety and diversity of thin-film materials used in MEMS manufacturing. These include materials such as semiconductor, dielectrics, metals, and others. The methods employed for depositing these materials usually depend on the material type but often are similar to those used in IC manufacturing. This section reviews thin-film materials used in MEMS fabrication that are performed on IC equipment. The material properties of deposited thin-film layers, which are extremely important in MEMS fabrication, will be covered in a later chapter.

4.3.1.1 Thin-Film Semiconductors

MEMS manufacturing uses a number of deposited semiconductor thin-film material layers including silicon; silicon-germanium; silicon carbide; diamond; and others that are reviewed in this section.

4.3.1.1.1 Silicon (Si)

A polycrystalline form of silicon, often called polysilicon, is extensively used in MEMS manufacturing, most commonly as a structural layer in surface micromachining process sequences.[2] This is due to the excellent mechanical properties that this material exhibits combined with its cost-effectiveness. There are a number of reported recipes for undoped LPCVD deposited of polysilicon with deposition temperatures ranging from 570 to 640 °C [1–11]. Reported recipes for doped LPCVD deposited polysilicon have ranges in the deposition temperatures ranging from 550 to 650 °C [12–16]. Importantly, the reported thicknesses of the polysilicon depositions were up to 3 microns, far more than would ever be used in IC fabrication. Typical gases used for doping include PH_3 for n-type material and B_2H_6 for p-type material. As a general rule, polysilicon is often used having material layer thicknesses that are significantly larger in MEMS compared to IC manufacturing. A polysilicon thickness more than 1 micron would normally be considered thicker than what would be used in IC manufacturing.

There are several reasons why thicker polysilicon layers often employed in MEMS microsystems.[3] There is often the need for the MEMS device to have a relatively large dimensional size compared to IC microsystems devices due to the device physics involved. For example, the application may require a substantial amount of mass for the MEMS sensor to have the desired level of sensitivity such as for an inertial-based MEMS device. Additionally, for many capacitive-based sensors, the device application may dictate a certain overall minimum capacitance value. If the capacitance sensor mechanism is through the effects of laterally spaced interdigitated electrodes (e.g., a comb-drive configuration), then the capacitance can only be increased by using thicker material layers without increasing the device lateral dimensions.

In surface micromachining processes, often the structural material layer used to make the micromachined MEMS devices is composed of polysilicon. The polysilicon is usually deposited using the same LPCVD equipment that would be employed in IC manufacturing. The major difference is that the time for the depositions is increased to allow for thicker material layers.

A few further comments about LPCVD polysilicon are in order. First, many of the recipes to deposit polysilicon using LPCVD result in material layers having a high level of stress.[4] A common technique of dealing with high stress is to perform a high-temperature anneal. Second, extremely thick polysilicon films (2–5 or more microns) can exhibit significant surface roughness that can impact the subsequent patterning of these material layers. Methods can be used to smoothen these surfaces [17].

PECVD is not as commonly used as LPCVD for the deposition of polysilicon material layers in MEMS manufacturing. Nevertheless, PECVD does offer a significant advantage to be able to deposit at much lower temperatures. Deposition temperatures as low as 100 °C have been reported using PECVD [18]; however, most of the reported deposition temperatures are in the range of 250–300 °C

[2]Surface micromachining is described in a later section of this chapter.

[3]This discussion about thicker polysilicon layers also pertains to many types of materials used in MEMS manufacturing.

[4]Thin-film residual stresses will be covered in a later chapter on material properties.

[19–22]. PECVD silicon films can also be in situ doped using the typical dopant gases to modify the films conductivity [19]. PECVD silicon has found some application in MEMS microsystems [23, 24].

Silicon can also be epitaxially deposited at temperatures below 1000 °C or onto a non-single-crystal silicon substrate, with a resultant silicon layer that is polycrystalline. The major advantage for using an epitaxial process to deposit a silicon layer is that the growth rates for epitaxial deposition are much higher than for LPCVD. Deposition rates for epitaxy are about 1 micron/min [25–29]. Epitaxial polysilicon can be in situ doped.

Silicon can also be deposited using PVD techniques. The advantage of sputtering deposition is the low temperatures that these processes can be performed [30]. Silicon layers deposited using sputtering are amorphous unless the substrate is heated substantially during deposition or the substrate is annealed after the deposition. It has been reported that annealing temperatures of at least 800 °C are needed in order to induce crystallization of the deposited silicon layers [31].

The uniformity of polysilicon material layer depositions for MEMS, regardless of thickness or method, including LPCVD, PECVD, and PVD, will be similar to the guidance given in Chap. 3.

4.3.1.1.2 Silicon-Germanium (SiGe)

The main advantage of LPCVD polycrystalline SiGe compared to LPCVD polysilicon is that it can be deposited at lower temperatures, while exhibiting material properties that make it suitable as a structural layer for surface micromachining applications [32]. Of particular interest is in the application where MEMS can be fabricated directly on CMOS wafers at temperatures that do not adversely affect the microelectronics [33, 34]. Deposition of SiGe at temperatures as low as 210 °C has been demonstrated; however, the residual strain in the as-deposited material were high and necessitated a high-temperature anneal thereby eliminating the advantages of the low-temperature deposition [35]. Therefore, most of the reported depositions of LPCVD SiGe were performed at temperatures ranging from 400 to 450 °C [38, 39]. LPCVD SiGe can be in situ doped [36] and can be deposited on most other materials including oxides, nitrides, and metals [32].

If the germanium content during deposition is above 70%, it has been found that the deposition can be performed at temperatures as low as 400 °C [37]. Unlike germanium (see below), SiGe can be deposited directly onto silicon dioxide, and this is an advantage for integrated MEMS fabrication [36].

The interest in PECVD SiGe is due to the ability to deposit material layers at lower temperatures. PECVD SiGe is either amorphous or polycrystalline in microstructure and can be deposited using source gas mixtures of SiH_4 and GeH_4. The deposited layers can be in situ doped. In addition to the lower deposition temperatures, the other advantages of PECVD are the higher deposition rates that are possible and potentially more flexibility in the resultant material properties based on the sensitivity of the material properties to the processing conditions [40–43].

PECVD SiGe has found application as a structural material layer for surface micromachined devices integrated onto CMOS substrates. SiGe can also be deposited using sputtering [44, 45].

The uniformity of deposited SiGe has not been widely reported but can be expected to be no better to that of other materials, such as polysilicon, using the same deposition equipment.

4.3.1.1.3 Germanium (Ge)

Polycrystalline germanium (Ge) can also be deposited using LPCVD methods. The advantage of LPCVD Ge is that it can be deposited at temperatures as low as 325 °C [46]. The mechanical properties of polycrystalline Ge are similar to those of polysilicon [47]. Residual stress in polycrystalline Ge thin-film layers can be reduced to nearly zero after performing an anneal at a relatively modest temperature of 600 °C. However, this temperature is still too high for most advanced integrated circuits. An important disadvantage of LPCVD Ge is that it cannot be deposited directly onto SiO_2 [46].

Polycrystalline Ge is chemically resistant to a variety of common micromachining etchants such as KOH, TMAH, and BOE, thereby allowing fabrication of micromachined devices [46]. Most of the interest in polycrystalline germanium is for MEMS and integrated MEMS devices [32]. Hydrogen peroxide, H_2O_2, has been demonstrated as an isotropic etchant for germanium. Using this result, combined with the near zero etch rate of polycrystalline SiGe in hydrogen peroxide, has allowed polycrystalline Ge to be used as a sacrificial layer and polycrystalline SiGe to be used as a structural layer for surface micromachining [48]. Ge can be also deposited at low temperatures using sputtering [49]. A similar statement to that above on SiGe uniformity can be made with regard to the uniformity of Ge.

4.3.1.1.4 Silicon Carbide (SiC)

SiC deposited using LPCVD methods is polycrystalline in microstructure and can be deposited on Si, SiO_2, and SiN [50–52] at temperatures from about 700 to 1200 °C. Most of the research performed on polycrystalline SiC deposited using LPCVD employed either a dual source gas mixture of acetylene, C_2H_2, and dichlorosilane, SiH_2Cl_2 [53, 54, 57, 58], or a single gas approach using 1,3 disilabutane, SiH_3-CH_2-SiH_2-CH_3 [55, 59–63]. The major application of polycrystalline SiC is for surface micromachined MEMS microsystems since the mechanical and chemical properties of SiC are attractive for many mechanical applications [56].

Polycrystalline SiC films deposited by LPCVD using dichlorosilane and acetylene source gases were reported to be textured with (111) columnar grain orientation [51], while layers deposited using disilabutane were found to have variable orientations [50]. Silicon carbide can be doped n-type or p-type, and high conductivity has been achieved by heavy doping with boron, aluminum, or nitrogen [56].

Silicon carbide deposited using PECVD can be performed at temperatures below 300–350 °C and therefore much lower than LPCVD [68, 69]. The typical source gases for the deposition of PECVD SiC are SiH_4 and CH_4, $C_6H_{18}Si_2$, and $(CH_3)_3SiH$. The as-deposited SIC thin films are amorphous in microstructure and can exhibit a relatively high compressive residual stress. The deposited layers are also electrically insulating and incorporate a considerable amount of hydrogen. For example, it was reported that SiC films deposited using PECVD had an as-deposited resistivity of 10^7 Ohm-cm [64]. A post-deposition anneal can be performed on the deposited SiC layers at temperatures above 450 °C to modify the residual stress from compressive to tensile [65, 66]. Annealing at temperatures above 800 °C can result in crystallization and densification of the material. SiC thin-film layers, like other forms of SiC, are highly resistant to chemical exposure. However, this chemical resistance also makes it difficult to etch this material for patterning. The electrical and mechanical properties of SiC, being inferior to those of single-crystal material, generally restrict its applications to mostly MEMS microsystems [67].

Polycrystalline silicon carbide material layers can also be deposited using epitaxy. The source gases used are usually a mixture of SiH_4 and C_3H_8. Epitaxially grown thin films of SiC having either p-type or n-type conductivity can be grown. The growth temperatures tend to rather high with most above 1280 °C. It should be noted that prevention of nitrogen doping of the grown SiC layers is a significant challenge since the amount of nitrogen that may be present in the chamber is sufficient to result in n-type doping of SiC. Of particular interest is epitaxial growth of 3C-SiC on silicon substrates. While both Si and SiC have the same lattice structure, there is a significant mismatch between these two materials. Nevertheless, it is possible to grow SiC on Si using a carbonization step that first converts the surface of the silicon substrate to a 3C-SiC layer by exposing the Si to a hydrocarbon and hydrogen gas mixture at a temperature of at least 1200 °C. The quality of the 3C-SiC material grown on silicon substrates is not sufficient for most microelectronic applications. However, it is adequate for MEMS devices [70–72].

Silicon carbide can be deposited using sputtering at room temperature. RF magnetron sputtering of SiC has been reported [73, 74]. Unlike PECV-deposited SiC thin films, sputtered SiC layers do not contain hydrogen. The deposited layers of SiC are amorphous in microstructure and electrically insulating.

There is not much information in the literature concerning the uniformity of SiC-deposited films, but it can be safely assumed that the uniformity is usually no better than 5% across the substrates.

4.3.1.1.5 Diamond

LPCVD-deposited diamond is polycrystalline. The main interest in diamond for MEMS microsystems is due to the appealing mechanical properties of this material [75, 76], which include: a very high value of Young's modulus; a very high hardness; a high thermal conductivity; and it is chemically resistant to almost nearly every chemical etchant. Although diamond is thought of as an electrical insulator, it is actually a semiconductor with a large energy band gap of 5.5 eV. It can be doped p-type with boron and n-type with nitrogen.

Polycrystalline diamond is often classified by the size of the crystallites in the film. Diamond films are usually deposited by CVD methods using a hydrocarbon precursor such as methane, CH_4, combined with hydrogen gas, H_2. The hydrogen is responsible for suppressing the growth of sp^2-bonded carbon. The grain size of deposited diamond layers is largely determined by the diameter and density of the seed particles. Seed particles or a negative biasing are needed to initiate growth. Polycrystalline diamond can be deposited on Si or SiO_2 material surfaces that are properly seeded. CVD diamond films are usually deposited using a process technology called microwave plasma chemical vapor deposition, or MPCVD, at temperatures above 400 °C and using CH_4.

Polycrystalline diamond thin films tend to have rough surfaces due to the presence of faceted, large grains. The rough surface topography limits the resolution that can be obtained in subsequent photolithography and etching steps to pattern these material layers.

An alternative diamond deposition capability is ultra-nanocrystalline diamond (UNCD). UNCD films are particularly attractive for MEMS since they can be deposited at lower temperatures making them more compatible with MEMS fabrication. UNCD has been reported to have grown rates as high as 0.4 micron/h [76] at a deposition temperature of 400 °C and exhibiting a smooth surface [77]. UNCD films have been demonstrated for use as conformal coatings on silicon surfaces, as well as surface micromachining process sequences [78–83]. The electrical properties of UNCD can be tailored from highly insulating to highly conductive by controlling the incorporation of dopants. Patterning of polycrystalline diamond thin films is challenging due to the chemical inertness of the material, but several methods have been reported [76, 84–87].

PECVD can also be used to deposit carbon-based thin-film layers that are often called diamond-like carbon (DLC) films. These films are deposited using a hydrocarbon source gas such as methane or ethane. The resultant residual stresses in these types of material layers exhibit are very high, and this limits their applications. Nevertheless, these layers sometimes are useful for coating MEMS devices undergoing wear or as an anti-stiction coating [88, 89].

There is not much reported data on the uniformity of diamond films. Moreover, many of the results reported were performed on custom-made deposition systems. Therefore, it is unlikely that the uniformity will be better than 5% across the substrate.

4.3.1.2 Metals

There are a variety of different metals used in MEMS fabrication including noble, transition, and refractory metal types. Often the application involved is the primary determinant of which type of metal is most useful. Since most metals are excellent electrical conductors, metal thin films are often used for electrical interconnections. Many metals also are good thermal conductors and therefore are also used as thermal conductors and heat sinks.

It is not uncommon for MEMS process sequences to use noble metals, such as gold, silver, and platinum. These metals do not readily form oxides on their surfaces and therefore can make excellent materials for use on electrical bonding pads, electrical interconnections, and as an intermediary bonding layer for wafer bonding. Platinum is used as a sensing layer in certain biomedical applications. Some of these materials, such as gold, are considered a significant contamination threat in an IC manufacturing and therefore would be completely barred from use in those facilities.

Metals can also be used as structural material layers for mechanical functionality in MEMS devices or as coatings on MEMS devices in order to make their surfaces optically reflective. For other MEMS applications where the devices may be exposed to harsh environment conditions, such as high temperatures and/or corrosive gases, the transition or refractory metals such as nickel, tantalum, molybdenum, and tungsten can be useful.

Most metals can be deposited using physical vapor deposition (PVD) techniques, either sputtering or evaporation. Some metals, such as platinum, titanium, and tantalum, can also be deposited using atomic layer deposition (ALD), and some other metals can be deposited using electrochemical deposition. There is a far too vast array of different metals and process recipes reported for them to be covered here, and therefore the interested reader is encouraged to seek additional information in the literature [90–95].

Table 4.1 presents some qualitative information about some of the properties of metal thin-film layers including whether they form a self-passivating oxide on the surface when exposed to air or whether they are wire bondable and solderable [96–99].

Depositing metals onto some materials present special challenges in microfabrication [95], but the adhesion can be improved by using intermediate layers [96–99]. The patterning of metal thin-film layers can be done conventionally by performing photolithography followed by either wet or dry etching. Wet etches

Table 4.1 Qualitative properties of thin-film metals [96–99]

Metal material	Self passivation oxide	Wire bondable	Solderable
Ag	No	Al	Yes
Al	Yes	Au, Al	Very challenging
Au	No	Au, Al	Yes
Cr	Yes	No	No
Cu	No	Au	Yes
Ni	Yes	No	No
Pt	No	Al	Yes
Ta	Yes	Al	Yes
Ti	Yes	No	No
W	Yes	No	No

are usually isotropic in nature and therefore tend to exhibit poor dimensional control. A good reference source for metal etchants can be found in Vossen and W. Kern [100]. Dry etching of metals is more complicated and when performed using reactive ion etching (RIE) often involves highly toxic chlorine gases. In cases where there is not an available method for RIE etching, then ion milling can be used. Many metals thin-film layers can also be patterned using the lift-off technique, which will be described separately below.

When using the same types of deposition equipment, it can be assumed that the uniformity when depositing these metal thin-film layers will usually be about the same ranges as the IC materials that were previously covered in Chap. 3. As always, it is best to measure the uniformity of a given deposition process, equipment, and material type rather than assume the uniformity is the same as depositions for other materials.

4.3.1.3 Thin-Film Metal Oxides

Metal oxides are also often used in MEMS manufacturing for a variety of purposes. For example, titanium oxide, TiO_2, is commonly used as an optical coating [90, 100]. Vanadium oxide is frequently used as a material for long-wavelength infrared thermal imaging devices due to its large change in resistance with temperature [101]. Aluminum oxide, Al_2O_3, is an extremely robust, nonconductive, and optically transparent material that has many uses [90, 100].

Like metals, metal oxides are also usually deposited using evaporation or sputtering PVD techniques [90, 100]. This can be performed by depositing the metal oxide directly from a single crucible or target containing the metal oxide, co-sputtering or co-evaporating from multiple crucibles or targets or causing the metal from a crucible or target to form a vapor in the presence of oxygen, and depositing the metal oxide. The latter approach is termed reactive PVD deposition. Additionally, some metal oxides, such as HfO_2, ZnO, and Al_2O_3, can also be deposited using atomic layer deposition (ALD). Similar to metal thin-film depositions, there are far too many metal oxides and recipes for the deposition of metal oxides to cover here.

Most metal oxides can be etched using wet chemical immersion into solutions of hydrofluoric acids [100]. However, these etchants are isotropic, and therefore the dimensional control will not be good. Dry etching and ion milling of metal oxide can also be performed with recipes that vary depending on the material type, equipment, and process technology. Lift-off is also an option for patterning metal oxide thin-film layers. It can be assumed that the uniformity when depositing these metal oxide thin-film layers will usually be about the same ranges as the IC materials that were previously covered in Chap. 3.

4.3.1.4 Dielectrics

Dielectrics, such as deposited SiO_2 and SiN, are employed in IC manufacturing mostly for their electrical properties as insulating materials. In MEMS manufacturing, dielectrics have a much wider range of uses. For example, these materials are routinely used as etch masks in micromachining processes; as sacrificial layers in surface micromachining process sequences; and as mechanically functional layers. These materials and the methods for their deposition were mostly covered in Chap. 3. Therefore, the coverage here will be limited to only aspects of these materials and their deposition processes that are different from what is used in IC manufacturing.

4.3.1.4.1 Silicon Nitride (SiN)

The material properties of stoichiometric SiN allows its use as an electrical insulator, an etch masking material, a mechanical structural material, and a mechanical biasing material for MEMS manufacturing. One of the more critical material properties of stoichiometric SiN thin films in microsystems manufacturing is the residual stress. The residual stress in this material has been reported to have a tensile value of approximately 1 GPa [102]. As a practical matter, this very high value of residual stress means that this material is susceptible to cracking if more than a few hundred nanometers in thickness.

While stoichiometric SiN has some uses in MEMS, the stress in this material constrains its applications. Therefore, the MEMS community has developed a different formulation of SiN that is compositionally rich in silicon that is commonly called "silicon-rich silicon nitride" or "low-stress silicon nitride," having a considerably lower residual stress. In fact, the residual stress can be reduced to close to zero with the proper amount silicon incorporated into the material layer [103–107][5]. The lowering of the residual stress allows SiN layers with thicknesses of 1 micron or more to be deposited without the cracking seen in the stoichiometric materials. The ability to deposit nearly stress-free silicon nitride layers has been an important advancement for MEMS [108, 109]. Low-stress SiN is widely used in MEMS manufacturing as an electrical isolating layer, a masking layer, a membrane material, and as a mechanical structural layer. The relationship between ratio of dichlorosilane to ammonia and the residual stress is complicated [103, 104, 106, 110–112].

PECVD SiN is rarely used in IC microsystems but has found some application in MEMS as a structural material layer or as a passivation layer. The material properties

[5]Material properties of most MEMS material layers will be discussed in a later chapter. However, low-stress silicon nitride is a special example of a thin-film material that is used very often in MEMS manufacturing, and its stress level is directly related to the gas chemistry ratio. Therefore, the residual stress is given for this material here.

of PECVD SiN thin film can vary greatly ranging from moderately compressive to moderately tensile depending on the process recipe used [113–116].

The uniformity of silicon nitride, regardless of stress level, will be about the same as the guidance provided in Chap. 3.

4.3.1.4.2 Silicon Dioxide (SiO_2)

SiO_2 can be thermally grown or deposited using LPCVD or PECVD. MEMS manufacturing uses the same equipment to grow or deposit SiO_2 layers as IC manufacturing with the major differences being that MEMS will often use much thicker layers and the residual stresses in the deposited SiO_2 films may be modified.

Thick oxides using LPCVD or PECVD processing steps are common as masking layers for deep etches as well as sacrificial layers in surface micromachining process sequences. Often the thick oxides used as sacrificial layers in surface micromachining process sequences are also doped with phosphorus since phosphosilicate glass (PSG has a much higher etch rate compared to undoped oxide layers).

Silicon dioxide layers have been reported to be deposited at temperatures ranging from 425 to 700 °C and deposition rates from 7.5 to 70 nm/min [117–119]. The dielectric constant of low-temperature oxide (LTO) using a LPCVD process has been reported to be 4.3 and a dielectric strength about 80% of that for thermal oxide [120].

PECVD is a popular process choice for depositing oxides for MEMS microsystems again due to the low deposition temperatures. Additionally, another advantage of PECVD SiO_2 is that the mechanical properties can be modified over a fairly wide range to suit the requirements for many device applications [121, 122].

Silicon dioxide thin-film layers can also be deposited using sputter PVD. The big advantage of sputtering over other methods is that the deposition can be performed at temperatures as low as room temperature. The deposited layers are amorphous in microstructure and electrically insulating, and the properties will tend to be lower in magnitude than the bulk material equivalents. Like other sputtered deposited material types, the residual stress in sputter deposited silicon dioxide can exhibit either tensile or compressive stresses depending on the processing conditions [123].

Silicon dioxide can be deposited using atomic layer deposition (ALD). However, the maximum layer thickness using this technique will be limited to tens of nanometers.

The uniformity of thick oxides like other thin films that are thicker than what is typical in IC manufacturing will be no better than the guidance provided for uniformity of oxides in Chap. 3.

4.3.1.5 Polymers

The uses of polymers in IC manufacturing are mostly limited to photoresists and some packaging processes, but they are more commonly employed in MEMS fabrication including as an intermediary material layer in fabrication processes as well as a functional material in certain applications. Major interests in polymer for MEMS devices relate to the fact that many types are biocompatible, most involve low-temperature processing steps, and the manufacturing costs are relatively low.

Some of the more commonly used polymer materials for MEMS are SU-8, PDMS, and polyimide. Readers interested in more details on other polymer materials used in MEMS manufacturing, such as benzocyclobutene, BCB, as well as conducting polymers, such as polyacetylene, are referred to a more comprehensive review [124].

4.3.1.5.1 SU-8

SU-8 is a negative polarity resist sensitive to near-ultraviolet radiation. The lithographic process for SU-8 is very similar to a typical photoresist (e.g., spin coat, soft-bake, exposure, hardbake, and development). It is also a thermoset polymer, and a post-baking step is used to cross-link the polymers. Once cross-linked, SU-8 is an extremely robust material. SU-8 is commonly used for high-aspect ratio MEMS structures [125–134]. It can be exposed using X-ray radiation and electron-beam writing [135]. SU-8 is very also useful as an etch masking material for both wet and dry etch processes [136, 137].

The most common applications of SU-8 are for microfluidic MEMS devices including microchannels [138–141]; microreactors [142–144]; micronozzles [144]; microneedles [145, 146]; microcheck valves [147, 148]; and micromixers [131].

Probably the most limiting fabrication issue of SU-8 is the difficulty of removal of this material once it is cross-linked. There have been a number of different methods reported for removing SU-8 including elevated temperature N-methyl-2-pyrrolidone (NMP) [149, 150]; sulfuric acid, H_2SO_4 [136]; sulfuric acid mixed with hydrogen peroxide, H_2SO_4/H_2O_2 [150]; hot nitric acid, HNO_3 [150]; and others [151, 152]. Additionally, SU-8 has also been reported to be removed using pyrolysis and using a plasma etcher with an O_2 and CF_4 chemistry [150, 152]. For purposes of lift-off, NANOTM RemoverPG (80 °C) has been reported [153], as well as the use of thin or thick sacrificial material layers, such as metals [128, 147, 154], LTO [155], photoresist [156], polysilicon [157], and others [158–160].

Cross-linked SU-8 layers usually have a large tensile residual stress and stress gradients on silicon substrates [161]. This is thought to be mostly due to the large difference in thermal expansion coefficient of SU-8 and silicon.

SU-8 is usually spun onto substrates. The uniformity of SU-8 will be better than ±5%.

4.3.1.5.2 PDMS

Polydimethylsiloxane (PDMS) is a viscoelastic silicon-based organic polymer used in MEMS fabrication, particularly for microfluidic applications. It has a low density, is transparent over the visible spectrum, has a low mechanical stiffness, and is considered biocompatible [162–166]. Moreover, since it has a low curing temperature, PDMS can be used with microsensors integrated with electronics without adverse effects on the electronics. PDMS has emerged as an attractive choice for the implementation of MEMS microfluidic devices [167, 168].

The process for shaping PDMS is called soft lithography, which is a process whereby the pre-cured material is forced into a reverse image mold, heated, and then removed or released from the mold, resulting in the PDMS having the desired shape [169]. PDMS is made by mixing the pre-polymer with a curing agent in a ratio of about 10 to 1. The PDMS mixture is then placed in a vacuum environment for a few minutes to remove any air that may be entrapped into the mixture during stirring. The PDMS is then spun onto a substrate to create a uniform coating. Subsequently, the PDMS is pre-cured at an elevated temperature of about 65 °C. This pre-cured is timed carefully to ensure that the material does not completely cross-link. Next, the inverse image mold is placed onto the surface of the PDMS with the mold side in contact with the top surface of the PDMS, and a small amount of force loading is typically applied to the backside of the mold. The PDMS and mold stack are then placed into a vacuum for a few minutes. Then the PDMS and mold are exposed to atmospheric pressure and then completely cured at 65 °C. The mold is then removed from the PDMS thereby resulting in the mold shape being impressed into the PDMS.

Additionally, new photo-definable PDMS formulations have emerged onto the market [170, 171] thereby allowing the material to be processed like a photoresist into the desired patterns. These materials are made photo-definable by mixing a photo-initiator with the PDMS. This eliminates the requirement of a mold.

To first order, the dimensional variation of PDMS is dependent on the dimensional variation in the mold, which in turn will depend on the method used to fabricate the mold. Molds for PDMS are often made in silicon using DRIE.

PDMS spun onto substrates have a uniformity of better than ±5%

4.3.1.5.3 Polyimide

Polyimide is a polymer of imide monomers that has been in commercial production since the 1950s. It is extensively used in the microelectronics industry, mostly for packaging applications. It is an excellent electrical insulating material and also is resistance to heat. Currently, polyimide is available in a variety of forms, including film and tapes, as well as liquid formulations that are photo-definable or non-photo-definable that can be spun onto substrates. Polyimides exhibit either thermoset or thermoplastic behaviors. Some of the attractive material properties of polyimide for MEMS applications include a low mechanical stiffness, biocompatibility, chemical and biological inertness, and low cytotoxicity [172]. Other important characteristics of polyimides include a high glass transition temperature, high thermal and chemical stability, low dielectric constant, high mechanical strength, low moisture absorption, and high solvent resistance [173].

The processes for use of liquid forms of polyimide are very simple. The material is usually spun onto a wafer to obtain a uniform coating. For non-photo-definable polyimides, the material is then cured at a temperature from about 300–500 °C depending on the formulation. Other processing may be performed on the polyimide if the layer is to be patterned. If the material is photo-definable, then after spinning onto the substrate, it may then be pre-cured to drive off some of the solvents depending on the polyimide formulation type. This is followed by an exposure and development, followed by a cure at a temperature from about 300–500 °C depending on the formulation. Polyimide layers can be removed using a variety of reported methods including hot acids and bases [174]; a mixture of sulfuric acid and hydrogen peroxide, H_2SO_4 and H_2O_2 [175]; O_2 plasmas [176]; and O_2 combined with CF_4 plasma [177]. Polyimide layers can also be selectively released from substrates using a number of reported methods including using a thin layer of oxide under the polyimide as a sacrificial layer with the release performed by immersion into hydrofluoric acid, HF [178], and using an aluminum sacrificial layer under the polyimide with the release performed by immersion into a mixture of phosphoric–acetic–nitric acids and water [176].

Polyimides are usually spun onto substrates and therefore will have a uniformity of better than ±5%.

4.3.1.6 Ceramics

A ceramic is a nonmetallic solid material composed of an inorganic mixture of elements held in ionic and covalent bonds. Ceramics exhibit varying material properties depending on the exact chemical makeup. The types of ceramics used in MEMS devices include a number of materials that can be co-categorized as metal oxides, metal nitride, dielectrics, etc. Examples of some of the ceramic materials used in MEMS include nitrides of silicon (e.g., Si_3N_4 or other stoichiometries that can also be classified as dielectric materials); aluminum (e.g., AlN which is also classified as a piezoelectric and pyroelectric material); titanium (e.g., TiN which is also classified as a highly electrically conductive material); silicon carbide (e.g., SiC which is also classified a semiconductor); zinc oxide (e.g., ZnO, which also classified as a piezoelectric material); aluminum oxide (Al_2O_3, which is also classified as a dielectric material); and others.

Ceramics can be deposited using a variety of methods depending on the specific material to be deposited including LPCVD, PECVD, PVD, and ALD. Since most of the important ceramic materials are covered in other sections of this chapter, no further discussion on them is needed here.

4.3.1.7 Special MEMS Materials

In this section several special thin-film material types will be reviewed including piezoelectrics, shape-memory alloys, and magnetic materials. These are considered special due to their unique properties. Most of these materials are deposited using the same equipment as used in IC fabrication.

4.3.1.7.1 Piezoelectric Materials

Piezoelectric materials are of interest in MEMS microsystems for use as both sensing and actuation applications. There are a wide variety of materials that exhibit piezoelectric effects[6]; and those used in MEMS include aluminum nitride (AlN), zinc oxide (ZnO), and lead zirconate titanate (PZT). AlN and ZnO are examples of polar materials, and PZT is an example of a ferroelectric material.

Aluminum nitride (AlN) can be deposited using sputtering and ALD [179–184]. The deposition processes are devised so that the material layers have a crystalline microstructure since this is important for them to exhibit piezoelectric properties.

ZnO is usually deposited using sputtering methods. The piezoelectric properties of ZnO are typically not as attractive as those of AlN, and therefore AlN is usually a preferred piezoelectric material compared to ZnO [185, 186].

PZT is a piezoelectric ceramic with high piezoelectric constants often used in MEMS sensors and actuators. Deposition of PZT, which is a ternary compound material, is considerably more complicated compared to AlN and ZnO. It is difficult to control the stoichiometry in ternary materials, and the resultant stoichiometry is extremely important in PZT [187, 188]. Among the methods demonstrated for deposition of PZT include sputtering and chemical solution deposition (CSD) techniques [189–191]. Importantly, thin-film-deposited PZT usually requires an anneal in order to crystallize the material layer for use as a piezoelectric. A preferred electrode material layer is platinum since its use results in (111) oriented PZT films that have relatively attractive material properties. Titanium is commonly used as an adhesion layer for the deposition of platinum on silicon dioxide. However, it has been shown that a stack of silicon dioxide/titanium/titanium oxide/platinum works better since the titanium oxide is a diffusion barrier between the titanium and platinum and allows improved PZT film quality [192].

Sputtering of PZT has been demonstrated using separate targets (Pb, Zr, and Ti) in a reactive sputtering process (reaction with oxygen) as well as RF sputter with

[6]There are 20 known piezoelectric materials. Of these 20 materials, 10 of them are called polar materials in which an electric dipole moment can be present even when no electrical field is present. This effect is called spontaneous polarization. Spontaneous polarization changes with temperature, which is also referred to as pyroelectricity. The other remaining piezoelectric materials exhibit a spontaneous polarization that can be permanently reoriented between crystallographically defined states by the application of an electrical field. These are commonly referred to as ferroelectrics.

composite targets [193]. The major challenge with sputtering is control of the stoichiometry of the resultant material layers.

Chemical solution deposition (CSD) methods for PZT thin-film layers have been reported using a mixture of lead, titanium, and zirconium precursors in a solvent-based solution [194]. The highest reported piezoelectric coefficients were for a slight variation that used lead acetate trihydrate, titanium isopropoxide, and zirconium n-propoxide as the precursors in a solvent-based solution [191, 195, 196]. The material is deposited onto the substrates by spinning at high speeds to create a uniform coating. Following the deposition, the material is heated to about 350 °C to drive off the solvents and stabilize the material. The thickness of the layer in each spinning is limited to less than 1 micron. Therefore, it is common practice to spin coat the substrate multiple times, with a pyrolysis after each spin deposition, in order to build up a total thin-film thickness of the desired amount. Once the PZT layer thickness is sufficient, the material is then subjected to a crystallization anneal at about 700 °C in oxygen [197].

Both wet and dry etch processes are available for patterning AlN and ZnO [198–209]. The etch rates for both AlN and ZnO are related to the crystalline microstructure of the layers. PZT can be wet etched using mixtures of HCl and HF [210, 211] with the addition of water. However, a major issue with wet etchants for PZT is the high lateral etch rates that can be several times higher than the vertical etch rate. This results in a large amount of variation in the dimensional control of the features and renders wet etching an unattractive option for patterning of this material. Chemical and physical etching methods using either a reactive ion etch (RIE) or an ion-milling removal are better able to provide improved anisotropy and dimensional control necessary for implementation of MEMS devices. Similar to the chemistries used in wet etching of PZT, RIE techniques of PZT mostly employ fluorine and chlorine containing reactive gases (i.e., CF_4 and CCl_4, etc.) [188, 212–218]. However, both Zr and Ti etching results in low volatility by-products, such as ZrF_4, $ZrCl_4$, TiF_4, and $TiCl_4$, that make a more chemical-based etch prone to problems. The amount of etch reaction by-products redepositing onto the substrate can be reduced by ensuring that the etch process has sufficient mechanical energy (i.e., ion bombardment).

For most MEMS applications, the PZT must be poled, which is the aligning of the ferroelectric domains in a preferred direction. To perform poling, an electric field is applied across the material that has a magnitude of two to three times the coercive field of the material for a duration of at least 10–15 min. The poling field can be reduced by heating the sample during the poling process [197].

AlN and ZnO deposited using sputtering would expect to have variations similar to those of any thin-film layer deposited using this technique as covered in Chap. 3. PZT can be deposited a number of ways. CSD PZT will have a uniformity similar to spin coating, that is, usually better than a few % across the wafer. However, the problem is that CSD is usually deposited onto pre-patterned wafers with existing topography and therefore this can degrade the uniformity of spun-on thin-film layers. Sputtering of PZT will have a value of thickness uniformity essentially equivalent to the sputtering of any thin-film layer. The stoichiometry of this complex material may have variation of $\pm 1\%$ or more.

4.3.1.7.2 Shape-Memory Alloys (SMAs)

Shape-memory alloys (SMAs) are an attractive material choice for use in MEMS actuator applications. The major advantages that SMAs offer over other means of actuation are the large forces and displacements that are possible with these materials [219]. By far the most commonly used SMA material type is titanium nickel (TiNi). TiNi SMA can be deposited as a thin-film layer using physical vapor deposition (PVD), namely, sputtering and evaporation [220–227]. As-deposited TiNi thin-film layers are amorphous, and therefore a subsequent anneal must be performed at temperatures typically above 540 °C to crystallize the material [228]. The sputter deposition of TiNI thin-film layers has been reported using a single target as well as multiple targets [221, 222, 224–226, 229]. Evaporation of TiNi is challenging since the vapor pressure of nickel is higher than that of titanium and therefore will result in a gradient in the stoichiometry through the thickness of the deposited layer. This problem can be partly remedied by using what is called flash evaporation [230].

A major issue with TiNi thin-film material layers is that the phase transformation temperature of the material is highly dependent on the exact stoichiometry of the layer [231, 232]. Therefore, the control of the stoichiometry of these materials is extremely important.

With regard to patterning, TiNi can be wet etched using a mixture of hydrofluoric acid and nitric acid [233]. In one report, TiNi thin film deposited on Si substrates using evaporation was wet etched in a mixture of a part HF, one part HNO3, and four parts H2O using negative photoresist as a masking material with a reported etch rate that was about 30 nm/s [234]. The lateral etch rates were not reported, but assuming that this etch is mostly isotropic, the dimensional control and resolution of this etching technique will be very limited.

Reactive ion etching (RIE) of TiNi thin films using a chlorine-based reactive gas has been reported. Due to the low vapor pressure of Ni, the etch rate was only about 10 nm/min which is impractical for production environments [233].

Another patterning technique reported for TiNi is lift-off [226]. For example, using a polyimide layer as the mask, a 7–10 micron thick layer of TiNi was deposited using DC sputtering and then patterned using lift-off. The polyimide was removed using immersion in an elevated temperature KOH solution. Of course, thinner layers of TiNI can be patterned using lift-off techniques with photoresist.

TiNi thin-film actuators can exhibit fatigue at higher strain levels, and therefore this is a very important design issue with SMA materials [235]. Specially, the maximum number of reversible cycles of an SMA actuator is related to the maximum strain levels in each cycle. Therefore, in applications the usable strain for these materials is kept to lower levels.

As noted above, the phase transformation temperatures of TiNi SMAs are extremely sensitive to the stoichiometry of the alloy, and with only a 1% variation, the transformation temperature can vary over tens of degrees [236]. This places an incredibly high demand for accurate control and reproducibility in the deposition of this material in a manufacturing environment.

The addition of copper (Cu) to TiNi alloys to form a ternary compound, namely, TiNiCu, has been shown to affect the phase transformation temperature and thermal hysteresis [221, 237]. Specifically, the addition of Cu acts to decrease the thermal hysteresis, which is desirable for applications requiring repeated actuation cycles and a narrow range of temperatures for the phase transformations [238]. The Cu also has some beneficial effects on the mechanical properties, such as lowering the elastic limit of the material in the martensite phase.

Laser micromachining can be used to pattern TiNi thin films [239]. However, care must be taken to ensure that the material does not heat up too much when radiated by the laser. TiNi SMA thin-film layers can also be etched using electrochemical etching techniques [240].

TiNi thin films have been reported to have a number of uses in MEMS devices, particularly for microfluidic applications due to the large displacements and forces that are possible [241, 242]. The large displacements, due to the large strains possible with SMAs, are important in microfluidic components since the flow resistances can be reduced significantly. And, the high forces allow microfluidic devices to be realized that can control fluids at higher operating pressures. Among the devices to be realized using TiNi SMAs include microvalves [243]; micropumps [244–246]; and optical components [247–249].

The thickness uniformity of deposited SMA thin-film layers will be determined by the sputtering equipment used and therefore will be similar to the guidance given for sputtering in Chap. 4. A more difficult issue is the control of the uniformity of the stoichiometry of deposited SMA layers. The content can vary of ±0.5–1% or more.

4.3.1.7.3 Magnetic Materials

Certain metal alloys are known to exhibit magnetic behavior. Magnetic materials are typically classified as either hard or soft depending on whether they exhibit a strong magnetization when no external magnetic field is present. These are so-called hard magnetic materials. Soft magnetic materials exhibit a small level of remanent magnetization even without the presence of an external magnetic field [250, 251]. Among the soft magnetic materials that have been deposited in thin-film form include $Ni_{80}Fe_{20}$; Co–Fe–Cu; Co–Ni–Fe; and Co–Fe [250]. Among the hard magnetic materials in thin-film form include Co–Ni–P; $FePt-L1_0$; Sm–Co; and Nd–Fe–B [251].

Magnetic materials are useful for implementation of motors, actuators, energy harvesters, transformers, etc. Both soft and hard magnetic materials can be used in these types of devices.

Magnetic materials are usually deposited using either electroplating[7] or sputtering. Fabrication of MEMS using magnetic materials is often very complex. Many of the stronger hard magnetic materials require high-temperature deposition

[7] Electroplating is discussed in the next section.

methods or high-temperature anneals in order to optimize their magnetic performance. Electroplating of ferromagnetic materials that are metal alloys is especially challenging. Details on process recipes for electrodeposition can be found elsewhere [252, 253].

The selection of magnetic materials for a given application requires considerable examination of both the possible materials and their properties as well as the fabrication issues related to the use of each particular material and device design [250, 251].

The uniformity of thickness with either sputter deposition was covered in the previous chapter. The stoichiometry control will likely be no better than about ±0.5–1%. The uniformity of thickness of electroplated magnetic materials is discussed later.

4.3.2 *MEMS Specific Processing Steps*

This section reviews MEMS processing steps that do not use the equipment that is commonly employed in IC manufacturing. Most of these processing steps are unique to MEMS fabrication. When possible, variations in the device dimensional parameters will be given. Many MEMS process sequences also use what are called "micromachining technologies." Since most micromachining methods are really process modules that entail more than one processing step, these will be discussed in the next section.

4.3.2.1 Electrochemical Deposition

Electrochemical deposition methods are wet chemical processes that involve the reduction of metal ions from an electrolyte solution resulting in the deposition of thin-film layer of metal atoms onto a surface [252, 253]. There are essentially two different types of electrochemical deposition processes: the first is electroplating where an externally generated electric current is passed between two electrodes placed in the electrolyte solution to supply the charges for the oxidation and reduction reactions and electroless deposition wherein a reducing agent in the electrolyte solution provides the charges for these reactions. A variety of metals can be electroplated including Au; Ag; Cu; Cr; Pt; Ni; Zn; S; Cd; and Pb, as well as number of different alloys composed of metals that can be electrochemically deposited [252, 253].

Figure 4.2 illustrates the setup for electroplating. As noted, there are two electrodes placed into the electrolyte solution, namely, a cathode and an anode. The substrate on which the metal is to be plated from the solution is the cathode and is where the applied electrical current causes a reduction (addition of an electron) to take place resulting in the deposition of the metal atoms. The anode is where an oxidation (removal of an electron) takes place and can be soluble. A current source is

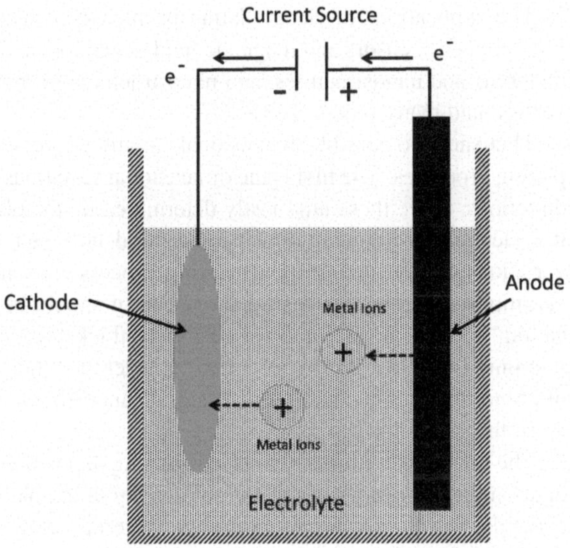

Fig. 4.2 Setup for electroplating of metals onto substrates

placed between the cathode and anode. The solution is usually stirred and temperature controlled and may also be heated.

There are a number of different compositions for the electrolyte solutions for use in electroplating different metals. Often wetting agents (i.e., surfactants), buffering agents, and other agents are added to the electrolyte to enhance the plating process and/or the plated materials. Also, the applied currents can be DC or have a specific waveform that can also enhance the plating process depending on the materials being plated as well as other factors [252].

The other type of electrochemical deposition is electroless and thereby does not require an externally applied current source [254]. There are three separate processes involved in electroless plating: (1) a galvanic displacement wherein there is an electron exchange that takes place of the surface substrate that results in the reduction of metal ions; (2) a substrate-catalyzed process wherein the surface is modified to make it more reactive for reduction and oxidation; and (3) an autocatalytic process wherein a metal salt and a reducing agent react continuously in the presence of a catalyst. The third process is critical to sustaining the plating process once the substrate is initially covered by a monolayer of the metal.

Because not all metals are self-catalytic, the number of types of metals that can be deposited using electroless plating are more limited than compared to electroplating. One advantage of electroless plating is that is not constrained to only the deposition onto conductive surfaces. Electroless plating can be performed on polymers as well as other electrically insulating materials. However, electroless plating is rarely used in industry due to challenges of reproducibility [255].

Electrochemical plating for MEMS is most often performed using a prefabricated mold on the surface of the substrate into which the metal is plated. This enables the plating of the features desired.

The applications of electroplating in microsystems manufacturing include thin-film magnetic sensors and magnetic field generators at the nanoscale, thermoelectric materials and nanostructures, and nanostructures for photovoltaics and energy conversion and many others [255].

There are two possible dimensional variations of consideration in electrochemical plating processes. The first is the dimensional variations that may occur in the lateral directions. Since these are mostly determined by the plating mold, the variations in the electroplated lateral dimensions are determined by the lateral dimensional variations of the plating mold making process (i.e., lithography). These types of variations have been discussed above, but it should be noted that in MEMS fabrication, it is not uncommon for relatively thick layers to be plated, up to tens of microns. Therefore, it is important to consider the lithography process used since use of photolithography could result in significant diffraction of the features in resists of these thickness levels.

The thickness uniformity of electroplating deposition processes are highly dependent on the uniformity of the current density in the plating process, which in turn is dependent on the plating mold design, material being plated, loading effects, solution depletion effects, temperature uniformity, and many other factors. In general, electrochemical plating processes are not considered as having particularly good uniformity. Moreover, the surfaces of plated material layers can be rough and may require polishing for use in some applications thereby adding more difficulty in obtaining good uniformity in the plated material thickness. It has been reported that with the use of a multiple electrode plating configuration, plating uniformities of about $\pm 8\%$ were possible with copper on 200 mm diameter wafers [256]. However, it should be noted that in this work, an unusual level of development was incurred to obtain this level of thickness uniformity, and most electroplating processes will likely exhibit higher levels of non-uniformity.

4.3.2.2 MEMS Lithography

We discussed photolithography processes that are employed in IC microsystems manufacturing in Chap. 3. As noted in that discussion, projection photolithography in the form of step-and-repeat (i.e., steppers) systems is almost universally used in IC production today. In MEMS manufacturing, both contact aligners and steppers are used, but contact aligners are more common. Since stepper photolithography was covered in Chap. 3, it will not be further discussed in this chapter. In addition to contact and stepper photolithography, MEMS fabrication also uses some unique lithographic capabilities that are not used in IC manufacturing that will be reviewed in this section. These include front-to-back contact alignment; lithography on curved surfaces; grayscale lithography; e-beam lithography; X-ray lithography; direct-write laser lithography; and mixed tool set lithography.

4.3.2.2.1 Contact Photolithography

Contact photolithography was briefly mentioned in Chap. 3 wherein it was noted that stepper photolithography has displaced contact photolithography for use in IC manufacturing. Contact photolithography is used extensively in MEMS manufacturing since most MEMS devices do not need to pattern features below a few microns. With regard to resolution and alignment accuracy using contact photolithography, both will depend on the amount of topography on the substrate surface, the materials on the substrate, the skill of the person performing the alignment, and other factors. Nevertheless, using contact photolithography, it is possible with some effort to be able to obtain feature resolution of below 1 micron, although many foundries specify a minimum resolution of about 1–2 microns. The lateral dimensional variation in a defined feature and the layer-to-layer registration alignment accuracy that can be obtained are both about ±0.5 microns [257].

4.3.2.2.2 Front-to-Back Contact Photolithography

Often in MEMS manufacturing, there is the need to perform photolithography processes on both sides of the substrate. Specifically, there may be a series of processing steps completed on both the top and bottom sides of the wafer, and these features on either side of the wafer must accurately align to one another. One of the most common examples of the use of a front-to-back aligner is in the manufacturing of pressure sensors. An example is shown in Fig. 4.3 wherein photolithography is required on both sides of the substrate in order to able to perform an etch from the backside (i.e., bulk micromachining) of the wafer to implement the thin silicon membrane that is needed for the pressure sensor. The pressure sensor shown in Fig. 4.3 follows a standard bulk micromachining process sequence that will be covered in more detail later in this chapter, and therefore only a cursory explanation will be given here. Specifically, the piezoresistive elements are first fabricated on the top surface of the substrate, and then using an etch mask on the backside of the wafer, a front-to-back alignment is performed to align a mask to open up regions in the backside etch mask to expose the silicon substrate. Then an anisotropic wet etch

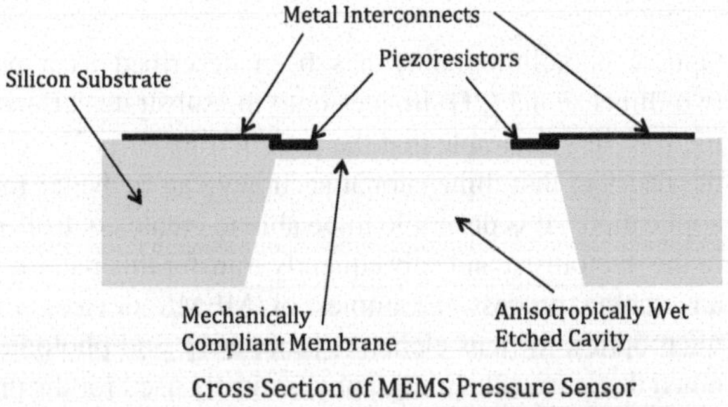

Fig. 4.3 An example of a MEMS device fabrication requiring front-to-back alignment for contact photolithography

is performed on the silicon wafer to etch from the backside of the wafer and make a thin silicon membrane that will act as the mechanically compliant element in the pressure sensor. It is important to be able to accurately align the piezoresistors on the top surface of the wafer to the opening in the mask on the backside of the wafer.

As a general rule, the resolution on front-to-back photolithography is the same as for front–side exposures or about 1–2 microns. The alignment registration accuracy for front-to-back alignment is specified by some tool manufacturers as about ±1.0 micron in both x and y directions although in practice this accuracy of alignment is very difficult to obtain [257]. In practice most foundries give an alignment registration accuracy of at least ±3 microns, a resolution of 1–2 microns, and a dimensional variation of ±0.5 microns.

4.3.2.2.3 Direct-Write Laser Photolithography

In direct-write laser photolithography, a wafer having a pre-deposited layer of photoresist is scanned using a computer-controlled laser beam to expose the resist according to a pre-defined mask pattern. Since the laser is electronically controlled, the need for a photomask plate is eliminated. Instead, a computer file with the mask pattern directly controls the laser scanning. Some of these systems are able to write features having a size of 0.2 microns, on an address grid as small as 5 nm, a write speed of 3 mm^2/min, an edge roughness at 3σ of 50 nm, a critical dimension uniformity at 3σ of ±60 nm, and a layer-to-layer alignment accuracy at 3σ of ±100 nm [258, 259]. These systems can be used to make patterns directly onto the substrate and make masks for use in conventional photolithography.

While these systems were mostly developed for rapid prototyping in research and development applications, they can also be used in small-scale MEMS production markets as well. Additionally, new capabilities have been introduced on these systems to allow direct writing of patterns over 3-D surfaces with feature sizes down to 1 micron. These systems are capable of writing patterns on both concave and convex surfaces.

4.3.2.2.4 Grayscale Photolithography

Optical photolithography has been described as a pattern transfer technique of two-dimensional (2D) images onto the substrate surface. In fact, for IC manufacturing it is very desirable that the pattern transfer is 2D with sharp vertical sidewalls in the resist so that dimensional accuracy can be better maintained. For some MEMS applications, it is desirable to be able to create controlled three-dimensional patterns in the photoresist and subsequently transfer this pattern into a material layer through an etching process. Examples of MEMS devices where this is useful include microoptics, for lens elements [260, 261], and photonic crystals [262]. It should be noted that grayscale lithography is mostly used for the production of very specialized optical elements for niche markets where no other method for making these sorts of structures is available.

The photomask used in conventional photolithography is a glass plate with regions on the surface having a nontransparent material that represents the image to be transferred. These regions block the light from the exposure to reach the photoresist, whereas the open regions allow the exposure radiation to reach the resist. Consequently, a photomask is a binary optical element with regions that are either "on" or transparent and other regions that are "off" or blocking. The resolution of these photomasks is diffraction limited.

The purpose of grayscale lithography is to create a pattern in the photoresist that is 3D, that is, having a thickness that is variable. This is achieved by pixelating the mask with pixel dimensions that are sized below the diffraction limit of the optical photolithographic system being used for the pattern transfer [263]. This allows the pixel pattern to modulate the amount exposure radiation transmission through the photomask. Specifically, the pixel sizes and spacing patterns are varied so as to allow different levels of transmission depending on the location on the grayscale mask.

A grayscale mask can be made in two different methods [263, 264]. The first is to use a stepper photolithography system to create a pixelated pattern on a mask plate, and the second uses a commercial process called high-energy beam-sensitive (HEBS) glass. The HEBS process is more expensive but would be recommended for manufacturing since the resolution and quality are better. HEBS has a resolution of below 0.25 microns [264]. The positional accuracy of grayscale lithography has been reported to be $\leq$150 nm [258]. Figure 4.4 illustrates a version of the grayscale photolithography process. The photomask is pixelated in the central region with a linear gradient in transparency from zero to near 100% from left to right, with the ends completely nontransparent. The radiation from the exposure tool passes through the grayscale mask to impinge onto the photoresist on the top surface of the substrate with a gradient in the intensity mimicking the mask gradient. The result is a gradient in the resist exposure. Once developed, the photoresist has a linearly sloped profile as shown in the bottom.

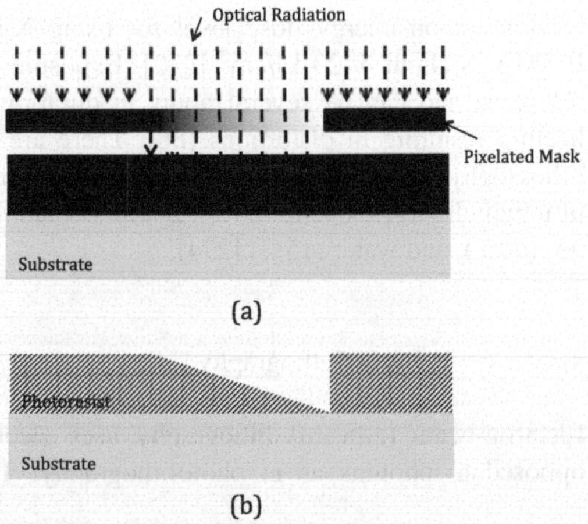

Fig. 4.4 An illustration of performing grayscale lithography in the photoresist on a wafer. (**a**) Exposure of photoresist using pixelated mask. (**b**) After exposure and development showing resist profile

4.3.2.2.5 X-Ray Lithography

X-ray lithography uses X-ray radiation instead of optical radiation to expose the photoresist on a substrate. The wavelengths of the X-rays used in this technique are about 0.7–1 nm. While the theoretical resolution of this technique is 20 nm, in practice the resolution is about 50 nm [265]. X-ray lithography is almost always a single-layer process. X-ray lithography requires a suitable light source for generating highly collimated X-ray beams. For this purpose a synchrotron is often used as the source since it is capable of emitting high-power, highly collimated X-ray radiation. This allows exposure through thick resist layers as well as creation of high-aspect ratios in the resist.

While X-ray lithography is used as a research tool, it is not often used for pattern transfer in MEMS manufacturing since it is too costly and complicated. Instead, it is used for the production of high-aspect ratio molds in polymers that are used in a MEMS process called LIGA (acronym for Lithographie, Galvanoformung, Abformung) [265, 266] that will be described below.

X-ray lithography requires making of very specialized masks. The mask is usually made from a substrate material that is X-ray transparent (low-Z material) with a patterned X-ray absorbing material (high-A material) to define the features on the mask. The substrate is commonly silicon with membranes made from silicon nitride or silicon carbide [267, 268]. The absorbing materials on the mask are usually either copper or gold [269].

Since the pattern transfer process from the mask to the wafer is 1–1, the mask-making process usually sets the tolerances on the resultant image on the wafer. E-beam lithography (described below) can be used for small critical dimensions but is very costly for large area masks. Poly(methyl methacrylate) (PMMA) is typically employed as the resist for X-ray lithography. SU-8 is also sometimes used [270, 271]. PMMA or SU-8 are applied to the substrates by spinning. For thicker mold layers such as commonly seen in LIGA processes, a precast sheet of PMMA can be attached to the substrate using glue. If the PMMA sheet thickness is not exactly the thickness required for a given application, it can be milled to reduce its thickness prior to use.

A common energy dose level for using X-ray radiation for the exposure of PMMA is about 4–20 kJ/cm^3 [272]. Exposure times depend on the thickness of the resist and can be several hours in duration. Long exposure times can cause heating resulting in distortions [86]. There are several developers used in X-ray lithography, including a mixture of diethothxy and morpholine [273] and a mixture of tetrahydro-1,4-oxazine (20%), 2-aminoethanol-1 (5%), 2-(2-butoxyethoxy) ethanol (60%), and water (15%) [274].

4.3.2.2.6 E-Beam Lithography

Electron-beam (e-beam) lithography uses electrons to expose a resist layer as opposed to photons as in photolithography. The dosage level of the electrons modifies the solubility of the resist. An e-beam lithography system creates an electron beam that is electrically focused onto the substrate surface and scans this beam across the surface of the wafer.

An example of a high-end e-beam lithographic system is the Vistec EBPG 5000+ system. The specifications for this system are as follows: spot size <2.5 nm (at 100 keV); beam current, 103 pA to 200 nA; acceleration voltages, 50 kV and 100 kV; pixel size, 2.5 nm; scan field size, 1 mm by 1 mm; field stitching accuracy, <20 nm; alignment accuracy, <20 nm; scan speed, 50 MHz; and minimum reproducible feature size, <10 nm. This system uses a high current density thermal field emission gun as the electron beam source.

Due to the need to scan a single electron beam across the substrate to write a pattern, e-beam lithography is very slow and costly and mostly used as a research tool and rarely in production for MEMS microsystems. However, it can be used cost-effectively for mask making since the cost of the mask can be amortized over many wafers.

Generally the electron beam can only be scanned over a certain area of the substrate at a time. The amount of area that can be scanned depends on the e-beam settings. To scan larger areas, the areas must be stitched together by a translation of the stage with the wafer on it.

Polymethylmethacrylate (PMMA) is the most commonly used e-beam resist. It has a positive polarity. Poly (2,2,2-trifluoroethyl-chloroacrylate) (EBR-9) is an acrylate-based resist that is commonly used in mask-writing applications since it has faster exposure times [275].

4.3.2.2.7 Lithography on Large Topography

Often in MEMS manufacturing, there is the need to perform lithography on substrates having large topography. This requires the ability to deposit a uniform coating of photoresist over the substrate surface having large topography. For example, the substrate may have been bulk etched to create deep depressions into the substrate, and there is the need to provide metal interconnects from the top surface of the substrate to features located with the bulk micromachined cavity as shown in Fig. 4.5.

There are two methods that have been reported for depositing resist onto surfaces with large topography. The first is to spray the photoresist [276]. The method uses an ultrasonic spray nozzle to create droplets of resist that are microns in diameter and that stick once they contact the substrate surface. During the spraying process, the substrate is rotated at a slow speed while the sprayer is moved across the surface. A resist having a viscosity of less than 20 cSt is used. AZ4562 properly diluted with a solvent to reduce its viscosity has been reported [277]. This system has been shown to be able to deposit AZ452 photoresist over features having a depth of 375 microns with a uniformity of about $\pm$10%. While resist spraying provides much improved uniformity compared to spinning, there may still be variations in the resist thickness

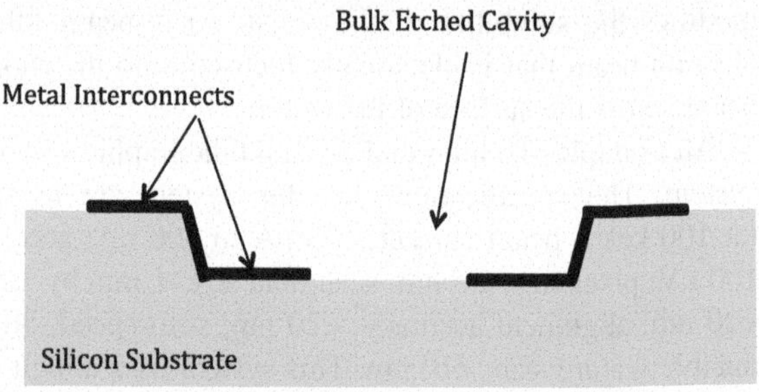

Fig. 4.5 Illustration of a simple MEMS structure with large topography that would benefit from being able to deposit a uniform layer of photoresist

across the substrate topography depending on the circumstances. Usually, this process must be optimized for a given substrate topography.

Another method for depositing photoresist is electrodeposition [278]. This technique requires special equipment, a cataphoretic resist emulsion, and the wafer surface must first be coated with an electrically conductive material over the surface. Therefore, it is very useful for patterning on metal lines but is not suitable for any type of process. It has a demonstrated improved uniformity compared to spray resists [278].

It should be noted that since the photoresist is covering large topography over the substrate surface, the resolution of the lithography will be limited. For example, if using contact lithography, there will be a large gap between the mask and the substrate surface at the depths of the features.

4.3.2.2.8 Lift-Off Patterning

Lift-off is an alternative method of patterning thin-film layers that is particularly useful when the etching of a material layer cannot be performed or is difficult. Due to the fact that MEMS fabrication uses so many different materials, often there are not suitable methods for etching available. Lift-off is not material dependent and can be performed for most any material type.

The lift-off patterning technique is illustrated in Fig. 4.6. In Fig. 4.6a, a layer of photoresist is deposited onto the substrate surface. Next, the photoresist is patterned using conventional photolithography as shown in Fig. 4.6b. The metal to be patterned is then deposited as seen in Fig. 4.6c. Then the resist with the metal on top is removed. This leaves behind the metal deposited in the open areas in the photoresist as shown in Fig. 4.6d.

Lift-off has a few requirements for it to be performed successfully. First, it is desirable that the resist be considerably thicker than the metal to be deposited and patterned. A good rule of thumb is that the resist should be about 2× thicker than the metal to be patterned. Second, it is preferably that the resist have a retrograde (or overhanging) profile since this makes it more different for the deposited metal to be continuous over the photoresist edges. Third, the deposition method used for the metal should not be conformal. As such, evaporation is commonly used for performing the deposition of the material that is patterned using lift-off.

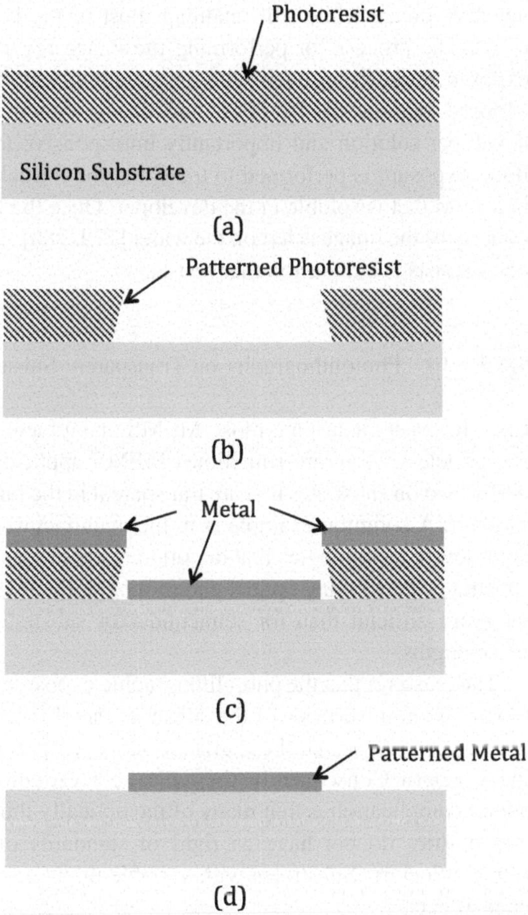

Fig. 4.6 The process of lift-off to pattern a metal thin-film material layer. (**a**) Deposition of photoresist on substrate. (**b**) Perform photolithography to pattern resist. (**c**) Deposition of metal layer. (**d**) Perform lift-off

It should be noted that lift-off patterning is not restricted to only patterning metals. It can be used to pattern virtually any thin-film-deposited material layer as long as the deposition temperatures do not adversely affect the photoresist.

The thickness uniformity of lift-off patterning of deposited thin-film material layers will be determined by the uniformity of the evaporation process employed as discussed in Chap. 3.

4.3.2.2.9 Image Reversal Photoresists

Positive resists are used more often than negative resists in MEMS fabrication due to their better overall performance and lower toxicity. However, positive resists commonly exhibit a resist sidewall angle that makes lift-off processes difficult. An image reversal resist allows a normally positive polarity photoresist to be transformed into a negative image while still retaining most of the beneficial attributes of a positive resist. The process for performing the image reversal involves a post-bake of the exposed resist in an oven with an ammonia vapor. The ammonia reacts with the exposed regions of the resist and changes them from solution to insoluble in developer solution and importantly unresponsive to further UV exposure. Then a flood exposure is performed to transition the formally unexposed areas of the resist in a form that is soluble in the developer. Once the resist is developed, the negative polarity of the image is left on the wafer [279, 280]. The uniformity of image reversal photoresists is similar to other resists.

4.3.2.2.10 Photolithography on Transparent Substrates

Like IC manufacturing, most MEMS manufacturing is performed on silicon . Nevertheless, there are sometimes MEMS applications requiring fabrication to be performed on substrates that are transparent in the range of optical wavelengths, such as glass. A common example is in the manufacturing of microfluidic and lab-on-a-chip device technologies that are often made in glass or optically transparent plastic substrates. Performing optical photolithography on optically transparent substrates is far more difficult than for semiconductor substrates, which are opaque at visible wavelengths.

The reason is that the photolithographic exposure radiation can reflect off both the top and bottom surfaces of the substrate thereby resulting in unpredictable interference effects. At worst, these effects may even prevent the resist from receiving the mask pattern. Consequently, the achievable resolution will likely be limited. Another added complication is that many of the optically transparent substrates such as glass and plastics do not have as rigid of standards on the thicknesses as seen with semiconductors. Substrates with variable thicknesses will exhibit different interference effects.

Therefore, when performing photolithography on transparent substrates, it must be assumed that the photolithography process will have to have a higher level of rework that would be the case for silicon wafers.

There are a number of complicated issues involved in the dimensional variations that will result in performing photolithography on transparent substrates that can only be determined by performing experiments and measuring the results. Nevertheless, the dimensional variations will likely be larger than if the same photolithography were performed on nontransparent substrates.

4.4 MEMS Micromachining Methods

This section reviews most of the micromachining methods used for MEMS fabrication. Micromachining methods are used to make mechanical and electromechanical structures out of materials for the implementation of MEMS devices. Since micromachining is such an important topic in MEMS fabrication, an entire section is dedicated to the subject. Moreover, many of the micromachining methods consist of multiple processing steps and therefore are better classified as process modules rather than processing steps. Among the micromachining methods covered herein include bulk micromachining; deep reactive ion etching (DRIE); surface micromachining; wafer bonding; LIGA, molding, and replication processes; laser micromachining; electro-discharge micromachining; and electrochemical fabrication.

4.4.1 Bulk Micromachining

Bulk micromachining is one of the oldest and most widely employed MEMS fabrication process used in the production of MEMS microsystems. Bulk micromachining involves the selective removal of substrate material to fabricate three-dimensional shapes and structures. Specifically, a masking layer is first deposited and patterned to expose areas on the substrate surface, and then the substrate is etched in the exposed regions. Due to the fact that bulk micromachining involves at least the processing steps of a mask deposition and patterning combined with etching, it is considered a process module. Since bulk micromachining implements the MEMS devices directly from the substrate material, it offers some specific advantages due to the fact that the substrate is usually a single-crystal material (e.g., single-crystal silicon). The superior mechanical and electrical material properties of substrate materials such as single-crystal silicon enable the obtainment of excellent device performance with known and predictable material properties.

Bulk micromachining can be accomplished using chemical and/or physical means. Purely chemical techniques involve a chemical reaction between a reactive species (usually in the form of either a liquid solution or gas) and the substrate material. These can be very selective and either isotropic or anisotropic depending on the materials and chemistries involved. Purely physical techniques usually involve removal of substrate material using momentum transfer of energetic chemically inert particles and are not selective but can be highly anisotropic. The combination of chemical and physical techniques combines reactive species with momentum transfer and can exhibit some degree of selectivity and anisotropic behavior.

This section will cover the more commonly used methods of bulk micromachining including isotropic wet etching; anisotropic wet etching; and dry plasma techniques including deep, high-aspect ratio etching (DRIE).

4.4.1.1 Wet Chemical Etchants

A widely used bulk micromachining technique is chemical wet etching. This is performed by immersing the substrate into a solution of reactive chemicals that etch the exposed regions of the substrate. Chemical wet etching provides relatively high etch rates, with good etch selectivity, and has a very low cost (in terms of both operational and setup costs). The etch rates and etch selectivity of wet chemical etching can be modified by adjusting the etch solution chemical composition; varying the temperature of the etch solution; or altering the substrate dopant concentration. Additionally, the etch selectivity can be adjusted depending on the etch mask material used. The etch rate can also depend on the substrate crystallographic planes exposed to the etchant for certain types of etchants. There are two general types of chemical wet etching processes used in bulk micromachining: isotropic wet etching and anisotropic wet etching [281].

4.4.1.1.1 Isotropic Wet Chemical Etchants

In isotropic wet etching, the etch rate is not dependent on the crystallographic orientation, and ideally the etching proceeds at equal rates in all directions of the substrate. A very common isotropic etchant for single-crystal silicon is a mixture of HNO_3, HF, and H_2O. The etch rate varies depending on the concentrations of HF and HNO_3. The etch rate decreases if the concentration of either HF or HNO_3 becomes low. The etch rate decreases with low concentrations of HNO_3 since SiO_2 formation is restricted. Similarly, the etch rate decreases with low concentrations of HF since the SiO_2 etch rate is restricted. Using this etchant, there is a very large range of etching rates possible, from nearly a zero etch rate to over 500,000 nm/min depending on the concentrations of HF and HNO_3 [281–284].

Accurate control of the etching rate and therefore the etched feature dimensions require that the temperature be controlled to within ±0.5 °C. The addition of acetic acid to the mixture has been shown in improving the uniformity of the etch rate across the exposed silicon surface [281–284].

In theory, lateral etching under the masking layer of an isotropic etchant etches at the same rate as the etch rate in normal direction into the substrate. However, in practice lateral etching can be slower without stirring, and consequently isotropic wet etching is almost always performed with vigorous stirring of the etchant solution.

Any etching process requires a masking material to be used with preferably a high selectivity relative to the substrate material. Common masking materials for isotropic wet silicon etching include silicon dioxide and silicon nitride. However, the presence of HF in the mixture of HNO_3, HF, and H_2O will result in the fast erosion of an oxide mask layer. Therefore, silicon nitride is more commonly used as a masking layer with isotropic wet silicon etchants since it has a much lower etch rate.

The etch rate of some isotropic wet etchant solution mixtures also depends on the concentration of dopants in the substrate. For example, the mixture of $HC_2H_3O_2$:HNO_3:HF in the ratio of 8:3:1 etches highly doped silicon (>5 × 10^{18} atoms/cm^3) at a rate of 50–200 microns/h but etches lightly doped silicon material at a rate about 150 times less. This attribute can be used to create regions in the silicon substrate where the etching is selective. It is noteworthy that the etch rate selectivity with respect to dopant concentration is highly dependent on solution mixture [281, 284].

Isotropic wet etchants for bulk micromachining of substrates are rarely used in MEMS manufacturing due to the fact that they are isotropic and thereby do not provide very good control over the etched feature dimensions.

There is not much information available about the dimensional variations that result from isotropic wet etchants. This is mostly due to the fact that the dimensional control is so comparatively poor and dependent on many factors (e.g., temperature, concentration, loading, stress, etc.). We can estimate the magnitude of the variations. If the etch is to be x units deep, then it is possible that the lateral undercut of the mask will be x units on each side or $2\times$ units total based on an etchant that is adequately stirred. Therefore, the total lateral undercut of the mask could be estimated as about $2\times$. Additionally, the etch rate uniformity will vary by at least 5% across the substrate. Therefore, the etch depth variation will be at least $\pm 5\%$, and the lateral dimension variation will be also at least 5% combined with the $2\times$ mask undercut.

4.4.1.1.2 Anisotropic Wet Chemical Etchants

Anisotropic wet etchants are much more commonly used in MEMS fabrication compared to isotropic wet etchants. Anisotropic wet etching involves the immersion of the substrate into a wet chemical solution where the etch rate is dependent on crystallographic orientation of the substrate. This capability enables three-dimensional shapes and structures to be made in the substrate.

Wet anisotropic chemical etching is commonly defined in terms of etch rates according to the different major crystallographic planes, specifically <100>, <110>, and <111>. In general, silicon anisotropic etching etches more slowly along the <111> planes than all the other planes with a difference in etch rates as high as 1000 to 1. It is thought that the reason for the slower etch rate of the <111> planes is that these planes have the highest density of exposed silicon atoms in the etchant solution, as well as three silicon bonds below the plane, thereby leading to some amount of chemical shielding of the surface from the etchant.

Anisotropic etching techniques have been around for over 30 years and are commonly used in the manufacturing of silicon pressure sensors, microfluidic devices, bulk micromachined accelerometers, and other important MEMS devices.

Figure 4.7 illustrates anisotropic etching of a <100> orientation of single-crystal silicon. A masking layer has been deposited and patterned on the top surface of the substrate to expose a square area of the silicon substrate (see: before etch, Fig. 4.7a). If the substrate is immersed into the anisotropic etchant for a brief amount of time, the etchant etches along the <100> crystallographic faster than the other planes, particularly the <111> planes as shown in Fig. 4.7b. It is noted that the bottom of the etched feature has a flat surface. If the substrate is etched for a long period of time, the etching along the <100> planes continues until the <111> planes are only exposed thereby resulting in an etched feature of an inverted pyramid as seen in Fig. 4.7c. The etch rate of the <111> planes is slow, and therefore the etching once the <111> planes are exposed reduces to a very small rate. Therefore, the etch rate

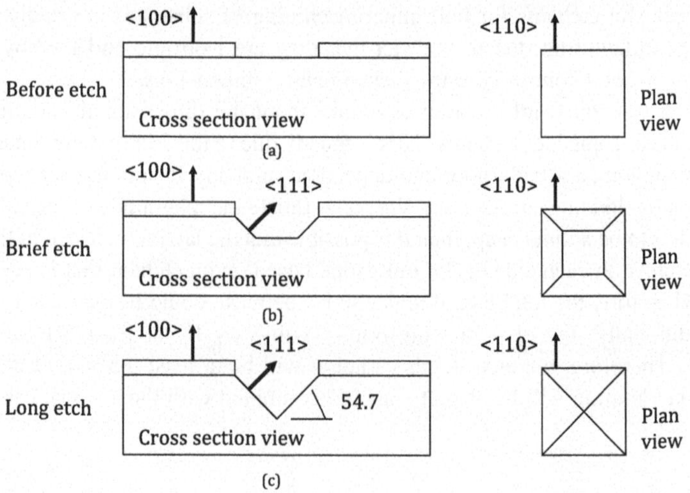

Fig. 4.7 Illustration of shape of the etch profiles of a <100> oriented silicon substrate before immersion in Fig. 4.7a; after immersion in an anisotropic wet etchant solution for a brief time in Fig. 4.7b; and a long time immersion in Fig. 4.7c; to demonstrate that the etch is self-limiting on the <111> planes

has a self-limited characteristic that is useful for dimensional control of the etched features. It is also noted in Fig. 4.7c that the sidewall angle of the etch pyramid etch pits is 54.7°.

Figure 4.8 is an illustration of some of the shapes that are possible using anisotropic wet etching of silicon single-crystal substrate. In this figure, the silicon substrate has a <100> orientation. Among the shapes shown include an inverted pyramidal; a flat bottomed trapezoidal etch pit; and a thin membrane. The thin membrane is created using an etch stop that will be explained below. It is important to note that the slower etching <111> planes have a large impact on determining the resultant shapes of the etched features.

There are two possible issues that can affect the dimensional control that is obtainable with anisotropic wet etching. The first is that although the <111> planes do have a slower etch rate, it does not mean that the actual etch rate is zero. A typical ratio of <100>/<111> etch rate would be about 200–400. Therefore, if an etch is performed through the thickness of a 100 mm diameter wafer with a nominal thickness of 525 microns, the lateral etch of the <111> planes would be 1.3–2.6 microns. Note that this amount of expansion of the lateral dimension would be present on all four sides of the etched feature. Depending of the overall size of the etched feature, this amount of lateral etching may or may not be important. This expansion of the dimensions of the etched feature can be reduced by biasing the mask for the etch and controlling the thicknesses of the wafers.

The second issue related to dimension control concerns how well the mask for the anisotropic etch is aligned to the crystallographic orientation of the wafer. This is

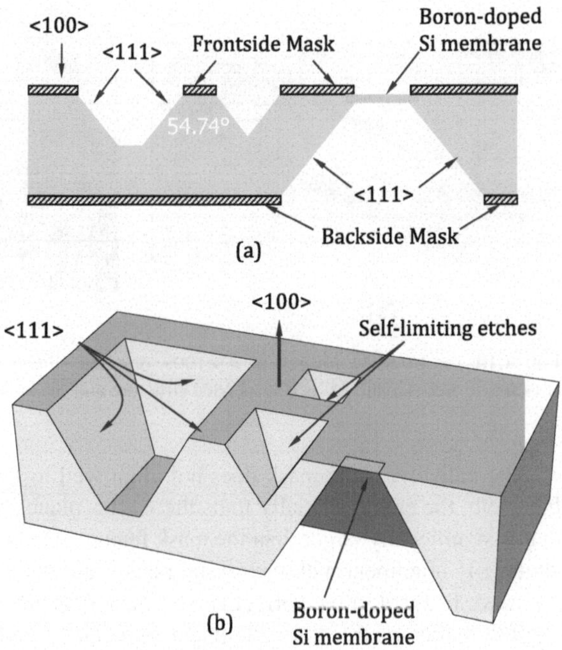

Fig. 4.8 Illustration of shape of the etch profiles of a <100> oriented silicon substrate after immersion in an anisotropic wet etchant solution. In Fig. 4.8a is shown the cross-section of the substrate and in Fig. 4.8b is a 3-D image of the same substrate

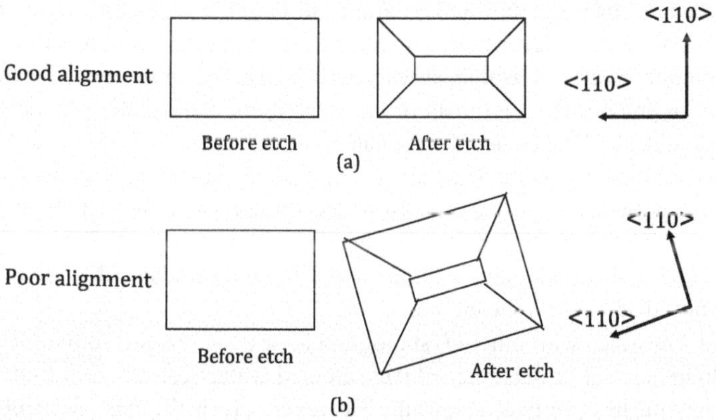

Fig. 4.9 Illustration of how the alignment of the mask to the planes in the substrate crystal affects the resultant dimensions of the etched features. In Fig. 4.9a, the alignment of the mask with the crystal orientation of the substrate is good and the etched feature has roughly the same orientation as the mask feature; whereas in Figure 4.9b, the alignment of the mask with the crystal orientation is not good and the orientation and dimensions of the etched feature is not the same as on the mask

demonstrated in Fig. 4.9. Shown in Fig. 4.9a is a rectangular mask that is well aligned to the crystal planes of the wafer before etching on the left and the result after etching on the right. If the mask openings align with the <110> planes accurately, the etched feature will have dimensions fairly close to those of the mask.

(a) (b)

Fig. 4.10 (**a**, **b**) SEM images of a <100> oriented silicon substrate after immersion in an anisotropic wet etchant. (**a**) is a suspended structure and (**b**) is an trapezoidal etched pit

Alternatively, if the mask does not align well to the <110> planes as shown in Fig. 4.9b, the etch essentially finds the <110> planes resulting in an etched feature that is significantly larger than the mask feature. The wafer flats are typically within about $\pm 1°$ alignment with the crystal planes and therefore can be used for aligning the mask features to the correct crystal planes. If better accuracy is needed, then a short duration small feature alignment etch can be performed in areas on the wafer surface using the anisotropic wet etchant to expose the <110> planes. Then these small etched features can be used to align the larger features on the mask to the correct crystal planes. Nevertheless, it can be estimated that even a $\pm 1°$ alignment angle variation results in less than about a $\pm 1\%$ expansion of the etched feature size. Therefore, in practice the alignment of the mask to the flat usually provides reasonably good control of the etched feature lateral dimensions.

Shown in Fig. 4.10a,b are SEM images of a silicon substrate after an anisotropic wet etching. Show in Fig. 4.10a is a suspended structure held by a small width tether on both ends. The material that the suspended structure was made from was also the masking layer in the field regions. Figure 4.10b is a trapezoidal etch pit that has been subsequently diced across the etch pit.

Silicon substrates with other crystallographic orientations are sometimes used in MEMS fabrication. For example, <110> oriented wafers can be used to implement features having near-vertical sidewalls. However, given the cost, lead times, and availability, the overwhelming majority of silicon wafers used in wet anisotropic processes bulk micromachining have <100> orientation.

There are three types of anisotropic wet etchants that are commonly used. By far the most popular anisotropic etchants are aqueous alkaline solutions such as KOH, NH_4OH, NaOH, CsOH, and TMAH, with KOH being the most frequently used etchant [281, 285]. The characteristics of aqueous alkaline solutions are high etch rates and a relatively high etch rate ratios between the (100) and (111) planes. TMAH is sometimes preferred for use on preprocessed microelectronic wafers for integrated MEMS technologies,[8] due to the fact that this etchant does not

[8]Integrated MEMS is defined as combining MEMS and integrated circuits onto the same substrate.

appreciably etch aluminum under certain conditions. The major drawback of alkaline solution etchants is that they have a relatively high etch rate of silicon dioxide, which is often employed as a masking material layer. Additionally, there is also the potential for alkali contamination of the wafers, although there are cleaning procedures that can be employed to minimize these risks.

Another anisotropic wet etchant of silicon used for bulk micromachining is ethylenediamine and pyrocatechol or EDP. This etchant has a higher etch rate ratio of the (100) and (111) planes which is attractive for better feature dimensional control [281]. Another benefit of EDP is that there is a larger variety of masking materials that can be used with this etchant compared to aqueous alkaline solutions. The major drawback of EDP is that it is a carcinogenic material. Another problem with EDP is that it is difficult to observe the wafer when immersed into this solution, which is a thick, nearly opaque, orange-yellow solution. Also, EDP is more difficult to clean off etched wafers and the associated etching apparatus.

The last type of anisotropic etchant for bulk micromachining of silicon is a mixture hydrazine and water (N_2H_2:H_20) [286]. Although hydrazine is well known as a rocket fuel, the concentrations used for anisotropic etching are below the threshold for combustibility. An advantage of this etchant is that it has a very low silicon dioxide etching rate thereby allowing relatively thin oxide masking layers to be used. However, this etchant does have a relatively poor etch rate ratio of the (100) and (111) planes. This makes dimensional control of the etched features less accurate. Hydrazine is a comparatively hazardous material, and while it was used in the early days of wet chemical bulk micromachining of silicon, it is now rarely used since its disadvantages outweigh its advantages.

For anisotropic etching to be useful, it requires the capability to mask certain areas of the substrate. Therefore, an important criterion for selecting an etchant is the availability of suitable masking material layers. Silicon nitride has a very low etch rate in most wet chemical etchant solutions and therefore is a commonly used masking material for anisotropic wet etchants. Thermally grown SiO_2 is also frequently used as a masking material, although it is important to ensure that the oxide thickness is sufficient for long etches in KOH. Photoresists are unusable in any anisotropic wet etchant. Many metals including Ta, Au, Cr, Ag, and Cu withstand immersion in EDP, and Al has been shown to be a suitable masking material in TMAH under certain conditions [281, 287].

In general, the etch rate, etch rate ratios (100)/(111), and etch selectivity of anisotropic wet etchants are strongly dependent on chemical composition and temperature of the etchant solution. The etch rate [*R*] obeys an Arrhenius law given by:

$$[R] = R_o e^{\frac{-E_a}{KT}} \quad \text{(micron/hour)} \tag{4.1}$$

where R_o is a constant, E_a is the activation energy, k is Boltzman's constant, and T is temperature in degrees Kelvin [281, 288]. Both R_o and E_a vary depending on the type of etchant, the etchant composition, and the crystallographic orientation of material being etched. Experimentally determined activation energies and rate constants for commonly used anisotropic silicon etchants for different crystal orientations can be found in the literature [288].

Bulk anisotropic micromachining can be used to make thin silicon membranes employing a timed etching process. Timed etches are a process wherein the etch depth and thereby the membrane thickness are determined by multiplying the etch rate by the etch time. These fabrication methods are difficult to control since the resultant etch depth is very dependent on a number of factors including etchant species diffusion effects; loading effects; etchant aging; surface preparation; etc. The use of timed etches to get a desired membrane thicknesses requires a properly selected etchant and solution chemistry; adequate stirring of the etchant solution; the use of substrates having low TTVs; and the frequent inspection of the substrate etching depths (and/or the membrane thickness) during the etching process. The latter obviously makes timed etching a labor-intensive operation. If the etchant is well stirred and the temperature well controlled, from Eq. (4.1), it can be determined that the etch rate will vary by approximately ±5–10% or more. For a wafer that is 525 microns thick, this translates into a range in etch depth of ±26 microns or more. To allow a higher level of dimensional control of membrane thickness and repeatability using anisotropic wet etching, etch stops have been developed to address this issue.

Etch stops are very useful to control the dimensions of etched features across the wafer. Etch stops are often used to make relatively thin diaphragms in single-crystal silicon wafers. There are three basic types of etch stop methods that are used in micromachining: dopant etch stops; electrochemical etch stops; and material etch stops.

Dopant etch stops can be implemented by the introduction of impurity dopants into the silicon substrate [281, 289]. The most commonly used etch stop is heavy doping of silicon with boron concentrations typically above 5×10^{19} cm^{-3}. The lightly doped region of the wafer will etch at the normal rate, while the highly doped region of the silicon will have a very low rate of etching. The dopants can be introduced into the silicon using the techniques of diffusion or ion implantation. These are often called "boron etch stops." These methods provide for well controlled depth and reasonably good uniformity of dopants in the substrate. While boron etch stops are a useful and inexpensive method to control the resultant dimensions of etched features using anisotropic etching, it does have some disadvantages. Specifically, high dopant concentration of boron can lead to slip planes in the single-crystal material and high residual stress levels [286]. Also, silicon that is highly doped boron will have degraded piezoresistive properties.

Another useful etch stop method used in silicon bulk micromachining is the electrochemical etch stop [285, 290, 291]. Electrochemical etching of silicon using an anisotropic etchant is useful since it provides very good dimensional control (e.g., diaphragm thickness is reproducible) and can make diaphragms with lightly doped material, which is required for high-quality piezoresistive devices. The disadvantage of electrochemical etch stops is that they require special fixtures to each wafer in order to make electrical contacts and an electronic control system is needed to control and apply the correct voltage potential to the wafer during the etch.

Material etch stops involve the use of the material layer that has high selectivity compared to the substrate. The most common example of a material etch stop is the use of a SOI wafer. An anisotropic etch performed from the backside of a silicon wafer will essentially stop once the buried oxide layer is reached. Often the buried oxide is then removed to leave the suspended single-crystal silicon device layer. Since the device layer in SOI wafers can have a thickness that is reasonably well controlled, this process can provide a means for an accurate membrane thickness.

There are two different dimensional variations of interest for bulk anisotropic wet chemical micromachining processes using etch stops: the etch depth and the lateral dimensions. With respect to the etch depth, the dopants for boron etch stops are commonly introduced using solid source diffusion that was covered in Chap. 3. Therefore, the expected non-uniformity of the boron etch stop is estimated as $\pm 5\%$ across the substrate and from wafer-to-wafer. This will be the largest contribution to the etch depth (and resultant membrane thickness) variation. Better control can be obtained with ion implantation with a variation of less than 1.5%. The lateral dimensional variations are mostly due to the lateral etch rate and the orientation of the mask with the crystal planes. As noted above, the lateral etch rate is given by the ratio of the etching rate ratio of the <111> and <100> planes and results in a lateral etch of between 1.3 and 2.6 microns on each side of the feature for a 525 micron thick wafer. The crystallographic orientation alignment using the substrate flat results in about a 1% lateral variation. The combination of these two can be used to estimate the total lateral variation. For most applications the etched feature size is hundreds of microns or more, and therefore most the lateral dimensional variation is due to the crystallographic misalignment. The lateral etch variation can be reduced by performing a pre-etch on the substrates to identify the precise crystal planes. The electrochemical etch stop method has been reported to have a membrane thickness uniformity of about $\pm 1.3\%$ across the wafer [292].

The lateral etch variation will be the same as the boron etch stop method. The use of a buried oxide of SOI wafers will have the same lateral variations as discussed above for other etch stop methods. The membrane thickness using a SOI substrate will be determined by the variation in the device layer thickness, which can be as low as ± 0.25 microns.

4.4.1.2 Gas-Phase Isotropic Chemical Etchants

Xenon difluoride (XeF_2) is a vapor state isotropic etchant for silicon [293]. This etchant is highly selective with respect to other materials commonly used in microelectronics fabrication including silicon nitride, SiO_2, aluminum, and others, which can be used as masking layers. Since this etchant is a completely dry release process, it does not suffer from the stiction problems that can be an issue in wet release processes, such as surface micromachining (described below). Figure 4.11 below is a

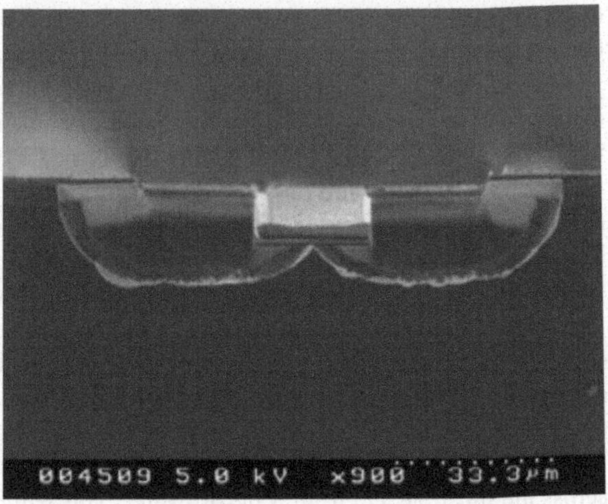

Fig. 4.11 SEM of a beam undercut using XeF_2 isotropic etching of silicon

cross-sectional SEM of a cantilever beam made by isotropic etching the silicon substrate using XeF_2 to partially undercut the metal material of the cantilever.

Like wet isotropic etchants discussed above, the amount of undercut of the mask will be about the same as the etch depth, and therefore it is difficult to control the dimensions of features etched with XeF_2. Therefore, in most circumstances of MEMS fabrication using this etchant, other layers that can provide better control are used to determine the device dimensions. For example, in the end-on SEM of the beam in Fig. 4.11, the critical dimension would not be based on the etched trench into the silicon substrate to release the beam. Instead, it would set by the beam itself, which can be controlled reasonably well using standard photolithography and RIE methods.

Vapor hydrofluoric (HF) acid is another gas phase bulk isotropic etchant. However, it is rarely used to etch bulk material. Instead it is mostly used for surface micromachining process (discussed below) to remove the sacrificial oxide layer(s). The advantage of vapor HF is that since the wafer with devices is not immersed into a liquid solution, there are no capillary forces that can cause the released structure to come into contact with the substrate, which can result in the device being stuck due to stiction effects. Like XeF_2 above, vapor HF is not used to control the dimensions of the device but instead to remove the sacrificial oxide layers.

4.4.1.3 Deep Reactive Ion Etching (DRIE) of Silicon

Wet anisotropic etching of silicon has been used for decades to micromachine single-crystal silicon substrates for a variety of MEMS devices. While wet etching is very inexpensive to perform, it has some limitations with respect to the dimensional control of the etched features. As discussed in Chap. 3, dry plasma reactive ion etching (RIE) usually provides much better dimensional control over wet etching techniques; however, it had not been possible to perform plasma etching into silicon more than a few microns in depth until relatively recently.

In the 1990s several new plasma etching technologies were developed that offered the possibility of etching tens or even hundreds of microns into single-crystal silicon [294]. This technology has been given the moniker of deep reactive ion etching (or DRIE) and was very quickly adopted by the MEMS industry as the bulk silicon micromachining method of choice. It enables implementation of very deep and very high-aspect ratio etches to be performed into silicon substrates. The sidewalls of the etched holes are nearly vertical, and the depths of the etches can be hundreds of microns or more into the silicon substrate [294, 295].

Figure 4.12 illustrates how deep reactive ion etching is accomplished in one variation of the DRIE process, called the Bosch™ process [294]. The etch is a cyclical dry plasma etch process that alternates between a high-density plasma to etch the silicon in one cycle and then deposit an etch-resistant polymer layer on the sidewalls in the other cycle. The etching of the silicon is performed using a SF_6 chemistry, whereas the deposition of an etch-resistant polymer layer on the sidewalls uses a C_4F_8 chemistry.

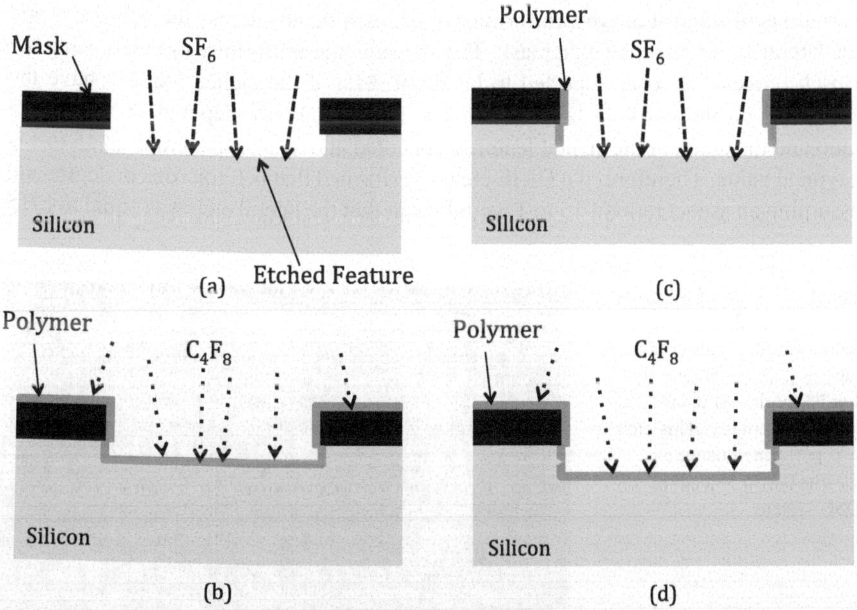

Fig. 4.12 An illustration of the mechanism of DRIE etching of silicon. In Fig. 4.12a, the etch cycle uses SF_6 process gas to etch into the silicon through an opening in the masking layer; in 4.12b, C_4F_8 process gas is used to deposit a protective polymer layer on the substrate surface and no etching occurs during this phase; in 4.12c, the gas is switched back to SF_6 and etching continues again, wherein the polymer on the surfaces orthogonal to the reactive ions in the plasma is removed quickly whereas the polymer on the sidewalls of the etched features is not removed as quickly; and in Fig. 4.12d, the polymerization continues again with C_4F_8 process gas

The sidewalls of features made with DRIE etching are not perfectly (or optically) smooth, and if the sidewall is magnified under SEM inspection, a characteristic washboard or scalloping pattern is seen in the sidewalls. It should be noted that there are other types of RIE processes that enable the etching of deep and high-aspect ratio features into silicon, such as the cryogenic silicon etch process [296]. However, the vast majority of DRIE systems today employ the Bosch™ process.

Photoresist and silicon dioxide are commonly used masking layers for DRIE etching. The selectivity with photoresist and oxide can be as high as about 75 to 1 and 150 to 1, respectively. With process optimization to the features being etched, the aspect ratio of the etch can be as high as 30 to 1 [297] but, in practice, tends to be closer to 15 to 1.

Figure 4.13 is a cross-sectional SEM of a silicon microstructure fabricated using DRIE technology. This etch was performed using a commercially available PlasmaTherm Versaline Deep Silicon Etch (DSE) system. This tool has etch rates as high as 20 microns/min, virtually no scalloping on the sidewalls, and minimal lateral etch (i.e., undercut). These characteristics are important to accurately control the dimensions of the etched features.

The achieved dimensional variations in DRIE depend on the etch depth, tool, recipes, mask design, loading effects, and other factors. The principal dimensional variations of interest are the uniformity of the etch depths across the substrates and the lateral undercut of the etch mask. The across-wafer uniformity of DRIE using the Bosch process has been reported to be about ±1% if the etched features have the same size on the masking layer [298]. The lateral etch rate depends on the aspect ratio and the depth of the etched features. As noted above, an aspect ratio of 15 to 1 is a typical value. Therefore, if a DRIE etch is performed that is L microns in depth, and assuming an aspect ratio of 15 to 1 would mean that the lateral etch was equal to L/15

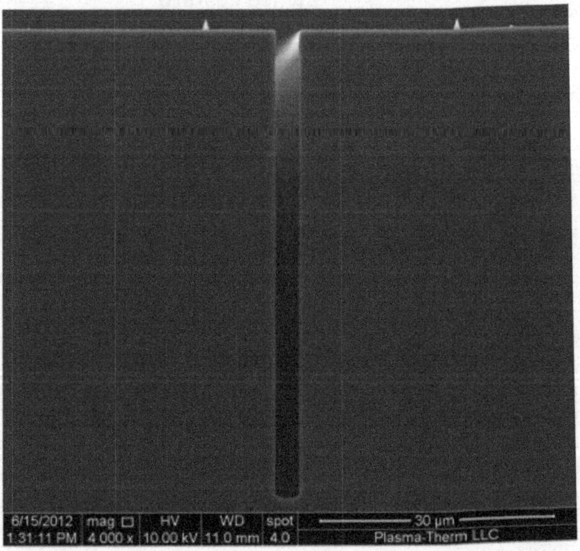

Fig. 4.13 SEM of the cross section of a silicon wafer demonstrating high-aspect ratio and deep trenches that can be fabricated using DRIE technology. This etch was performed using a PlasmaTherm Versaline DSE system

or 0.067 L. As seen from this example, the lateral etch can be substantial. It should be noted that the masking layer for the DRIE etch can biased by the lateral etch amount to result in etched feature dimensions more closely aligned with the desired dimensions. The variation in lateral etched dimensions will be less than $\pm 1\%$.

4.4.1.4 Deep, High-Aspect Ratio RIE of Fused Silica, Quartz, and Glass

Various forms of silicon dioxide materials such as fused silica (fused quartz) have many very desirable material properties for applications in MEMS devices. For example, fused silicon exhibits a high-quality (Q) factor; high stiffness; chemical inertness; good thermal stability; small viscoelastic losses; low thermal expansion; good thermal shock resistance; low dielectric constant and low dielectric losses; excellent optical transparency ranging from deep ultraviolet to the far infrared; and low thermal conductivity. Additionally, quartz, a crystalline form of silicon dioxide, is a piezoelectric material making it an excellent material choice for certain sensor, actuator, and electronic applications. Similarly, plain amorphous glass also has many desirable properties for MEMS device applications such as low thermal conductivity, low electrical conductivity, transparency, good stability, and biocompatibility [299]. Despite the usefulness of these types of silicon dioxide materials, the fabrication technologies to perform bulk micromachining have been mostly limited to old macroscale machining techniques such as crystal cutting, grinding, and wet etching. The dimensional control of these fabrication methods has been a severe constraint.

While dry plasma etching of silicon dioxide has been available in IC manufacturing for a number of years, the technology had been limited to depths of a few microns or less, very limited aspect ratios, and typically non-vertical sidewalls of the etched features [300]. The capability to make very deep (more than tens of microns), small-dimensioned devices and device features with high-aspect ratios and vertical etched sidewalls in these important materials has not been previously available to device developers.

Recently, the ability to etch very deep and high-aspect features into fused silica and glasses using a reactive ion etching process has been reported [301]. Etches over 200 microns deep, with the capability of etching completely through 500 microns or thicker fused silica material layers or substrates, with high-aspect ratio features of over 10 to 1 and with nearly vertical sidewalls in the etched features have been demonstrated.

Plasma etching of silicon dioxide can be summarized as follows: the etch process employs a continuous polymerization–dissolution process based on the ionization of C_xF_y fluorocarbons (including CF, CF_2, C_2F_3, and C_2F_4); the polymers deposited onto the silicon dioxide surface partly dissolve the silicon dioxide in a complex interaction of fluorination, desorption, and passivation. The dissolution of the fused silica occurs as a complex interaction between ions in the plasma and the solid fused silica, and in addition to removal of glass, the process forms a complex fluorocarbon polymer on the substrate. Due to the directionality of the incoming ions, there is a preferential removal direction of the polymer, which leads to anisotropy in the etched feature shape. This is similar to the behavior seen in silicon deep reactive ion etching (DRIE), with the important difference that the polymer in the fused silica etch process is formed continuously during etch. The addition of oxygen or argon to the plasma affects the polymerization rate in the process, since oxygen ions chemically attack the polymer and argon ions physically sputter the polymer. Therefore, changes in etch rate, mask selectivity, and anisotropy (sidewall shape) can be realized with the introduction of these gases into the process chamber during etching of silicon dioxide.

This etch process is different compared to deep reactive ion etching (DRIE) of silicon using the Bosch™ process since it is a continuous process and not switched or cycled between two process chemistries such as in the DRIE Bosch process. As a consequence, there is no scalloping of the sidewalls of the etched features. Additionally, there is also a reduced lateral under-etching of the etch mask in this process than seen in silicon DRIE.

Commercially available tools for performing deep, high-aspect ratio etching of silicon dioxide include the ULVAC Technologies Neutral-Loop Discharge (NLD) 6000 etch system. This is an inductively coupled plasma (ICP) etch tool that has a unique feature of three separate, independently controlled, electromagnetic neutral coils positioned external and completely encircle the etch chamber. This allows the magnetic field shape and strength to modify the plasma and spatial redistribution of the ions, which in turn impacts the etch rate, the resulting sidewall angle, and the etch uniformity. The magnetic neutral loop discharge configuration is used to simultaneously create a high-density plasma and low operating pressure (e.g., 10^{11} cm^{-3} at 10^{-1} Pa) to enable higher etch rates in hard to etch materials as well as enabling more directional, or anisotropic, etching to be performed.

Etch rates of about 500 nanometers/min into fused silica have been reported; however, the etch rate tends to decrease as the deep of the etch increases and the gap between etched features decreases [301]. Figure 4.14 shows a cross-sectioned substrate of fused silica after performing a deep high-aspect ratio etch. The deep

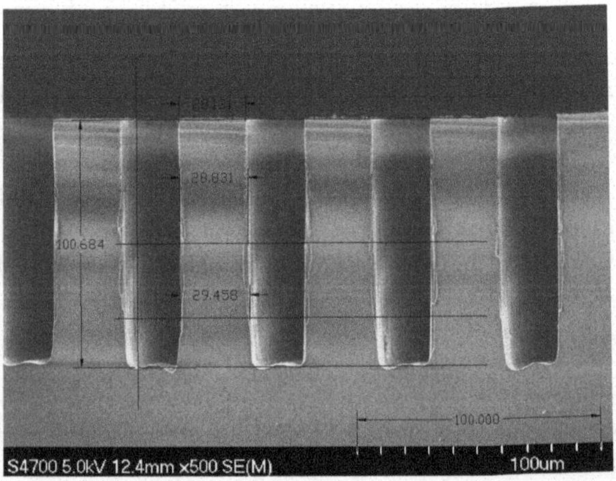

Fig. 4.14 Scanning electron microscope (SEM) image of a cross section of fused silica sample after performing etch

of this etch is between 100 and 125 microns, and the aspect ratio shown in this figure is about 5 to 1. Deeper etches (up to 500 microns or more) or having higher aspect ratios (up to 10 to 1) are possible with this etch.

The uniformity of this etch process was reported to be 2.41% in the lateral features across the substrate. The uniformity on the etch depths was 1.51% for an average etch depth of 132 microns [302, 303].

4.4.1.5 Deep, High-Aspect Ratio RIE of Silicon Carbide (SiC)

Silicon carbide (SiC) is a refractory compound semiconductor material that has several very unique and desirable electrical, mechanical, and chemical material properties including large energy band gap; high mechanical stiffness; high electron and hole mobility; chemical resistance; extreme hardness; high-quality factor (Q); high thermal stability; good thermal shock resistance; high thermal conductivity and electroluminescence; high abrasion resistance; and others. As a result of these unique material properties, silicon carbide is an excellent material choice for several MEMS applications such as MEMS sensors for harsh environments.

Many important MEMS device applications are in need of a bulk etching technology that can be used to implement deep, high-aspect ratio features into bulk SiC substrates. However, due to its chemical resistance, silicon carbide is an extremely challenging material to etch using any technique, particularly dry plasma etching processes.

While deep reactive ion etching (DRIE) of silicon has become a mainstay of the MEMS industry in the formation of deep and high-aspect ratio microstructures and features into silicon (Si) substrates, no truly comparable technology providing similar etch results in silicon carbide (SiC) has been available until recently when the ability to etch very deep and high aspect features into 4H oriented silicon carbide substrates using an inductively coupled plasma (ICP) reactive ion etching (RIE) process conducted on an ULVAC NLD-6000 etch tool was reported [304].

Using this etch process, etches of over 150 microns deep, with the capability of etching completely through 500 microns or thicker silicon carbide material layers or substrates, were demonstrated. A SEM image of the cross section of an etched SiC substrate is shown in Fig. 4.15. As can be seen, the recipe performed an etch that was deep and high in aspect ratio. The depth of the etched features was nearly 160 microns, thereby resulting in an etch rate of about 1 μm/min. Measurement of the sidewall angles indicated that they were 87.5° or nearly vertical. Moreover, the mask etch selectivity was measured to be approximately 130:1, and the aspect ratio was 12 to 1. No etch defects, such as pillars, were exhibited in the etched samples. The recipe for performing deep etches into silicon carbide can be found in the literature [304].

The measured etched feature lateral dimensional repeatability for an average etch depth of more than 150 microns into SiC substrates was found to depend on the lateral dimensions of the feature size and averaged less than 1%. Also, for an average

Fig. 4.15 Scanning electron microscope (SEM) images of array of vias and posts after performing plasma etch using ULVAC NLD-6000. 50 um via array (top left), 25 um post array (top right), 25 um via array (bottom left), and 25 um post array (bottom right) are shown

etch depth of 150 microns, the measured repeatability of the depth of the etched features from substrate-to-substrate was found to be less than 1% [305].

4.4.2 *Surface Micromachining*

Surface micromachining is one of the most widely used fabrication methods for the manufacturing of microsystems. This method employs thin-film depositions of various types of material layers combined with selective removal of underlying thin-film material layers to form free-standing structures that can undergo mechanical movements. Since it is composed of multiple processing steps, it is best categorized as a process module. Given the wide latitude of choices of materials that can be deposited using thin-film deposition techniques, etchants, and possible process sequences, surface micromachining enables the implementation of a large diversity of possible MEMS devices. Moreover, there are also a number of design choices that enable MEMS microsystems to be integrated with existing microelectronics using surface micromachining. As a general rule, surface micromachining

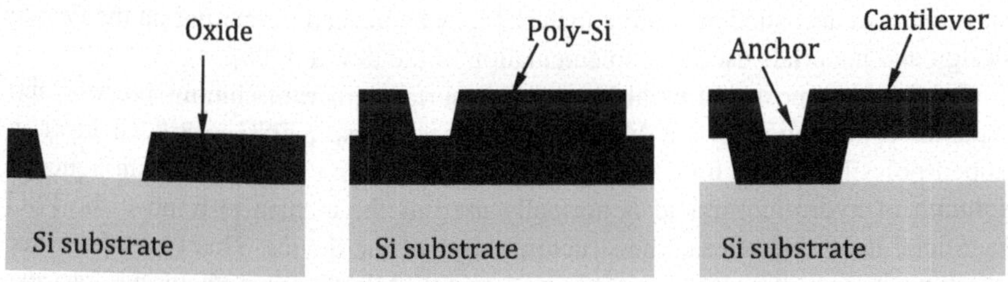

Fig. 4.16 Illustration of a surface micromachining process

can be performed while consuming less substrate real estate area thereby making it potentially a lower-cost manufacturing approach compared to bulk micromachining.

The common theme of surface micromachining is that the fabrication processes only involve thin-film layers deposited onto the substrate and not the micromachining (i.e., etching) of the substrate itself. The typical surface micromachining process involves a sequence of processing steps beginning with the deposition and patterning of a thin film that functions as a temporary mechanical layer on which the actual device layers are built (the temporary layer is commonly referred to as the sacrificial layer), followed by the deposition and patterning of a functional thin-film layer made of a different material (commonly referred to as the structural layer), and followed by the removal of the sacrificial material layer. This last step is usually called the release step since it is used to release the mechanical structural layer from the constraint of the underlying sacrificial layer and thereby allowing the device layer to move. An illustration of a surface micromachining process shown in cross section is given in Fig. 4.16.

Surface micromachining is attractive for manufacturing of MEMS microsystems for several important reasons. First, it usually provides for better dimensional control of the device features compared to bulk micromachining. Second, in general surface micromachining processes can be more compatible with microelectronics devices to enable integrated MEMS devices compared with bulk micromachining. Third, surface micromachining frequently exploits the good conformal properties of thin-film depositions. Fourth, surface micromachining is relatively simple and uses single-sided wafer processing. In general, this allows smaller device sizes and thereby higher integration densities, as well as lower resultant per die cost compared to bulk micromachining.

Perhaps the most important major disadvantage of surface micromachining fabrication processes is that the mechanical properties of deposited structural thin films are often unknown and therefore must be determined by measurement. Additionally, the release of the structural layer can be challenging due to stiction effects. Stiction is a phenomenon whereby the structural layer is pulled down and stuck to the underlying substrate due to van der Waals attractive forces between the bottom of the thin-film structural layer and the top surface of the substrate. Stiction effects can be induced by the capillary forces of retreating liquids used to release or rinse the substrates. Stiction can also occur during use if the structural layer touches the substrate. An anti-stiction coating material may be needed depending on the device design and materials used in implementation of the device [306].

By far, the most commonly employed surface micromachining process and material combination in the MEMS industry is to use a PSG sacrificial layer, a doped polysilicon structural layer [307]. In this material combination, an aqueous solution of hydrofluoric acid is normally used as the etchant to remove the PSG sacrificial layer and release the structural layer of the device. This type of surface micromachining process is used in several MEMS manufacturing process technologies.

Figure 4.17 is a SEM image of an example of a surface-micromachined polysilicon MEMS device using this material combination. It is a plan view image of an electric micromotor. The polysilicon in this image is the material with the lighter contrast and is composed of multiple layers of LPCVD deposition. The movable rotor can be seen in the center. The drive elements to electrostatically move the rotor element are located around the perimeter. These are electrically connected to bonding pads by polysilicon interconnects. Voltages that are offset in phase are applied to the stator elements located around the perimeter of the rotor, and these create electrostatic fields that attract the closest electrodes on the rotor. If the voltages are phased and timed correctly, the rotor spins in the plane of the substrate surface at a speed proportional to the timing of the voltages to the stators. The sacrificial layers used in this process were LPCVD deposited PSG.

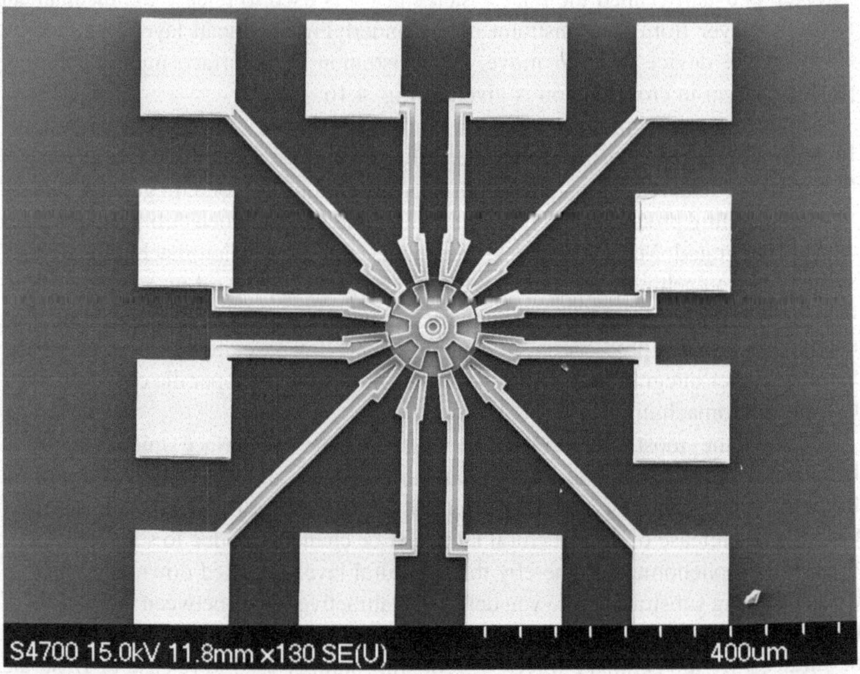

Fig. 4.17 Polysilicon micromotor fabricated using a surface micromachining process

Importantly, the layering of the alternative depositions of polysilicon and PSG enables the realization of a rotor element that is free to move in a circular direction (either clockwise or counterclockwise) while being constrained by a center-bearing element from separating from the substrate surface. Moreover, this layering also allows the stator elements to be placed into a position whereby they are in close proximity to, but do not touch, the rotor electrodes. This layering of multiple thin films allows a large amount of design freedom to implement a wide variety of MEMS microsystems. Other material combinations can be used for implementing devices using surface micromachining process modules.

There is a large variety and diversity of possible surface micromachining processes wherein the materials and fabrication process sequences will differ greatly. The resultant dimensional variations of any specific surface micromachining process will depend on the various materials and processes used in the fabrication of the devices. Similarly, the material properties will also depend on the materials and processes employed. Therefore, it is not possible to provide guidance on the dimensional or material property variations that can result from any surface-micromachined process. Nevertheless, some guidance on one of the most commonly used surface micromachined process technologies will be given in a later chapter on process integration.

4.4.3 *Wafer Bonding*

Wafer bonding is a technique that allows substrates to be bonded to one another. The bonding process involves a chemical attachment of the wafer pair. Some methods allow the wafers to be bonded without the use of a bonding agent or an intermediate layer. Instead, the substrate surfaces are pre-treated so as to make them to stick to one another, followed by an elevated temperature anneal, resulting in the wafer pair chemically fusing. Wafer bonding can be used to create multi-wafer (i.e., two or more bonded wafers) stacks and SOI wafers.

Beyond the ability of adding thicker material layers than would be possible using conventional deposition techniques, wafer bonding also has another distinctly attractive benefit in that the bonded material layers or substrates can be single crystal. This is in contrast to thin-film depositions whereby the deposited layers are not only restricted to material layers a few microns or less but also material layers that are usually either polycrystalline or amorphous.[9]

There are three basic types of wafer bonding including direct or fusion bonding; field-assisted or anodic bonding; and bonding using an intermediate layer. In general, all bonding methods require substrates that are very flat, smooth, and clean, in order for the wafer bonding to be successful and free of voids.

[9]Epitaxy does allow single-crystal material layers to be grown but only under special circumstances.

Direct or fusion bonding is typically used to mate two silicon wafers together. One or both of the wafers may have a pre-existing silicon dioxide or silicon nitride layer on the surface(s) prior to the bonding as well. The direct wafer-bonding process consists of the following five steps [308, 309]:

1. Hydration and cleaning of the wafer surfaces (RCA clean, piranha immersion, etc.)
2. Physical contacting and pressing together of wafers (must be done quickly and in a clean environment since silicon will be charged and attract particles)
3. Infrared inspection of pre-anneal bond quality
4. High-temperature anneal, typically 1000 °C
5. Infrared inspection of post-anneal bond quality

Figure 4.18 illustrates some of the steps involved in the direct wafer-bonding process. The wafer pair is first cleaned and the surfaces hydrated (Fig. 4.18a). This hydration can be done simply with immersion in water or other agents such as immersion in piranha etch (three parts sulfuric acid and one part hydrogen peroxide) or other solutions [309]. The cleaning process is important to ensure that the wafer surfaces have all particulate matter and residual material completely removed that would otherwise interfere with the bonding. The hydration process allows the surfaces of the wafers to develop electrical charges on the surfaces that will facilitate the bonding. These charges coupled with the van der Waals attractive forces on the surfaces are important to establishing a pre-anneal bond condition, whereby the bond strength between the two wafers is sufficiently strong for the wafer pair to be handled and not separate from one another.

The wafer pair is then physical contacted as shown in Fig. 4.18b. A force is usually applied to the wafers once in contact to ensure that the surfaces have bonded

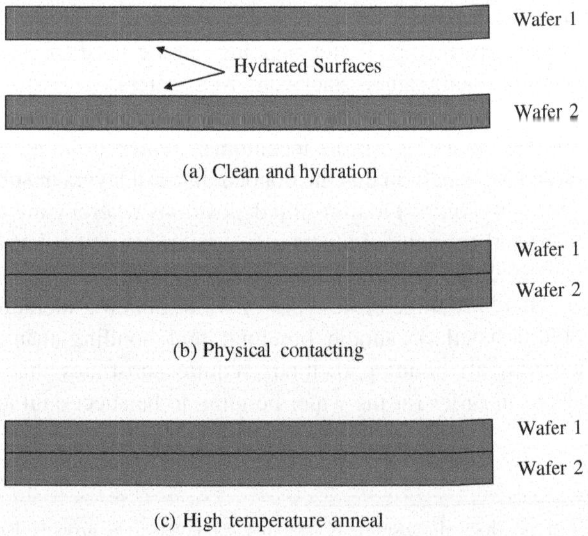

Fig. 4.18 Direct or fusion wafer bonding

to one another. One or both of the bonded wafers can be preprocessed and then aligned during the bonding procedure so as to register features on the top wafer to features on the bottom wafer. The bonding of the wafer pair can be performed in a controlled ambient, such as nitrogen gas, as well as other gases, or even a vacuum. Vacuum enclosures made using wafer-scale packaging are attractive for certain kinds of MEMS devices that require a reduced-pressure environment such as resonators. Next, the wafer pair is then annealed as shown in Fig. 4.18c. This is commonly performed at a high temperature, such as around 1000 °C. After the high-temperature anneal, the bond strength between the two wafers can be as high as that of single-crystal silicon. The high-temperature essentially chemically fuses the two wafer surfaces together. There is relatively low residual stress in the bonded layers after anneal.

Although the high-temperature anneal can be performed at a lower temperature than 1000 °C in direct wafer bonding, the bond strengths tend to be weak when the anneals are performed below 800 °C [309]. However, it has been shown that the surfaces can be prepared with plasma activation to allow lower temperature anneals and still maintain high bond strengths [310].

Another popular wafer-bonding technique is anodic bonding, which is illustrated in Fig. 4.19 below. In anodic bonding a silicon wafer is bonded to a Pyrex™ 7740 wafer using an electric field and elevated temperature [311]. The two wafers can be preprocessed prior to bonding and can be aligned during the bonding procedure [309].

The mechanism by which anodic bonding works is based on the fact that Pyrex™ 7740 has a high concentration of Na+ ions and a positive voltage applied to the silicon wafer drives the Na+ ions from the Pyrex™ glass surface, thereby creating a negative charge at glass surface nearest the silicon substrate.[10] The elevated temperature during the bonding process allows the Na+ ions to migrate in the glass substrate with relative ease. When the Na+ ions reach the interface, an extremely high electrical field results between silicon and glass since the silicon can create a

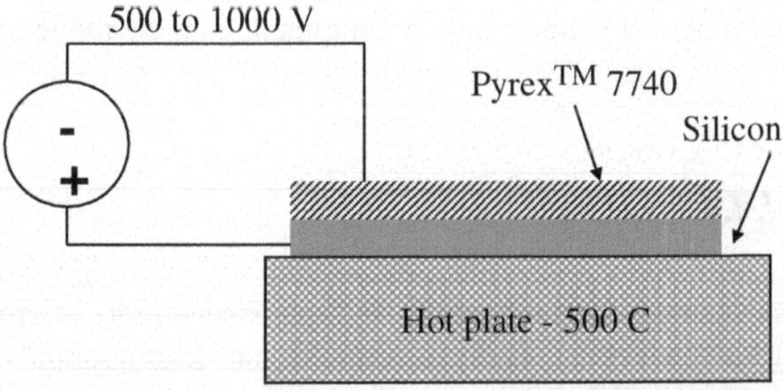

Fig. 4.19 Illustration of the setup for anodic wafer bonding

[10]The voltage potential shown in Fig. 4.19 is for electrons and therefore has the opposite polarity for a potential for positive charges.

negative image charge and the distance between the two opposed charges is very small. This combined with the elevated temperatures chemically fuses the two wafers together. As with direct wafer bonding, it is imperative that the wafers are flat, smooth, and clean and that the anodic bonding process is performed in a very clean environment. Any particles on the wafer surfaces will result in voids between the two wafers. One advantage of this process is that Pyrex™ 7740 has a thermal expansion coefficient nearly equal to silicon and therefore there is a low value of residual stress in the layers. Anodic bonding is a widely used technique for MEMS packaging [309, 311].

Direct and anodic are wafer-bonding techniques that do not use an intermediate layer between the wafers to facilitate the bonding. There are other wafer-bonding techniques that do use an intermediate layer. One method is called thermocompression bonding and involves the bonding of a silicon substrate to another silicon substrate at an elevated temperature using an intermediate layer of gold on the surface of one or both of the wafers [312]. It is a preferred method of wafer bonding at relatively low temperatures since the thermocompression bonding temperature is typically below 400 °C. Thermocompression bonding is commonly used in MEMS packaging.

Another wafer-bonding technique used in MEMS is glass frit bonding. In this process a glass is spun or screen-printed onto one substrate surface. Subsequently this wafer is physically contacted to another wafer, and the composite is annealed to flow the glass intermediary layer and bond the two wafers.

Lastly, various polymers can also be used as intermediate layers to bond wafers including epoxy resins, photoresists, polyimides, silicones, etc. [313]. Included in this category are tapes with one or both sides having adhesive coatings. This technique is commonly used during various fabrication steps in MEMS such as when the device wafer becomes too fragile to handle without mechanical support.

The dimensional variations of importance regarding wafer-bonding processes concern how well the alignment of the wafers. This of course assumes that there is preprocessing on the substrates prior to the bonding since if two blank wafers are bonded the alignment will not be important. The alignment registration accuracy of commercial wafer-bonding equipment is in the range of ± 3 microns or larger [314].

4.4.4 LIGA

A method for fabrication of high-aspect ratio micromachined devices is called LIGA, which is a German acronym for "LIthographie Galvanoformung Abformung [315, 316]." This is a non-silicon-based technology and requires the use of synchrotron generated X-ray radiation. It is used to form high-aspect ratio parts in metal that have near-vertical sidewalls and extremely smooth sidewalls. The basic process is outlined in Fig. 4.20 and starts with the cast of an X-ray radiation-sensitive PMMA

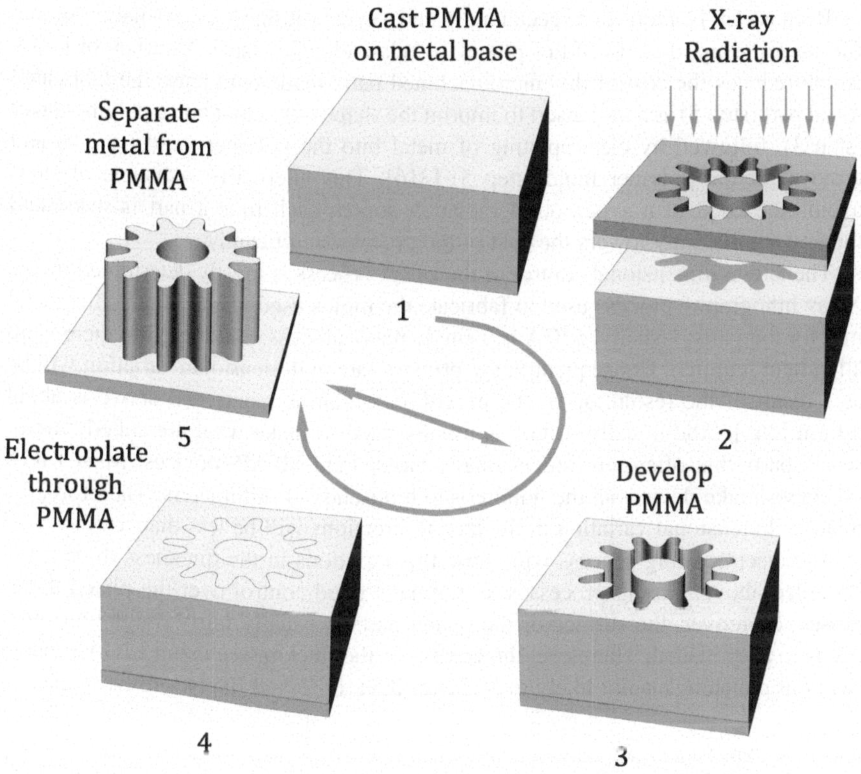

Fig. 4.20 An illustration of the steps involved in the LIGA process to fabricate high-aspect ratio MEMS devices

onto a suitable substrate.[11] The PMMA can be spun onto a substrate for thickness of less than about 1 mm, or sheet material of PMMA can be applied to a substrate and thin-back using grinding to the desired thickness, including thicknesses of 1 mm or more.

A special X-ray mask is used for the selective exposure of the PMMA layer using X-rays from the synchrotron. The PMMA is then developed and will be defined with extremely smooth and nearly perfectly vertical sidewalls. Also, the penetration depth of the X-ray radiation into the PMMA layer is quite deep and allows exposure through very thick PMMA layers, up to and exceeding several millimeters. After the development, the patterned PMMA acts as a polymer mold and is placed into an electroplating bath and nickel, or some other desired electroplated metal is plated into the open areas of the PMMA. The PMMA is then removed, thereby leaving the metallic microstructure (Fig. 4.20).

[11]A metal substrate can be used as well as a silicon substrate. It is important that the substrate is sufficiently electrically conductive for a subsequent electroplating deposition process.

Because LIGA requires a special mask and a synchrotron (X-ray) radiation source for the exposure, the cost of this process is relatively expensive. A variant of LIGA which reduces the cost of the micromachined parts made is to reuse the fabricated metal part (step 5) as a tool insert to imprint the shape of the tool into a polymer layer (step 3), followed by electroplating of metal into the polymer mold (step 4) and removal of the polymer mold (step 5) [316]. This alternative sequence of steps avoids the need of a synchrotron radiation source each time a part is made and thereby significantly lowers the cost of the process significantly.

The lateral dimensional control of the LIGA process is mostly determined by the X-ray lithography process used to fabricate the molds used to electroplate the metal into for the parts. Usually, LIGA is a single mask process, and therefore there is no alignment required. Consequently, the primary lateral dimensional variation will be as a result of the resolution of the exposure, which as mentioned above is about 50 nm [265]. Additionally, LIGA is usually used to make what are called "meso-scale" parts that often are dimensionally larger than MEMS devices. Most LIGA processes make devices in the hundreds to thousands of millimeters. Therefore, the relative dimensional variation in the lateral directions will be less than 1%.

The electroplating process will cause the variations in the thickness dimension. As noted above, plating processes do not have good control over the plated thicknesses. Moreover, the surface of the plated parts is usually fairly rough and often needs to be polished. Therefore, the control of the thicknesses using LIGA process (without polishing) is not likely to be better than ±8% and likely will be worse.

4.4.5 Hot Embossing

A variation of the LIGA process is to fabricate the metal tool insert using LIGA or an alternative technology and then to emboss the tool insert pattern into a polymer substrate, which is then used as the part. This process can make imprints into a polymer of over 200 microns deep with very good dimensional control. The advantage of this process is that the cost of the individual polymer parts can be very low compared to the same structures made using other technologies. Because of the overwhelming cost advantages combined with very good performance, this polymer molding process is very popular for producing microfluidic components [124]. The mold can be made using standard lithography and a plating process as well. While this approach will not provide as good of control over the dimensions as LIGA, it does not require a synchrotron and therefore is less expensive.

The dimensional variations of hot embossing will mostly be due to the dimensional variations of the mold inserts used, which have been discussed above. Additionally, there is the issue of how well the embossed polymer properly replicates the mold insert and the shrinkage that may occur with the polymer as it is cooled after embossing that may also affect the dimensional variations. These will need to be determined based on the mold insert fabrication process, embossing tool, polymer used, and embossing process.

4.4.6 *Other MEMS Micromachining Technologies*

In addition to bulk micromachining, surface micromachining, wafer bonding, and high-aspect ratio micromachining technologies, there are a number of other techniques that have been demonstrated for the fabrication of MEMS devices. We shall review a few of the more notable methods in this section. Importantly, with the exception of focused ion beam (FIB) micromachining, each of the other technologies in this section is not known for either high precision or accuracy. Additionally, they are mostly used to make relatively large MEMS devices. Most of these technologies are rarely used in MEMS fabrication.

4.4.6.1 Electro-Discharge Micromachining

Electro-discharge micromachining or micro-EDM is a process used to machine a conductive material using electrical breakdown discharges to remove material. A working electrode is made from a metal material onto which high voltage pulses are applied. The working electrode is brought into close proximity to the material to be machined, which has been immersed in a dielectric fluid. The minimum-sized features that can be made with micro-EDM are dependent on the size of the working electrode and how it is fixtured, but holes as small as tens of microns have been made using this method. Among the issues of micro-EDM include: it is a slow serial process; it is not performed by machining devices on substrates, but rather on piece parts, and therefore it is difficult to manipulate the workpiece during machining; it is costly; and the machined surfaces can be relatively rough.

4.4.6.2 Laser Micromachining

Lasers can generate an intense amount of energy in very short pulses of light and direct that energy onto a selected region of material for micromachining. Among the many types of lasers now in use for micromachining include CO_2, YAG, excimer, etc. Each has its own unique properties and capabilities suited to particular applications. Factors that determine the type of laser to use for a particular application include laser wavelength, energy, power, and temporal and spatial modes; material type; feature sizes and tolerances; processing speed; and cost. The action of CO_2 and Nd:YAG lasers is essentially a thermal process, whereby focusing optics are used to direct a predetermined energy/power density to a well-defined location on the workpiece to melt or vaporize the material. Another mechanism, which is nonthermal and referred to as photo-ablation, occurs when organic materials are exposed to ultraviolet radiation generated from excimer, harmonic YAG, or other UV source.

Among some of the issues with laser micromachining is that the laser beam can cause extreme local thermal gradients that can result in cracking of the workpiece. Additionally, there is a significant amount of redeposited material around the laser-micromachined features that can be difficult to remove. Often a masking layer is placed on the workpiece prior to performing laser micromachining. It also can be difficult to obtain extremely high-aspect ratios using laser material removal. Laser micromachining can produce features on the order of tens of microns and is a serial writing process.

A new technology for laser micromachining uses extremely short duration pulses to perform ablation of material and not result in significant heating or thermal gradients, which is called femtosecond laser micromachining. This technology offers promise for precision micromachining and trimming applications on MEMS and NEMS devices [317].

4.4.6.3 Focused Ion Beam (FIB) Micromachining

Another technology for performing micromachining is the focused ion beam (FIB). The accelerating voltages are adjustable from few keV to several hundred keV. The spot sizes can be focused down to 50 nm, making it capable for making extremely small structures. The user can input a 3D CAD solid model of desired etching topology, and the computer-controlled stage with submicron positional accuracy allows very precise registration of sample. The removal rate depends on beam energy and diameter as well as dwell time. Additionally, the tool allows control of the pixel spacing, which is the distance between adjacent beam exposure positions. Roughness of the etched features in FIB can be minimized by using small pixel spacing, but this will increase machining time. The FIB beam is modeled as a Gaussian with a beam diameter standard deviation of σ. Smooth features have been reported to be obtained when the pixel spacing or beam standard deviation has a ratio equal to 1.5 [318]. Most FIB systems are dual-column tools, which mean that in addition to a focused ion beam, there is the capability for imaging using scanning electron microscopy (SEM).

FIB technology is mostly used to perform cross-sectioning of microfabricated devices for metrology and diagnostic purposes. But it can be used for machining purposes as well. The speed of material removal is slow and therefore is not usually cost effective for wafer-scale micromachining. Instead it is most useful as a technology for trimming purposes but can be combined with other higher removal rate technologies for precision micro- and nanomachining application [317].

In addition to material removal, the FIB can also be used to perform ion-induced deposition, lithography, implantation doping, mask repair, device repair, and device diagnostics. Many of these tools can also be outfitted with a secondary column for mass analysis of particles removed from the substrate using uSIMS.

Figure 4.21 are two SEM images of a (a) triangular feature made into the surface of a silicon substrate using a FIB and (b) a metal line of platinum deposited over the previously machined triangular feature using a FIB.

Variations: FIB systems typically have spot sizes down to 50 nm. However, the actual resultant etched feature has sloping sidewalls and redeposited material at the edges of the machined features. Therefore, multiple passes are usually required in order to obtain higher aspect ratios and remove redeposited materials. The variation

Fig. 4.21 Very small features made with FIB including the deposition of a metal across a triangular features made into the substrate

in dimension for a 50 nm feature could be on the order of 50% or more [318]. However, the amount of variation will stay relatively constant for larger dimensional machined feature sizes.

4.4.6.4 Electrochemical Fabrication (EFAB)

EFAB is a micromachining process technology that is in many ways similar to stereolithography. Using this technology, three-dimensional microstructures are created through a sequential series of depositions of thin layers of the structural material. Each thin layer can be independently patterned so as to provide flexibility of the resultant device shape. The process easily allows overhanging components and components that bridge across supports. Individual layer thicknesses range between about 1 micron and over 10 microns, and the minimum lateral feature size is a few microns.

4.5 Summary

This chapter has reviewed a number of individual processing steps and process modules commonly used in the fabrication of MEMS microsystems devices. MEMS processing steps such as LPCVD, PECVD, etc. that are commonly performed in IC manufacturing but modify the one or more important parameters, such as the thickness of a deposited layer or deposition of a different material type, were first reviewed. Subsequently, processing steps and process modules that are unique to MEMS manufacturing were reviewed. As in Chap. 3, general guidance as the best-case expected dimensional variations that can be obtained in performing these MEMS processes were provided. Table 4.2 summarizes these best-case dimensional variations.

Table 4.2 Best-case dimensional variations for various MEMS fabrication technologies

Technology	Typical best case variation
Substrates (commercial-grade)	
Silicon	Total thickness variation (TTV) less than 1 micron
Silicon-on-insulator (SOI)	Device layer thickness ±0.25 micron
Glass	TTV >20 microns
Metal	No data available
Ceramic	No data available
MEMS processing steps	
Materials deposited using IC equipment	See Chap. 3 for processing step technology type thickness uniformity for LPCVD, PECVD, PVD, etc.
Special MEMS materials	
Piezoelectric	PVD and CSD thickness uniformity see Chap. 3; PVD stoichiometry variation ±1%: CSD stoichiometry variation <1%
Shape-memory alloy	PVD thickness uniformity see Chap. 3; stoichiometry variation ±1%
Magnetic materials	PVD thickness uniformity see Chap. 3; PVD stoichiometry variation ±1%: Electroplating see below
MEMS specific processing steps	
Electrochemical deposition	Thickness uniformity of ±8%; lateral dimension variations determined by plating mold; electroplating stoichiometry variation ±1%
MEMS lithography	
Contact photolithography	Resolution 1–2 microns; lateral feature variation ±0.5 micron; alignment accuracy ±0.5 micron
Front-to-back contact photolithography	Resolution 1–2 microns; lateral feature variation ±0.5 micron; alignment accuracy ±3 micron
Direct write photolithography	Resolution 0.2 microns, lateral feature variation ±60 nm, alignment accuracy ±100 nm
Grayscale photolithography	Resolution 0.25 microns; alignment accuracy less than ±150 nm
X-ray lithography	Resolution 50 nm
e-beam lithography	Resolution 10 nm; alignment accuracy less than ±20 nm
Photoresists for large topography	Depends on topography; requires individual determination
Lift-off patterning	Depends on feature specifics; requires individual determination
Image-reversal photoresists	Depends on feature specifics; requires individual determination
Photolithography on transparent substrates	Depends on many factors; requires individual determination

(continued)

Table 4.2 (continued)

Technology	Typical best case variation
MEMS micromachining methods	
Bulk micromachining	
Isotropic wet chemical etchants	Lateral variation (i.e., mask undercut) 2x, where x is the etch depth; variation of etch depth 10% or more
Anisotropic wet chemical etchants	
Without etch stop	Lateral variation about 1%; variation of etch depth ±5–10% or more
With etch stop	Lateral variation about 1%; variation of etch depth ±5% with diffusion; 1.5% with ion implantation; ±1.3% with electrochemical; ±0.25 microns SOI
Gas-phase isotropic chemical etchants	
Xenon difluoride	N/A, not used to control device dimensions
Vapor HF	N/A, not used to control device dimensions
Deep reactive ion etching (DRIE) of silicon	Etch depth uniformity ±1%; lateral etch uniformity ±1%
Deep reactive ion etching of fused silica, quartz and glass	Etch depth uniformity 1.5%; lateral etch uniformity 2.4%
Deep reactive ion etching of silicon carbide	Etch depth uniformity 1%; lateral etch uniformity 1%
Surface micromachining	Depends on materials, photolithographic, and etching methods used
Wafer bonding	Wafer-to-wafer alignment accuracy ±3 microns
LIGA	Lateral dimension <1%; thickness variations ±8%
Hot embossing	Determined by mode used and type of polymer
Other MEMS micromachining technologies	
Electro-discharge micromachining	Resolution of tens microns; dimension variation at resolution ±10% or more
Laser micromachining	Resolution of tens microns; dimensional variation at resolution of ±10% or more
Focus ion beam micromachining	Resolution of 50 nm; dimension variation at 50 nm 50% or more

References

1. J. Yang, H. Kahn, A.Q. He, S.M. Phillips, A.H. Heuer, A new technique for producing large-area as-deposited zero-stress LPCVD polysilicon films, the multipoly process. J. Microelectromech. Syst. **9**, 485–494 (2000)
2. D. Poenar, R. Wolffenbuttel, Optical properties of thin-film silicon-compatible materials. Appl. Opt. **36**, 5122–5128 (1997)
3. P.J. French, Polysilicon: a versatile material for microsystems. Sensors Actuators A **99**, 3–12 (2002)

4. L. Elbrecht, R. Catanescu, J. Zacheja, J. Binder, Highly phosphorus-doped polysilicon films with low tensile stress for surface micromachining using POCl3 diffusion doping. Sens. Actuators A **61**, 374–378 (1997)
5. G.M. Dougherty, A.P. Pisano, T.D. Sands, Processing and morphology of permeable polycrystalline silicon thin films. J. Mater. Res. **17**, 2235–2242 (2002)
6. M. Madou, *Fundamentals of Microfabrication*, 2nd edn. (CRC Press, Boca Raton, 2002), p. 274
7. X. Zhang, T.Y. Zhang, M. Wong, Y. Zohar, Rapid thermal annealing of polysilicon thin films. J. Microelectromech. Syst. **7**, 356–364 (1998)
8. Y.B. Gianchandani, M. Shinn, K. Najafi, Impact of high-thermal budget anneals on polysilicon as a micromechanical material. J. Microelectromech. Syst. **7**, 102–105 (1998)
9. O. Tabata, K. Kawahata, S. Sugiyama, I. Igarashi, Mechanical property measurements of thin films using load-deflection of composite rectangular membrane, in *Proceedings of Micro Electro Mechanical Systems*, Salt Lake City, UT, pp. 152–156, 1989
10. H. Guckel, D.W. Burns, Planar processed polysilicon sealed cavities for pressure transducer arrays. Proc. Electron Device Meet. **30**, 223–225 (1984)
11. R.T. Howe, R.S. Muller, Polycrystalline silicon micromechanical beams. J. Electrochem. Soc. **130**, 1420 (1983)
12. G. Fresquet, C. Azzaro, J.P. Couderc, Analysis and modeling of in situ boron-doped polysilicon deposition by LPCVD. J. Electrochem. Soc. **142**, 538 (1995)
13. M. Biebl, G.T. Mulhern, R.T. Howe, In situ phosphorus-doped polysilicon for integrated MEMS, technical digest, in *8th International Conference on Solid-State Sensors and Actuators, Eurosensors IX*, Stockholm, Sweden, pp. 198–201, 1995
14. J.G.M. Mulder, P. Eppenga, M. Hendriks, J.E. Tong, An industrial LPCVD process for in situ phosphorus-doped polysilicon. J. Electrochem. Soc. **137**, 273 (1990)
15. S. Bouwstra, E.L. de Weerd, M.C. Elwenspoek, In situ phosphorus-doped polysilicon for excitation and detection in micromechanical resonators. Sens. Actuators A **24**, 227–235 (1990)
16. R.C. Anderson, R.S. Muller, C.W. Tobias, Porous polycrystalline silicon: A new material for MEMS. J. Microelectromech. Syst. **3**, 10–18 (1994)
17. H. Kahn, S. Stemmer, K. Nandakumar, A.H. Heuer, R.L. Mullen, R. Ballarini, M.A. Huff, Mechanical properties of thick, surface micromachined polysilicon films, in *Proceedings of the 9th IEEE International Workshop on Microelectromechanical Systems, MEMS 96*, San Diego, CA, pp. 343–348, 11–15 Feb 1996
18. J. Gaspar, T. Adrega, V. Chu, J.P. Conde, Thin-film paddle microresonators with high quality factors fabricated at temperatures below 110 °C, in *Proceedings of the 18th International Conference on Microelectromechanical Systems*, Miami, FL, pp. 125–128, 2005
19. S. Chang, S. Sivoththaman, Development of a low temperature MEMS process with a PECVD amorphous silicon structural layer. J. Micromech. Microeng. **16**, 1307–1313 (2006)
20. P. Alpuim, V. Chu, J.P. Conde, Amorphous and microcrystalline silicon films grown at low temperatures by radio-frequency and hot-wire chemical vapor deposition. J. Appl. Phys. **86**, 3812–3821 (1999)
21. C.-K. Chung, M.-Q. Tsai, P.-H. Tsai, C. Lee, Fabrication and characterization of amorphous Si films by PECVD for MEMS. J. Micromech. Microeng. **15**, 136–142 (2005)
22. S. Chang, W. Eaton, J. Fulmer, C. Gonzalez, B. Underwood, Micromechanical structures in amorphous silicon, in *Technical Digest, International Conference on Solid State Sensors and Actuators*, San Francisco, CA, pp. 751–754, 1991
23. S.B. Patil, V. Chu, J.P. Conde, Surface micromachining of a thin film microresonator using dry decomposition of a polymer sacrificial layer. J. Vac. Sci. Technol. B **25**, 455–458 (2007)
24. S. Chang, S. Sivoththaman, A tunable RFMEMS inductor on silicon incorporating an amorphous silicon bimorph in a low-temperature process. IEEE Electron Device Lett. **27**, 905–907 (2006)
25. P. Gennissen, M. Bartek, P.J. French, P.M. Sarro, Bipolar-compatible epitaxial poly for smart sensors: stress minimization and applications. Sens. Actuators A **62**, 636–645 (1997)

26. M. Kirsten, B. Wenk, F. Ericson, J.A. Schweitz, W. Reithmuller, P. Lang, Deposition of thick doped polysilicon films with low stress in an epitaxial reactor for surface micromachining applications. Thin Solid Films **259**, 181–187 (1995)
27. P. Lange, M. Kirsten, W. Riethmuller, B. Wenk, G. Zwicker, J.R. Morante, F. Ericson, J.A. Schweitz, Thick polycrystalline silicon for surface-micromechanical applications: deposition, structuring, and mechanical characterization. Sens. Actuators A **54**, 674–678 (1996)
28. S. Greek, F. Ericson, S. Johansson, J.A. Schweitz, In situ tensile strength measurement and Weibull analysis of thick film and thin film micromachined polysilicon structures. Thin Solid Films **292**, 247–254 (1997)
29. O. De Sagazan, M. Denoual, P. Guil, D. Gaudin, O. Bonnaud, Microelectromechanical systems fast fabrication by selective thick polysilicon growth in epitaxial reactor. Microsyst. Technol. **12**, 953–958 (2006)
30. T. Abe, M.L. Reed, Low strain sputtered polysilicon for micromechanical structures, in *Proceedings of the 9th International Workshop on Micro Electro Mechanical Systems*, San Diego, CA, pp. 258–262, 1996
31. P. Pal, S. Chandra, RF sputtered silicon for MEMS. J. Micromech. Microeng. **15**, 1536–1546 (2005)
32. A.E. Franke, Y. Jiao, M.T. Wu, T.J. King, R.T. Howe, Post-CMOS modular integration of poly-sige microstructures using poly-ge sacrificial layers, in *Technical Digest, Solid State Sensor and Actuator Workshop,* Hilton Head, SC, pp. 18–21, 2000
33. C.W. Low, T.J.K. Liu, R.T. Howe, Characterization of polycrystalline silicon-germanium film deposition for modularly integrated MEMS applications. J. Microelectromech. Syst. **16**, 68–77 (2007)
34. M.-A. Eyoum, T.-J. King, Low-resistance silicon-germanium contact technology for modular integration of MEMS with electronics. J. Electrochem. Soc. **151**, J21–J25 (2004)
35. S. Sedky, A. Bayoumy, A. Alaa, A. Nagy, A. Witvrouw, Optimal conditions for micromachining Si1–xGex at 210 °C. J. Microelectromech. Syst. **16**, 581–588 (2007)
36. S. Sedky, P. Fiorini, M. Caymax, S. Loreti, K. Baert, L. Hermans, R. Mertens, Structural and mechanical properties of polycrystalline silicon germanium for micromachining applications. J. Microelectromech. Syst. **7**, 365–372 (1998)
37. S. Sedky, A. Witvrouw, K. Baert, Poly SiGe, a promising material for MEMS monolithic integration with the driving electronics. Sens. Actuators A **97–98**, 503–511 (2002)
38. A.E. Franke, J.M. Heck, T.J. King, R.T. Howe, Polycrystalline silicon-germanium films for integrated microsystems. J. Microelectromech. Syst. **12**, 160 171 (2003)
39. T.-J. King, J.P. McVittie, K.C. Saraswat, Electrical properties of heavily doped polycrystalline silicon-germanium films. IEEE Trans. Electron Device **41**, 228–232 (1994)
40. M. Gromova, A. Mehta, K. Baert, A. Witvrouw, Characterization and strain gradient optimization of PECVD poly-SiGe layers for MEMS applications. Sens. Actuators A **130–131**, 403–410 (2006)
41. S. Kannan, C. Taylor, D. Allred, PECVD growth of Six:Ge1-x films for high speed devices and MEMS. J. Non-Cryst. Solids **352**, 1272–1274 (2006).
42. C. Rusu, S. Sedky, B. Parmentier, A. Verbist, O. Richard, B. Brijs, L. Geenen, A. Witvrouw, F. Larmer, F. Fischer, S. Kronmuller, V. Leca, B. Otter, New low-stress PECVD poly-SiGe layers for MEMS. J. Microelectromech. Syst. **12**, 816–825 (2003)
43. A. Mehta, M. Gromova, C. Rusu, R. Olivier, K. Baert, C. van Hoof, A. Witvrouw, Novel high growth rate processes for depositing poly-sige structural layers at CMOS compatible temperatures, in *Proceedings of the 17th International Conference on Micro Electro Mechanical Systems*, Maastricht, The Netherlands, pp. 721–724, 2004
44. I. Nakamura, T.A. Hirokazu, D. Hoshi, M. Isomura, Formation of polycrystalline SiGe thin films by the RF magnetron sputtering method with Ar–H2 mixture gases. Vacuum **80**(7), 31 (2006)
45. G.-P. Ru, G.-W. Wang, Y.-L. Jiang, W. Huang, X.-P. Qu, S.-Y. Zhu, B.-Z. Li, Rectifying characteristics of sputter-deposited SiGe diodes. J. Vac. Sci. Technol. B: Microelectron. Nanometer Struct. Process. Meas. Phenom. **21**, 1301 (2003)

46. B. Li, B. Xiong, L. Jiang, Y. Zohar, M. Wong, Germanium as a versatile material for low temperature micromachining. J. Microelectromech. Syst. **8**, 366–372 (1999)
47. A. Franke, D. Bilic, D.T. Chang, P.T. Jones, T.J. King, R.T. Howe, C.G. Johnson, Post-CMOS integration of germanium microstructures, in *Proceedings of the 12th International Conference on Microelectromechanical Systems, IEEE*, Piscataway, NJ, pp. 630–637, 1999
48. J.M. Heck, C.G. Keller, A.E. Franke, L. Muller, T.J. King, R.T. Howe, High aspect ratio polysilicon-germanium microstructures, in *Proceedings of the 10th International Conference on Solid State Sensors and Actuators*, Sendai, Japan, pp. 328–334, 1999
49. I.H. Khan, Low-temperature epitaxy of Ge films by sputter deposition. J. Appl. Phys. **44**, 14 (1973)
50. C.R. Stoldt, C. Carraro, W.R. Ashurst, D. Gao, R.T. Howe, R. Maboudian, A low temperature CVD process for silicon carbide MEMS. Sens. Act. A **97–98**, 410–415 (2002)
51. C.A. Zorman, S. Rajgolpal, X.A. Fu, R. Jezeski, J. Melzak, M. Mehregany, Deposition of polycrystalline 3C-SiC films on 100 mm-diameter (100) Si wafers in a large-volume LPCVD furnace. Electrochem. Solid State Lett. **5**, G99–G101 (2002)
52. L. Behrens, E. Peiner, A.S. Bakin, A. Schlachetzski, Micromachining of silicon carbide on silicon fabricated by low-pressure chemical vapor deposition. J. Micromech. Microeng. **12**, 380–384 (2002)
53. X.A. Fu, R. Jezeski, C.A. Zorman, M. Mehregany, Use of deposition pressure to control the residual stress in polycrystalline SiC films. Appl. Phys. Lett. **84**, 341–343 (2004)
54. X.-A. Fu, J.L. Dunning, C.A. Zorman, M. Mehregany, Polycrystalline 3C-SiC thin films deposited by dual precursor LPCVD for MEMS applications. Sens. Actuators A **119**, 169–176 (2005)
55. D. Gao, M.B. Wijesundara, C. Carraro, R.T. Howe, R. Maboudian, Recent progress toward and manufacturable polycrystalline SiC surface micromachining technology. IEEE Sensors J. **4**, 441–448 (2004)
56. Properties of SiC, Ioffe Institute. http://www.ioffe.ru/SVA/NSM/Semicond/SiC/ Accessed 2018–06–06
57. C.S. Roper, R.T. Howe, R. Maboudian, Stress control of polycrystalline 3C-SiC films in a large-scale LPCVD reactor using 1,3-disilabutane and dichlorosilane as precursors. J. Micromech. Microeng. **16**, 2736–2739 (2006)
58. Y. Yamaguchi, H. Nagasawa, T. Shoki, N. Armaka, H. Mitsui, Properties of heteroepitaxial 3C-SiC films grown by LPCVD. Sens. Actuators A **54**, 695–699 (1996)
59. D. Gao, C. Carraro, V. Radmilovic, R.T. Howe, R. Maboudian, Fracture of polycrystalline 3C-SiC films in microelectromechanical systems. J. Microelectromech. Syst. **13**, 972–976 (2004)
60. E. Hurtos, J. Rodriguez-Viejo, Residual stress and texture in poly-SiC films grown by low pressure organometallic chemical-vapor deposition. J. Appl. Phys. **87**, 1748–1758 (2000)
61. K. Murooka, I. Higashikawa, Y. Gomei, Improvement of the Young's modulus of SiC film by low pressure chemical vapor deposition with B2H6 gas. Appl. Phys. Lett. **69**, 37–39 (1996)
62. J. Chen, J. Scofield, A.J. Steckl, Formation of SiC SOI structures by direct growth on insulating substrates. J. Electrochem. Soc. **147**, 3845–3849 (2000)
63. G. Sun, J. Ning, X. Liu, Y. Zhao, J. Li, L. Wang, W. Zhao, L. Wang, Heavily doped polycrystalline 3C-SiC growth on SiO2/Si (100) substrates for resonator applications. Mater. Sci. Forum **556–557**, 179–182 (2007)
64. H. Zhang, H. Guo, Y. Wang, G. Zhang, Z. Li, Study on a PECVD SiC-coated pressure sensor. J. Micromech. Microeng. **17**, 426–431 (2007)
65. P.M. Sarro, C.R. deBoer, E. Korkmaz, J.M.W. Laros, Low-stress PECVD SiC thin films for IC-compatible microstructures. Sens. Actuators A **67**, 175–180 (1998)
66. C. Iliescu, B. Chen, D.P. Poenar, Y.Y. Lee, PECVD amorphous silicon carbide membranes for cell culturing. Sens. Actuators B **129**, 404–411 (2008)

67. L.S. Pakula, H. Yang, H.T.M. Pham, P.J. French, P.M. Sarro, Fabrication of a CMOS compatible pressure sensor for harsh environments. J. Micromech. Microeng. **14**, 1478–1483 (2004)
68. A. Klumpp, U. Schaber, H.L. Offereins, K. Kuhl, H. Sandmaier, Amorphous silicon carbide and its application in silicon micromachining. Sens. Actuators A **41–42**, 310–316 (1994)
69. J. Du, C.A. Zorman, Low temperature a-SiC/Si direct bonding technology for MEMS/NEMS, in *Technical Digest of the 14th International Conference on Solid State Sensors, Actuators and Microsystems*, Lyon, France, pp. 2075–2078, 2007
70. M. Eickhoff, H. Moller, G. Kroetz, J. von Berg, R. Ziermann, A high temperature pressure sensor prepared by selective deposition of cubic silicon carbide on SOI substrates. Sensors Actuators **74**, 56–59 (1999)
71. G. Krotz, H. Moller, M. Eickhoff, S. Zappe, R. Ziermann, E. Obermeier, J. Stoemenos, Heteroepitaxial growth of 3C-SiC on SOI for sensor applications. Mater. Sci. Eng. B **61**, 516–521 (1999)
72. D. Young, J.D. Du, C.A. Zorman, W.H. Ko, High-temperature single crystal 3C-SiC capacitive pressure sensor. IEEE Sensors J. **4**, 464–470 (2004)
73. N. Ledermann, J. Baborowski, J.P. Muralt, N. Xantopoulos, J.M. Tellenbach, Sputtered silicon carbide thin films as protective coating for MEMS applications. Surf. Coat. Technol. **125**, 246–250 (2000)
74. S. Inoue, T. Namazu, H. Tawa, M. Niibe, K. Koterazawa, Stress control of a-SiC films deposited by dual source dc magnetron sputtering. Vacuum **80**, 744–747 (2006)
75. M.A. Huff, D.A. Aidala, J.E. Butler, MEMS applications using diamond thin films. Solid State Technol. **49**(4), 45–48 (2006)
76. O. Auciello, J. Birrell, J.A. Carlisle, J. Geri, X. Xiao, B. Peng, H.D. Espinosa, Materials science and fabrication processes for a new MEMS technology based on ultrananocrystalline diamond thin film. J. Phys. Condens. Matter **16**, R539–R552 (2004)
77. A.R. Krauss, O. Auciello, D.M. Gruen, A. Jayatissa, A. Sumant, J. Tucek, D.C. Mancini, N. Moldovan, A. Erdemir, D. Ersoy, M.N. Gardos, H.G. Busmann, E.M. Meyer, Q. Ding, Ultrananocrystalline diamond thin films for MEMS and moving mechanical assembly devices. Diam. Relat. Mater. **10**, 1952–1961 (2001)
78. D.K. Reinhard, T.A. Grotjohn, M. Becker, M.K. Yaran, T. Schuelke, J. Asmussen, Fabrication and properties of ultranano, nano, and microcrystalline diamond membranes and sheets. J. Vac. Sci. Technol. B **22**, 2811 (2004)
79. X. Xiao, B.W. Sheldon, Y. Qi, A.K. Kothari, Intrinsic stress evolution in nanocrystalline diamond thin films with deposition temperature. Appl. Phys. Lett. **92**, 131908 (2008)
80. H. Li, B.W. Sheldon, A. Kothari, Z. Ban, B.L. Walden, Stress evolution in nanocrystalline diamond films produced by chemical vapor deposition. J. Appl. Phys. **100**, 094309 (2006)
81. D.R. Wur, J.L. Davidson, W.P. Kang, D.L. Kinser, Polycrystalline diamond pressure sensor. J. Microelectromech. Syst. **4**, 34–41 (1995)
82. H.D. Espinosa, B.C. Prorok, B. Peng, K.H. Kim, M. Moldovan, O. Auciello, J.A. Carlisle, D.M. Gruen, D.C. Mancini, Mechanical properties of ultrananocrystalline diamond thin films relevant to MEMS/NEMS devices. Exp. Mech. **4**, 256–268 (2003)
83. H.D. Espinosa, B. Peng, B.C. Prorok, N. Moldovan, O. Auciello, J.A. Carlisle, D.M. Gruen, D.C. Mancini, Fracture strength of ultrananocrystalline diamond thin films – identification of Weibull parameters. J. Appl. Phys. **94**, 6076 (2003)
84. H. Bjorkman, P. Rangsten, P. Hollman, K. Hjort, Diamond replicas from microstructured silicon masters. Sensors Actuators **73**, 24–29 (1999)
85. M. Aslam, D. Schulz, Technology of diamond microelectromechanical systems, in *Proceedings of the 8th International Conference on Solid State Sensors and Actuators, Eurosensors IX*, Stockholm, Sweden, pp. 222–224, 1995
86. R. Ramesham, Fabrication of diamond microstructures for microelectromechanical systems (MEMS) by a surface micromachining process. Thin Solid Films **340**, 1–6 (1999)
87. Y. Yang, X. Wang, C. Ren, J. Xie, P. Lu, W. Wang, Diamond surface micromachining technology. Diam. Relat. Mater. **8**, 1834–1837 (1999)

88. S.A. Smallwood, K.C. Eapen, S.T. Patton, J.S. Zabinski, Performance results of MEMS coated with a conformal DLC. Wear **260**, 1179–1189 (2006)
89. J.K. Luo, Y.Q. Fu, H.R. Le, J.A. Williams, S.M. Spearing, W.I. Milne, Diamond and diamond-like carbon MEMS. J. Micromech. Microeng. **17**, S147–S163 (2007)
90. M. Ohring, *The Materials Science of Thin Films* (Academic Press, London, 1993)
91. J.R. Davis (Ed.), Base-metal selection, in *Metals Handbook Desk Edition*, 2nd edn. (ASM International, Materials Park, 1998)
92. S.K. Prasad, *Advanced Wirebond Interconnection Technology* (Kluwer, Bangalore, 2004)
93. G.G. Harman, *Wire Bonding in Microelectronics*, 2nd edn. (McGraw Hill, New York, 1997)
94. S. D. Cramer, B. S. Covino (eds.), *ASM Handbook, Vol. 13B: Corrosion: Materials* (ASM International, Materials Park, 2005)
95. K. L. Mittal (ed.), *Adhesion Aspects of Thin Films*, vol 2 (VSP, Utrecht, 2005), p. 125
96. C.K. Hu, M.B. Small, F. Kaufman, D.J. Pearson, in *Tungsten and Other Advanced Metals for VLSI/ULSI Applications*, ed. by S. S. Wong, S. Furuka, (Materials Research Society, Pittsburgh, 1989), p. 369
97. M. K. Ghosh, K. L. Mittal (eds.), *Polyimides: Fundamentals and Applications, Ch. 20–21* (Marcel Dekker, New York, 1996)
98. F.K. LeGoues, B.D. Silverman, P.S. Ho, The microstructure of metal-polyimide interfaces. J. Vac. Sci. Technol. A **6**, 2200–2204 (1988)
99. N. Bowden, S. Brittain, A.G. Evans, J.W. Hutchinson, G.M. Whitesides, Spontaneous formation of ordered structures in thin films of metals supported on an elastomeric polymer. Nature **393**, 146–149 (1998)
100. J.L. Vossen, W. Kern, *Thin Film Processes* (Academic Press, London, 1978)
101. P.W. Kruse, *Uncooled Thermal Imaging* (SPIE Press, Bellingham, 2002)
102. J. Yang, O. Paul, Fracture properties of LPCVD silicon nitride thin films from the load deflection of long membranes. Sens. Actuators A **97–98**, 520–526 (2002)
103. J.M. Olson, Analysis of LPCVD process conditions for the deposition of low stress silicon nitride. Part 1: preliminary LPCVD experiments. Mater. Sci. Semicond. Process. **5**, 51–60 (2002)
104. M. Sekimoto, H. Yoshihara, T. Ohkubo, Silicon nitride single-layer x-ray mask. J. Vac. Sci. Technol. **21**, 1017–1021 (1982)
105. J.G.E. Gardeniers, H.A.C. Tilmans, C.C.G. Visser, LPCVD silicon-rich silicon nitride films for applications in micromechanics studied with statistical experimental design. J. Vac. Sci. Technol. A **14**, 2879–2892 (1996)
106. P. Temple-Boyer, C. Rossi, E. Saint-Etienne, E. Scheid, Residual stress in low pressure chemical vapor deposition SiNx films deposited from silane and ammonia. J. Vac. Sci. Technol. A **16**, 2003–2007 (1998)
107. C. Mastrangelo, Y.-C. Tai, R. Muller, Thermophysical properties of low-residual stress, silicon-rich, LPCVD silicon nitride films. Sens. Actuators **856–860**, A21–A23 (1990)
108. P.P. Tsai, I.-C. Chen, C.J. Ho, Ultralow power carbon monoxide microsensor by micromachining techniques. Sens. Actuators B **76**, 380–387 (2001)
109. P.J. French, P.M. Sarro, R. Mallee, E.J.M. Fakkeldij, R.F. Wolffenbuttel, Optimization of a low-stress silicon nitride process for surface micromachining applications. Sens. Actuators A **58**, 149–157 (1997)
110. J. Yang, J. Gaspar, O. Paul, Fracture properties of LPCVD silicon nitride and thermally grown silicon oxide thin films from the load-deflection of long Si3N4 and SiO2/Si3N4 diaphragms. J. Microelectromech. Syst. **17**, 1120–1134 (2008)
111. S. Hong, T.P. Weihs, J.C. Bravman, W.D. Nix, Measuring stiffness and residual stresses of silicon nitride thin films. J. Electron. Mater. **19**, 903–909 (1990)
112. E.I. Bromley, J.N. Randall, D.C. Flanders, R.W. Mountain, A technique for the determination of stress in thin films. J. Vac. Sci. Technol. B **1**, 1364–1366 (1983)
113. M. Martyniuk, J. Antoszewski, C.A. Musca, J.M. Dell, L. Faraone, Dielectric thin films for MEMS-based optical sensors. Microelectron. Reliab. **47**, 733–738 (2007)

114. H. Huang, K.J. Winchester, A. Suvorova, B.R. Lawne, Y. Liud, X.Z. Hud, J.M. Dell, L. Faraone, Effect of deposition conditions on mechanical properties of low-temperature PECVD silicon nitride films. Mater. Sci. Eng. A **435–436**, 453–459 (2006)
115. M.-C. Jo, S.-K. Park, S.-J. Park, A study on resistance of PECVD silicon nitride thin film to thermal stress-induced cracking. Appl. Surf. Sci. **140**, 12–18 (1999)
116. C. Iliescu, F.E.H. Tay, J. Wei, Low stress PECVD-SiNx layers at high deposition rates using high power and high frequency for MEMS applications. J. Micromech. Microeng. **16**, 869–874 (2006)
117. D. Poenar, P.J. French, R. Mallee, P.M. Sarro, R.F. Wolffenbuttel, PSG layers for surface micromachining. Sens. Actuators A **41–42**, 304–309 (1994)
118. Y. Shioya, M. Mamoru, Comparison of phosphosilicate glass films deposited by three different chemical vapor deposition methods. J. Electrochem. Soc. **133**, 1943–1950 (1986)
119. R.M. Levin, A.C. Adams, Low pressure deposition of phosphosilicate glass films. J. Electrochem. Soc. **129**, 1588–1592 (1982)
120. S.K. Ghandhi, *VLSI Fabrication Principles – Silicon and Gallium Arsenide* (Wiley, New York, 1983)
121. J. Leconte, F. Iker, S. Jorez, N. Andre, J. Proost, T. Pardoen, D. Flandre, J.P. Raskin, Thin films stress extraction using micromachined structures and wafer curvature measurements. Microelectron. Eng. **76**, 219–226 (2004)
122. X. Zhang, K.-S. Chen, R. Ghodssi, A.A. Ayon, S.M. Spearing, Residual stress and fracture in thick tetraethylorthosilicate (TEOS) and silane-based PECVD oxide films. Sens. Actuators A **91**, 373–380 (2001)
123. V. Bhatt, S. Chandra, Silicon dioxide films by RF sputtering for microelectronic and MEMS. J. Micromech. Microeng. **17**, 1066–1077 (2007)
124. E. Meng, X. Zhang, W. Benard, Additive processes for polymeric materials, Chapter 4, in *MEMS Material and Process Handbook*, (Springer, New York, 2011)
125. K.Y. Lee, N. LaBianca, S.A. Rishton, et al., Micromachining applications of a high resolution ultrathick photoresist. J. Vac. Sci. Technol. B **13**, 3012–3016 (1995)
126. J.M. Shaw, J.D. Gelorme, N.C. Labianca, et al., Negative photoresists for optical lithography. IBM J. Res. Dev. **41**, 81–94 (1997)
127. P. Abgrall, V. Conedera, H. Camon, et al., SU-8 as a structural material for labs-on-chips and microelectromechanical systems. Electrophoresis **28**, 4539–4551 (2007)
128. M. Despont, H. Lorenz, N. Fahrni, et al., High-aspect-ratio, ultrathick, negative-tone nearuv photoresist for MEMS applications, in *Proceedings of the IEEE Microelectromechanical Systems Conference*, Nagoya, Japan, pp. 518–522, 1997
129. L.J. Guerin, M. Bossel, M. Demierre, et al., Simple and low cost fabrication of embedded micro-channels by using a new thick-film photoplastic, in *Technical Digest, International Conference on Solid State Sensors and Actuators*, Chicago, IL, pp. 1419–1422, 1997
130. B. Bohl, R. Steger, R. Zengerle, et al., Multi-layer SU-8 lift-off technology for microfluidic devices. J. Micromech. Microeng. **15**, 1125–1130 (2005)
131. R.J. Jackman, T.M. Floyd, R. Ghodssi, et al., Microfluidic systems with on-line UV detection fabricated in photodefinable epoxy. J. Micromech. Microeng. **11**, 263–269 (2001)
132. A. Mata, A.J. Fleischman, S. Roy, Fabrication of multi-layer SU-8 microstructures. J. Micromech. Microeng. **16**, 276–284 (2006)
133. M. Han, W. Lee, S.K. Lee, et al., 3D microfabrication with inclined/rotated UV lithography. Sens. Actuators A Phys. **111**, 14–20 (2004)
134. Y.K. Yoon, J.H. Park, M.G. Allen, Multidirectional UV lithography for complex 3-DMEMS structures. J. Microelectromech. Syst. **15**, 1121–1130 (2006)
135. W.H. Wong, E.Y.B. Pun, Exposure characteristics and three-dimensional profiling of SU8C resist using electron beam lithography. J. Vac. Sci. Technol. B **19**, 732–735 (2001)
136. J. Zhang, K.L. Tan, G.D. Hong, et al., Polymerization optimization of SU-8 photoresist and its applications in microfluidic systems and MEMS. J. Micromech. Microeng. **11**, 20–26 (2001)

137. N.C. LaBianca, J.D. Gelorme, High-aspect-ratio resist for thick-film applications, in *Proceedings of SPIE – The International Society for Optical Engineering*, Santa Clara, CA, USA, pp. 846–852, 1995
138. Y.J. Chuang, F.G. Tseng, J.H. Cheng, et al., A novel fabrication method of embedded microchannels by using SU-8 thick-film photoresists. Sens. Actuators A Phys. **103**, 64–69 (2003)
139. B.E.J. Alderman, C.M. Mann, D.P. Steenson, et al., Microfabrication of channels using an embedded mask in negative resist. J. Micromech. Microeng. **11**, 703–705 (2001)
140. A. Heeren, C.P. Luo, G. Roth, et al., Diffusion along microfluidic channels. Microelectron. Eng. **83**, 1669–1672 (2006)
141. Y.T. Chen, D. Lee, A bonding technique using hydrophilic SU-8. J. Micromech. Microeng. **17**, 1978–1984 (2007)
142. D.J. Strike, G.C. Fiaccabrino, M. Koudelka-Hep, et al., Enzymatic microreactor using Si, glass and EPON SU-8. Biomed. Microdevices **2**, 175 (2000)
143. J. West, B. Karamata, B. Lillis, et al., Application of magnetohydrodynamic actuation to continuous flow chemistry. Lab Chip **2**, 224–230 (2002)
144. S. Li, C.B. Freidhoff, R.M. Young, et al., Fabrication of micronozzles using low-temperature wafer-level bonding with SU-8. J. Micromech. Microeng. **13**, 732–738 (2003)
145. J.H. Park, Y.K. Yoon, S.O. Choi, et al., Tapered conical polymer microneedles fabricated using an integrated lens technique for transdermal drug delivery. I.E.E.E. Trans. Biomed. Eng. **54**, 903–913 (2007)
146. S. Rajaraman, S.O. Choi, R.H. Shafer, et al., Microfabrication technologies for a coupled three-dimensional microelectrode, microfluidic array. J. Micromech. Microeng. **17**, 163–171 (2007)
147. V. Seidemann, J. Rabe, M. Feldmann, et al., SU8-micromechanical structures with in situ fabricated movable parts. Microsyst. Technol. **8**, 348–350 (2002)
148. V. Seidemann, S. Butefisch, S. Buttgenbach, Fabrication and investigation of in-plane compliant SU8 structures for MEMS and their application to micro valves and micro grippers. Sens. Actuators A Phys. **97–98**, 457–461 (2002)
149. H. Lorenz, M. Despont, N. Fahrni, et al., SU-8: A low-cost negative resist for MEMS. J. Micromech. Microeng. **7**, 121–124 (1997)
150. H. Lorenz, M. Despont, N. Fahrni, et al., High-aspect-ratio, ultrathick, negative-tone near-UV photoresist and its applications for MEMS. Sens. Actuators A Phys. **64**, 33–39 (1998)
151. S.J. Kim, H. Yang, K. Kim, et al., Study of SU-8 to make a Ni master-mold: Adhesion, sidewall profile, and removal. Electrophoresis **27**, 3284–3296 (2006)
152. P.M. Dentinger, W.M. Clift, S.H. Goods, Removal of SU-8 photoresist for thick film applications. Microelectron. Eng. **61–2**, 993–1000 (2002)
153. C.-H. Ho, K.-P. Chin, C.-R. Yang, et al., Ultrathick SU-8 mold formation and removal, and its application to the fabrication of LIGA-like micromotors with embedded roots. Sens. Actuators A Phys. **102**, 130–138 (2002)
154. A. Johansson, M. Calleja, P.A. Rasmussen, et al., SU-8 cantilever sensor system with integrated readout. Sens. Actuators A Phys. **123–124**, 111–115 (2005)
155. P. Basset, A. Kaiser, D. Collard, et al., Process and realization of a three-dimensional gold electroplated antenna on a flexible epoxy film for wireless micromotion system. J. Vac. Sci. Technol. B **20**, 1465–1470 (2002)
156. L. Dellmann, S. Roth, C. Beuret, et al., Fabrication process of high aspect ratio elastic and SU-8 structures for piezoelectric motor applications. Sens. Actuators A Phys. **70**, 42–47 (1998)
157. S. Mouaziz, G. Boero, R.S. Popovic, et al., Polymer-based cantilevers with integrated electrodes. J. Microelectromech. Syst. **15**, 890–895 (2006)
158. G. Kim, B. Kim, J. Brugger, All-photoplastic microstencil with self-alignment for multiple layer shadow-mask patterning. Sens. Actuators A Phys. **107**, 132–136 (2003)
159. S. Tuomikoski, S. Franssila, Free-standing SU-8 microfluidic chips by adhesive bonding and release etching. Sens. Actuators A Phys. **120**, 408–415 (2005)

160. C. Luo, A. Govindaraju, J. Garra, et al., Releasing SU-8 structures using polystyrene as a sacrificial material. Sens. Actuators A Phys. **114**, 123–128 (2004)
161. H. Lorenz, M. Laudon, P. Renaud, Mechanical characterization of a new high-aspect-ratio near UV-photoresist. Microelectron. Eng. **41–42**, 371–374 (1998)
162. S.J. Clarson, J.A. Semlyen, *Siloxane Polymers* (Prentice Hall, Englewood Cliffs, 1993)
163. J.C. Lotters, W. Olthuis, P.H. Veltink, et al., The mechanical properties of the rubber elastic polymer polydimethylsiloxane for sensor applications. J. Micromech. Microeng. **7**, 145–147 (1997)
164. ABCR, *Research Chemicals and Metals Catalogue* (ABCR, Karlsruhe, 1994)
165. J.C. Salamone, *Polymeric Materials Encyclopedia* (CRC Press, Rochester, 1996)
166. D.W. van Krevelen, P.J. Hoftyzer, *Properties of Polymers, Their Estimation and Correlation with Chemical Structure* (Elsevier Scientific, Amsterdam, 1976)
167. Y.H. Kim, K.S. Shin, J.Y. Kang, et al., Poly(dimethylsiloxane)-based packaging technique for microchip fluorescence detection system applications. J. Microelectromech. Syst. **15**, 1152–1158 (2006)
168. M.A. Eddings, B.K. Gale, A PDMS-based gas permeation pump for on-chip fluid handling in microfluidic devices. J. Micromech. Microeng. **16**, 2396–2402 (2006)
169. Y.N. Xia, J.J. McClelland, R. Gupta, et al., Replica molding using polymeric materials: A practical step toward nanomanufacturing. Adv. Mater. **9**, 147–149 (1997)
170. H.L. Cong, T.R. Pan, Photopatternable conductive PDMS materials for microfabrication. Adv. Funct. Mater. **18**, 1912–1921 (2008)
171. K. Tsougeni, A. Tserepi, E. Gogolides, Photosensitive poly(dimethylsiloxane) materials for microfluidic applications. Microelectron. Eng. **84**, 1104–1108 (2007)
172. A.M. Wilson, Polyimide insulators for multilevel interconnections. Thin Solid Films **83**, 145–163 (1981)
173. M. Gad-el-Hak, *MEMS: Design and Fabrication* (CRC/Taylor & Francis, Boca Raton, 2006)
174. M. Pedersen, W. Olthuis, P. Bergveld, High-performance condenser microphone with fully integrated CMOS amplifier and DC-DC voltage converter. J. Microelectromech. Syst. **7**, 387–394 (1998)
175. D. Memmi, V. Foglietti, E. Cianci, et al., Fabrication of capacitive micromechanical ultrasonic transducers by low-temperature process. Sens. Actuators A Phys. **99**, 85–91 (2002)
176. M.A. Schmidt, R.T. Howe, S.D. Senturia, et al., Design and calibration of a microfabricated floating-element shear-stress sensor. IEEE Trans. Electron Device **35**, 750–757 (1988)
177. A.B. Frazier, C.H. Ahn, M.G. Allen, Development of micromachined devices using polyimide-based processes. Sens. Actuators A Phys. **45**, 47–55 (1994)
178. P.J. Rousche, D.S. Pellinen, D.P. Pivin, et al., Flexible polyimide-based intracortical electrode arrays with bioactive capability. I.E.E.E. Trans. Biomed. Eng. **48**, 361–371 (2001)
179. E. Valcheva, J. Birch, P.O.A. Persson, S. Tungasmita, L. Hultman, Epitaxial growth and orientation of AlN thin films on Si(001) substrates deposited by reactive magnetron sputtering. J. Appl. Phys. **100**, 123514 (2006)
180. H.Y. Shin, W.H. Lee, W.C. Kao, Y.C. Chuang, R.M. Lin, H.C. Lin, M. Shiojiri, M.J. Chen, Low-temperature atomic layer epitaxy of AlN ultrathin films by layer-by-layer, in-situ atomic layer annealing. Sci. Rep. **7**, 39717 (2017)
181. F. Medjani, R. Sanjinés, G. Allidi, A. Karimi, Effect of substrate temperature and bias voltage on the crystallite orientation in RF magnetron sputtered AlN thin films. Thin Solid Films **515**, 260–265 (2006)
182. B. Abdallah et al., Deposition of AlN films by reactive sputtering: effect of radio frequency substrate bias. Thin Solid Films **515**, 7105–7108 (2007)
183. A.K. Chu, C.H. Chao, F.Z. Lee, H.L. Huang, Influences of bias voltage on the crystallographic orientation of AlN thin films prepared by long-distance magnetron sputtering. Thin Solid Films **429**, 1–4 (2003)

184. M.A. Dubois, P. Muralt, Stress and piezoelectric properties of aluminum nitride thin films deposited onto metal electrodes by pulsed direct current reactive sputtering. J. Appl. Phys. **89**, 6389–6395 (2001)
185. C. Lin, Y. Chen, K.-S. Kao, Two-step sputtered ZnO piezoelectric films for film bulk acoustic resonators. Appl. Phys. A Mater. Sci. Process. **89**, 475–479 (2007)
186. S.J. Chang, Y.K. Su, Y.P. Shei, High quality ZnO thin films on InP substrates prepared by radio frequency magnetron sputtering. I. Material study. J. Vac. Sci. Technol. A **13**, 381–384 (1995)
187. R.D. Klissurska et al., Effect of Nb doping on the microstructure of sol-gel derived PZT thin films. J. Am. Ceram. Soc. **78**, 1513–1520 (1995)
188. P. Muralt, J. Baborowski, N. Lederman, Piezoelectric micro-electro-mechanical systems with PbZrxTi1-xO3 thin films: integration and application issues, in *Piezoelectric Materials and Devices*, (EPFL Swiss Federal Institute of Technology, Lausanne, 2002), pp. 231–260
189. R.N. Castellano, L.G. Feinstein, Ion-beam deposition of thin films of ferroelectric PZT. J. Appl. Phys. **50**, 4406–4411 (1979)
190. M. Okada et al., Preparation of c-axis oriented PbTiO3 thin films byMOCVD. Ferroelectrics **91**, 181–192 (1989)
191. K.D. Budd, S.K. Dey, D.A. Payne, Sol-gel processing of PT, PZ, PZT, and PLZT thin films. Br. Ceram. Proc. **36**, 107 (1985)
192. G. Fox, K. Suu, High temperature deposition of Pt/TiOx for bottom electrodes, US Patent 6,682,772, 27 Jan 2004
193. S. Hiboux, P. Muralt, N. Setter, Orientation and composition dependence of piezoelectric dielectric properties of sputtered Pb(ZrxTi1-x)O3 thin films. Mater. Res. Soc. Symp. Proc. **596**, 499–504 (2000)
194. R.W. Whatmore, Q. Zhang, Z. Huang, R.A. Dorey, Ferroelectric thin and thick films for microsystems. Mater. Sci. Semi-Solid Proc. **5**, 65–76 (2003)
195. Q.F. Zhou, E. Hong, R. Wolf, S. Trolier-McKinstry, Dielectric and piezoelectric properties of PZT 52/48 thick films with (100) and random crystallographic orientation, in *Ferroelectric Thin Films IX, Material Research Society Proceedings*, vol. 655, pp. CC11.7.1–CC11.7.6, 2001
196. N. Ledermann, P. Muralt, et al., {100}-textured, piezoelectric Pb(Zrx, Ti1−x)O3 thin films for MEMS: Integration, deposition and properties. Sens. Actuators A **105**, 162–170 (2003)
197. R.G. Polcawich, J.S. Pulskamp, Additive processes for piezoelectric materials; piezoelectric MEMS, Chapter 5, in *MEMS Materials and Process Handbook*, (Springer, New York, 2011)
198. C.B. Varlul et al., Wet chemical etching of AlN and InAlN in KOH solutions. J. Electrochem. Soc. **143**, 3681–3684 (1996)
199. D. Zhuang, J.H. Edgar, Wet etching of GaN, AlN, and SiC: a review. Mater. Sci. Eng. **48**, 1–46 (2005)
200. D. Chen, J. Wang, D. Xu, Y. Zhang, The influence of the AlN film texture on the wet chemical etching. J. Microelectron. **40**, 15–19 (2009)
201. S. Saravanan, E. Berenschot, G. Krijnen, M. Elwenspoek, A novel surface micromachining process to fabricate AlN unimorph suspensions and its application for RF resonators. Sens. Actuators A **130–131**, 340–345 (2006)
202. T.Y. Sheng, Z.Q. Yu, G.J. Collins, Disk hydrogen plasma assisted chemical vapor deposition of aluminum nitride. Appl. Phys. Lett. **52**, 576–578 (1988)
203. S.M. Tanner, V.V. Felmetsger, Microstructure and chemical wet etching characteristics of AlN films deposited by ac reactive magnetron sputtering. J. Vac. Sci. Technol. A **28**, 69–76 (2010)
204. M.J. Vellekoop, C.C.O. Visser, P.M. Sarro, et al., Compatibility of zinc oxide with silicon IC processing. Sens. Actuators A **23**, 1027–1030 (1990)
205. D. Devoe, Piezoelectric thin film micromechanical beam resonators. Sens. Actuators A **88**, 263–272 (2001)
206. S.S.J. Pearton, C.C.R. Abernathy, et al., Dry etching of thin-film InN, AlN and GaN. Semicond. Sci. Technol. **8**, 310–312 (1993)
207. R.J. Shul, G.A. Vawter, et al., Comparison of plasma etch techniques for III-V nitrides. Solid-State Electron. **42**, 2259–2267 (1998)

208. H. Cho, C.B. Vartuli, et al., Comparison of inductively coupled plasma Cl2 and Cl4/H2 etching of III-nitrides. J. Vac. Sci. Technol. A **16**, 1631–1635 (1998)
209. G.-K. Lee, J.-H. Moon, B.-T. Lee, Inductively coupled plasma reactive ion etching of ZnO using C2F6 and NF3-based gas mixtures. Semicond. Sci. Technol. **21**, 971–974 (2006)
210. S. Mancha, Chemical etching of thin film PLZT. Ferroelectrics **135**, 131–137 (1992)
211. R. Miller, J. Bernstein, A novel wet etch for patterning PZT thin films. Integr. Ferroelectr. **29**, 225–231 (2000)
212. D.P. Vijay, S.D. Desu, W. Pan, Reactive ion etching of lead zirconate titanate (PZT) thin film capacitors. J. Electrochem. Soc. **140**, 2635–2539 (1993)
213. W. Pan et al., Reactive ion etching of PZT and RuO2 films by environ-mentally safe gases. J. Mater. Res. **9**, 2976–2980 (1994)
214. J. Baborowski et al., Mechanisms of PZT thin film etching with ECR/RF reactor, Int. Ferroelectrics **31**, 261–271 (2001)
215. C.W. Chung, Reactive ion etching of Pb(ZrxTi1-x)O3 thin films in an inductively coupled plasma. J. Vac. Sci. Technol. B **16**, 1894–1900 (1998)
216. S. Koo, D. Kim, K. Kim, S. Song, C. Kim, Etching properties of Pb(ZrxTi1-x)O3lead–zirconate–titanate thin films in an inductively coupled plasma Cl2/Ar and BCl3/Ar gas chemistries. J. Vac. Sci. Technol. B **1622**, 1519–1523 (2004)
217. R.G. Polcawich, Design, fabrication, test, and evaluation of RF MEMS series switches using lead zirconate titanate (PZT) thin film actuators, Ph.D. Thesis, Pennsylvania State University, 2007
218. K. Zheng, J. Lu, J. Chu, A novel wet etching process of Pb(Zr,Ti)O3 thin films for applications in microelectromechanical system. J. Appl. Phys. **43**, 3934–3937 (2004)
219. T. Mineta, Y. Haga, Materials and process in shape memory alloys, Chapter 6, in *MEMS Materials and Processes Handbook*, (Springer, New York, 2011)
220. P. Krulevitch, A.P. Lee, P.B. Ramsey, J.C. Trevino, J. Hamilton, M.A. Northrup, Thin film shape memory alloy microactuators. J. Microelectromech. Syst. **5**(4), 270–281 (1996)
221. K. Kuribayashi, M. Yoshitake, S. Ogawa, Reversible SMA actuator for micron sized robot, in *Proceedings of the IEEE Micro Electro Mechanical Systems. MEMS*, Napa Valley, CA, USA, pp. 217–221, 1990
222. A.D. Johnson, Vacuum-deposited TiNi shape memory film: characterization and applications in microdevices. J. Micromech. Microeng. **1**, 34–41 (1991)
223. S. Miyazaki, A. Ishida, Martensitic transformation and shape memory behavior in sputter deposited TiNi-base thin films. Mater. Sci. Eng. A **273–275**, 106–133 (1999)
224. J.D. Busch, A.D. Johonson, A shape-memory properties in Ni-Ti sputter-deposited film. J. Appl. Phys. **68**(12), 6224–6228 (1990)
225. A. Ishida, S. Miyazaki, Microstructure and mechanical properties of sputter-deposited Ti-Ni alloy thin films. J. Eng. Mater. Technol. **121**, 3–8 (2004)
226. Y. Nakamura, S. Nakamura, L. Buchaillot, M. Ataka, H. Fujita, Tow thin film shape memory alloy microactuators. Trans. IEE Jpn. **117-E**(11), 554–559 (1997)
227. S. Miyazaki, K. Nomura, Development of perfect shape memory effect in sputter-deposited Ti-Ni thin films, in *Proceedings of the IEEE Micro Electro Mechanical Systems (MEMS-94)*, Oiso, Japan, pp. 176–181, 1994
228. K.R. Collen, A.B. Ellis, J.D. Busch, A.D. Johnson, Structural and electrical properties of titanium-nickel films deposited onto silicon substrates, ONR Technical Report UWIS/DC/TR-91/3
229. A. Ohta, S. Bhansali, I. Kishimoto, A. Umeda, Novel fabrication technique of TiNi shape memory alloy film using separate Ti and Ni targets. Sensors Actuators **86**, 165–170 (2000)
230. E. Makino, M. Uenoyama, T. Shibata, Flash evaporation of TiNi shape memory thin film for microactuators. Sens. Actuators A **71**, 187–192 (1998)
231. F.E. Wang, W.J. Buehler, S.J. Pickart, J. Appl. Phys. **36**, 3232 (1965)
232. T.W. Duering, K.N. Melton, D. Stockel, C.M. Wayman, *Engineering Aspects of Shape Memory Alloys* (Butterworth-Heinemann, London, 1990)

233. S. Nakamura, Y. Nakamura, M. Ataka, H. Fujita, A study on patterning method of TiNi shape memory thin film. Trans. IEE Jpn. **117-E**(1), 27–32 (1997)
234. E. Makino, T. Mitsuya, T. Shibata, Micromachining of TiNi shape memory thin film for fabrication of micropump. Sensors Actuators A Phys. **79**(3), 251 (2000)
235. D. Stockel, Status and trends in shape memory technology, in *Proceedings of Actuators'92*, Bremen, Germany, pp. 79–84, 1992.
236. K. Otsuka, C.M. Wayman, *Shape Memory Materials* (Cambridge University Press, Cambridge, 1998)
237. S. Miyazaki, T. Hashinaga, A. Ishida, Thin Solid Films **364**, 281–282 (1996)
238. T.H. Nam, T. Saburi, K. Shimizu, Cu-content dependence of shape memory characteristics in Ti-Ni-Cu alloys. Mater. Trans. JIM **31**, 959–967 (1990)
239. D. Reynaerts, J. Peirs, H.V. Brussel, Shape memory micro-actuation for a gastro-intestinal intervention system. Sens. Actuators A **77**(2), 157–166 (1999)
240. T. Mineta, Electrochemical etching of NiTi shape memory alloy sheet using new electrolyte solutions. J. Micromech. Microeng. **14**, 76–80 (2004)
241. H. Kahn, W. Benard, M. Huff, A. Heuer, The TiNi shape-memory alloy and its application for MEMS. J. Micromech. Microeng. **8**, 213–221 (1998)
242. M. Huff, W. Benard, Thin-film titanium-nickel shape-memory alloy microfluidic devices, in *Proceedings Third International Symposium Microstructures and Microfabricated Systems, Electrochemical Society Meeting*, pp. 26–38, 1997
243. K. Nandakumar, A. Parr, M. Huff, S. Phillips, A smart SMA actuated microvalve with feedback control, in *ASME MEMS*, 1998
244. W.L. Benard, H. Kahn, M.A. Huff, TiNi shape-memory alloy actuated micropump with fluid isolation, in *Smart Structures and Materials '97, Smart Electronics and MEMS*, International Society for Optics and Photonics, pp. 156–164, 1997
245. W.L. Benard, H. Kahn, A.H. Heuer, M.A. Huff, A titanium nickel shape-memory alloy actuated micropump, in *Proceedings of the 9th IEEE International Conference on Solid-State Sensors and Actuators, Transducers 97*, vol. 1, Chicago, IL, pp. 361–364, 16–19 June 1997
246. W. Benard, H. Kahn, A. Heuer, M. Huff, Thin-film shape-memory alloy actuated micropumps. IEEE/ASME J. Microelectromech. Syst. **7**(2), 245 (1998)
247. B. Sutapan, M. Tabib-Azar, M. Huff, Applications of shape memory alloys in MOEMS devices and in optics, in *Micromachining and Microfabrication, Microelectronic Structures and MEMS for Optical Processing IV, International Society for Optics and Photonics*, pp. 223–232, 1998
248. B. Sutapan, M. Tabib-Azar, M. Huff, Applications of TiNi thin films shape-memory alloys in micro-opto-electro-mechanical systems. Sens. Actuators A Phys. **77**(1), 34–38 (1999)
249. B. Sutapan, M. Tabib-Azar, M. Huff, Applications of shape memory alloys in optics. J Appl Opt **37**(28), 6811–6815
250. E.I. Cooper, C. Bonhôte, J. Heidmann, Y. Hsu, P. Kern, J.W. Lam, M. Ramasubramanian, N. Robertson, L.T. Romankiw, H. Xu, Recent developments in high-moment electroplated materials for recording heads. IBM J. Res. Dev. **49**, 103–126 (2005)
251. D.P. Arnold, N. Wang, Permanent magnets for MEMS. J. Microelectromech. Syst. **18**, 1255–1266 (2009)
252. M. Schlesinger, M. Paunovic, *Modern Electroplating* (Wiley, New York, 2000)
253. D.R. Crow, *Principles and Applications of Electrochemistry* (Stanley Thornes (Publishers) Ltd., Cheltenham, 1998)
254. A. Brenner, G.E. Riddell, Nickel plating on steel by chemical reduction, United States Bureau of Standards. J. Res. **37**, 31–34 (1946)
255. G. Zangari, Electro-deposition of alloys and compounds in era of microelectronics and energy conversion. Coatings **5**, 195–218 (2015)
256. Copper Damascene Plating B. Brooks, Semitool Presentation at University of Utah. See: https://www.che.utah.edu/.../Lecture12.1%20Copper%20Deposition%20(SemiTool)/. Retrieved Aug 2018

257. Datasheet on MA/BA6+ Mask and Bond Aligner for SussMicroTec. See: https://www.suss.com/brochures-datasheets/mask-aligner-ma-ba-6.pdf. Accessed Sept 2018
258. Datasheet on Heidelberg Instruments, Model DWL 66+ system. See: https://www.himt.de/. Accessed Sept 2018
259. D. Radtke, U.D. Zeitner, Laser-lithography on non-planar surfaces. Opt. Express. Optical Society of America **15**(3), 1168 (2007)
260. E. DiFabrizio, F. Romanato, M. Gentili, S. Cabrini, B. Kaulich, J. Susini, R. Barret, High efficiency multilevel zone plates for keV X-rays. Nature **401**, 895–898 (1999)
261. P. Ehbets, H.P. Herzig, D. Prongue, M.T. Gale, High-efficiency continuous surface-relief gratings for two-dimensional array generation. Opt. Lett. **17**, 908–910 (1992)
262. P. Yao, G.J. Schneider, D. Prather, E. Wetzel, D. O'Brein, Fabrication of three-dimensional photonic crystals with multilayer photolithography. Opt. Express **13**, 2370–2376 (2005)
263. Y. Olinger, P. Sixt, J.M. Stauffer, J.M. Mayor, P. Regnault, G. Voirin, One-step 3D shaping using a gray-tone mask for optical and microelectronic applications. Microelectron. Eng. **23**, 449–454 (1994)
264. V.P. Korolkov, A.I. Malyshev, A.G. Poleshchuck, V.V. Cherksshin, H.J. Tiziani, C. Pruss, T. Schoder, J. Westhauser, C. Wu, Fabrication of gray-scale masks and diffractive optical elements with LDW glass, in *Proceedings of SPIE – Lithographic and Micromachining Techniques for Optical Component Fabrication*, vol. 4440, pp. 73–84, 2001
265. Y. Chen, R.K. Kupka, F. Rousseaux, F. Carcenac, D. Decanini, M.F. Ravet, H. Launois, 50-nm x-ray lithography using synchrotron radiation. J. Vac. Sci. Technol. B **12**, 3959–3964 (1994)
266. W. Ehrfeld, P. Bley, F. Götz, P. Hagmann, D. Münchmeyer, J. Mohr, O.H. Moser, D. Münchmeyer, W. Schelb, D. Schmidt, E.W. Becker, Fabrication of microstructures using the LIGA process. Presented at Micro Robots and Teleoperators Workshop, IEEE, 1987
267. A.R. Shimkunas, P.E. Mauger, L.P. Bourget, R.S. Post, L. Smith, R.F. Davis, G.M. Wells, F. Cerrina, R.B. McIntosh, Advanced electron cyclotron resonance chemical vapor deposition SiC coatings and x-ray mask membranes. J. Vac. Sci. Technol. B: Microelectron. Nanometer Struct. **9**, 3258 (1991)
268. G.M. Wells, S. Palmer, F. Cerrina, A. Purdes, B. Gnade, Radiation stability of SiC and diamond membranes as potential x-ray lithography mask carriers. J. Vac. Sci. Technol. B **8** (6), 1575–1578 (1990)
269. V. Saile, U. Wallrabe, O. Tabata, G.K. Fedder, *LIGA and Its Applications, Volume 7 of Advanced Micro and Nanosystems* (Wiley-VCH, Weinheim, 2008)
270. J. Goettert, *From Design to LiGA Polymer Chips: A Customer Experience, Presentation* (Louisiana State University, Baton Rogue, 2005)
271. M. Gad-el-Hak, *MEMS: Design and Fabrication*, 2nd edn. (CRC Press, Boca Raton, 2006)
272. H. Guckel, Deep x-ray lithography for micromechanics and precision engineering. Rev. Sci. Instrum. **67**, 3357 (1996)
273. Y. Liu, P. Sheng, Modeling of X-ray fabrication of macromechanical structures. J. Manuf. Process. **4**, 109 (2002)
274. H. Guckel, T.R. Christenson, K.J. Skrobis, D.D. Denton, B. Choi, E.G. Lovell, J.W. Lee, S.S. Bajikar, T.W. Chapman, Deep X-ray and UV lithographies for micromechanics. Presented at Solid-State Sensor and Actuator Workshop. 4th Technical Digest, IEEE, 1990
275. M.A. McCord, M.J. Rooks, *Electron Beam Lithography, PM39* (SPIE Publications, Bellingham, 1997)
276. N.P. Pham, T.L.M. Scholtes, R. Klerks, E. Boellaard, P.M. Sarro, J.N. Burghartz, Direct spray coating of photoresist- a new method for patterning 3-D structures, in *Proceedings of EUROSENSORS 2002*
277. V.G. Kutchoukov, Fabrication technology for through-wafer interconnects, PhD thesis, Delft 2002, ISBN 90-5166-897-X
278. P. Kersten, S. Bouwstra, J.W. Petersen, Photolithography on micromachined 3D surfaces using electrodeposited photoresists. Sensors Actuators **A51**, 51–54 (1995)

279. E. Alling, C. Stauffer, Image reversal of positive photoresist, in *SPIE Proceedings, 539 Advances in Resist Technology & Processing*, 1985
280. M.L. Long, J. Newman, Image reversal techniques with standard positive photoresist. Proc. Soc. Photo Opt. Instrum. Eng. **469**, 189–193 (1984)
281. M. Madou, *Fundamentals of Microfabrication and Nanotechnology* (CRC Press, Boca Raton, 2011)
282. S. Sze, *Semiconductor Sensors* (Wiley, New York, 1995)
283. M. Gad-el-Hak, *The MEMS Handbook* (CRC Press, Boca Raton, 2002)
284. D.W. Burns, MEMS wet etch processes and procedures, in *MEMS Materials and Processes Handbook*, (Springer, New York, 2011)
285. W.K. Zwicker, S.K. Kurtz, Anisotropic etching of silicon using electrochemical displacement reactions, in *Semiconductor Silicon*, ed. by H. R. Huff, R. R. Burgess, (Electrochemical Society Press, Princeton, 1973), p. 315
286. M. Huff, A thermally isolated microstructure for gas sensing applications, S.M. thesis, Dept. EECS and MS&E, MIT, 1988
287. G.-Z. Yan, P.C.H. Chan, I.-M. Hsing, R.K. Sharma, J.K.O. Sin, An improved TMAH Si-etching solution without attacking exposed aluminum, in *IEEE Proceedings 13th Annual International Conference on MEMS*, Miyazaki, JP, Jan. 2000
288. H. Seidel, L. Csepregi, A. Heuberger, H. Baumgartel, Anisotropic etching of crystalline silicon in alkaline solutions, I. Orientation dependence and behavior of passivation layers. J. Electrochem. Soc. **137**(11), 3612–3626 (1990)
289. H. Seidel, L. Csepregi, A. Heuberger, H. Baumgartel, Anisotropic etching of crystalline silicon in alkaline solutions, II. Influence of dopants. J. Electrochem. Soc. **137**(11), 3626–3632 (1990)
290. B. Kloech, S.D. Collins, N.F. de Rooij, R.L. Smith, Study of electrochemical etch-stop for high precision thickness control of silicon membranes. IEEE Trans. Electron Devices **ED-36**, 663 (1989)
291. V.M. McNeil, S.S. Wang, K.Y. Ng, M.A. Schmidt, An investigation of the electrochemical etching of (100) silicon in CsOH and KOH, in *Tech. Dig. IEEE Solid-State Sensor and Actuator Workshop*, Hilton Head, SC, p. 92, 1990
292. G.-S. Chung, Anisotropic etching and electrochemical etch-stop properties of silicon in TMAH:IPA:Pyrazine solutions. Met. Mater. Int. **7**(6), 643 (2001)
293. P.B. Chu, J.T. Chen, R. Yeh, G. Lin, C.P. Hunag, B.A. Warneke, K.S.J. Pister, in *Proceedings of international solid state sensors and actuators conference (Transducers' 97)*, p. 665, 1997
294. F. Larmar, P. Schilp, Method of anisotropically etching of silicon, German Patent DE 4,241,045, 1994
295. J. Bhardwaj, H. Ashraf, Advanced silicon etching using high density plasmas, in *Proc. SPIE, Micromachining and Microfabrication Process Technology Symp.*, vol. 2639, Austin, TX, p. 224, 23–24 Oct 1995
296. M.D. Henry, C. Welch, A. Scherer, Techniques of cryogenic reactive ion etching in silicon for fabrication of sensors. J. Vac. Sci. Technol. A **27**(5), 1211 (2009)
297. J. Yeom, Y. Wu, J.C. Selby, M.A. Shannon, Maximum achievable aspect ratio in deep reactive ion etching of silicon due to aspect ratio dependent transport and the microloading effect. J. Vac. Sci. Technol. B.: Microelectron. Nanometer Struct. Process. Meas. Phenom. **23**, 2319 (2005)
298. T.F. Hill, Analysis of DRIE uniformity for microelectromechanical systems, S.M. Thesis, MIT, Department EECS, May 2004
299. B.H.W.S. De Jong, R.G.C. Beerkens, P.A. Van Nijnatten, Glass, in *Ullmann's Encyclopedia of Industrial Chemistry*, (Wiley-VCH, Weinheim, 2000)
300. S. Wang et al., Optimized condition for etching fused-silica phase gratings with inductively coupled plasma technology. Appl. Opt. **44**(21), 4429–4434 (2005)
301. M. Pedersen, M. Huff, Plasma etching of deep, high-aspect ratio features into fused silica. IEEE J. MEMS (JMEMS) **26**(2), 448 (2017)
302. M. Pedersen, M. Huff, Based on research results that have not yet been reported.

303. M. Pedersen, M. Huff, Across substrate lateral dimensional repeatability using a highly-anisotropic deep etch process on fused silica material layers. IEEE J. MEMS (JMEMS) **27** (1), 31 (2018)
304. M. Ozgur, M. Huff, Plasma etching of deep, high-aspect ratio features into silicon carbide (SiC). IEEE J. MEMS (JMEMS) **26**(2), 456 (2017)
305. M. Ozgur, M. Huff, Based on research results that have not yet been reported.
306. N. Tas, T. Sonnenberg, H. Jansen, R. Legtenberg, M. Elwenspoek, Stiction in surface micromachining. J. Micromech. Microeng. **6**, 385–397 (1996)
307. M. Huff, A.F. Bart, P. Lin, MEMS process integration, Chapter 14, in *MEMS Materials and Processes Handbook*, (Springer, New York, 2011)
308. M. Huff, M.A. Schmidt, Fabrication, packaging, and testing of a wafer-bonded microvalve, in IEEE Solid-State Sensor and Actuator Meeting, Hilton Head, SC, June 22–25, 1992
309. M. Huff, Silicon micromachined wafer-bonded valves, Ph.D. Thesis, MIT, Department EECS, Feb 1993
310. G.-L. Sun et al., Cool plasma activated surface in silicon wafer direct bonding technology. J. Phys. **9**, 79–82 (1988)
311. Y. Kanda et al., The mechanism of field-assisting silicon-glass bonding. Sens. Actuators **A21–A23**, 939 (1992)
312. A.L. Tiensuu, J.A. Schweitz, S. Johansson, In situ investigation of precise high strength micro assembly using Au-Si eutectic bonding, in *8th International Conference on Solid-State Sensors and Actuators, Transducers 95*, Stockholm, Sweden, p. 236, June 1995
313. C. den Besten, R.E.G. van Hal, J. Munoz, P. Bergveld, Polymer bonding of micromachined silicon structures, in *Proceedings of the IEEE Micro Electro Mechanical Systems, MEMS 92*, Travemunde, Germany, p. 104, 1992
314. S.J. Cunningham, M. Kupnik, Wafer bonding, Chapter 11, in *MEMS Materials and Processes Handbook*, (Springer, New York, 2011)
315. W. Ehrfeld, P. Bley, F. Gotz, P. Hagmann, A. Maner, J. Mohr, H.O. Moser, D. Munchmeyer, W. Schelb, D. Schmidt, E.W. Becker, Fabrication of microstructures using the LIGA process, in *Proc. IEEE Micro Robots and Teleoperators Workshop*, Hyannis, MA, Nov 1987
316. W. Menz, W. Bacher, M. Harmening, A. Michel, The LIGA technique – a novel concept for microstructures and the combination with Si-technologies by injection molding, in *IEEE Workshop on Micro Electro Mechanical Systems, MEMS 91*, p. 69, 1991
317. M. Huff, System and method for precision fabrication of micro- and nano-devices and structures, United States Patent No. US8790534B2, granted 29 July 2014
318. A.A. Tseng, Recent developments in micromilling using focused ion beam technology. J. Micromech. Microeng. **14**, R15–R34 (2004)

Chapter 5
Metrology for Microsystems Manufacturing

5.1 Introduction

Microsystems fabrication utilizes a number of metrology techniques and instruments during development and manufacturing. These techniques are used in development to find and properly diagnose design and fabrication problems that may be present. In manufacturing, metrology is employed to measure important process results and maintain quality control of the microsystems being made and thereby increase production yields and lower costs. Metrology also enables the process engineers to monitor the status of the processing equipment used in production.

Although metrology tools and methods are useful in performing forensics in fabrication development activities, it should also be noted that they often do not always provide clear or definitive information. Often, the determination of the cause (s) of a problem requires considerable inference. Therefore, experienced fabrication engineers use these techniques judiciously.

There are a wide variety of metrology tools that are available, including inspection of the devices during and after fabrication; measurement of the dimensions of various important elements of the microsystems; and chemical analysis of materials used in fabrication.

The most important metrology techniques in MEMS microsystems manufacturing will be reviewed in this chapter. We will also provide some guidance as the accuracy of each of the methods discussed. It is important to note that there are basic metrology techniques that have been developed for the IC industry that are also widely used in the manufacturing of MEMS microsystems. In addition to the techniques commonly used in the IC industry, there have emerged several more specialized metrology tools and techniques that are employed for development and manufacturing of MEMS microsystems that will also be covered.

M. Huff, *Process Variations in Microsystems Manufacturing*, Microsystems and Nanosystems, https://doi.org/10.1007/978-3-030-40560-1_5

5.2 Fabrication Metrology Equipment and Methods

In this section we cover the widely employed metrology methods and tools that are mostly used during production.

5.2.1 *Optical Microscopy*

The most commonly used metrology tool in MEMS manufacturing is the optical microscope. These systems use white light illumination that is either reflected from or transmitted through samples during inspection and magnified using an optical lensing system. Optical microscopes allow for very fast, simple, and nondestructive inspection of devices and substrates during and after fabrication and, if properly instrumented and calibrated, can also be used to make dimensional measurements in the x-, y-, and z-directions.

There are several considerations that are important for optical microscopes including magnification; resolution; and numerical aperture. Additionally, there are several types of optical microscope capabilities that are used in different situations that include bright field; dark field; transmitted field; and differential interference contrast or Nomarski.

The resolution of an optical microscope is given by the ability of the instrument to discern closely spaced features. In general, the resolution, *r*, is given by:

$$r = \frac{\lambda}{2(Na)}, \tag{5.1}$$

where λ is the wavelength of the illumination and *Na* is the numerical aperture of the objective lens of the microscope used [1].

The resolution of a typical optical microscope using white light illumination and having a *Na* of 0.95 will be about 0.25 microns. The magnification of a microscope is defined as the ratio of the size of the object's image after magnification compared to the actual size of the object. As a general rule, the maximum magnification of an optical microscope is about 1000×. The maximum magnification is set by the resolution of the human eye, which is about 0.25 mm, and therefore, at 1000×, the smallest feature is about 0.25 microns. However, in practice the minimum resolution of an optical microscope in a manufacturing environment is usually specified as about 0.5 microns.

Most commercially available microscopes are compound microscopes, which are illustrated in Fig. 5.1. There are two basic lens elements: the first is the objective lens that collects light from the object being observed and magnifies the image, and the second is the eyepiece lens that provides for additional magnification than the objective lens would provide alone. These lens elements working in conjunction are able to provide a higher level of magnification that is possible with a single-lens

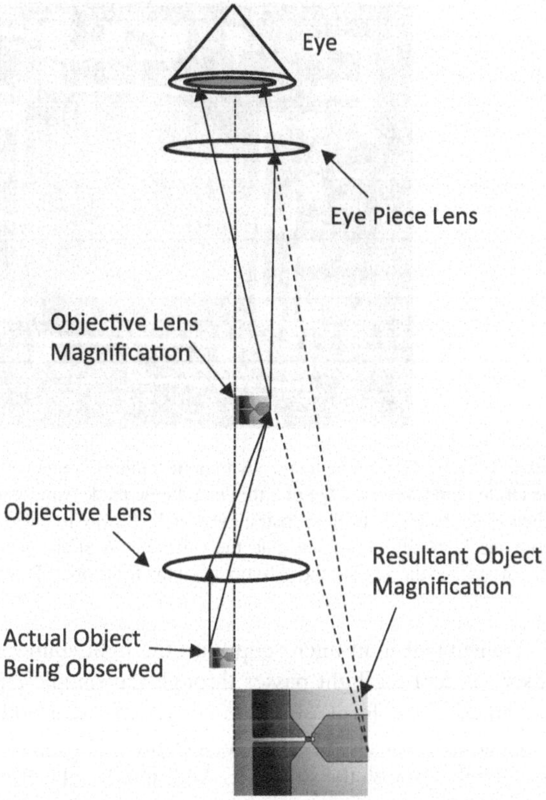

Fig. 5.1 Illustration of the optical setup for a compound microscope that is commonly used in microsystems manufacturing. The total magnification is given by the product of the magnification of the objective and eyepiece lenses

microscope. Additionally, a compound microscope suffers from less chromatic aberration than a single-lens microscope. A compound microscope configuration also allows the magnification to be easily and quickly changed by swapping the lens elements in the optical path.

Compound microscopes in microsystems development and manufacturing facilities often are outfitted with multiple objective lens (typically three or four objective lenses) in some combination of the increments of 2×, 10×, 20×, and 50×. The eyepiece lenses are typically 10× or 20×. The total magnification is given by the product of the magnification of the objective and eyepiece lenses and usually ranges from 20× to 1000×.

There are several possible modes of a compound microscope. In bright-field mode of microscopy, the substrate is observed using reflected white light. A light source illuminates the area on the substrate under observation wherein the light travels down the optical path and through the objective lens to impinge on the sample surface. The light that strikes the sample surface is reflected back along the same optical path wherein the image is magnified. This is the most commonly used method of optical microscopy.

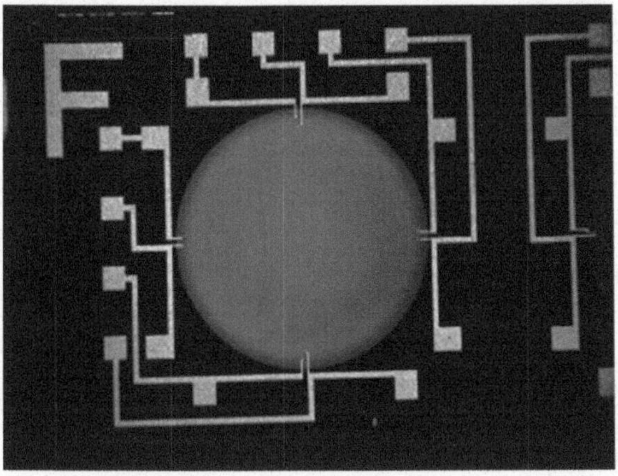

Fig. 5.2 Optical photograph taken with optical microscope in transmission field mode. The edge of the circle represents the edge of a thin membrane made using micromachining fabrication methods wherein the backside of the substrate was etched away to leave a thin membrane of silicon that is more mechanically compliant than the substrate. A strain gauge, with electrical connections, is shown made at the edge of the membrane on the front side. This device was fabricated by the MNX

Transmitted light microscopy has the light source underneath the sample being observed, and the light passes through the sample and into the optical path to be magnified. This form of microscopy is very useful for inspection of optically transparent substrates as well as substrate having features made in it that extend completely through the substrate. Additionally, if a thin membrane on a substrate has been fabricated, transmitted light microscopy may be useful for observation if the membrane is sufficiently thin to allow adequate light to pass through (Fig. 5.2).

Dark-field microscopy uses a light source to illuminate the sample surface at an oblique angle that is outside the optical path. Only light that strikes surface features is reflected back into the microscope's objective lens and therefore seen by the operator. The magnified image is a black background with very visible surface features. Dark-field microscopy enables some features to be visible that otherwise would not be able to be seen in bright-field microscopy.

Differential interference contrast (DIC), or Nomarski microscopy, employs a light source that is split into two separate optical illumination beams that impinge on the substrate surface. These two beams impinge on the surface and are reflected back through the optical path where they are combined. If one of the beams strikes a feature on the substrate surface having a different height from the height of the surface where the other beam strikes, then each beam will have a different optical path length, thereby resulting in a slight phase differential between the two beams. Once these two beams are combined, the differential phases will produce interference effects that result in differing contrasts (different colors and shades of gray) for the features having different heights across the substrate surface. DIC microscopy allows adjustment of the polarizer, analyzer, and prism in the optical path to maximize the contrast effects. Using DIC, differences in heights across the surface as small as 50 Angstroms can be seen. This mode of microscopy is very useful for viewing very small features and defects on the substrate surface that would be otherwise unable to be seen using bright-field microscopy.

Optical microscopes are used not only for inspecting substrates but also for measuring the dimensions of features. The precision of performing dimensional measurements depends on the skill of the operator and the size of the object being measured. For larger objects (e.g., tens of microns or more), the precision can be quite good with a spread in measured values of less than a few %. For smaller objects (e.g., a few microns), the precision will not be as good and the spread in measured values may be tens of %. Importantly, taking more measurements and using statistics to analyze the measurements can increase the precision.

The accuracy of measuring the dimensions of features using an optical microscope depends on a number of issues. First, the microscope must be equipped with a reticule (i.e., measurement or ocular scale) in the eyepiece. Second, the reticule must be properly calibrated. Calibration is performed using a stage micrometer, which is a form of standard. Third, the calibration will need to be done with the magnification to be used for the measurement. Lastly, the accuracy of a measurement is also highly dependent on the sharpness of the edges of the features and the contrast of the features being measured.

The reticule (or ocular scale) is a gauge built into the eyepiece of microscopes and allows the measurement of lateral dimensions in the field of view of the microscope. It can be rotated in order to measure the dimension at any angle, such as the x- and y-directions. The dimension of an object in the field of view of the microscope is determined by the number of ocular divisions on the scale spanned by the object multiplied by a calibration factor. The standard used for calibrating the ocular scale is called a stage micrometer scale, which is a microscope slide having a measurement scale etched into the surface of the slide. A micrometer scale may be a few millimeters in length with equally spaced divisions along the length.

With respect to the accuracy of measurements using an optical microscope, it is first important to note that the accuracy cannot be better than the resolution. Therefore, if the microscope has a resolution of 0.5 micron, then the accuracy is limited to being no better than that. Since the resolution improves with the higher magnification, more accurate measurements are taken using the highest possible magnification.

5.2.2 *Fluorescence Microscopy*

A fluorescence microscope is usually composed of a modified optical compound microscope setup that uses an ultraviolet light source to illuminate the substrate sample rather than a white light source. A fluorescence microscope configuration is illustrated in Fig. 5.3. The illumination from the light source passes through an excitation filter to narrow the range of wavelengths illuminating the sample, particularly if the light source is broadband. After passing through the excitation filter, a

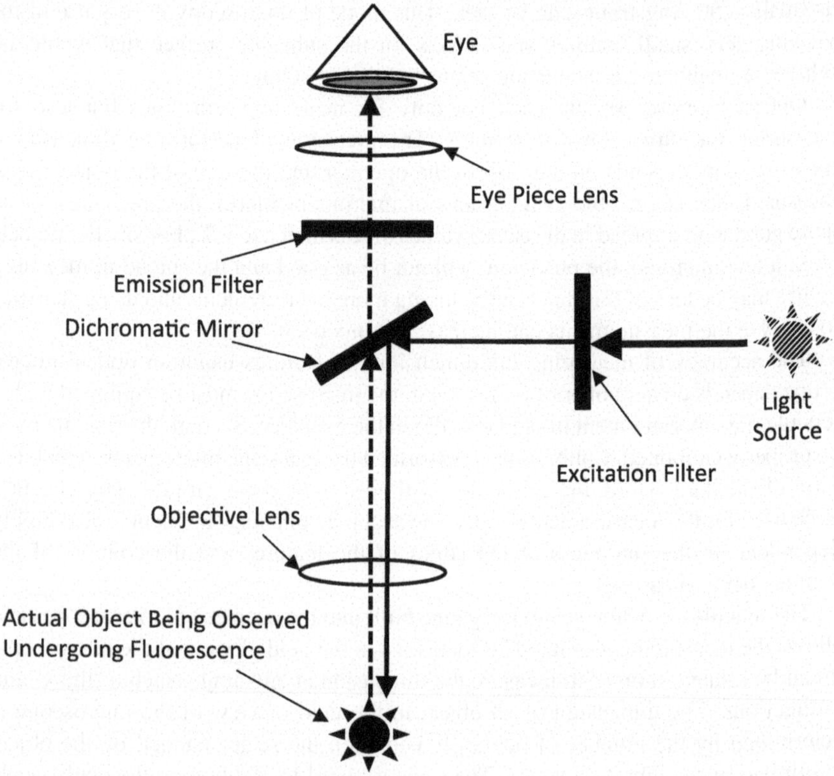

Fig. 5.3 Illustration of the optical setup for a fluorescence microscope that is used to detect the presence of photoresist, as well as other polymers that are used during microsystems manufacturing

dichromatic mirror reflects the illumination into the objective lens and then onto the object. The object may undergo a fluorescence wherein the illumination light wavelength causes the object to emit radiation at a different wavelength. The radiation emitted from the sample passes through the objective lens and the dichromatic mirror, through an emission filter and through the eyepiece lens where it can be observed.

Organic material when illuminated with an ultraviolet source will emit light in the visible portion of the spectrum that can be detected through the microscope. Fluorescence microscopy is very useful for inspecting substrates after lithography has been performed on photoresist to ensure that the resist is completely gone from the exposed and developed regions on the substrate surface. It can also be used to determine if masking photoresist is left after a long etching process [2].

Similar to optical microscopy, the resolution of fluorescence microscopy depends on the lens and eyepiece magnification, the calibration, and the skill of the operator; however, fluorescence microscopy is more often used for inspection rather than for measurement of the dimensions of features.

5.2.3 Confocal Microscopy

Confocal microscopy is an optical magnifying method that allows for increased resolution and contrast. An issue with microscopy is that the entire sample is illuminated with the light source, and therefore the entire surface reflects light backward. As a result, not only will the sample under observation at the microscope's focal point reflect, but the entire sample will also reflect, thereby resulting in a significant amount of unfocused background light from across the sample also being observed. The result is an image with reduced resolution and contrast. The confocal microscope is an optical setup that avoids this problem by inserting a screen at the focal point that has a spatial pinhole that eliminates the emitted light that is outside of the focal point from being observed. Most confocal microscopes employ a set of scanning mirrors that allow the entire sample to be scanned and a photodetector to collect images.

An illustration of a confocal microscope setup is shown in Fig. 5.4. As can be seen, there are two spatial pinholes inserted into the optical paths, one in the

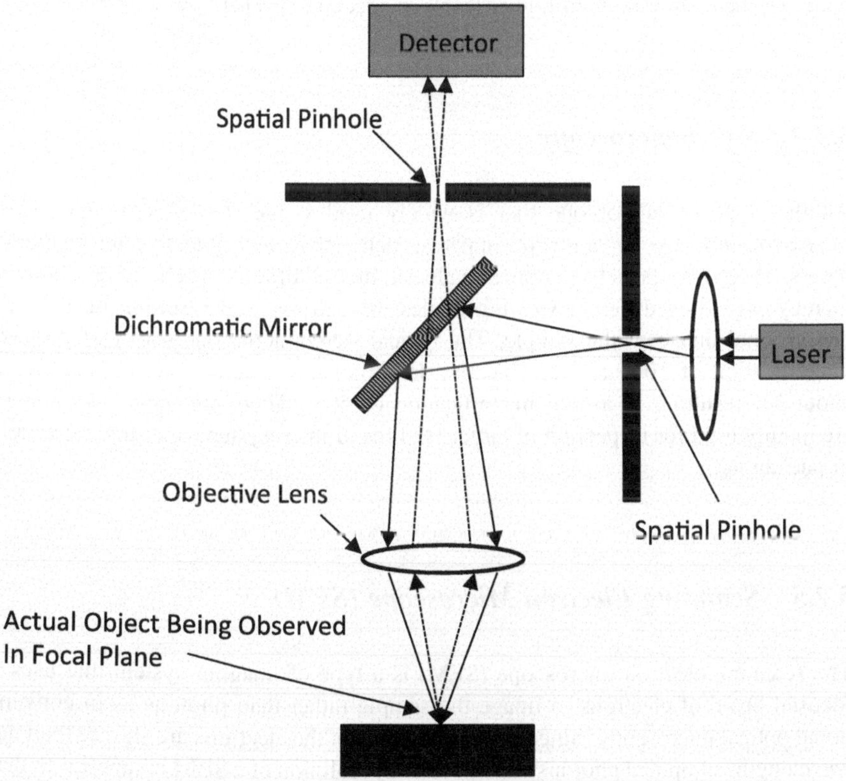

Fig. 5.4 Illustration of the optical setup for confocal microscopy that is used in microsystems manufacturing. The important feature of the confocal microscopy is that any light reflected from the object being observed that is not in the focal plane will be prevented from reaching the detector and therefore background light is eliminated from the image

illumination optical path between the light source (labeled as a "laser" in Fig. 5.4) and the dichromatic mirror and the other between the detector and the dichromatic mirror. The emitted radiation from the laser passes through a lens element and then through the spatial pinhole that prevents any out-of-focus radiation from passing through the screen. This radiation that is able to pass through the spatial pinhole reaches the dichromatic mirror and is reflected toward the object being observed. Before reaching the object, this radiation is focused using an objective lens. The focused radiation impinges onto the object, and any reflected light (or fluorescence emitted) passes through the objective lens, through the dichromatic mirror, and, if in focus, will pass through the second spatial pinhole onto the detector. Any light or emitted fluorescence from the object under observation out of focus will be prevented from reaching the detector.

Confocal microscopy is used to inspect very small point defects on the surfaces of substrates and for device failure analysis in microsystems manufacturing. The resolution of confocal microscopy depends on the system being used and the operator skill. In general, the actual resolution is lower than the theoretical limit. A typical commercial equipment vendor quoted resolution is around 150 nm [3], although there are reports of higher levels in special cases [4].

5.2.4 *Stereomicroscopy*

Another type of microscope that is widely used is the stereomicroscope. This microscope uses two separate optical paths, each with its own objective and eyepiece lenses. These two paths have optical paths slightly at different angles to the sample, thereby providing different viewing angles to each eye and resulting in a three-dimensional image of the sample. The typical stereomicroscope uses an eyepiece magnification of 10× and an objective lens that has a range with maximum of about 5× resulting in a total magnification of 50×. Therefore these instruments are mainly used for inspection of larger features of microsystems and for inspection of packaging.

5.2.5 *Scanning Electron Microscope (SEM)*

The scanning electron microscope (SEM) is a type of imaging system that uses a focused beam of electrons to image the sample rather than photons as in conventional optical microscopy. Since the wavelengths of the electrons are shorter than the wavelengths of optical photons, the maximum resolution of a SEM is superior to that of an optical microscope. Additionally, the SEM provides considerably enhanced depth of focus capability compared to an optical microscope, thereby allowing the imaging of features having depth without having to adjust the focusing.

For imaging purposes, the electron scanning beam is synchronized with the detector signal, in both time and position, so as to form a mapping of the intensity of the signal emitted and thereby form an image of the surface. A SEM provides a three-dimensional perspective of the surface with the entire image field in focus since there is not the depth of focus issues as in optical microscopy. Additionally, the sample can be tilted and rotated along multiple axes that can be used to greatly enhance the quality of the resultant image. In metrology applications where the dimensions are quite small, critical dimension (CD) inspections are performed using a SEM placed orthogonally to the feature being examined.

SEMs are used for both development and production. As a development tool, the SEM enables inspection and measurement with a resolution that is not possible with optical microscopy. Additionally, the larger depth-of-focus capability of the SEM enables the examination of how various layers are aligned and stacked onto one another, device elements are positioned, etc. that is extremely valuable and cannot be obtained using any other means. For production, the SEM can be invaluable for inspection and/or measurement of critical features. Given the feature sizes of critical dimensions of advanced ICs, the SEM is the only capability for process inspection since optical microscopy does not have adequate resolution. Therefore, the automated SEM is a standard in-process inspection tools in these production environments. In MEMS and other microsystems where the feature sizes can mostly be imaged using optical microscopy, SEMs are not as commonly used for production monitoring.

In general, SEMs are capable of magnification from about 10× to 500,000× or more, and resolutions of up to about 1 nm are possible [5]. Nevertheless, in practice and more realistic manufacturing environments, the resolution may be a few nanometers.

The accuracy of performing dimensional measurements using a SEM is a very complicated issue with a number of effects that can impact the result. Similar to many metrological technologies, the accuracy of the SEM will not be better than the resolution. Therefore, if the resolution is 1 nm, then the accuracy of a measurement will be constrained by this value.

Additionally, as a general rule, the accuracy of a SEM for use in dimensional measurement will be improved at higher magnifications compared to lower magnifications. More exactly, the maximum resolution will only be obtained at high magnification levels. Nevertheless, it may not be possible when performing measurements on some microsystems, such as technologies with larger dimensioned features as seen in MEMS, to use a higher magnification and still maintain the entire feature in the image. Therefore, it will be important to know the relationship between the resolutions at a usable magnification level rather than at the maximum magnification since this will impact the achievable measurement accuracy.

The images formed using a SEM are prone to several types of distortions, including spatially varying distortions and time-varying distortions. Most modern SEMs have digital image correlation software to enable these distortions to be partially corrected if used properly. However, proficient use of these software packages requires significant time commitments, and even then it has been argued that these capabilities introduce other sources of errors [6].

Additionally, it is important to note that there are different types of SEMs, some of which are specifically designed for automated accurate measurements of dimensions (often called "CD-SEMs" with "CD" representing "critical dimension") and others that are more multipurpose systems that allow a large degree of flexibility. We will discuss the CD-SEM capability in the next section.

Most SEMs include the capability of a measurement scale that can be superimposed onto the image of the feature being observed. However, in order to be accurate, this scale must be calibrated for the specific SEM parameters to be used in the actual measurement. There are calibration standards available for this purpose (e.g., NIST SRM 484).

The calibration procedures provided by the manufacturer are typically specific to a specific tool. Importantly, the accuracy of these standards obviously has an impact on the resultant calibration accuracy that can be obtained. A typical uncertainty of a standard is between +/−1 and 3%. Therefore, any measurement of dimensions using a SEM will not be better than this level of accuracy.

A standard multipurpose SEM (i.e., not an automated CD-SEM) allows adjustment of a number of parameters including viewing angle, accelerating voltage, and magnifying level over a large range of values that may impact the accuracy of a dimensional measurement. For example, attempting to measure a feature at an off-angle requires the user to accurately know the angle of viewing (which can be difficult to know with high precision) and employ geometry to calculate the actual dimension of the feature. Consequently, when attempting to make accurate dimensional measurements, it is advisable to orient the sample to be orthogonal to the beam.

Measurement of dimensions also depends on being able to correctly identify the edges of the features. This can be difficult since the feature may have a rounded topography, rough edges, edges that are not straight, etc.

The materials and material combinations imaged with the SEM also have an impact on the accuracy. Some materials and material combinations provide a sharper image than others.

Consequently, given all of the factors involved, it is very difficult to provide a generic number of the accuracy of a SEM. Suffice it to say that if everything is perfect, a measurement performed at the highest magnification, with the calibration performed correctly, and a sample oriented normally to the beam, the accuracy can be on par with the resolution of the SEM [7].

Also, the interaction of the impinging electrons with the surface results in the characteristic X-rays to be produced. The resultant X-rays will have energy and intensity levels that are a function of the elements in the surface impinged upon and the relative amount of these elements. Therefore, SEMs can be equipped with an energy-dispersive X-ray spectroscopy (EDS) that can be used to provide information about the chemical composition of the material surface as a function of location. This is extremely useful for the detection of the presence of elemental materials, and many SEMs are equipped with this analysis capability. This is

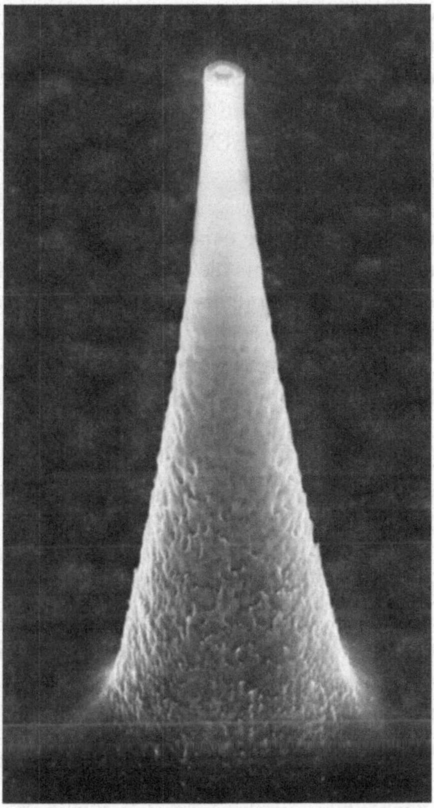

Fig. 5.5 An example of a SEM image of a nano-dimensional-scale field emitter device. This is a tall, high-aspect ratio pillar fabricated from silicon using a deep reactive ion etching (DRIE) into a single-crystal silicon substrate. The pillar is approximately 10 microns in height and has a diameter at the top of approximately 500 nm. On top of the silicon pillar is a thin-film layer of patterned titanium nitride, having a diameter and thickness of approximately 300 nm and 50 nm, respectively. Additionally, on top of the patterned TiN layer is a thin-film layer of patterned nickel having a diameter of 100 nm and a thickness of approximately 10 nm

described in more detail in a later section. Figure 5.5 is a SEM image of a field emission tip that was fabricated by the MNX.

5.2.6 Automated Scanning Electron Microscope

The use of scanning electron microscopes for inspection and measurement of microsystem devices is an extremely valuable tool for performing forensics and diagnostics on patterned substrates during development, especially when the features to be inspected are too small for optical microscopy. However, for routine process monitoring and inspection during integrated circuit microsystems manufacturing, the multipurpose SEM capability described above wherein the inspections and measurements are performed manually is often too time-consuming and expensive. Therefore, SEM manufacturers have responded to this need by developing automated SEMs for high-throughput inspection of patterned substrates. The systems work by imaging a known and quantified pattern on the substrate at multiple locations and using specialized software to recognize defects in these patterns and classify these defects. The analysis software in these tools can be used to map out these defects on the substrate surfaces, run statistics on these defects, and even classify the defects as short circuits, open circuits, dimensional violations, etc. These are sometimes called "defect SEMs."

The automated SEM must be pre-programmed so as to recognize what constitutes a pattern violation. Additionally, each time a new pattern or process is to be inspected using these systems, the programming must be revised accordingly, and these revisions are expensive in terms of both time and cost. Therefore, due to the up-front costs of these automated systems, they are most useful when the manufacturing foundry is producing large volumes of devices using a limited number of process technologies and designs.

Commercially available automated SEM defect review and classification systems have the capability of detecting and classifying defects at the $1\times$ nm design node. The accuracy of CD measurements depends on the calibration of the system as well as the size of the sampling areas and number of sampling points within the area, particularly when the lines being measured have line-edge roughness (LER) and line-width roughness (LWR) [8].

Additionally, SEM manufacturers have also developed specialized tools called CD-SEMs that are used for the measurement of critical dimensions with high substrate throughputs. For example, these tools could be used for measuring and ensuring that the gate lengths on transistors in an advanced integrated circuit manufacturing process were fabricated according to the desired design values. The CD-SEM uses specialized software to analyze the image and determine the dimensions.

These achieve a high level of precision (i.e., reproducibility in the measured values) by a number of constraints including having the beam aligned normally to the substrate and the use of a previously set of tool parameters and a proper alignment and calibration of the tool. Most CD-SEMs specify a CD measurement repeatability of about 1% of the measurement dimension within 3σ.

Importantly, precision in the measurements is not the same as accuracy. If the tool is able to provide measurement precision, the accuracy can be adjusted by using a measurement standard to ensure that the measurement value coincides with the true value. Assuming all of the above procedures have been performed, then the accuracy of the CD-SEM can approach the resolution of the tool.

It is important to note that these automated SEMs (whether defect or CD), while useful for high-throughput manufacturing environments, do not have the flexibility of conventional manual and multipurpose SEMs and therefore would not be useful in process development activities, when the microsystems being produced have critical features that vary over a large range of values, when the critical features are very large (as compared to a typical gate length), or when the types of devices being made are not amenable to inspection or measurement from a normal incident beam angle. Defect and CD-SEMs are highly automated (and expensive) systems that are designed specifically for IC manufacturing and would only be appropriate for MEMS manufacturing when the production volumes are large.

5.2.7 *Thin-Film Thickness*

The measurement of the thicknesses of thin films is one of the most often-used metrology methods in microsystems manufacturing. As discussed in Chaps. 3 and 4, thin-film deposition and growth processes, such as thermal oxidation; CVD; PVD; and others, are used extensively for implementation of microsystems devices. The techniques used for these depositions and growths do not result in thin-films having perfectly uniform thicknesses across the substrates, as well as from substrate-to-substrate and from lot-to-lot. Therefore, it is important to determine if the film thickness has the correct value and amount of non-uniformity that is present after each deposition is performed. Measurements of the thicknesses of thin films are a standard practice after nearly every thin-film growth or deposition. Additionally, it is common to take a number of measurements across the substrate surface in order to determine how the thickness varies over the surface. If a sufficient number of measurements are taken, a mapping of the distribution of the thickness across the substrate can be generated to provide extremely useful information about the process, the equipment, and how to improve the performance of the process. In some process steps, every substrate will be measured, while in other processes some representative number of sample substrates will be measured. The process engineers must determine the number of measurements that are required for maintaining quality while also balancing the costs involved.

5.2.7.1 Interferometry

Interferometry is a fast, nondestructive, noncontact, and accurate method that can be used to measure the thicknesses of thin films. Typically, a light source is focused onto the surface of the thin film on a substrate, and the reflected optical signal is captured in a photodetector (see Fig. 5.6). Multiple reflections will occur inside the thin film that cause some of these light waves to recombine with other light waves reflecting from the surface at different locations to experience either constructive or destructive interference according to the phase differences in these waves [1].

The phase differences and resultant constructive or destructive interference effects are determined by the wavelength of the light source and optical path length (given by the distance that the light moves back and forth in the thin film multiplied by the film's refractive index). The reflected signal from the sample therefore becomes a particular spectrum depending on the film thickness. The angle and/or the wavelength of the light source may be varied in order to take multiple measurements and improve measurement accuracy. Most often, the wavelength of the incident light is varied in the range from about 480 nm to 790 nm, and the incident angle remains constant.

These maximum and minimum reflected energies can be calculated or by the use of a fringe chart. A fringe chart plots the thin-film thickness as a function of the

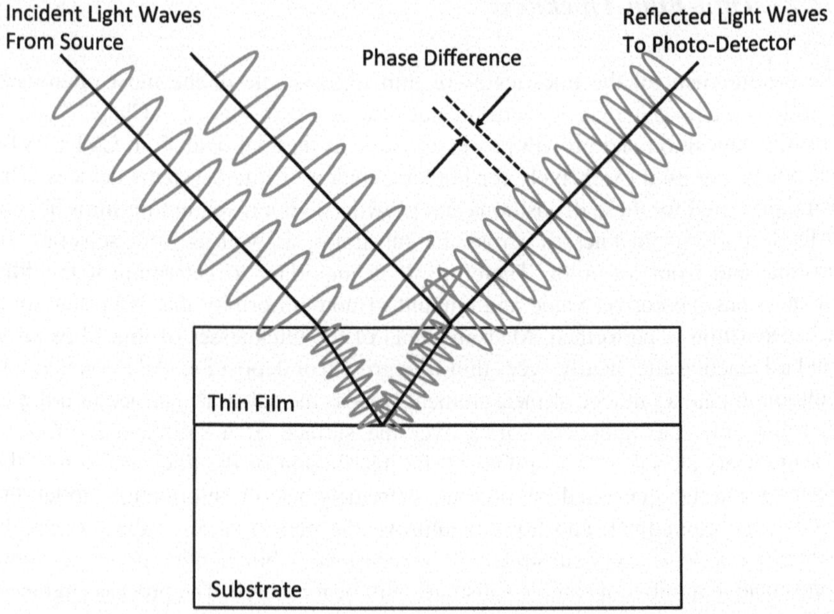

Fig. 5.6 Illustration of the use of interferometry to measure the thickness of a thin film on the surface of a substrate

reflected wavelength minima for the integral order of reflection, and oxide layers can be measured over a range of thicknesses from 500 Angstroms to 1 micron with an accuracy of ±25 Angstroms. However, the use of calculations or fringe charts is rarely used in modern manufacturing facilities since they are difficult and time-consuming and the accuracy is dependent on the experience of the person performing the measurement.

Instead, most manufacturing sites use highly automated interferometers wherein a substrate is placed onto a chuck, a measurement is taken, and the instrument employs internal software that quickly calculates the thin-film thickness at a specific location on the substrate surface.

It is common to take a number of measurements of the thin films on the substrates according to a pre-defined grid in order to calculate the resultant uniformity of the thin-film thickness and indices across the substrate. Common thin-film materials (and substrates) measured using this technique include silicon dioxide on a silicon substrate; Si_3N_4 on a silicon substrate; photoresist on a silicon substrate; and polysilicon on a layer of silicon dioxide on a silicon substrate.

Software upgrades allowing other variations of types of materials and films stacks that can be measured using interferometers are available from vendors. Importantly, this technique is suitable for thin films that are transparent to the optical energy, and therefore interferometry would not be used for opaque thin-film material layers such as metals.

Although the accuracy of these instruments depends on many factors, interferometry when used to measure the thickness of a thin film of silicon dioxide on a silicon substrate can measure over a range of oxide thickness from about 500 Angstroms to 10 microns with an accuracy of ±2%. Similar accuracies are obtained for other thin-film layer material types. For oxide layers thinner than 500 Angstroms, special algorithms can be used. For layers less than 100 Angstroms, ellipsometry should be used as an alternative as discussed below. It should be noted that these accuracies are dependent on a well-calibrated instrument. Most of the interferometry instrument makers offer calibration standards [1].

5.2.7.2 Ellipsometry

Ellipsometry is another fast, accurate, and nondestructive method that is widely used to measure the thickness of thin films and their indexes of refraction. It can be used to measure other material properties as well. This method is based on changes in the polarization as optical light reflects or passes through thin films on a substrate. The basic setup for performing ellipsometry is shown in Fig. 5.7, wherein a light source

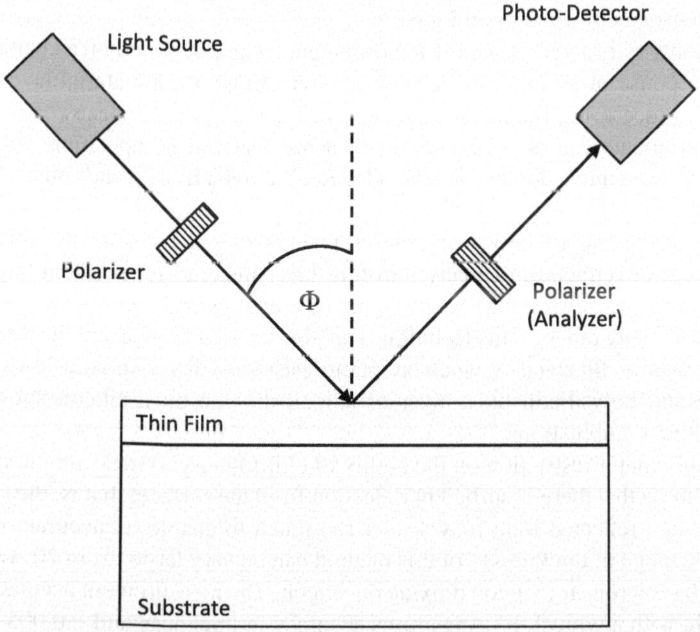

Fig. 5.7 Illustration of an ellipsometry measurement instrument using reflected light to measure the thickness of a thin film on the surface of a substrate. A light source is used to produce light waves at an angle Φ to the surface that is near the Brewster angle. The light from the source passes through a polarizer and then is reflected from the surface, thereby passing through another polarizer, called an analyzer, and into a photodetector. The light that reflects from the surface of the substrate undergoes changes in the polarization that are detected and analyzed to measure the thickness and index of refraction

is placed at an angle Φ to the sample that is usually near the Brewster angle of the sample to be scanned. The impinging light waves pass through a polarizer before reaching the sample surface, whereupon it interacts with the surface of the thin film. This interaction of the light wave and the material layer can be reflection, absorption, scattering, and transmission. As shown in Fig. 5.7, this experimental setup uses the reflected light, and after the light is reflected from the surface of the substrate, it passes through another polarizer, also called an analyzer, and into a photodetector. The light interacting with the material layer undergoes a change in the polarization that can be related to the thickness and index of refraction on the material layer. The polarization change is quantified by the complex reflectance ratio that is related to the amplitude and phase differences. Therefore, the determination of the measured characteristics of the thin films scanned involves complex calculations, curve fitting, and modeling. Most often the makers of ellipsometer instruments develop the algorithms and material models and integrate them into the software of the tools, resulting in instruments that are fairly automated for quick and accurate measurements on microsystems manufacturing lines.

Additionally, the instrument makers can develop optional models for specialized materials and applications. Because the instrument's received signal depends on the thickness as well as the materials properties, ellipsometry can be used as a multiple material property measurement tool for contact-free determination of thickness and optical constants of films of all kinds.

Ellipsometry is a very sensitive measurement technique based on a change of the optical response of an incident light wave interacting with a material layer, and in addition to measuring the thicknesses of thin films and their indexes of refraction, these instruments can also be used to measure material composition, roughness, crystalline structure, doping levels, electrical conductivity, and other material properties.

Automated commercially available ellipsometers can measure the thicknesses and indexes of refraction of many thin-film layer material types, including silicon dioxide on silicon; silicon nitride on silicon; silicon oxynitride on silicon; alumina on silicon; and many others. This technique can also be used to measure the thicknesses of films in thin film stacks, such as photoresist on silicon dioxide on a silicon substrate and polysilicon on a layer of silicon dioxide on a silicon substrate, as well as other combinations.

The only major restriction on the ability of ellipsometry to measure the thickness of thin films is that there is sufficient reflection from the surface, that is, the surface is too rough the reflected light may scatter too much to enable an accurate measurement. The range of thicknesses of this method can be very large, from 20 Angstroms to over 10+ microns for silicon dioxide on silicon. The measurement accuracy is also very high, with a typical ±3 Angstroms accuracy in thickness and ±0.005 or better in index of refraction value. Ellipsometry instrument makers offer standards for calibrating these tools. Since the accuracy of ellipsometry is dependent on the properties of light, this technique is often used as a standard for other instruments as well [1]. Ellipsometry can also be performed in situ by integrating the instrument into a process tool, such as for atomic layer deposition (ALD) to determine the thickness during the actual thin-film layer deposition [9].

5.2.7.3 Stylus Profilometry

While interferometry and ellipsometry can both be used for measuring thicknesses of thin films (and in the case of ellipsometry, can be used to measure other material properties), these techniques are not well suited for measuring the thicknesses of thin films that are not transparent to optical illumination, such as metals. Additionally, it is often very useful to be able to measure the resultant topology of the surface of the substrate during or after an etching step in order to measure and monitor the etch rates or whether an etch has been completed or not. It is also often useful to be able to measure the width of features made in thin-film layers, such as to ensure that metal interconnect lines have the desired width after etching. Also, it is sometimes useful to have the capability of measuring the roughness of surfaces after certain processing steps have been performed. In these cases, a metrology tool called a stylus profilometer can be very useful.

To measure thicknesses of thin films, the stylus profilometer requires that the thin film be patterned and etched so as to create a step in the material layer having a height equal to the thin-film layer's thickness. This step can be created by masking certain pre-defined areas on the substrate prior to deposition or alternatively by performing a photolithography and etch of the thin-film layer after deposition. Therefore, a shortcoming of this technique is that either it requires additional processing to be performed on dedicated substrates or it may be performed on the actual production substrates later in the manufacturing process. That is, the layer thickness can be measured once the photolithography and etching have been performed on actual device wafers. The former will result in lower yields for this production step since some substrates will be used to obtain the layer thickness, while the latter will mean that the layer thickness will only be known at a later manufacturing step, and this may result in requiring the layer be stripped and all manufacturing work lost if the layer thickness is found to be out of the required tolerance specification.

The functional aspect of a stylus profilometer is illustrated in Fig. 5.8, wherein a very small diameter stylus is placed into contact with the surface of the substrate. Once in contact, the stylus is then physically dragged over the surface, and its vertical displacement, both upward and downward, is measured along a line drawn across the substrate of a pre-defined scan length. The stylus does make physical contact with the substrate, but depending on the materials being scanned, it usually does not result in any damage to the surface elements or features. This type of profilometer can only perform linear surface scans of the topology, and therefore multiple scans must be performed in order to obtain a wider perspective of the thin-film layer's thickness variations.

The resolution of a stylus profilometer is dependent on the diameter of the tip, the scan speed, the contacting forces employed, and the surface topography being

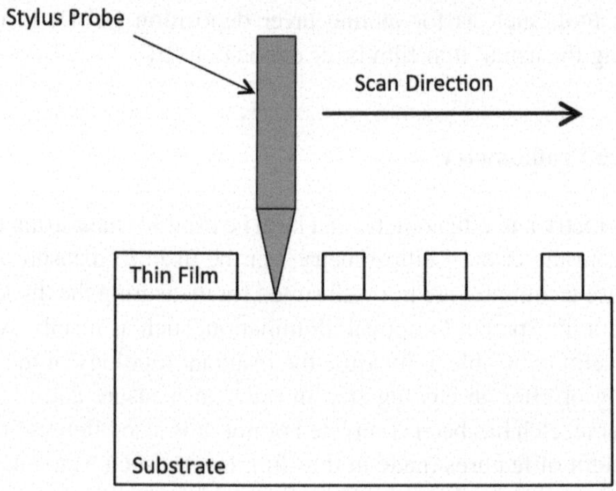

Fig. 5.8 Illustration of stylus profilometer that is employed to measure the thickness of a thin-film layer that was deposited onto the surface of a substrate. A stylus is brought into physical contact at a pre-defined location and scanned over the surface, and the vertical motions (both upward and downward) of the stylus probe are measured using a transducer element (usually capacitive or inductive-based) as a function of the location along a linear scan path

Table 5.1 Uncertainty of calibration standards used for stylus profilometers

Nominal step value (μm)	Typical uncertainty (μm)	Typical uncertainty (%)
0.1	±0.0021	±2.2
2	±0.018	±0.95
10	±0.070	±0.7
25	±0.151	±0.6
50	±0.266	±0.55

scanned. The scanning tip of the profilometer has a radius that ranges from about 20 nm to a few tens of microns. The typical vertical resolution is specified as about 1 nm on modern instruments, with a repeatability of less than 0.5 nm to 6σ. The dynamic ranges of surface heights ranges from about 5 nm to 1 mm. The lateral resolution is dependent on the features being scanned. For example, if scanning a surface with trenches having widths of nearly the size or smaller than the tip radius, the tip will not be able to resolve these trenches.

The accuracy of the stylus profilometer is usually dictated by the uncertainty of the calibration standard used. As a general rule, the error in the calibration standards is in the range from 0.5% to 2.5%, with the uncertainty in the standard being larger for standards having smaller step heights (e.g., 100 nm) and smaller for standards having larger step heights. The uncertainty or error in the calibration standards is given in Table 5.1 [10].

The scanning can be performed by moving the tip across the surface or, more commonly, translating the substrate in a linear direction while holding the tip steady. The contact force on the stylus can be adjusted from 0.1 mN to about 150 mN.

The scan lengths are adjustable over a dynamic range from about 50 microns to over 50 mm, and multiple scans can be stitched together along the length (i.e., end-to-end scan stitching) or side by side (i.e., side-by-side stitching). The latter can be used to create a three-dimensional topology mapping of the surface. The accuracy will be no better than the resolution, and high accuracy will require that the instrument is calibrated correctly.

5.2.8 Four-Point Probe

The four-point probe is a commonly used instrument for measuring the sheet resistance and resistivity of material and material layers, which are important electrical material properties. This metrology tool is often used to make sure that the dopant impurities that have been introduced into the material are of sufficient quantity to result in a desired resistivity (or inversely conductivity) level. Additionally, this tool can be used to determine if the conductivity of deposited metals layers is adequate, or alternatively if the conductivity is assumed constant, it can be used to indirectly measure a conductive material layer's thickness. Multiple measurements can be taken across a substrate or on multiple substrates to derive information about the uniformity of sheet resistance or resistivity as well.

Figure 5.9 is an illustration to describe the four-point probe performed on a substrate having either a diffused layer of dopant impurities of thickness, t, that

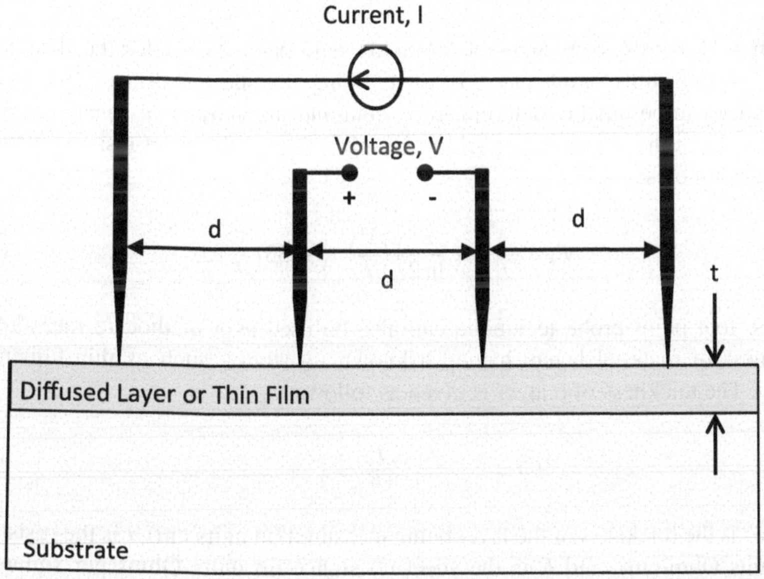

Fig. 5.9 Illustration of the four-point probe technique for measuring the resistivity and sheet resistance of material layers

has been introduced into the top surface of the substrate to change the resistivity of the shaded layer or a material layer of thickness, t, that has been deposited having a certain resistivity. A set of four probes are placed into contact with the surface of the substrate with sufficient mechanical force to ensure that a good electrical contact is made. The two outer probes have a constant current, I, applied between them as shown. The other set of inner probes measures the resultant voltage, V, that is due to the current passing through the shaded material layer having a certain resistivity value. As shown, the probes are equally spaced at a distance of d between them. The four-point probe configuration cancels out any effects of the varying contact resistances that may be present from one probe to another. Equation (5.2) defines the resistivity, ρ (in units of Ohms-cm), of the material layer being probed to the distance between the probes, d (in units cm), the applied current to the outer two probes, I (in units of Amps), and the resultant voltage, V (in units of Volts) on the inner two probes, as follows:

$$\rho = 2\pi d \frac{V}{I}. \tag{5.2}$$

Equation (5.2) is strictly valid if the thickness of the layer being probed is much greater than the distance between the probes, that is, when $t \gg d$. In cases, when the thickness of the layer being probed is much less than the distance between the probes, that is, when $t \ll d$, Eq. (5.3) would be used:

$$\rho = \left(\frac{\pi t}{\ln 2}\right)\left(\frac{V}{I}\right). \tag{5.3}$$

Often in microsystems manufacturing, a useful parameter called the sheet resistance, having units ohm/square, is used. Using the sheet resistance, the overall resistance can be quickly determined by counting the number of squares between two points on the substrate. Rewriting Eq. (5.3) above, the sheet resistance can be found as follows:

$$R_s = \frac{\rho}{t} = \left(\frac{\pi}{\ln 2}\right)\left(\frac{V}{I}\right) = 4.53\left(\frac{V}{I}\right). \tag{5.4}$$

The four-point probe technique can also be used as a method to measure the thickness of material layers having a known resistivity, such as thin-film metal layers. The thickness of a layer is given as follows:

$$t = \frac{\rho}{R}, \tag{5.5}$$

where t is the thickness of the layer being measured (in units cm), r is the resistivity (in units Ohm-cm), and R is the sheet resistance (in units Ohms per square). It should be noted that Eq. (5.5) assumes that the resistivity is constant through the thickness of the layer being measured. Therefore, Eq. (5.5) is not valid for

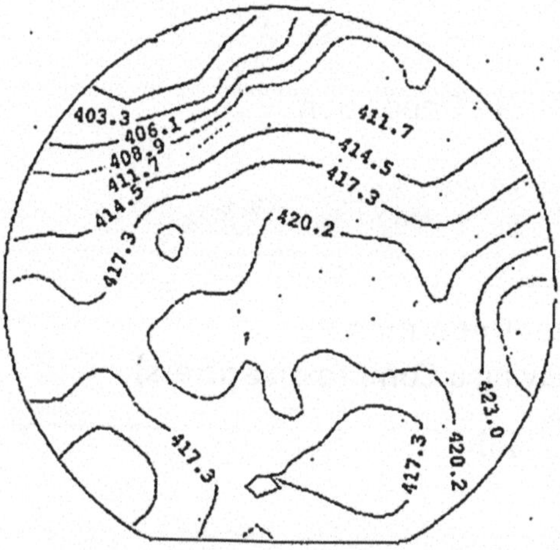

Statistical Summary

Number of Test Points	65
Average Value	416.4
Maximum Value	428.9
Minimum Value	399.2
Sample Spread(%)	7.14
Std Dev Value	5.8
Wafer Uniformity Value(%)	1.38
Contour Interval Value	2.816

Fig. 5.10 Contour-mapping profile output showing the mapping of the sheet resistance across a substrate. As can be seen, in this measurement, 65 tests were conducted across the substrate, and the average, maximum, minimum, sample spread, and standard deviation for this substrate are calculated

measuring the thicknesses of diffused layers of dopants into a semiconductor where the profile of dopants as function of depth is changing.

Microsystems manufacturing facilities will have modern sheet resistance measurement systems that are highly automated to make multiple measurements across the substrate and can be outfitted with robotic systems for cassette-to-cassette substrate measurement taking. Most commercially available sheet resistance measurement systems employ the ASTM standard F84-99 and can measure over a dynamic range from about 1 milli-Ohm per square to nearly 1 mega-Ohm per square. The measurement accuracy of these systems is usually better than 1%. Figure 5.10 shows a typical sheet resistance plot across a substrate [1].

5.2.9 *Thin-Film Stress Measurement*

Thin-film stress measurement systems are used for determining and monitoring the residual stresses in deposited thin-film layers. The measurement of the residual stress levels is important for thin films used in microsystems (both ICs and MEMS) since these stresses will cause the substrate to bow upward (i.e., like a bowl) or downward (i.e., like a dome) depending on whether the residual stresses are tensile or compressive in nature (see Fig. 5.11). Excessive bowing of the substrate can lead to a number of problems including the following: it can cause the photolithography process to be out of focus across a wafer; it can cause the thin films to crack or delaminate; it can result in strains being applied to electrically active devices, thereby changing their behavior; it can cause the gas flow characteristics across

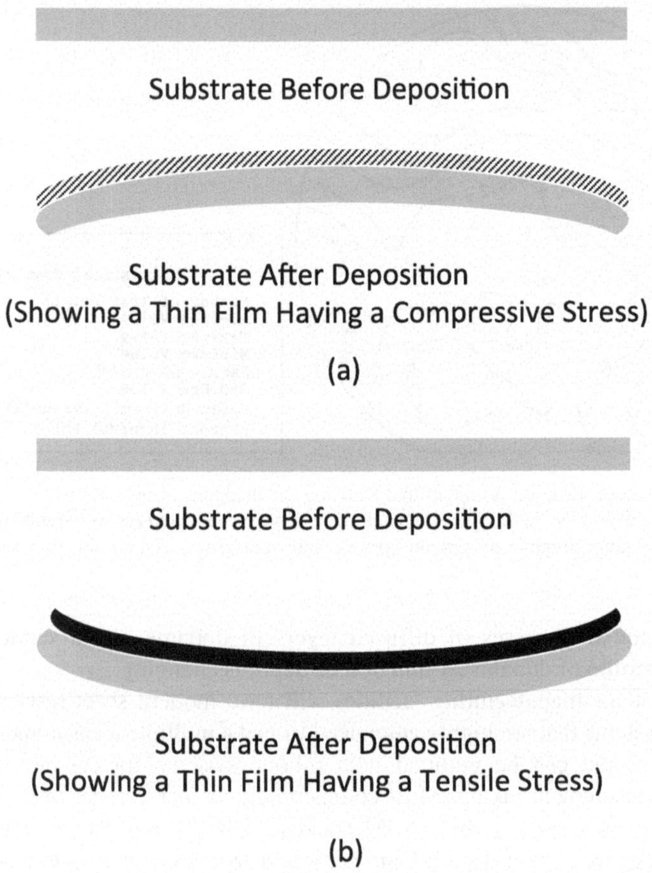

Fig. 5.11 Cross-sectional illustration of substrates before and after the deposition of a thin film having a residual stress. In (**a**) the substrate is flat before the deposition, and after the deposition, the substrate bows like a dome due to the compressive residual stress in the deposited layer. In (**b**), the substrate bows in the opposite direction after the deposition of the thin film due to the tensile residual stress in the deposited layer

the substrate to be modified so as to result in deposition or etching processes that are less uniform across the substrate; and, if these stresses are sufficiently large, they can cause the substrate to fracture.

Measurement of residual stress is particularly important; then there is some type of mechanical or electromechanical functionality of the device that is made from thin films, such as in MEMS. In addition to the problems cited above concerning microsystems in general, residual stresses in MEMS devices will change the resultant mechanical behavior and performance of these types of devices and therefore must be taken into consideration.

Some of the material properties of the thin-film layers commonly employed in microsystems manufacturing are highly variable depending on the processing conditions. Perhaps the most notable example of a material property that is process dependent is the residual stress of thin-film layers. This subject has been extensively studied for some thin-film layers, such as polysilicon, which is often used as a structural layer in many surface micromachining microsystem devices [11]. It is well known that the residual stress of a deposited thin-film layer varies depending on the exact recipe and tool employed. Therefore, the use of stress values from look-up tables or the literature has limited value, and the developers are forced to rely on performing measurements of the residual stresses for their specific processing conditions. MEMS developers and manufacturers require the capability for a fast, simple, inexpensive, nondestructive routine measurement of thin-film stresses. The most common method for measuring the residual stresses in thin films is to measure the resultant curvature of the substrate before and after the deposition of the thin film. This is portrayed in Fig. 5.11.

As seen in Fig. 5.11a, the substrate is shown at the top without a thin film deposited onto its surface, and the substrate is relatively flat. Once a thin film has been deposited as shown in the bottom of Fig. 5.11a, the thin film's resultant residual stress causes the substrate to develop a curvature. In this case, the residual stress is compressive in nature, and the substrate with the thin film results in the curvature of the substrate to be dome-like. The elastic properties of the substrate provide a force to attempt to push the substrate back to its previous flat state and the thin film wants to expand and therefore provides an opposite force. An equilibrium situation will result, wherein the two forces balance as shown in the dome-like substrate in the bottom of Fig. 5.11a. It should also be noted that in Fig. 5.11, the thin film is only deposited onto the top surface. If the thin film was deposited on both sides of the substrate and the thicknesses and material properties were identical, then it would be expected that the substrate after the thin-film deposition would remain flat since the residual stress forces on both sides of the substrate would cancel each other out. Another scenario of residual stress is shown in Fig. 5.11b. In this case, the substrate that starts out as flat in the top of this figure without a thin film and develops a bow-like curvature after the thin film is deposited onto the surface, as shown at the bottom of Fig. 5.11b. In this case, the residual stress is tensile in nature and the biaxial forces of the deposited film as such to make the film want to contract.

In practice, substrates without thin films on their surface may have some amount of curvature. Therefore, the measurement of the film stress is performed by first measuring the curvature of the substrate before the thin film has been deposited and then after the film has been deposited to remeasure the substrate curvature to determine the differential curvature of the substrate due to the presence of the deposited film.

The substrate curvature can be directly related to the residual stress by knowing the elastic properties of the material and thin film using the Stoney equation, given as follows:

$$\sigma_f = \frac{(E_s t_s^2)}{[6 t_f (1 - \upsilon_s) R]}, \tag{5.6}$$

where the subscripts f and s refer to the thin film and substrate, respectively, E_s is the Young's modulus of the substrate, t_s is the thickness of the substrate, t_f is the thickness of the thin film, ν_s is Poisson's ratio of the substrate, and R is the radius of curvature of the substrate. This equation assumes that the stresses in the film and substrate have isotropic elastic properties, the film thickness is uniform, the temperature is uniform, and the stress is uniform throughout the film thickness.

Of these assumptions, the one most problematic in typical situations is the assumption about the substrate being isotropic in its elastic mechanical properties. In most situations, the thin-film residual stresses to be measured have been deposited onto semiconductor materials that are usually anisotropic. Therefore, there is a more complex form of Stoney equation that is used in most practical measurements that takes the anisotropic material properties into account. For example, in the case of a thin film deposited onto a single-crystal silicon substrate having an orientation of (100), the modified Stoney equation is given by:

$$\sigma_f = \frac{(t_s^2)}{\left[6t_f\left(s_{11}^{si} + s_{12}^{si}\right)R\right]}, \tag{5.7}$$

where s_{11}^{si} and s_{12}^{si} are values from the compliance tensor of silicon. The factor $1/\left(s_{11}^{si} + s_{12}^{si}\right)$ is the biaxial modulus of (100) silicon and has a numerical value of approximately 180 GPa. Different versions of the Stoney equation are available for other crystallographic orientations of silicon as well as the most common orientations of other commonly used anisotropic semiconductor materials. Other appropriately modified versions of the Stoney equation are available that account for other situations as well [12].

The actual measurement of the substrate curvature is usually performed by scanning a laser across the substrate surface to measure the resultant change in the radius of curvature before and after the thin film has been deposited. A typical residual stress measurement tool has a dynamic range of stresses that it can measure from about 1 MPa to 4 GPa, with both a repeatability and resolution on the measurement of about 1 MPa. Calibration standard is available. The accuracy is typically less than 2.5% or 1 MPa depending on which is larger [13]. It is important to point out that this method measures the average film stress across the substrate and the stress at any location on the substrate may be different from this average value.

Most thin-film stress measurement systems have a cooled and heated substrate chuck capability, typically allowing the temperature of the substrate to be varied from about -65°C to 500°C. This capability allows the residual stresses to be measured at various temperatures of operation as well as enables the thermal expansion coefficient to be measured.

While the substrate curvature method is fast and nondestructive, there are some disadvantages of this technique that users should be aware of. First, this technique measures the average stress over the entire substrate. That is, there may be variations in the residual stresses across the substrate, and this method only provides an average value over the substrate and not information about specific values at different locations across the sample. Second, it assumes that the elastic properties of the thin film are known. Third, while the measurement of the stress includes any effects of a stress gradient that may be present in the thin film, it does not allow the stress gradient to be quantified.

5.2.10 *Particle Measurements*

Microsystems manufacturing entails the fabrication of microminiaturized devices having very small features. Any particulates that may exist in the environment and deposit on the surfaces of substrates during fabrication can result in a degradation of the performance of the devices or even making them not work at all. Since normal ambient air in an urban environment may contain upward of 1 M particles that are 0.5 micron or larger per cubic foot of air, there is a requirement in microsystems manufacturing for methods to control the particulate count in the air. Therefore, microsystems manufacturing is mostly performed in cleanrooms, wherein the amount and size of the particulate matter are well controlled to very small levels.

There are a number of classifications for cleanrooms including US FED STD 209 E, wherein a class 1 cleanroom is defined as having 1 or less particles per cubic foot or air with a size of 0.5 micron or larger, a class 10 cleanroom is defined as having 10 or less particles per cubic foot or air with a size of 0.5 micron or larger, a class 100 cleanroom is defined as having 100 or less particles per cubic foot or air with a size of 0.5 micron or larger, and so on. However, this standard has now been superseded by ISO 14644 standard for cleanrooms [14].

Although microsystems manufacturing is performed in cleanrooms, where there is an implicit assumption that everything is "clean," in practice there is a constant effort to ensure that the particle count is properly managed in a cleanroom and its associated equipment set. The particle count in any cleanroom is a function of the level of air filtration employed as well as the operational practices of the cleanroom staff. While a cleanroom may be designed to operate at a specified cleanliness level, the actual particle count can be increased if the equipment is not properly maintained and particle-generating activities are not properly managed. Nevertheless, no matter how good the cleanroom air filtration systems and clean room practices are, no clean room will be free of particles. One method of determining the cleanliness of the cleanroom is to install a particle counting measurement system that monitors the air in a defined area. However, it is also useful to measure the actual particle count on substrates in the cleanroom.

For this purpose, there are commercially available substrate particle inspection tools. These tools typically use one or more lasers to scan the surface of the substrates combined with integrating photodetectors to sense detect the presence of particles. On a clean smooth substrate surface, an impinging laser beam will be reflected at predictable angles. However, if there are any particles present on the surface, the laser beam will be scattered, and this scattered laser radiation can be sensed by photodetectors.

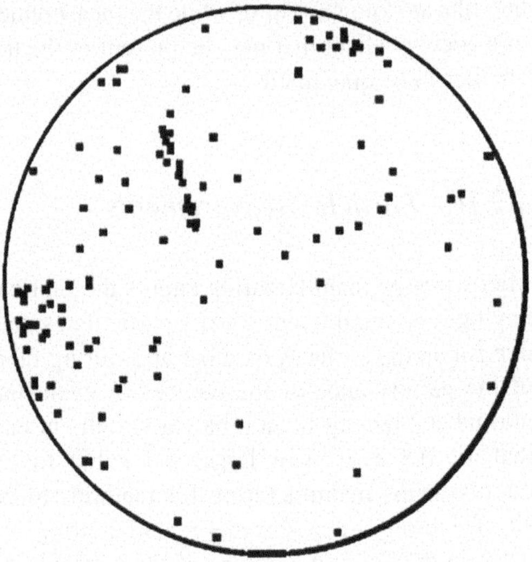

Fig. 5.12 A plot of the particle count showing the distribution of particles across the substrate surface detected by a particulate scanning system

The output of a particle scanning system is shown in Fig. 5.12. As can be seen, a plan view of the substrate is presented, and black dots represent particles that were detected as well as their location. These systems are usually equipped to provide information about the location and size distributions on the particles in a wafer mapping format or using statistical graphs for each substrate as well as over a defined group of substrates and over time. This capability allows the particle count to be measured and monitored as a function of processing steps performed in each tool to determine which tools and process steps may be responsible for increasing the particle counts. Armed with this information, the equipment engineers can identify the equipment and process steps that are the source of particle generation. Modern systems can detect particles down to sizes of 30 nm diameter with reasonable wafer throughputs, wherein each wafer can be scanned in a few seconds.

5.2.11 Noncontact Optical Profilometry

Noncontact surface profilometers are very useful metrology tools for measuring the topology of substrate surfaces without having to make physical contact to the surface such as when using stylus profilometers. This is important when the surface features have fragile structures, such as in the case of MEMS, that could be easily damaged by having a stylus come into contact with these structures during the measurement. These systems typically use white light interferometry to measure surface features, and most systems employ two types of measurements, phase-shifting interferometry (PSI) and vertical scanning interferometry (VSI) [15].

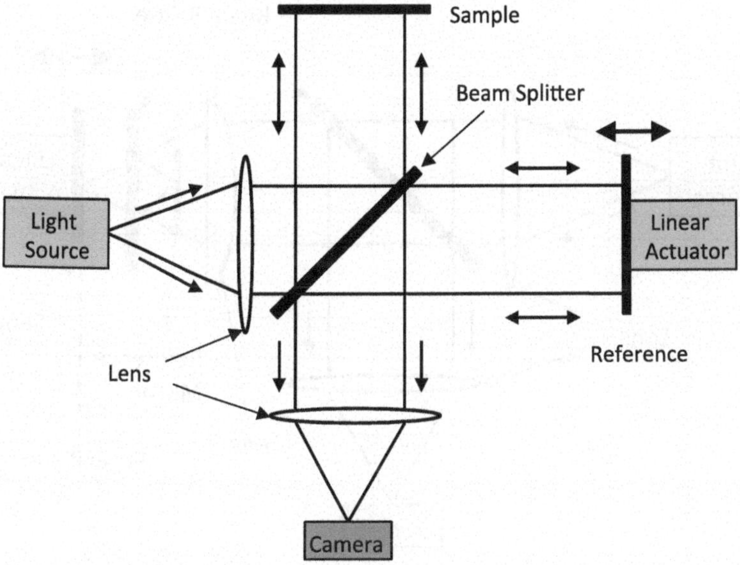

Fig. 5.13 Illustration of the configuration for performing phase-shifting interferometry (PSI). A monochromatic light source is used is illuminate the sample and reference through a beam splitter and the reference is linearly displaced by pre-defined displacements. The resultant interference pattern is captured by a camera and processed to measure surface roughness on the sample

In PSI, as illustrated in Fig. 5.13, a monochromatic light source is used in conjunction with a mechanical translation system to change the optical path length of both the test and reference beams in a series of pre-defined shifts, and these optical path length changes result in a lateral shift of the resultant interference fringes patterns. The mechanical translation is typically performed using a linear actuator, such as a piezoelectric actuator, on which the reference is placed. A camera to produce a series of interferograms that can be analyzed to produce a profile of the substrate surface captures the moving fringe patterns. PSI enables very accurate measurements but is only usable for surfaces that are relatively smooth and therefore is more appropriate for measuring surface roughness than performing metrology on actual features fabricated on a substrate surface.

VSI, illustrated in Fig. 5.14, uses an internal translator to scan the sample sideways during the measurement as the camera records single-intensity patterns, called frames. As each point on the surface comes into focus, the modulation on that point reaches a maximum and then tapers off as the translator passes through the focus. By recording the height of the translator at maximum modulation, the system can determine the height of each pixel. It should be noted that VSI gives less precise measurements than PSI but is better suited for measurements of actual features fabricated on sample surfaces. Additionally, VSI is sometimes called scanning white light interferometry (SWLI) and coherence scanning interferometry (CSI). There are a number of types of CSI including Mirau interferometry, Linnik interferometry, and Twyman–Green interferometry that are tailored to specific measurement

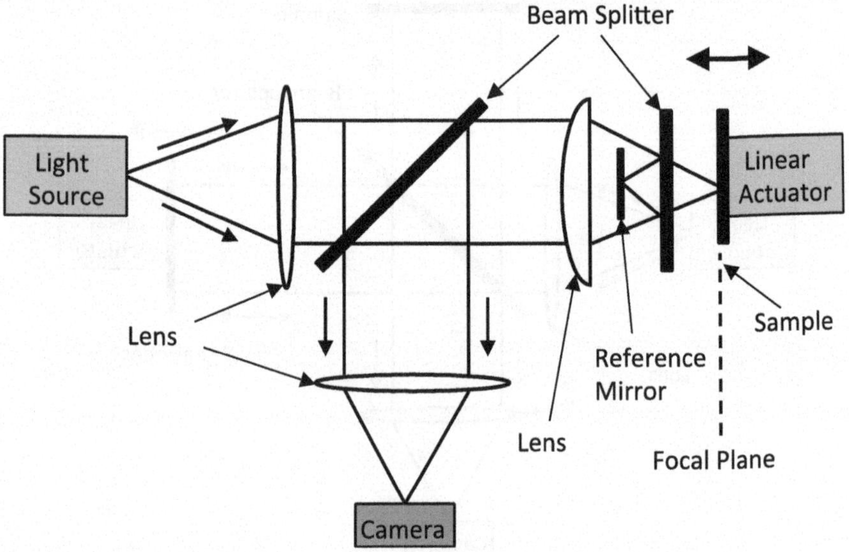

Fig. 5.14 Illustration of the configuration for performing vertical scanning interferometry (VSI). A white light source used illuminates the sample that is placed on a linear actuator inside the optical system shown above. The camera captures frames and looks for the maximum fringe contrast for determining step heights on sample

variations. A limitation of VSI for performing metrology is that enough light must be able to reflect from the features of the sample. Therefore, while this tool is very useful for measuring microsystem device features that have limited aspect ratios, insufficient light sometimes is not reflected in high-aspect ratio features commonly seen in MEMS devices.

Commercially available noncontact profilometer tools that are fully automated are available. The dynamic measurement ranges of a typical commercially available noncontact profilometer using the CSI mode, which is more appropriate for measurement of features, are less than 1 nm to over 20,000 microns with a surface topography repeatability of about 3.5 nm. If used in the PSI mode, which is appropriate for surface roughness measurements, the repeatability can be as low as 0.08 nm. The scan speeds are relatively fast on these tools, which is important for routine metrology on production lines.

5.2.12 *Wafer Bonding Inspection*

It is useful to have a method to inspect the quality of the wafer bonding. Wafer bonding defects can be caused by a number of reasons, including dust particles on the wafer surfaces prior to bonding; non-uniformities due to improper polishing of the wafer surfaces; excessive bowing of the wafers due to highly stressed films on the wafer surface(s); chemical contamination of surfaces; epitaxial spikes on the wafer surfaces; and clusters of material that protrude from the wafer surface after a thin-film deposition process. These defects result in diminished bond strength between the bonded wafer pair and are exhibited as voids between the bonded wafer pair. If the defects are sufficiently large or numerous, the wafers do not bond together.

The ability to be able to perform inspection of the bond quality is important after the wafers have been physically contacted (this is sometimes called a pre-bonding operation) since if there are visible defects between the bonded wafer pair, they can be separated, re-cleaned, and re-pre-bonded to attempt to reduce the defects to obtain better bonding results. Additionally, it is also useful to inspect bonded wafers after the annealing has been performed since any voids may result in yield loss in those locations.

If the substrates being bonded together are made of a semiconductor material, it is not possible to see any defects in the bonding that may exist between the wafers since semiconductor substrates are opaque to optical wavelengths. A technique around this problem for silicon wafers is to use infrared imaging to view the voids.

Figure 5.15 illustrates a simple infrared inspection system [16]. The bonded wafer pair is placed between an infrared source illumination and a charge-coupled device (CCD) or a CMOS imaging camera. The infrared illumination source can be as simple as a common incandescent lamp. Since silicon has an energy bandgap of 1.1 eV, electromagnetic radiation having an energy (and wavelength) greater than 1.1 microns will be partially transmitted through the material.

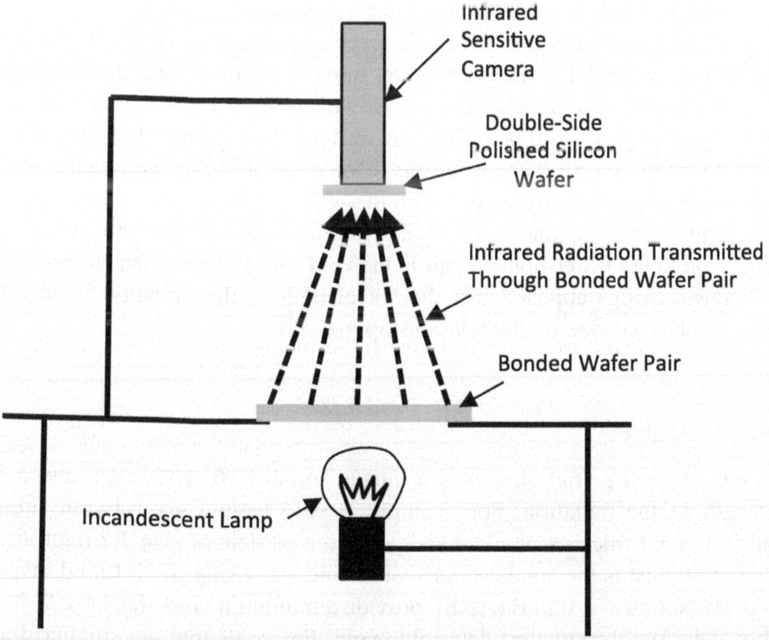

Fig. 5.15 An infrared inspection system for identifying voids between bonded wafer pairs [16]

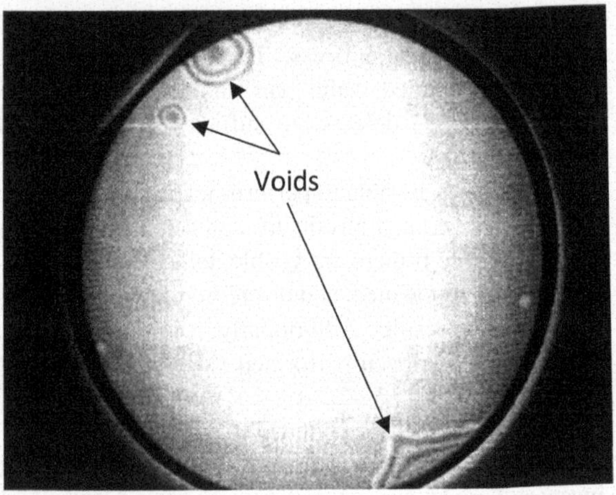

Fig. 5.16 The image of the infrared camera used to inspect a bonded wafer pair to inspect for voids. As can be seen, several voids are visible at the edges of the bonded wafers

The total amount of transmittance will be determined by the wavelength of the illumination source, the doping level of the silicon, the roughness of the wafers as well as any patterning that may exist, and the thickness of the wafers. The CMOS imaging camera has sufficient sensitivity in the infrared to develop an image. In order to prevent room light from saturating the camera, it is recommended that a double-sided polished silicon wafer be placed over the lens of the camera. This allows infrared radiation into the camera but prevents any electromagnetic radiation with wavelengths below 1.1 microns to reach the imager.

A resultant image from the infrared inspection system of a silicon wafer-bonded pair is shown in Fig. 5.16. As can be seen, most of the field has a uniform contrast indicating a good bond in these regions. However, in the top-left and bottom-right regions of the bonded wafer pair, there are multiple fringe rings that are visible that indicate poor bonding in these locations. Moreover, these fringe patterns indicate that the voids are caused by particles on the wafer surface prior to bonding.

The fringe rings are the result of the interference between infrared radiation passing through the wafer bonded pair in the void area and are often called "Newton rings." Newton rings appear when the wavelength of the radiation is an integral multiple of the void size by the relationship given by:

$$t = \left(\frac{1}{2} + n\right)\lambda, \tag{5.8}$$

where t is twice the void size, n is a integer number (0, 1, 2, ...), and λ is the wavelength of the radiation. For example, a void having six Newton rings and assuming $\lambda = 1.1$ microns would correspond to a particle of size 7.2 microns [16].

Also of interest is the smallest detectable void size using the infrared inspection method. By setting $n = 0$ in Eq. (5.8), provide a minimum void size of 0.27 microns which is therefore the smallest detectable void. For voids that are smaller than this dimension, there are other methods of inspection including X-ray topographical and ultrasonic graphical imaging that can improve the resolution. For example, an ultrasonic wafer bonding inspection system can detect voids as small as 0.1 micron [17]. However, both of these methods require specialized and expansive equipment and measurements can take considerably longer times.

5.3 Specialized Metrology Equipment and Methods

The previous section reviewed the metrology methods commonly used during microsystems manufacturing. As noted, many of these capabilities are performed on a routine basis on substrates after performing processing steps and therefore are primarily used for quality control purposes. This section provides an overview of more specialized capabilities for microsystems metrology. Specifically, the capabilities reviewed in this section are used sparingly in MEMS microsystems manufacturing since they are too time-consuming and too expensive and may result in damage to the substrate.

5.3.1 Focused Ion Beam

The focused ion beam (FIB) technology was mentioned in the last chapter as a tool technology for performing micromachining operations. The FIB tool is also an extremely useful tool for cross-sectioning and imaging portions of device structures that would otherwise be invisible due to coverage of thin-film or other material layers. The FIB performs nanometer dimensional-scale machining using a focused ion beam (FIB). Conventional FIB machining uses a liquid-metal ion source, usually a gallium ion source, whereby the gallium wets a tungsten needle that is heated and has a large electrical potential to cause ionization and field emission of the gallium ions (see Fig. 5.17). The source ions are accelerated to energies usually between 5 and 50 KeV and focused to a small spot size using a special electrostatic lens. As the focused beam of Ga^+ ions impinges on the surface of a sample, material on the surface is sputtered away. This sputtered material leaves the surface as either secondary ions or neutral atoms. Secondary electrons are also produced from this process. The sputtered ions or secondary electrons can be collected to form a high-resolution image of the surface (similar to the way a SEM works).

Conventional FIB technology can machine features down to about 5–10 nm and can remove material at rates up to nearly 100 μ^3/s. Newer technology using a helium ion source has recently been introduced into the commercial market and has a resolution below 1 nm [18]. Helium is also less damaging to the surface of the sample compared to gallium ions.

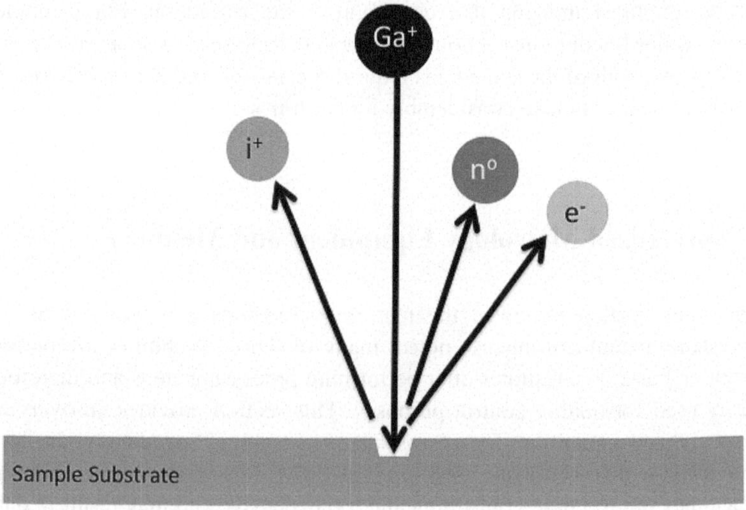

Fig. 5.17 Illustration of focused ion beam nanometer dimensional-scale machining whereby a gallium ion beam is directed at the surface of the sample to be machined and removes material from the sample with a high resolution and high level of positional accuracy. Secondary ions (labeled i^{+}), neutral atoms (labeled n^{o}), and secondary electrons (labeled e^{-}) are emitted from the sample surface, as shown

Using a FIB system, the user can input a 3D CAD solid model of the desired topology of the machining process and the computer-controlled stage allows very precise registration of sample with the ion beam with submicron positional accuracy.

FIB systems can be used to machine both conductive and nonconductive materials. Specifically, for the machining of nonconductive materials, a low energy beam of electrons is directed at the surface of the sample where the FIB is located to provide neutralization of the charge.

FIBs are most commonly used in the microsystems manufacturing industry as a forensic and diagnostic tool. For example, Fig. 5.18 is a scanning electron microscope (SEM) image of a FIB-cross-sectioned dynamic pattern generator device that is used to create a multiplicity of individual steerable electron beams for e-beam lithography applications. This device was fabricated by the MNX. The darker and thicker regions are made of silicon dioxide, and the lighter and thinner regions are made of a metal. A deep, high-aspect ratio etch technology was used to bore the cylindrical trenches through the dielectric and metal layers. There is a conductive layer at the bottom of the trenches. A surgical cross-sectioning and imaging like shown in Fig. 5.18 are extremely useful to literally "see into" the microstructure of the device and determine that the layers align as required, the cylindrical trench features are cleared during etching, and no residual resists are present. This sort of cross-sectioning would be extremely difficult to obtain with cleaving the sample since the location of the cleave is unpredictable at best, and often cleaving results in debris to be located at the broken surfaces that would interfere with the imaging of the structure.

Fig. 5.18 Scanning electron microscope (SEM) image of a FIB-cross-sectioned electron reflector or absorber structure showing various layers involved in the device. The darker and thicker regions are made of silicon dioxide, and the lighter and thinner regions are made of a metal. A deep, high-aspect ratio etch technology was used to bore the cylindrical trenches through the dielectric and metal layers. There is a conductive layer at the bottom of the trenches

Most commercial FIB tools have imaging capability wherein the system is combined with a SEM (these are often called "dual-column FIBs"), and many tools can also be outfitted with a secondary column for mass analysis of particles removed from the substrate using secondary ion mass spectrometry (SIMS). Elemental chemical analysis of the machining by-products can be enabled by incorporating energy-dispersive X-ray spectroscopy capability – something that is a common add-on feature on SEMs (and discussed later in this chapter).

Among the attributes of focused ion beam include the following: extremely high precision machining capability; resolution of between 5 and 10 nm; can be used to nano-polish sample surfaces; machining performed directly under computer control; secondary electrons can be used for scanning electron microscopy in dual-column FIB configuration; and energy-dispersive X-ray spectroscopy for materials analysis is available as an add-on. The major disadvantages of FIB are that it has slow machining rates of typically less than 100 microns^3/s and the tool is very expensive.

FIBs are mostly used in microsystems development purposes to be able to cross section areas in order to determine the cause of problems. Usually, given the slow machining rates and the destructive nature of the machining process, these tools are rarely used in manufacturing.

Additionally, FIBs are used in manufacturing for photolithographic mask repair. The cost of making masks at the nodes below 90 nm is very high, and it is not uncommon to have one or more defects present that would otherwise make the mask unusable. Depending on the type of defect, the FIB capability may be able to perform surgical machining on mask so as to make it useful and avoid the cost of expensive mask rework. On rare occasions, the FIB can also be used for device repair.

5.3.2 *Scanning Tunneling Microscopy (STM)*

A scanning tunneling microscope (STM) is an extremely precise metrology instrument that can be used to scan and image the surfaces of materials at the atomic scale [19]. The STM instrument employs a very small diameter needle-shaped tip (typically as small as one atom at the apex) that is made from a conductive metal material such as tungsten, gold, platinum-iridium, etc. that is fabricated using wet etching micromachining fabrication techniques.

The sharp tip of the STM is positioned in very close proximity to a surface of the material to be scanned using piezoelectric or electrostatic actuation. A biasing voltage, labeled "V_{Bias}" as shown in Fig. 5.18, is applied across the gap between the tip and material surface resulting in a decrease in the separation (labeled "z" in Fig. 5.19). If the tip is sufficiently near to the material surface, a quantum-mechanical tunneling current, labeled $I_{Tunneling}$ in Fig. 5.19, passes across the gap between the tip and the material surface. This distance between the tip and material for tunneling is typically within the dimensional range of about 5–10 Å.

The current, $I_{Tunneling}(d)$, through the tunneling tip to the substrate surface is given by:

$$I_{Tunneling}(d) = \text{Constant} \times qV_{Bias} \exp\left(\frac{-2d\sqrt{2m\Phi}}{\hbar}\right), \tag{5.9}$$

where d is the distance between the tunneling tip and the substrate surface, q is the electron charge, V_{Bias} is the voltage potential between the tip and the substrate, m is electron mass, $\hbar$ is Plank's constant, and Φ is the work function (energy barrier) of the material.

It should be noted that while a scanning tunneling microscope may appear to be similar to the stylus profilometer described above, the two techniques are quite different in how the measurements are taken. The stylus profilometer essentially places the tip into mechanical contact with the substrate surface, while the scanning

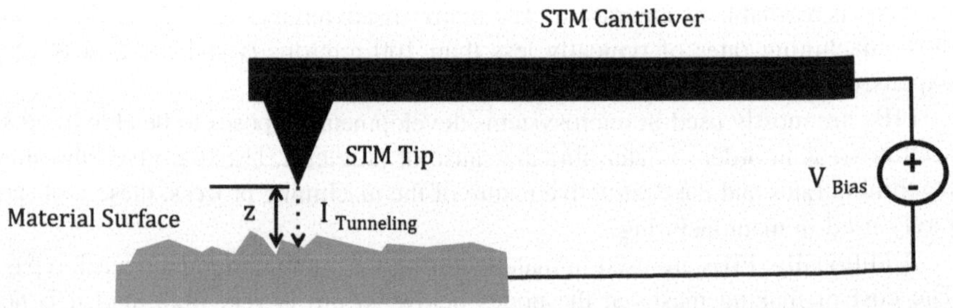

Fig. 5.19 Illustration of a scanning tunneling microscope (STM) showing the tip near the material surface and the resulting tunneling current $I_{Tunneling}$ that is indicative of the distance z between the tip and the material surface

tunneling microscope is based on a quantum mechanical electron tunneling effects. Consequently, the STM has a much higher level of resolution than the stylus profilometer.

It is important to point out that the STM does not probe the positions of nucleus of the atoms directly. Instead it actually probes the electron density surrounding the atoms. Also, the STM procedure is usually performed under ultrahigh vacuum conditions since the STM is probing the outermost atoms on the sample surface and any water molecules or other gas absorbents would otherwise be present on the material surface and interfere with the response.

A closed-loop feedback circuit is used to measure the tunneling current passing through the tip to the material surface, and this tunneling current is exponentially proportional to the distance between the tip and the material surface. The tip can be scanned linearly across the surface of the material to generate a spatially varying tunneling tip current that can be used to create an image of the material surface at the atomic dimensional size scale. Alternatively, the tip current is maintained at a constant value, and the actuator voltage to maintain a constant gap distance can be measured to create an atomic dimensional size-scale image.

The typical tip tunneling current is on the order of a few hundred pico-amps at a bias voltage of about 1 V. This instrument is extremely sensitive since the tunneling current is exponentially dependent on the gap between the tip and material surface. That is, the tunneling current increases by a factor of 10 for each Angstrom closer the tip gets to the surface. The resolution of the STM is approximately 0.1 nm in the lateral (i.e., x and y coordinate) directions and about 0.01 nm in the orthogonal (i.e., z coordinate) directions [20]. The repeatability of STM depends on the sample being scanned. If the sample has considered topography, then a number of scans may be needed in order to develop sufficient statistics in order to quantify the results. As with most highly sensitive metrology instruments, accuracy depends on having sufficient precision and then removing the bias error using calibration. Calibration standards for STMs are available from the instrument makers and other organizations. Nevertheless, an accuracy of below 1 nm is not uncommon.

While the STM can be used for imaging with atomic-level resolution the surfaces of materials as described above, it can also be used to perform spectroscopy. That is, the dependence of the current on the voltage can be used to provide information about the electronic-shell structure of the surface atoms. Specifically, the first derivative of the tunneling current with respect to the biasing voltage is proportional to the electronic states of the atoms on the surface immediately under the tip.

Figure 5.20 is an image created by a STM of the surface of gold material. As can be seen, the individual gold atoms are visible using this metrology technique.

STMs are mostly used in microsystems development environments for performing surface scans to determine and quantify the roughness of the substrate or material surface after a deposition or etching. The scanning process is relatively slow but is nondestructive.

Fig. 5.20 Image of the surface of a gold material layer that was created by a STM. The image is in grayscale

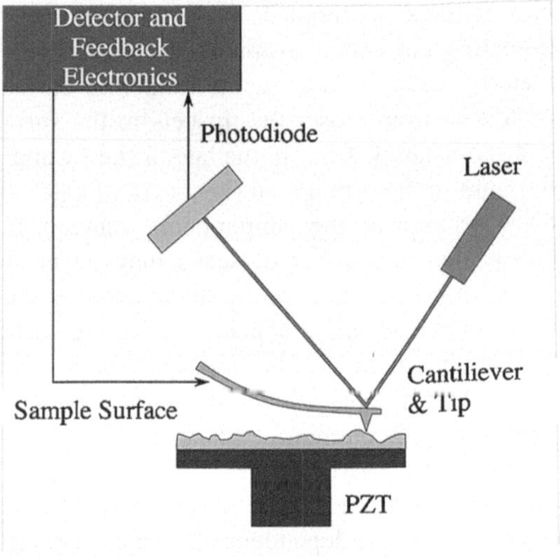

Fig. 5.21 Illustration of the setup for the atomic force microscope (AFM) that can be used to image the surface of a material at the atomic scale as well as measure other useful material properties

5.3.3 *Atomic Force Microscopy (AFM)*

The atomic force microscope (AFM) is similar to the STM described above since they both use sharp-tipped probes; however, the AFM differs from the STM in that it mostly uses a mechanical probe as opposed to an electrical probe to scan the surface of a material [21].

As shown in Fig. 5.21, the AFM uses a sharp tip at the end of a cantilever to scan across the material surface. The tip is typically made from silicon using micro-electromechanical systems (MEMS) fabrication techniques. Like the STM, the AFM

Fig. 5.22 SEM of an example AFM tip on the end of a cantilever

tip is brought into close proximity to the material surface usually using piezoelectric actuation, although electrostatic actuation can be used as well. As the tip comes into immediate nearness to the material surface, the cantilever will deflect based on the stiffness of the cantilever, that is, Hooke's law.

The mechanical forces that can be measured using the AFM include the mechanical contact forces; van der Waals attraction forces; chemical bonding forces; Casimir forces; capillary forces; etc.

The deflection of the AFM tip can be measured using a laser optical beam that is bounced off the top of the cantilever into a photodetector. Alternatively, the deflection can also be measured using capacitive sensors as well as other means. Using specialized probes, other quantities can be measured such as temperature and thermal conductivity. Figure 5.22 is a SEM of an example AFM tip and cantilever.

The AFM can be used in one of three imaging modes: contact mode; tapping mode; and the noncontact mode. The contact mode has the tip dragged across the surface, and a closed-loop feedback circuit is typically used in order to maintain the tip at a constant position while it scans the material surface. Contact mode imaging using AFM has several disadvantages. It is susceptible to noise and drift and also results in damage to the tip and the material surface, at least at the nanoscale. Additionally, many materials in ambient environmental conditions will develop a thin liquid layer of water on the surfaces that can cause the AFM tip to stick to the surface if brought into contact. Because of these reasons, contact mode AFM imaging is rarely used as a metrology method.

The tapping mode imaging method involves a cantilever with a tip at the end that is brought into oscillation at or near the cantilever's natural resonance frequency. The resonance is typically driven using a piezoelectric actuator. As the tip is brought into close proximity to the surface, certain atomic forces, such as van der Waals forces, will be exerted onto the apex of the tip that will result in a change in amplitude of the oscillation. Typically a closed-loop feedback loop circuit is used to maintain the oscillation amplitude, and the amount of signal required to maintain the constant amplitude is indicative of the height of the material surface at that specific location. As the oscillating tip is scanned over the surface, an image can be generated of the material surface. The tapping mode AFM imaging method does not result in same degree of damage as the contact mode imaging since the tip is only in contact in the vertical direction and for short periods of time.

Material surface imaging using AFM can also be performed using a completely noncontact mode. In this mode the tip never actually makes physical contact with the material surface. As with the tapping mode method described above, the noncontact mode involves driving the tip cantilever to its natural resonant frequency with a very small amplitude deflection, usually well below 10 nm. The surface van der Waals forces act over a distance of about 10 nm above the surface, and therefore as the tip is brought into close proximity, these van der Waals forces will exert forces on the oscillating tip and cantilever, thereby changing its resonant frequency. A closed-loop feedback circuit is used to maintain the amplitude or frequency and the signal required to maintain a constant amplitude or frequency that is indicative of the height of the material surface. As the oscillating tip is scanned over the surface, an image can be generated of the material surface. Noncontact mode AFM imaging is usually preferred over the contact mode for most applications.

The surface topology resolution of the AFM is the same as that as the STM or approximately 0.1 nm in the lateral (i.e., x and y coordinate) directions and about 0.01 nm in the orthogonal (i.e., z coordinate) direction. The resolution in force of the AFM is usually on the order of a few piconewtons. As with the STM above, calibration standards are available from the instrument makers and other organizations. Calibrated AFMs have been reported to have a step height inaccuracy of 0.4% and pitch inaccuracy of 0.1% [22].

The major application of AFM is for scanning non-electrically conducting surfaces. Unlike the STM which is based on a current passing through the gap between the tip and the material surface and thereby requires electrically conducting surfaces, the AFM can perform image surface scans of the surfaces of insulating materials. The AFM can also be used to measure van der Waals and Casimir forces on material surfaces.

A major disadvantage of these scanning probe methods, at least compared to conventional stylus profilometers, is that both STM and AFM have very slow scan rates and can only scan relatively small areas. Additionally, the typical maximum scanning area is only about 150 microns in the x- and y-directions and a maximum height on the order of 10 microns. Note, however, that if a square area image surface sample is to be taken, many scans would be required wherein each scan would take several minutes.

5.3.4 Energy-Dispersive X-Ray Spectroscopy (EDXS)

Energy-dispersive X-ray spectroscopy (EDXS) is a commonly used tool in microsystems manufacturing for the chemical analysis of samples [23]. This method is based on the interaction of the atomic structure of a material sample with an electron beam as a primary excitation that results in X-rays as secondary emission having a unique emission spectrum that can be detected.

In general, the X-rays are generated by impinging a sample with charged particles in the form of electrons. The impinging particle or X-ray beam stimulates electrons in the material sample from a ground state to higher energy states creating an unfilled electron energy state that is subsequently filled from electrons from higher energy states, thereby resulting in the emission of an X-ray. This technique has some similarity with Auger spectroscopy discussed below, except instead of Auger electrons being emitted, characteristic X-rays are emitted and analyzed. The emitted X-rays can be detected using a spectrometer that measures the energies and intensities of these emitted X-rays.

Figure 5.23 is an illustration of the interaction of an external stimulation, in the form of particles or X-rays that impinge on an electron state in an atom. This impinging stimulation results in an electron being ejected, thereby leaving an unfilled electron energy state as shown. Electrons from various higher energy levels can transition to the unfilled state. In Fig. 5.23, an electron from the L energy level fills the unfilled K level. This transition results in a Kα electromagnetic emission. These emissions provide information about the atomic energy levels of the material sample that can be used to determine its chemical makeup.

EDXS is in a widespread use in microsystems fabrication facilities since it is easily and inexpensively outfitted onto SEMs. It is used to determine the presence of chemical elements, both desired and undesired, and their amounts in material samples relatively quickly.

The accuracy of EDXS is dependent on a number of factors. For example, some elements have overlapping emission spectrums that make identification challenging. Also, the emission may be reabsorbed and not be able to escape the material sample depending on the energy of the emitted X-rays and the composition and density of the material sample. Due to this latter effect, most EDXS systems employ software correction procedures.

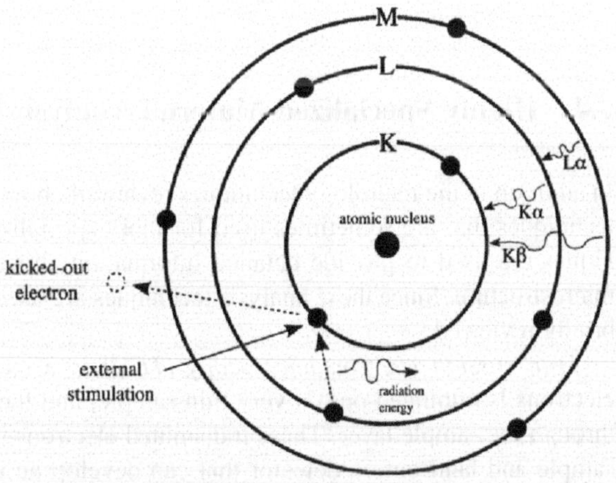

Fig. 5.23 Illustration of the interaction of an external stimulation and the energy levels of an atom in the material sample employed in the method of chemical analysis of energy-dispersive X-ray spectrometry

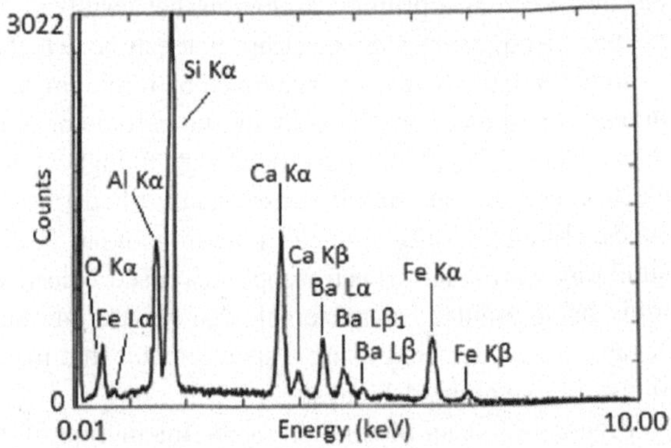

Fig. 5.24 An EDXS spectrum of NIST K309 glass showing its various elemental constituents

EDXS can be used to detect elements heavier than boron up to about uranium (atomic number 92). The limit of detection with EDXS is between 1000 and 3000 parts per million (ppm) or above 10% by weight percent. The spatial resolution varies with atomic number with low atomic number having a spatial resolution between 1 to 5 micron^3 and high atomic numbers having a range between 0.2 and 1 micron^3. The precision of EDXS is about ±0.1%. The accuracy depends on the sample analyzed and the calibration. In general, the accuracy is about ±10%. The spatial resolution is based on the fact that the X-rays may be generated deeply into the sample being analyzed. Therefore, EDXS is not considered a surface analysis technique. It can be used on insulators, but not on organics. EDXS does not provide information about the chemical state of the sample.

Figure 5.24 is an analysis performed using EDXS on a NIST sample of K309 glass. As can be seen, various elements are clearly identified including silicon, aluminum, calcium, etc. There is a different type of detector units available for use on SEMs that is called wavelength dispersive detectors (WDX) that provides lower levels of detection. WDX can provide detection limits between 30 and 300 ppm at 1% weight percent.

5.4 Highly Specialized Material Analysis Methods

In addition to the metrology techniques reviewed above, there are a number of other techniques that are sometimes used for more specialized applications. These techniques are used to provide detailed information about the material chemistry and microstructure. Since these analysis techniques are used sparingly, they will only be briefly reviewed.

Transmission electron microscopy (TEM) is a technique whereby a beam of electrons is impinged onto a very thin sample, and these electrons are able to pass through the sample layer. These transmitted electrons interact with the atoms in the sample and land onto a detector that can develop an image of the sample using a detector camera or equivalent. Most often the TEM system is able to scan and therefore is called a scanning transmission electron microscopy (STEM). TEMs are capable of atomic-level imaging and analysis of materials [24]. The typical applications for the TEM include atomic-level imaging and chemical analysis of materials; characterization of semiconductor devices and interfaces; and tomographic imaging and three-dimensional chemical spatial variations.

X-ray diffraction is a sophisticated technique that is used to measure the composition of materials, their crystal structure and phases, presence of strain, impurities, and defects in materials. It uses X-rays to irradiate a sample and observes the X-rays that penetrate through the sample and undergo diffraction effects. It is mostly used on crystalline materials [25]. X-ray diffraction has a lateral resolution of about 30 microns and a detection depth ranging from 0.1 micron to about 10 microns. The detection limit is typically about 1%. The accuracy of X-ray diffraction depends on what is being measured, but for determination of crystal structures, the accuracy is typically in the range of $\Delta 2\Theta < \pm 0.04°$ [26]. This is mostly due to the precision at which the angles of the sample and system can be aligned, the spread in the diffraction beam on the imaging plane, and other factors. X-ray diffraction can be used on crystalline forms of insulators, but not organics. It does not provide information about the chemical state of the sample.

Secondary ion mass spectrometry (SIMS) is a technique used to analyze the composition of substrates and thin films by sputtering the surface to be analyzed with a focused ion beam and collecting and analyzing the ejected (secondary) ions [27]. The mass and charge ratios of the secondary ions are measured using a mass spectrometer, which provides information about the atomic and molecular compositions of the region being analyzed. Typically, the depth of the ion beam sputtering is between 1 and 2 nm. The detection limits are between 10^{12} and 10^{16} atoms per cm^3 or in the parts per millions (ppm) to parts per billions (ppb) range. SIMS can be used to detect elemental materials with atomic numbers ranging from 1 to 92. The lateral resolution is about 60 microns, with a detection depth of up to 10 microns. The accuracy of SIMS is about +/−30% but can be improved using standards to calibrate the instrument. SIMS is commonly used in microsystems development and manufacturing for the analysis of trace elements in solid materials, especially semiconductors and thin films.

Auger electron spectrometry (AES) is an analytical technique for determining the elemental chemical composition of the surface of materials and used as an analytical technique in microsystems development [28–30]. In AES the primary form of excitation is an electron beam, and the detected secondary is an Auger electron. Auger spectroscopy is used in microsystems development and manufacturing for surface analysis. It is extremely useful if contamination is suspected and needs to be identified or the chemistry of certain materials used in production needs to be measured. This technique bombards the surface of a solid material with energetic electrons with energies ranging from about 2 keV to up about 10 keV. The Auger process essentially consists of three steps as follows: (1) the removal of a core electron from the material sample (i.e., atomic ionization) using a high every electron bombardment; (2) the emission of an electron which is called an Auger electron; and

(3) the analysis of the emitted Auger electrons. Auger spectrometry has a sensitivity of about 1% with a lateral resolution of about 200 Angstroms and a detection depth of 30 Angstroms. The instruments are usually equipped with an ion gun that removes material to allow the auger analysis to be performed deeper into the sample, thereby generating a depth profile. The elemental range of AES is all the elements with the exception of H and He but is usually used in the elemental range from 3 to 92. This method has an accuracy of about ±30% but can be improved with calibration using samples of known composition. Additionally, due to the charging effects due to an impinging electron beam, this method has not been considered appropriate for insulators. However, new techniques have recently been reported that enable Auger to be performed on insulating materials as well [30]. Auger is not suitable for analysis of organics and is useful for identifying the chemical state of a sample.

X-ray photoelectron spectroscopy (XPS), also sometimes called electron spectroscopy for chemical analysis (ESCA), impinges the sample material surface with monochromatic low-energy X-rays of accurately known energy magnitudes to cause photoelectron emission. XPS involves a primary excitation of the sample using X-rays, and the detected secondary is composed of a photoelectron [31]. XPS is considered a complementary analysis technique to AES. Some of the advantages of XPS include the following: some materials that dissociate under bombardment by electrons are undistributed by X-ray examination; XPS can be used on insulator materials where AES analysis can result in charging problems; and the energy resolution of XPS is higher than AES allowing analysis of chemical bonding of the material in addition to the elemental composition. XPS can be used to measure: the elemental composition of a material surface (confined to the top 0–10 nm unless outfitted with ion milling capability); the empirical formula of materials; identify any elements contaminating a material surface; the chemical and electronic state of elements on the material surface; the homogeneity of the elemental composition across the material surface; the homogeneity of the elemental composition as a function of depth into the material surface (if outfitted with ion milling capability). The range of materials that can be analyzed using XPS have atomic numbers from 3 to 92. The lateral resolution is about 10 microns, and the detection depth is around 30 Angstroms (however the analysis depth can be increased if outfitted with ion milling). The detection limit is about 1% with an accuracy of ±30%. XPS can be used to analyze insulators and some organics and can also provide information about the chemical state of the sample.

In X-ray fluorescence (XRF), incident X-rays (or gamma rays) are used to probe the material sample, and characteristic secondary (or fluorescent) X-rays are emitted and detected to analyze the sample. Specifically, XRF involves a primary excitation of the sample using X-rays, and the detected secondary is also composed of X-rays (at different energies and wavelengths). XRF can be used for elemental and chemical analysis [32]. Since XRF uses X-rays to probe the sample, as opposed to an incident electron beam, it offers some distinct advantages to analyze materials that would charge up if an electron beam was used. Therefore, it does not have the same problems with electronic charging as seen with EDXS, and XRF can be used for analyzing insulator materials. Additionally, some materials that decompose under an incident electron beam may be able to sustain X-ray probing with degradation. XRF can be used to identify elements in the range from at least carbon (element 6) to uranium (element 92). The lateral resolution is not as good as other methods (~1 cm), and large sample sizes (>1 g) are typically required for this technique. Since it uses incident X-rays to radiate the sample, it measures elemental species in the bulk sample and does not provide information about the depth profile. The accuracy and precision of XRF are quite good, with a typical accuracy of ±10% or better and a precision of at least 100 times better. Moreover, the detection limit of XRF is in the parts per billion range. It can be used to analyze insulating material types but is not appropriate for identifying organics. XRF is not appropriate for analysis of the chemical state of samples.

Laser ionization mass spectroscopy (LIMS) uses a highly focused laser beam to impinge the material surface being analyzed, and the laser beam results in material to laser desorb (LD) and laser ionize (LI) from the surface. The resultant material ions are then analyzed using a mass spectrometer [33]. Therefore, this method is very similar to SIMS discussed above. The detection limit of LIMS is quite good compared to other methods and is typically in the range of 10^{-4} to 10^{-2} atomic %. The laser beam probe diameter is typically around 1 micron, and all elements can be detected using this method.

Rutherford backscattering spectroscopy (RBS), also sometimes referred to as high-energy ion scattering (HEIS), is a technique that uses helium ions with an energy typically in the range of 1–4 MeV which are impinged onto a material surface that experiences elastic collisions with the atoms at or near the material surface. The collision of the high-energy ions with the material atoms results in backscattered ions that experience energy losses characteristic of the atom that was colliding with as well as the distance into which the collision occurred. Therefore, Rutherford backscattering spectroscopy provides information about the elemental composition of the material at the surface as well as how the elemental composition may vary as a function of depth into the material surface. The detection limit of RBS has a minimum of 5×10^{18} $atom^3/cm^3$, and therefore RBS is not one of the most sensitive methods of elemental composition analysis. Additionally, the spread of the incident ion beam is rather large, thereby limiting this techniques resultant lateral resolution to more than a few hundred microns. Furthermore, RBS does result in some small amount of surface damage. Lastly, the specificity of RBS varies with atomic number. For low atomic numbers, it has good specificity, while for the high atomic numbers, this quality factor degrades. The accuracy of has been reported as good as 1% [34].

Neutron activation analysis (NAA) is useful for detection of small concentration of some elements in materials. NAA uses a beam of thermal neutrons to impinge onto the material surface. This radiation by neutrons results in the creation of radioisotopes of the elements inside the material. The energy of the detected X-rays from these radioisotopes and their half-lives enables the identification of trace impurities with high sensitivity levels. The X-rays emitted by the radioisotopes are detected using a lithium-drifted germanium solid-state detector and analyzed using a multichannel analyzer system. This technique is most useful in determining the presence of trace levels of impurities in materials. NAA can detect concentrations of heavy elements such as gold, copper, tungsten, etc. down to levels of 10^{11} to 10^{12} atoms/cm^3. However, since lighter elements such as boron, carbon, and oxygen do not produce radioisotopes that are detectable using NAA, this technique is limited to elements heavier than oxygen (atomic number 8). NAA is very useful for detection of trace levels of impurities in materials, particularly as a result of processing steps that may be thought to cause contamination and evaluating the effectiveness of gettering processes and materials. The detection limits vary on the element and can range from 10 to 100 pg for gold (Au) to 1 to 10e^5 pg for niobium (Nb). The sample sizes needed for NAA range from 1 to 200 mg or more. Accuracy and precision of NAA are dependent on the amount of the element that is present, but a typical value is between 2% and 5% for each, assuming that the instrument has been properly calibrated [35].

Raman spectroscopy is a chemical analysis technique that uses monochromatic light in the visible, near-infrared, or ultraviolet regions of the spectrum to induce inelastic scattering, called Raman scattering, with the material sample. The incident laser radiation interacts with the molecular vibrations and phonons in the material that interacts with the incident light resulting in energy shifts in the photons, and these energy shifts provide information about the vibrational modes that are indicative of the molecular composition of the material [36]. Raman spectroscopy has a lateral resolution of about 2 microns and a detection depth of below 1 micron to up to 10 or more depending on the transparency of the material. The detection limit is about 0.1%, but the accuracy is poor with a typical range of ±100%. Raman spectroscopy can be used for the analysis of insulating materials and organics. It also can be used to provide the chemical state of molecular species in the sample.

Fourier-transform infrared spectroscopy (FTIR) involves the measurement of how much incident broadband infrared radiation composed of a predetermined combination of wavelengths is absorbed as a function of the wavelength. This measurement is repeated by modifying the incident IR radiation so as to be composed of a different combination of wavelengths and measuring the absorption of this modified beam as a function of wavelength. This process may be repeated many times. The received data from a detector located on the other side of the irradiated sample is then analyzed to using Fourier transforms to calculate the absorption at each wavelength. This spectrum of the absorption at each wavelength represents the unique molecular identity of the sample since different chemical molecules produce different spectral absorption curves. FTIR uses IR radiation as a primary excitation source and IR radiation as the detected secondary [37]. The lateral resolution of FTIR is between 10 and 100 microns, and it has a detection depth of range from 20 Angstroms to about 1 micron. The detection limit of FTIR is in the parts per million (ppm) range. The accuracy of FTIR is ±5%. It can be used to analyze insulating materials as well as organics. Also, FTIR can provide valuable information about the chemical bonding present in a sample. FTIR is useful in microsystems manufacturing for analyzing polymer and inorganic materials and their chemical bonding states. It has also been employed for use in analyzing and monitoring the precursors and active species in CVD and etching processes [38].

5.5 Electrical Material Properties Test Methods

This section reviews test structures and methods that are used to determine the electrical properties of materials, such as the junction depths of dopants introduced into semiconductor and the amount of dopants introduced into materials.

5.5.1 Junction Depth Measurements

Microsystems manufacturing often involves the introduction of dopant impurities into semiconductor materials in order to alter the electrical properties of the material. For silicon, the most often used dopants are boron and phosphorus, where the introduction of boron results in a silicon material that has an excess of holes and the introduction of phosphorus the excess of electrons. As discussed previously, these dopants can be introduced into the semiconductor using diffusion or ion implantation wherein implantation is usually followed by annealing to electrically activate the implanted elements.

Often, especially when creating electrical functionality, these dopants are introduced into a semiconductor material having an inherent opposite background dopant types so as to create electrical junctions in the material. For example, a silicon substrate with a background of phosphorus doping (with an excess of electrons and typically called n-type silicon) may be doped in surface regions with boron (with an excess of holes and called p-type silicon).

Frequently it is useful to know where the junction is located in the semiconductor material. The junction depth is defined as the boundary in a semiconductor substrate between two different dopant types. This section describes how to measure where in the semiconductor these junctions are located.

There are several basic techniques that have been developed over the years to measure the junction depths in semiconductors including the angle-lapping and stain method; the grove and stain method; and the tapered groove and stain method. These methods all rely on machining the semiconductor substrate to create an angle or groove in the substrate with accurately known dimensions that exposes the junction; followed by a staining procedure that clearly delineates the junction; and the performance of measurements of the exact location of the junction in relation to the angled or grooved machining that was performed and then calculations based on the geometry and measure junction in relation to the machined features.

For example, the angle-lapping and staining method involves attaching the semiconductor substrate onto an angled block (having a precisely machined angle of between 1- and 5-degree angle) and then grinding the semiconductor substrate using a slurry to create an angled feature into the substrate surface. Once the angle has been created and the substrate cleaned, a staining solution is applied that may be composed of 49% HF or 49% HF combined with a few drops of HNO_3 (or other chemical mixtures) and then the junction location measured using a strong illumination source. Illumination may need to be applied to the stained region for a few minutes before the junction is clearly noticeable. Ultraviolet lighting is sometimes used to illuminate the sample.

The angle-lapping and staining method has an ASTM standard (F-110-84), but this method has an accuracy that is limited to about ±0.5 microns. However, if performed with extreme care, the accuracy with this method can be as good as ±200 Angstroms. The groove and staining method has an accuracy of ±1% for junction depths as shallow as 500 Angstroms. The tapered groove and staining method is able to measure the junction depth to within ±200 Angstroms, and if a TEM is used in locating the junction, it can measure the junction depth to within ±50 Angstroms [39].

While all of these methods have been around and in use for many years, their accuracy highly depends on the skill and experience of the operator particularly with the machining and measuring processes that are used. Most often, microsystems manufacturing and process development engineers will not attempt to undertake these measurements themselves but send representative samples to organizations that routinely perform these services on a fee-for-service basis.

5.5.2 *Spreading Sheet Resistance*

While measurement of the location in the semiconductor substrate of the junction depth is of use, it is usually more useful to know the junction depth combined with the doping profile in the material. As discussed in Chap. 3, the introduction of dopants using diffusion and ion implantation into a semiconductor does not result in a uniform doping level as a function of depth into the material but instead has a complicated continuous and curved doping profile into the material that can be modeled using appropriate mathematical equations. In any case, most microsystems manufacturers will use doping profile measurements as opposed to junction depth measurements since the information provided by the doping profile is usually more informative.

Spreading sheet resistance measurements are widely used in microsystems manufacturing for determining the junction depths and dopant profiles in semiconductors. This technique illustrated in Fig. 5.25 involves machining the semiconductor substrate to create a precisely known beveled angle in the surface that exposes the junction(s) and then applying a two-point electrical probe multiple times across the beveled angle at further positions along the beveling to measure the sheet resistance, R_{SR}, according to the simple equation:

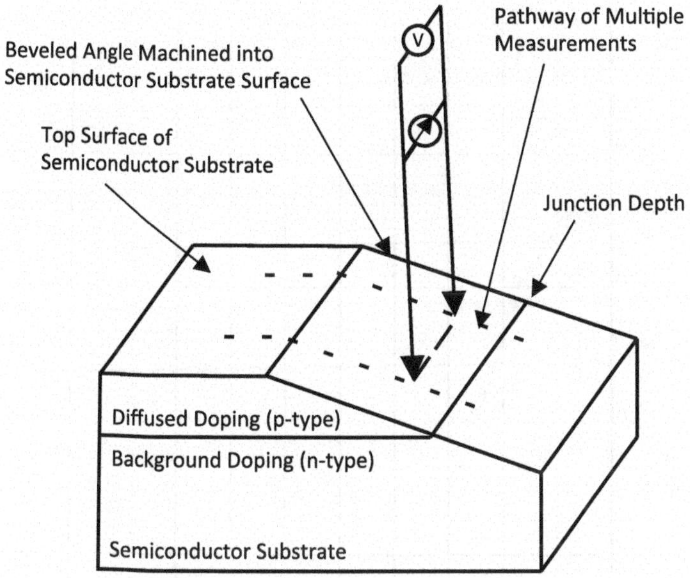

Fig. 5.25 An illustration of the spreading sheet resistance method for measuring the junction depth and doping profile as a function of depth in semiconductor substrates. As can be seen, the semiconductor substrate has been machined to define a known angle at the surface that exposes the junction. Subsequently, a two-point electrical probe is repeatedly placed into contact with the semiconductor with the probes precisely aligned along the beveling. Once in contact at each point, a known current is applied between the probes and the resultant voltage is measured. This measurement may be performed 500 times or more along a well-defined pathway

$$R_{SR} = \frac{V}{I}. \tag{5.10}$$

A known current is applied between the two probes, and the resultant voltage across the probes is measured at each location. The resistivity, ρ, of the material at each location is given by:

$$\rho = 2R_{SR}a, \tag{5.11}$$

where a is an empirical quantity associated to the effective electrical contact that determined by measuring the sheet resistance on a material with a known resistivity.

The electrical probes are moved along a defined pathway along the beveled surface of the substrate, keeping the probe positions parallel to intersection of the beveling with the top surface of the substrate, to measure the resistivity of the substrate as a function of position and therefore depth into the substrate surface. From the resistivity measurements, the doping levels for each measurement position can be determined as a function of depth.

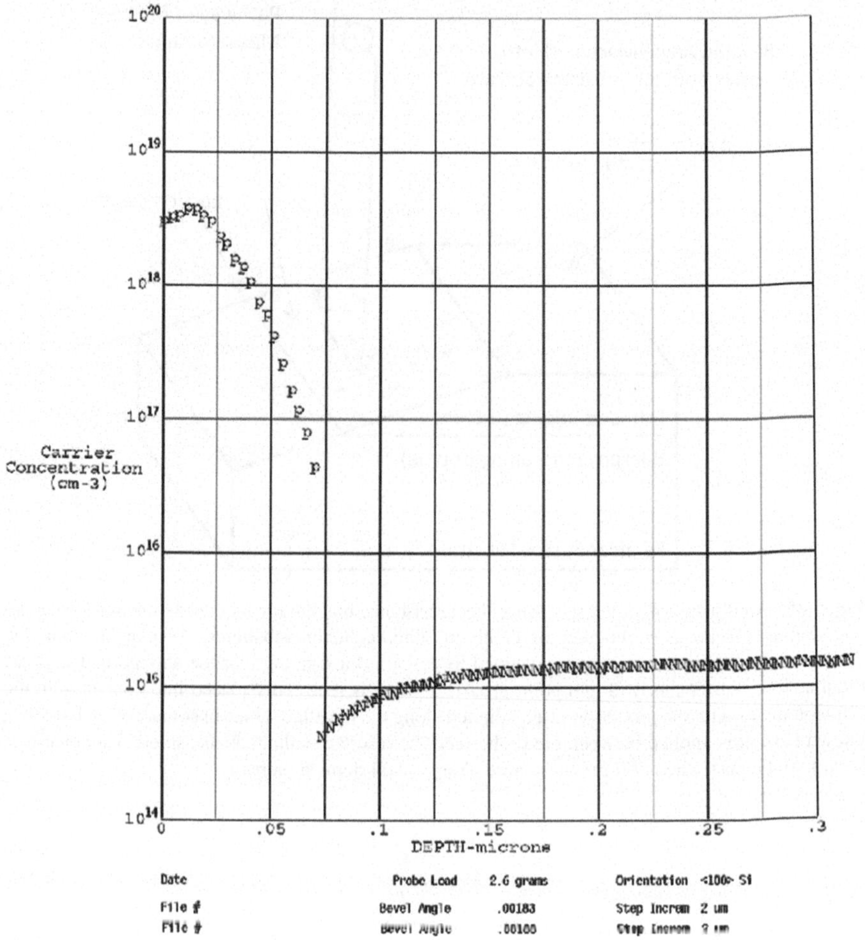

Fig. 5.26 A spreading sheet resistance profile measurement data for a silicon wafer having a background n type doping that has had the surface doped with a p-type dopant. As can be seen, the junction between the n- and p-type regions is about 0.75 microns deep into the substrate

A representative sheet resistance profile of a silicon wafer that had an n-type background carrier concentration that was doped at the surface with a p-type dopant is shown in Fig. 5.26. The curve in Fig. 5.26 plots the carrier concentration, in number of carriers per cubic centimeter, as a function of depth in the substrate (in microns) as measured from the top of the surface. As can be seen, the profile near the surface indicates p-type doping with a curved profile as a function of depth with a junction depth of slightly less than 0.75 microns deep in the silicon substrate.

It is important to point out that spreading sheet resistance data regarding concentrations in the substrate are limited to detection of only electrically active dopants. That is, this method relies on the measurement of the electrical resistivity of the sample material based on the introduction of dopants that create either electron or hole charge carriers. It does not measure electrically inactive dopant concentrations.

Spreading sheet resistance measurements, like junction depth measurements, rely on the ability to make accurately machined bevels into the substrate as well as electrical probes that are well conditioned and calibrated. Like junction depth profile measurements described above, most facilities send out the substrates to service organization that perform these measurements on a fee-for-service basis.

Spreading sheet resistance measurements are an attractive method due to the high resolution of depth measurements coupled with a high dynamic range. The resolution on depth has been reported to be with the nanometers, and the dynamic range of resistances is nine orders of magnitude. It has also been shown that spreading sheet resistance can provide submicron profiles of doping concentrations with an accuracy of 10% [40].

It should be noted that secondary ion mass spectroscopy (SIMS) that was discussed above in the section above can also be used as a method for measuring the impurity concentration profile in a substrate material. Unlike the spreading sheet resistance method discussed above, SIMS measures both electrically active and inactive impurity concentrations. To provide quantitative data on concentration levels, SIMS must be calibrated to a known standard. Nevertheless, SIMS is considered a more accurate method for measuring impurities than spreading sheet resistance.

5.6 Summary

This chapter has reviewed the methods and tools of metrology that are commonly used in microsystems manufacturing including those used during fabrication to monitor the quality of fabrication process steps; more specialized metrological methods that are used infrequently highly specialized materials analysis methods; and electrical property measurements. Tables 5.2, 5.3, 5.4, and 5.5 provide a quick look-up summary of these metrology methods along with their resolution, precision, and accuracy, if it is known. The numbers given in these tables should be considered as guidance since the resolution, precision, and accuracy should be independently determined or validated for each metrology tool and method.

Table 5.2 Fabrication metrology equipment and methods

Fabrication metrology	Typical best-case resolution, precision, and accuracy
Optical microscopy	Horizontal dimensions: resolution ≈ 1 um; precision 10–20%; and accuracy ≈ 1 um (assumes 1000× & calibrated)
	Vertical dimensions: height differences of 50 Angs. can be resolved with DIC; precision and accuracy ≈ 1 um with calibration
Fluorescence microscopy	Not used for dimensional measurements
Confocal microscopy	Resolution ≈ 150 nm; precision 10–20%; and accuracy ≈ ±50 nm (assumes calibration)
Stereomicroscopy	Not used for dimensional measurements.
Scanning electron microscopy (SEM)	Resolution ≈ 1 nm to a few nms; precision 1% at low mag.; accuracy ±1–3% for objects more than 1 um.
Automated scanning electron microscopy	Resolution ≈ 1 nm to a few nms; precision 1% at low mag.; accuracy ±1–3% for objects approx. 1 um.
Thin-film thickness	
Interferometry (reflectometry)	Dynamic range: 500 Angs. to 10 um; accuracy ±2%
Ellipsometry	Dynamic range: 20 Angs. to 10+ um; accuracy ±3 Angs. in thickness and ± 0.005% in index of refraction
Stylus profilometry	Dynamic range: 5 nm to 1 mm; vertical resolution 1 nm with precision of 0.5 nm; horizontal resolution >20 nm; accuracy 0.5–2.5%.
Four-point probe	Dynamic range: 1 m-Ohm per square to 1 M-Ohm per square; accuracy <1%
Thin-film stress measurement	Dynamic range: 1 MPa to4 GPa; resolution 1 MPa; accuracy 2.5% or 1 MPa whichever is greater
Particle analysis	Resolution of particles: 30 nm or larger
Noncontact optical profilometry	Dynamic range: 1 nm to 20,000 um; precision 3.5 nm in CSI mode and 0.08 nm in PSI mode. Accuracy ≈ 1 nm if calibrated
Wafer bonding inspection	IR resolution about 1.1 um; ultrasonic 0.1 um. Not used for dimensional measurements

Table 5.3 Specialized metrology equipment and methods

Specialized metrology	Typical best-case resolution (or detection limit), precision, and accuracy
Focused ion beam (FIB)	Beam resolution ≈ 5–10 nm; helium source resolution 1 nm; precision and accuracy of measurements ≈ SEM
Scanning tunneling microscopy (STM)	Lat. Res. ≈ 0.1 nm; Vert. Res. ≈ 0.01 nm; precision ≈ 0.1 nm; accuracy <1 nm
Atomic force microscopy (AFM)	Lat. Res. ≈ 0.1 nm; Vert. Res. ≈ 0.01 nm; precision ≈ 0.1 nm; step height inaccuracy ≈ 0.4%; pitch inaccuracy ≈ 0.1%
Energy-dispersive X-ray spectroscopy (EDXS)	Lat. Res. ≈ 10 um; detection depth ≈ 30 Angs.; detection limit ≈ 1%; accuracy ≈ ±30%

Table 5.4 Highly specialized metrology equipment and methods

Highly specialized metrology	Typical best-case resolution (or detection limit), precision, and accuracy
Transmission electron microscopy	Spatial resolution 0.136 nm at 300 kV in STEM mode; accuracy similar to Res. if calibrated
X-ray diffraction	Lat. Res. ≈ 30 um; detection limit ≈ 1%; accuracy of crystal struct. ≈ D2Q ± 0.04-degrees
Secondary ion mass spectrometry (SIMS)	Detection limit: ≈ $10e^{12}$ to $10e^{16}$ atoms/cm^3 (ppm to ppb range); lateral resol. ≈ 60 um; Detection Depth ≈ 10 um; Accuracy ±30%
Auger electron spectroscopy (AES)	Sensitivity ≈ 1%; Lat. Res. ≈ 200 Angs.; detection depth ≈ 30 Angs.; Accuracy ±30%
Energy-dispersive x-ray spectroscopy (EDXS)	Lat. Res. ≈ 10 um; detection depth ≈ 30 Angs.; detection limit ≈ 1%; Accuracy ≈ ±30%
X-ray photoelectron Spectroscopy (XPS)	Detection limit: ≈ 1000–3000 ppm; precision ≈ ±0.1%; accuracy ≈ ±10%
X-ray fluorescence (XRF)	Lat. Res. ≈ 1 cm; Detection Limit ≈ ppb; Precision ≈ ±0.1%; Accuracy ≈ ±10%
Laser ionization mass spectroscopy (LIMS)	Detection limit: $10e^{-4}$ to $10e^{-2}$ atomic%; Lat. Res. ≈ 1 um; accuracy ±30%
Rutherford backscattering spectroscopy (RBS)	Detection limit: 5e18 atoms3/cm^3; accuracy ≈ 1%
Neutron activation analysis (NAA)	Detection limit: 10–100 pg for Au; accuracy ≈ 2–5%
Raman spectroscopy	Lat. Res. ≈ 2 um; detection depth 1–10 um; detection limit ≈ 0.1%; accuracy ±100%
Fourier-transform infrared spectroscopy (FTIR)	Lat. Res. 10–100 um; detection depth 20 Angs. to 1 um; detection limit ppm range; accuracy ±5%

Table 5.5 Electrical material properties metrology methods

Electrical metrology	Typical best-case resolution (or detection limit), precision, and accuracy
Capacitance-voltage (CV) Measurements	Accuracy of CV is 0.1%
Electrical test structures for process control	Not typically used to make measurements but instead used to determine presence of electrical shorts and discontinuities

References

1. S. Wolf, R.N. Tauber, *Silicon Processing for the VLSI Era, Volume 1: Process Technology* (Lattice Press, Sunset Beach, 1986)
2. K.R. Spring, M.W. Davidson, Introduction to fluorescence microscopy, in *Nikon Microscopy U* (Retrieved 2017)
3. https://www.leica-microsystems.com/confocalsr-na/?nlc=2017419-SFDC-002728&utm_source=google&utm_medium=cpc&utm_term=confocal%20microscopy&utm_

campaign=Confocal_SP8_Generic&gclid=EAIaIQobChMI3Nayv-HZ1QIVj4p-Ch0AGg-4EAAYASAAEgJ35vD_BwE, Accessed Sep 2017
4. S.E. Mechels, M. Young, Scanning confocal microscope for accurate dimensional measurement, in *Proceedings SPIE/IS&T, symposium on electronic imaging: Science and technology*, vol. 1660 (1992)
5. http://www.hitachi-hightech.com/global/product_detail/?pn=em-su9000
6. A.D. Kammers, S. Daly, Digital image correlation under scanning electron microscopy: Methodology and validation. Exp. Mech. **53**(9), 1743 (2013)
7. A.C. Diebold, *Handbook of Silicon Semiconductor Metrology* (Marcel Dekker, New York, 2001)
8. R.N. Yamaguchi, H. Kawada, *CD-SEM Technologies for 65-nm Process Node, Hitachi Technical Note* (2005). See: http://www.hitachi.com/rev/pdf/2005/r2005_01_103.pdf
9. E. Langereis, S.B.S. Heil, H.C.M. Knoops, W. Keuning, M.C.M. van de Danden, In situ spectroscopic ellipsometry for atomic layer deposition. Presented at the 52nd SVC technical conference, optical coatings session, Santa Clara, 2009
10. E. Novak, Understanding the capabilities of your surface measurement system. Bruker, Technical Notes (2012). See: https://www.bruker.com/fileadmin/user_upload/8-PDF-Docs/SurfaceAnalysis/StylusProfilometry/Webinars/Understanding_the_capabilities_of_your_measurement_system-May_webinar-slides.pdf
11. L. Chen, J. Miao, L. Guo, R. Lin, Control of stress in highly doped polysilicon multi-layer diaphragm structure. Surf. Coat. Technol. **141**, 96–102 (2001)
12. M.A. Brown, A.J. Rosakis, X. Feng, Y. Huang, E. Ustundag, Thin film/substrate systems featuring arbitrary film thickness and misfit strain distributions. Int. J. Solids Struct. **44**(6), 1755 (2007)
13. http://www.tohotechnology.com/flx-series.php#Specification
14. https://www.iso.org/search/x/query/cleanrooms
15. K.J. Stout, L. Blunt, *Three Dimensional Surface Topography* (Elsevier, Amsterdam, 2000)
16. M. Huff, PhD thesis (MIT, 1988)
17. http://www.sonix.com/wafer-inspection/applications/bonded-wafer-inspection
18. Zeiss Orion Helium Ion Microscope Technical Data. See: http://www.smt.zeiss.com/C1256E4600307C70/EmbedTitelIntern/ORIONEssentialSpecificationPDF/$File/oriondata.pdf. Accessed 2017
19. C. Bai, *Scanning Tunneling Microscopy and its Applications* (Springer, New York, 2000)
20. C.J. Chen, Origin of atomic resolution on metal surfaces in scanning tunneling microscopy. Phys. Rev. Lett. **65**(4), 448–451 (1990)
21. G. Binnig, C.F. Quate, C. Gerber, Atomic force microscope. Phys. Rev. Lett. **56**, 930 (1986)
22. https://www.photonics.com/Article.aspx?AID=3115
23. J. Goldstein, *Scanning Electron Microscope and X-Ray Microanalysis* (Springer, New York, 2003)
24. B. Fultz, J. Howe, *Transmission Electron Microscopy and Diffractometry of Materials* (Springer, Berlin, Heidelberg, 2007)
25. H.P. Klug, L.E. Alexander, *X-Ray Diffraction Procedures for Polycrystalline and Amorphous Materials* (Wiley and Sons, New York, 1974)
26. H. Ress, L. Bruegemann, *Good Diffraction Practice* (Webinar, Bruker, 2010). See https://www.bruker.com/fileadmin/user_upload/8-PDF-Docs/X-rayDiffraction_ElementalAnalysis/XRD/Webinars/Bruker_AXS_Good_Diffraction_Practice_III_-_Powder_XRD.pdf
27. A. Benninghoven, F.G. Rüdenauer, H.W. Werner, *Secondary Ion Mass Spectrometry: Basic Concepts* (Instrumental Aspects, Applications, Trends, Wiley, New York, 1987)
28. T.A. Carlson, *Photoelectron and Auger Spectroscopy* (Plenum Press, New York, 1975). B. David, M.P. Seah, *Practical Surface Analysis by Auger and X-ray Photoelectron Spectroscopy* (John Wiley & Sons, New York, 1983)
29. M. Thompson, M.D. Baker, A. Christie, J.F. Tyson, *Auger Electron Spectroscopy* (John Wiley & Sons, New York, 1985)

30. G.S. Spasov, Method for auger analysis of thick insulators samples. Annu. J. Electron., 40 (2013). ISSN 1314-0078
31. J.F. Watts, J. Wolstenholme, *An Introduction to Surface Analysis by XPS and AES* (Wiley & Sons, Chichester, 2003)
32. L. Vincze, Confocal X-ray fluorescence imaging and XRF tomography for three-dimensional trace element microanalysis. Microsc. Microanal. **11**, 682 (2005)
33. S. Becker, *Inorganic Mass Spectrometry: Principles and Applications* (John Wiley & Sons, New York, 2008)
34. C. Jeynes, N.P. Barradas, E. Szilagyi, Accurate determination of quantity of material in thin films by Rutherford backscattering spectroscopy. Anal. Chem. **84**, 6061 (2012)
35. Z.B. Alfassi, *Chemical Analysis by Nuclear Methods* (John Wiley and Sons, New York, 1994)
36. D.J. Gardiner, *Practical Raman Spectroscopy* (Springer, Berlin/New York, 1989)
37. P.R. Griffiths, J.A. De Haseth, *Fourier Transform Infrared Spectrometry*, 2nd edn. (John Wiley & Sons, New York, 2007)
38. W.S. Lau, *Infrared Characterization for Microelectronics* (World Scientific Publishing, Singapore, 1999)
39. S. Prussin, Junction depth measurements for VLSI structures. J Electrochem Soc **130**(1), 184 (1982)
40. T. Clarysse, W. Vandervorst, E.J.H. Collart, A.J. Murrell, Electrical characterization of ultra shallow dopant profiles, in Analytical and Diagnostic Techniques for Semiconductor Material, Devices, and Processes, Joint Proceedings of the Symposia on ALTECH 99, The Electro-Chemical Society, Leuven, Belgium, ECS PV 99-16, 1999

References

[illegible] analysis of [illegible] and [illegible] samples. [illegible]
[illegible]
[illegible]
[illegible]
[illegible]
[illegible] in the [illegible] Anal Chem [illegible]
[illegible] Wiley and Sons, New York [illegible]
[illegible] New York [illegible]
[illegible] Wiley [illegible]
[illegible]
[illegible] Publishing [illegible]
[illegible]
[illegible] and [illegible] Techniques [illegible]
[illegible]
[illegible]

Chapter 6
Microsystems Material Properties

6.1 Introduction

This chapter discusses the properties of the materials used in the manufacturing of MEMS microsystems devices. This is an important topic due to the fact that the properties of thin-film material layers are highly dependent on the processing conditions. Specifically, the processing conditions, such as deposition temperature, gas rates, growth rates, tool type, etc., all have an impact on the material properties of deposited thin-film layers. Additionally, any subsequent processing steps performed after a thin-film layer has been deposited can impact the properties of the material layers as well. The processing condition usually having a significant impact on the material properties includes any processing steps performed at elevated temperatures.

Since material properties are dependent on the processing conditions, including the entire process sequence, and MEMS process sequences are often customized, prior knowledge of the properties may not be possible when the initial design is performed. Additionally, attempting to design a MEMS device without knowledge of the material properties can be very challenging. This is like a structural engineer designing a bridge without knowing the strength of the steel beams to be used in the construction of the bridge.

A method to overcome this problem that is commonly employed in MEMS development is to perform the process sequence (or some portion of it) and subsequently measure the resultant material properties. Test structures and techniques commonly used to measure some of the more important material properties are reviewed later in this chapter. Once the material properties have been determined, then these values can be used in the device designs. It is strongly recommended in any MEMS device development effort that the material properties and any associated variability in the values be experimentally determined to the extent possible.

M. Huff, *Process Variations in Microsystems Manufacturing*, Microsystems and Nanosystems, https://doi.org/10.1007/978-3-030-40560-1_6

Note that the discussion above was limited to thin-film material layers. This is due to the fact that semiconductor bulk materials, such as the substrates[1] used in MEMS manufacturing, usually do not have material properties that vary significantly from the bulk values published in material property databases. There are exceptions. Specifically, the electrical properties of bulk semiconductors can be altered radically by the type and amount of dopant introduced into the material. This would include the resistivity values and the majority carrier types. However, it is important to note that the processing steps involved in impurity doping of semiconductors are very well understood and can be accurately predicted and modeled. That is, microsystems designers and developers can usually control the resultant electrical properties in bulk semiconductors with reasonably good accuracy and predictability. The same is mostly true of the electrical properties of thin-film layers as well.

The same is not true of some of the mechanical properties of deposited thin-film material layers. Additionally, the mechanical material properties of deposited thin-film layers can be significantly different from the values in databases. Because many MEMS devices use deposited thin-film layers for a mechanical functionality, the mechanical properties of these material layers are very often of significant importance.

Control of the material properties in MEMS manufacturing is also important. In the present context, control can have two meanings. In the simpler instance, control is the ability to obtain the same values of material properties when performing the processing step (and process sequence) over and over again. This level of control is usually achievable to some extent. There is metrology and statistical process control (SPC) methods to monitor and even predict when the process drifts outside of an acceptable range in microsystems manufacturing. These techniques will be described in a later chapter. However, even when the property of a deposited material layer is known, it must be remembered that the processing steps performed after that thin film has been deposited can still modify the properties.

A much more difficult level of control is the ability to tailor the processing conditions to obtain a desired specific value of a material property. First, it is important to recognize that the range over which the properties can be modified by varying the process recipe is often limited for most materials. Therefore, it is useful for the MEMS engineer to be cognizant of what is realistic with regard to obtaining a specific desired material property value. Second, it is considerably easier to be able to tailor a single-layer material property to a desired value rather than the material property of multiple layers. This is due to the fact that different materials often respond differently to processing conditions.

Nevertheless, it usually is possible to develop deposition recipes to modify the material property values over a certain limited range of values. Oftentimes, to develop a recipe to obtain a desired material property value will require performance

[1]It should be noted that many bulk semiconductor materials exhibit anisotropic material properties whereby the material properties vary depending on the orientation of the crystal lattice of the semiconductor. However, this is different from what is being discussed in this chapter.

of an extensive experimental study, specifically a design of experiments (DOE), whereby the most important process parameters of the processing step are varied, the results are measured, and an optimal recipe can be derived from the data. However, it should be noted that performing a DOE is a time-consuming and expensive effort. The topic of process recipe design is complicated and will not be further discussed in this volume.

In many cases of MEMS microsystems device development, the engineers decide not to tailor the material properties to be a specific value. Instead, they use already existing process recipes that the foundry has available with the corresponding material property values in the device design. In these cases, it is common to modify the other parameters of the device design so as to accommodate these material property values.

In some cases, the MEMS engineers may be able to use an existing process sequence from a foundry for the manufacturing of their device. In this instance, the properties of the deposited material layers are usually known including the values of the material property after the entire process sequence has been performed. While this situation makes the design and development of a MEMS device much less costly, risky, and time-consuming, it should be noted that MEMS process sequences are notoriously specified to a specific type of device. Therefore, most MEMS devices necessitate the design and development of a fully or partially customized process sequence.

Often the material property that has the most interest and impact in MEMS manufacturing is the residual stresses in thin-film layers. Another material property that is of interest in MEMS manufacturing is Young's modulus. Therefore, this chapter will focus on these two properties. First, we will explain the basics of residual stress and Young's modulus of materials. Second, we will review some of the material properties, with special interest in the residual stress and moduli, of the most commonly used thin-film materials and processes in MEMS manufacturing. The objective is not to provide the reader with property values of these materials for use in their design endeavors, but instead to provide an appreciation on how much and over what ranges that these properties can vary. Importantly, the material properties for any given layer and process sequence should always be determined (or verified) by experimentation.

6.2 Residual Stress and Young's Modulus

The mechanical material properties that often are of the greatest interest in deposited thin-film layers are the residual stress and Young's modulus [1]. The interest in these two thin-film material properties is based on several reasons, the principal ones being many MEMS microsystems devices have a mechanical functionality wherein these two material properties have a large impact on the resultant device behavior; and, both of these material properties can vary over large range of values depending on the exact processing conditions used during fabrication.

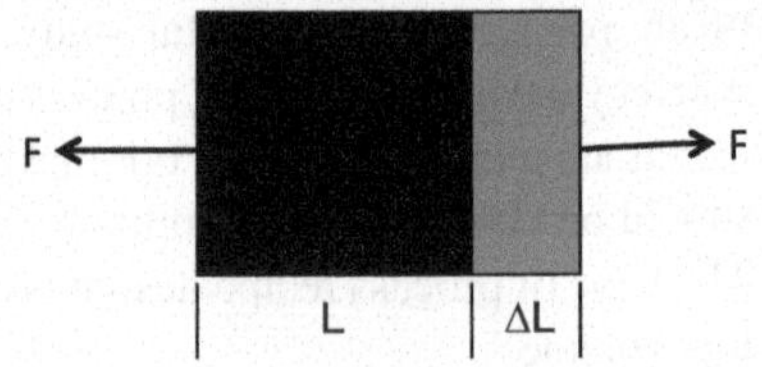

Fig. 6.1 An illustration of a bar of steel having a length L and area A with a force F applied along the length. As a result of this load, the bar lengthens by an amount of ΔL

In comparative terms however, the variations seen in the residual stresses are considerably larger than seen in Young's modulus. That is, it is not uncommon to see variations in residual stresses range over 100s of percent as well as change sign from compressive to tensile or vice versa depending on the processing conditions used in processing the thin-film layers.

This is not to imply that other material properties do not matter in MEMS microsystems. Indeed, depending on the device application and design, other mechanical properties, as well as the electrical and optical properties, may be more important. For example, the behavior of a MEMS thermal device will depend heavily on the thermal conductivity and thermal capacitance values of the material layers used in the device. However, the amount of variation typically seen in the thermal properties of most thin-film material layers will usually be limited to a relatively small amount (e.g., often less than $\pm 10\%$ of the nominal bulk value) depending on the processing conditions. It is useful to review the definition of Young's modulus and residual stress in deposited material layers.

6.2.1 Young's Modulus

Young's modulus (also called the elastic modulus) is an important mechanical material property and is a measure of the stiffness of a material in the linearly elastic region. A solid material having a load applied will undergo a deformation that is proportional to the magnitude of the load. For example, a steel bar having a uniformly load applied along the long axis as shown in Fig. 6.1 pulls the material along its length. This loading results in a small but measureable deformation in the amount of ΔL. The load applied is expressed by the stress, $\sigma(\varepsilon)$, given by the force, F, per unit area, A, as follows [2]:

$$\sigma(\varepsilon) = \frac{F}{A}. \tag{6.1}$$

Wherein the stress is a function of the strain given by e and the linear strain is given by:

$$\varepsilon = \frac{\Delta L}{L}. \tag{6.2}$$

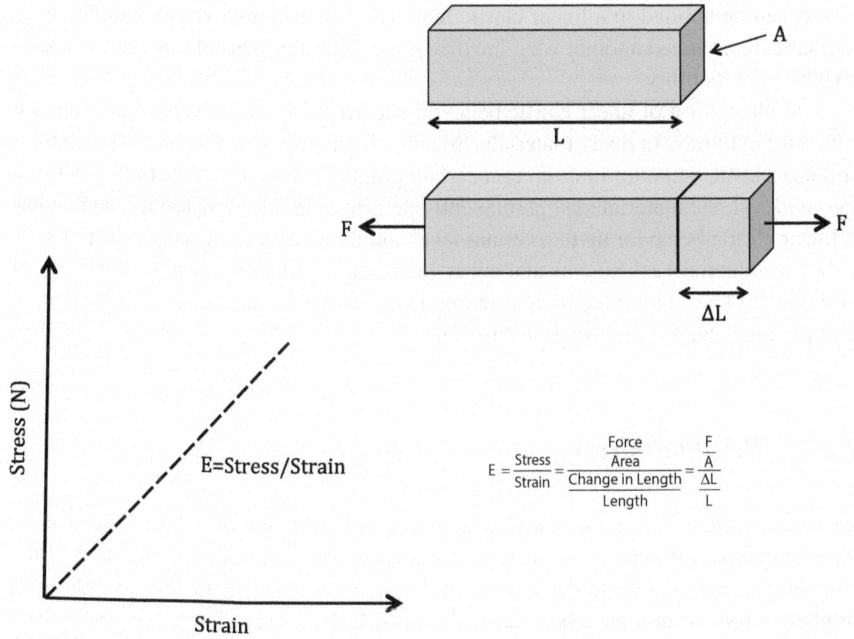

Fig. 6.2 Illustration of how Young's modulus is defined. A rectangular-shaped sample, with cross-sectional area, *A*, and length, *L*, is subjected to a force, *F*, on each of the ends. The sample lengthens by an amount of ΔL. As the force (or stress) level increases, the value of ΔL (or strain) increases linearly and proportionally to the applied force (or stress). The constant of proportionally between the stress and the strain is defined as Young's modulus, *E* [2]

In graphical terms, the stress-strain relationship in the linear elastic region, that is, where Young's modulus is determined, is illustrated in Fig. 6.2. The linear relationship between the stresses applied to the sample versus the strain levels for each value of stress is plotted in the figure in the bottom left. The slope of this line is the value of Young's modulus for the material the sample is made from. In equation form, this can be written as follows [2]:

$$E = \frac{\text{Stress}}{\text{Strain}} = \frac{\frac{\text{Force}}{\text{Area}}}{\frac{\text{Change in length}}{\text{Length}}} = \frac{\frac{F}{A}}{\frac{\Delta L}{L}}, \quad (6.3)$$

where E is Young's modulus (in units of Pascals, Pa), F is the force applied to the sample (in units of Newtons, N), A is the cross-sectional area of the sample (in units of square meters, m^2), ΔL is the change in length of the sample (in units of meters, m), and L is the original length of the sample (in units of meters, m).

Many materials, such as single-crystal silicon, other bulk semiconductors, and glasses, exhibit a linear elastic relationship such as shown in Fig. 6.2. A linear elastic material can undergo the stress-strain loading repeatedly over many cycles and perfectly maintain this relationship. Specifically, increasing and decreasing stress levels can be applied to a linear elastic material, and the strain values will be along the same line. Stated another way, the material exhibiting linear elastic behavior also exhibits reversibility.

The illustration of linear elastic behavior shown in Fig. 6.2 is simplified from the behavior exhibited in many materials. Specifically, many materials will often exhibit a linear elastic behavior only up to a certain point. That is, once the strain reaches a certain level, the material may permanently deform or break. A material type that has a linear elastic behavior up to a certain level and then breaks is usually referred to as a brittle material [2]. Silicon and other bulk semiconductors display this type of behavior.[2] Other materials, such as many metals and polymers, often exhibit a more complicated stress-strain relationship [2].

6.2.2 Residual Stress

The definition of Young's modulus involves the behavior of a thin-film layer to *externally applied stresses* so as to cause measurable strain levels. However, thin-film layers can also be under a state of stress even when there are no externally applied forces or strains. These are called residual stresses. There are generally a number of causes of residual stress. One is due to the thermal expansion coefficient difference between the thin-film layer and the substrate. These are called thermal stresses [3].

To understand the origins of thermal stresses, it will be recalled that most deposition processes involve the thin-film layer being deposited at an elevated temperature. If the deposited thin-film material has a higher thermal expansion coefficient than the substrate, then after the film is deposited and the temperature is reduced to room temperature, the result is that the thin film contracts more than the substrate. Since compatibility requires that the thin film and the substrate have the same length, once the film and substrate reach an equilibrium mechanical state, the film will be under a tensile state of stress. This is illustrated at the top in Fig. 6.3. Alternatively, if the substrate has a higher thermal expansion coefficient, then the situation at the bottom of Fig. 6.3 will be exhibited wherein the thin-film layer is under compressive stress.

Figure 6.3 helps to illustrate one cause of residual stresses in thin films due to temperature effects, but there are other causes as well. For example, if impurities, such as dopants, are introduced into a material, they may have atomic diameters different from the bulk host material, and this can alter the stress state. For example, the top surface layer of a substrate can be doped with boron that has a smaller atomic diameter than silicon, and this would cause the doped silicon layer to contract with

[2]It should be noted that this may not be true over any temperature. For example, silicon when loaded at elevated temperatures, typically above 600 °C, can display plastic deformation.

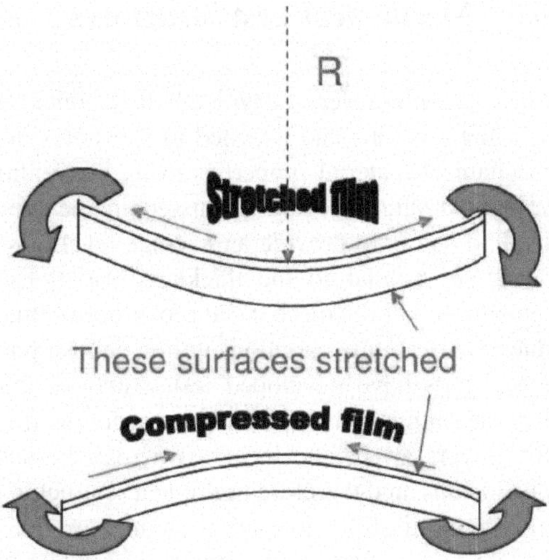

Fig. 6.3 An illustration of residual stresses in thin-film layers deposited onto substrates. The illustration at the top is a thin-film layer under tensile stress and the bottom a thin-film layer under compressive stress

respect to the rest of the substrate. The opposite would occur for phosphorus doping since it has a larger atomic diameter than silicon.[3]

Another cause of residual stresses in thin films is grain boundaries. Specifically, the interatomic spacing at the grain boundaries is not equivalent to that in the single-crystal material. When crystal grains are formed and grow in thin-film layers, they often exhibit a grain-like microstructure. Commonly the grains have a columnar shape with a diameter that changes through the thickness of the material layer, thereby resulting in a residual stress in the thin film. Atomic peening whereby ions are bombarded onto a thin-film layer, such as during a sputter deposition, can also result in a built-in state of stress. Gas entrapment into a thin-film layer during deposition can also alter the state of stress in thin-film layers. Voids that are present in thin-film layers and special arrangements of dislocations can also cause a built-in stress. Differences in lattice spacing in homogeneous material layers can also cause residual stresses [3, 4].

As a general rule, it is usually not possible to disentangle which of these causes may be present or their relative contributions to the total residual stress level in a thin-film layer. In any case, separating out the residual stress contributing factors is usually not of concern to MEMS developers. A more pressing concern is how to accurately measure the material properties of interest. The next section reviews some of the more commonly used test structures for measurement of important mechanical material properties.

[3] This assumes the dopants are substitutional in the silicon lattice. Interstitial doping may result in films with different stress states [4].

6.3 Mechanical Test Structures

Since thin-film layers are typically in the thickness range of a few tens of nanometers to a few microns and attached to substrates, it is difficult to measure some of the mechanical material properties using conventional methods. One exception is the wafer curvature method discussed in the previous chapter. However, even that method can only provide an average level of stress in the thin film over the entire substrate. Similar to the thickness variations of deposited thin films, it is not uncommon for the mechanical properties of thin films to also vary across a substrate due to temperature variations during depositions, as well as other effects. Therefore, researchers have developed test structures that are made using micromachining fabrication methods in order to measure the mechanical material properties of thin-film layers at specific locations across the substrate. These devices are small in dimensions, and therefore multiple test structures can be spread across the substrates to provide useful information about how the material properties vary across the wafers. Some of the most commonly used test structures are reviewed in this section.

6.3.1 *Test Structures for Young's Modulus*

An example of a test structure that was devised to measure the strain and elastic limit of a thin film is illustrated in Fig. 6.4. This test structure uses surface micromachining fabrication techniques for implementation. A thin-film layer is deposited and then patterned and etched to form the outline of a test structure. The thin film is released from the substrate by removal of the substrate from underneath the test structure pattern. If the thin-film layer is under tensile stress, it will result in the struts of length L_2 elongating by an amount given by δL. The strain in the struts is given by:

$$\varepsilon = \frac{\sigma}{E} = \frac{\delta L\left(\frac{W_1}{L_1} + \frac{W_2}{L_2}\right)}{W_1 - W_2}, \tag{6.4}$$

where the geometries are given in Fig. 6.4 [5]. The dimensions of the test structures can be varied over ranges of values to provide more precise measurements of strain and allow the determination of the dimension at the threshold of where the struts of length L_2 break, and this is the maximum strain limit of the thin-film material layer.

An issue with the test structure shown in Fig. 6.4 is that there must be a tensile residual stress in the thin film in order to result in a strain that can be measured, thereby allowing Young's modulus of the material layer to be calculated from Eq. (6.4). Additionally, the value of the stress in the film must already be known. Also, in many microsystems devices, it is desirable for the thin-film stress to be very low in value, in fact, near zero. Therefore, the test structure in Fig. 6.4 is not suitable

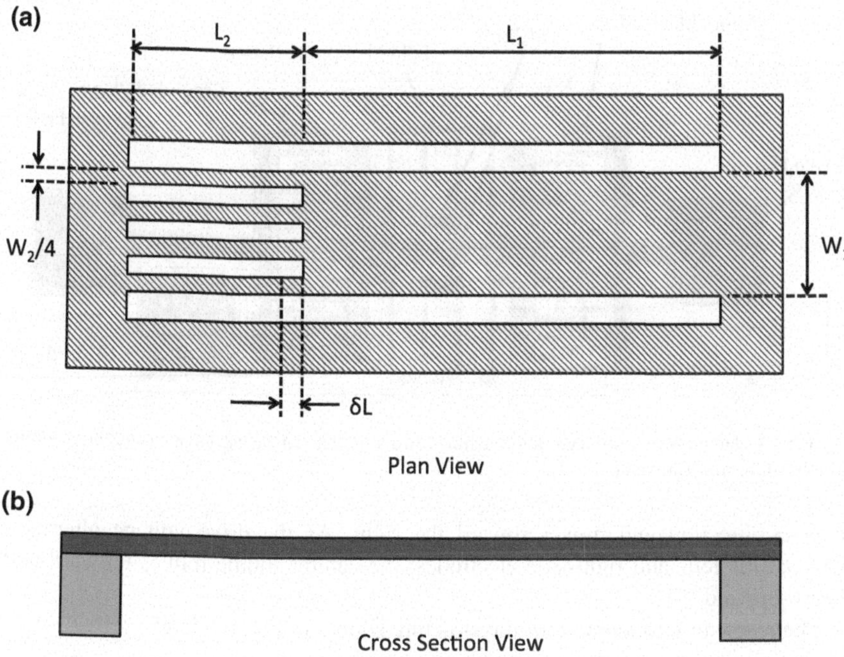

Fig. 6.4 An illustration of a micromachined test structure that can be used to determine Young's modulus

for determination of Young's modulus for a material when the stress is very low in magnitude or compressive.

Fortunately, there are other types of test structures that can be used for measuring Young's modulus of a material layer that do not have the requirement of a pre-existing residual stress. A commonly used device for this purpose consists of an electromechanical test structure with known dimensions that is electrically driven into a resonant operation as shown in Fig. 6.5. Young's modulus can be determined using the resonant frequency, dimensions, and material mass, in order to determine the modulus [6]. This type of resonant device can be used to determine the residual stress in polysilicon material layers.

A drive voltage signal is applied to the drive pad that creates an electrostatic force between the comb drive electrodes. There is a matching pair of comb electrodes on the structure with the outer pair attached to the substrate and unmovable. The inner pair is attached to the movable central shuttle. The voltages applied to the left comb drive electrodes are 180 degrees out of phase with the voltages applied to the right comb drive electrodes. As the voltage on the left-side comb drive electrodes increased in magnitude and the voltage on the right-side drive electrodes decreased in magnitude, an electrostatic force is generated between the left electrodes that causes the shuttle to move leftward. As the voltage of the right-side comb drive electrodes increases and the voltage on the left electrodes decreases, the shuttle

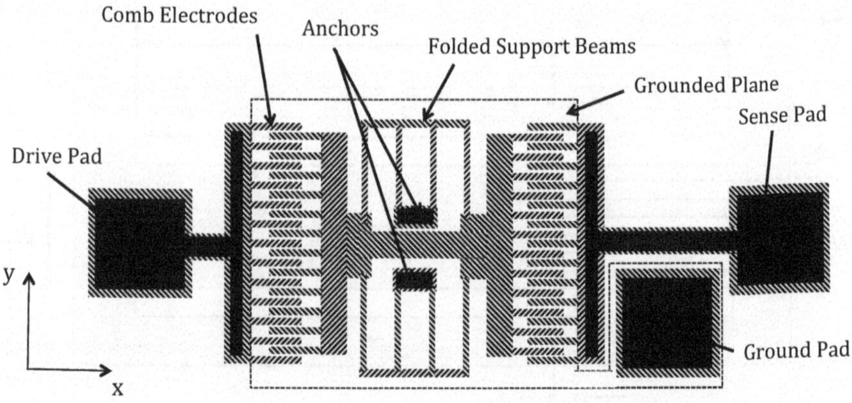

Fig. 6.5 A plan view of a surface micromachined structure that can be used to measure the modulus of a thin-film material layer

reverses direction and moves toward the right. As the drive voltage alternates between the left- and right-side electrodes, the central shuttle moves leftward and then rightward.

The resonant frequency, f, of the structure shown in Fig. 6.5 is given by:

$$f = \frac{1}{2\pi}\sqrt{\frac{k_x}{M}}, \tag{6.5}$$

where k_x is the mechanical stiffness (i.e., spring constant) of the structure in the x-direction and M is the mass. The structure mechanical stiffness, k_x, is given as:

$$k_x = \frac{24EI_z}{L^3}, \tag{6.6}$$

where E is Young's modulus of the material that the structure is made from, I_z is the moment of inertia of the beam cross section with respect to the z-axis (normal to x- and y-axes in Fig. 6.5), and L is the length of the folder beam elements. I_z is as:

$$I_z = \frac{w^3 t}{12}, \tag{6.7}$$

where w is the width of the beam and t is the thickness of the beam. Combining Eqs. (6.5), (6.6), and (6.7) allows the following to be written as an expression for the resonant frequency (in Hertz) of the structure in terms of the dimensions, modulus, and mass as follows:

$$f = \frac{1}{2\pi}\sqrt{\frac{24EI_z}{\left(M_p + \frac{1}{4}M_t + \frac{12}{35}M_b\right)L^3}}, \tag{6.8}$$

where M_p, M_t, and M_b are the masses of the plate, trusses, and beams of the structure [6]. Once the resonant frequency has been measured and the other terms determined, Eq. (6.8) can be solved for determining Young's modulus of the thin-film material layer that the structure is fabricated from.

6.3.2 *Thin-Film Residual Stress Test Structures*

Wafer curvature measurements are one technique that is commonly used to determine the stress in thin-film material layers and was discussed in Chap. 5. However, one of the issues with wafer curvature measurements is that they only provide information about the average stress levels and do not give any information about how the stress may vary across the substrate surface. That is, wafer curvature stress measurements do not indicate the specific value of stress at various locations of the substrates. Nevertheless, due to chemical concentration, chemical reaction, gas flow rate, plasma, and temperature variations that exist in many microsystems manufacturing processes, the resultant state of stress across the substrate can vary.

The measurement techniques employed in the macroscale dimensional domain for residual stress are not well suited for micro- or nanoscale dimensional elements. For example, removing a thin-film layer from the substrate to perform a measurement would change the stress state of the film. Therefore, more sophisticated techniques for measuring the films stresses at the localized level are needed.

Based on the requirement to measure the local stress level in thin films, a number of different test structures have been developed for this purpose. One of the most simple of these test structures is illustrated in Fig. 6.6. This technique uses a number of surface micromachined beams that are fabricated from the thin-film layer(s) each having a slightly different length, with the other dimensions held constant. The beams have anchor points affixed to the substrate surface at both ends and are freely suspended between the ends. This is commonly referred to as the "clamped-clamped beam" configuration.

The method of determining the stress in the thin film is to examine the array of beams having different lengths to determine the shortest beam that indicates the onset of buckling. The critical strain, ε_{cr}, of the thin-film layer can be calculated from the critical length of the beam, L_{cr}, at the onset of buckling, using the Euler formula for the elastic instability of a beam as follows:

$$\varepsilon_{cr} = \frac{\pi^2 t^2}{3L_{cr}^2}, \tag{6.9}$$

where t is the thickness of the thin-film material layer and the beam [7].

Once the strain has been calculated in the thin film using Eq. (6.9), the stress in the material layer can be determined by multiplying the strain by the modulus of the material. This of course assumes that the modulus of the thin-film material layer is

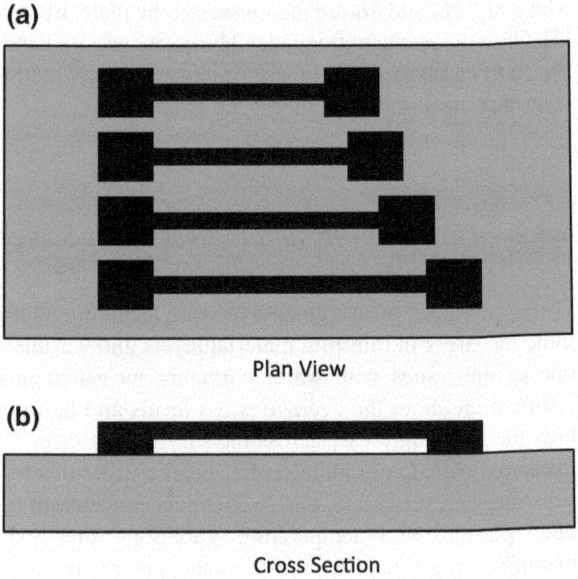

Fig. 6.6 (**a**) Plan view of the clamped-clamped beams having different lengths that are used to determine the critical stain level in the thin-film layer used to fabricate the beam. (**b**) Cross section of the longest beam. The beams are shown not in a buckled state

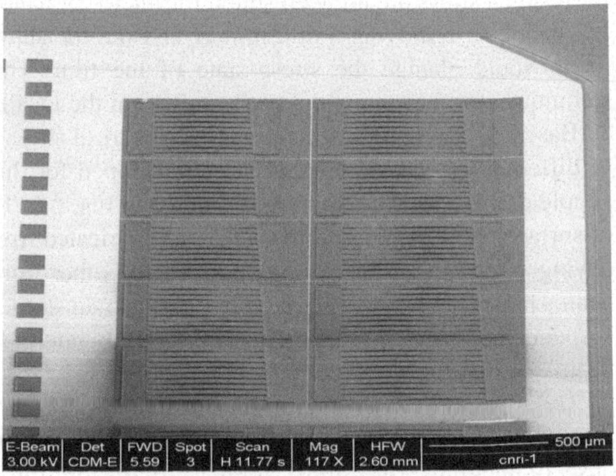

Fig. 6.7 SEM image of an array of clamped-clamped beams of different lengths wherein the beams having shorter lengths are not buckled, but longer beams are buckled

known. An example of clamped-clamped test structures is shown in the SEM image of Fig. 6.7. As can be seen, there are beams having shorter lengths that are not buckled and beams that are longer that are buckled.

While the clamped-clamped beams have proven to be a useful tool for determining the strain and stresses in thin-film layers, this method does have a few disadvantages. First, it can only measure compressive stresses. Second, in practice it can be difficult to accurately determine the length of beam of where the onset of buckling occurs. Third, the clamped-clamped beam method requires a large array of test structures in order to determine the threshold, and this can occupy a considerable

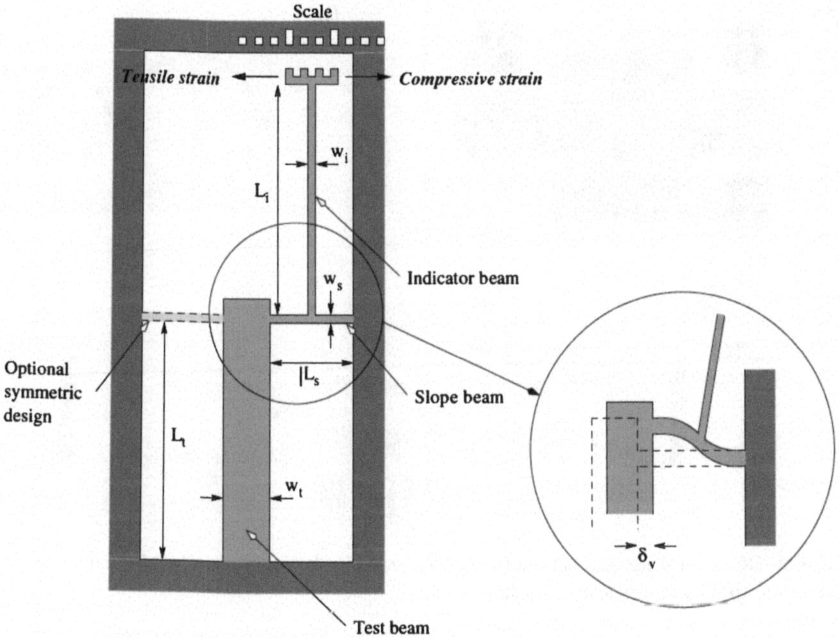

Fig. 6.8 Microscale test structure plan view

amount of the substrate area. Additionally, since the clamped-clamped beams are made using surface micromachining techniques, they can be prone to stiction effects, and if stiction is present, it can make identification of the buckling threshold very challenging.

There are other test structures for measuring the strain and stress in thin films. A popular test structure that can be used to determine either tensile or compressive stresses is shown in Fig. 6.8. This is a pointer structure that is surface micromachined from a thin-film layer. The indicator or pointer beam (oriented vertically in Fig. 6.8) is mechanically connected to the two struts on either side that are oriented horizontally, which are all connected to a test beam that is oriented vertically. All of these are freely released from the substrate. The frame surrounding the test structure is attached to the substrate. Depending on the strain in the test beam, it will expand or contract. An expansion of the test beam due to compressive stresses will cause the slope beam to obtain the shape as shown in the right of Fig. 6.8, thereby resulting in the indicator beam to displace toward the right. A microscale is fabricated on the end of the indicator beam and the frame to enable the amount of deflection to be measured. If the test beam contracts due to tensile stresses, the indicator beam will displace in the opposite direction.

As with the previous test structures, once the strain is measured, and assuming the modulus of the material of the thin film is known, the stress in the film can be determined. The strain in the thin film is found using the following equation:

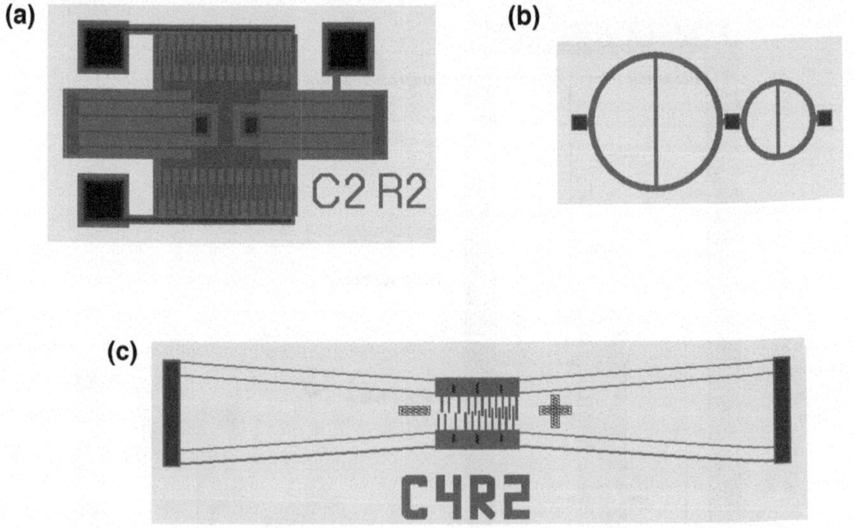

Fig. 6.9 Other test structures that can be used to determine the stress states in thin-film layers. (**a**) Resonator. (**b**) Guckel rings. (**c**) Bow tie

$$\varepsilon = \frac{2L_s\delta}{3L_iL_tC}, \tag{6.10}$$

where L_s, L_i, and L_t are the length of the slope, indicator, and test beams, respectively, d is the displacement of the indicator beam on the gauge scale, and C is a corrective factor [7].

One of the issues concerning the device shown in Fig. 6.8, beyond the stiction effects that can be a problem for all surface micromachined structures, is that if the stresses are compressive, the structure may buckle out of plane. Care must be exercised to ensure that the structure is not buckled since this would compromise the measurement of the film stress.

Additionally, there are other test structures that have been developed and can be used as well. Three are shown in Fig. 6.9 including (a) resonator structures; (b) Guckel rings; and (c) bow ties. The resonator structures depend on the change in the resonant frequency of a surface micromachined device depending on the state of stress. This device can be used for both compressive and tensile stresses. The Guckel rings are only applicable to thin films in tension. The rings under tension cause the center beam to buckle if the strain is sufficient to result in buckling. Lastly, the bow tie has arms on either side that result in the displacement along a gauge in the center and can be used for both compressive and tensile stresses. However, all of these structures require more sophisticated techniques to analyze [8].

6.3.3 Stress Gradients

The residual stresses in thin-film layers can be uniform, or it can have a gradient through the thickness of the material layer. Specifically, a stress gradient is when the stress varies through the thickness of the film from one side to the other in the direction normal to the substrate surface. It is common for polycrystalline materials, such as polysilicon, to have significant stress gradients. The presence of stress gradients can have a significant impact on the resultant device behavior. For example, the fabrication of cantilevers composed of material layers having stress gradients could result in the cantilevers not being straight and flat as would be desired for most applications [9]. It is important to note, but sometimes misunderstood, that it is possible for the average state of stress in a thin film to be near or at zero, but the film still have a significant stress gradient. Even devices that are clamped at both ends or around the periphery can have their performance impacted by the presence of a stress gradient.

The test structures described so far are not suitable for determining the gradient of the stress through the material layer. Appropriate test structures for measuring the stress gradients in material layers are the cantilevers as shown in Fig. 6.10. The left side shows a plan view and cross section of a cantilever wherein the unclamped free end of the cantilever has undergone a displacement due to the presence of a stress gradient. This type of curved displacement of the free end of the cantilever would be due to a stress gradient that is more tensile on the top compared to the lower part of the material layer.

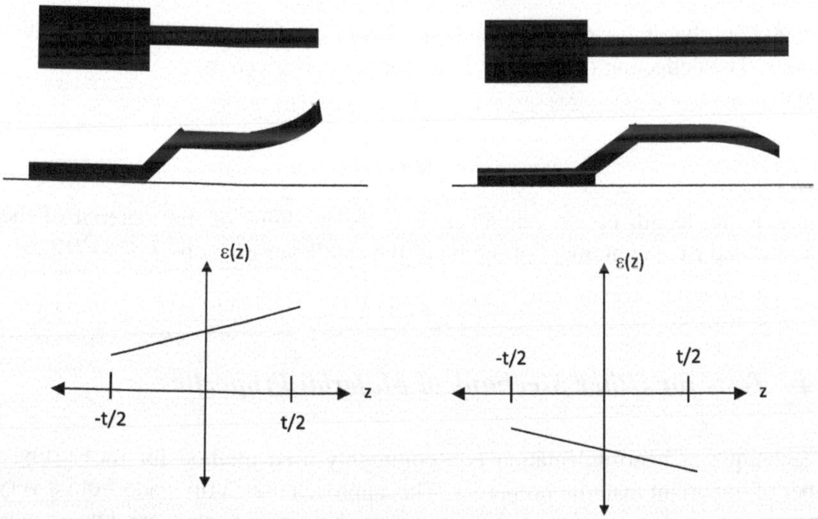

Fig. 6.10 Illustration of stress gradients in cantilevers that are exhibited as strain gradients that can be used to measure the stress gradient in material layers

Once the structure is released, the stress gradient is allowed to relax as the cantilever deforms and exhibits a strain. That is, as the cantilever relaxes, the stress goes to zero. The gradient in the strain is plotted in the left bottom, which shows the strain as a function in the normal direction, z, as a function of the thickness of the layer given by t. The strain has a larger magnitude in the top half of the material layer thickness and a lower magnitude in the lower half. The right side illustrates the effect of a stress gradient where it is compressive in nature, and once the cantilever relaxes, the stress goes to zero, and the end of the cantilever curving downward as shown in the bottom right exhibits the strain.

There may both be an average stress in the film and a stress gradient. For example, the average stress could be an extrinsic stress due to the mismatch in the thermal expansion coefficients of the material layer and the substrate. The stress gradient could be an intrinsic stress due to the columnar structure of the grains of a polycrystalline material layer. Once the cantilever is relaxed however, both of these stresses will relax and be exhibited as strains. The relaxation of the uniform average stress will be exhibited by a uniform strain that would be an expansion if the average stress is compressive or contraction if the average stress is tensile. However, the stress gradient will result in a curvature of the released cantilever. This curvature can be used to measure the stress gradient.

The bending moment, M, that causes the deflection of the cantilever is given by:

$$M = \int_{-t/2}^{t/2} \sigma(z) z w dz, \tag{6.11}$$

where $\sigma(z)$ is the stress gradient, w is the width of the cantilever, and t is the thickness. The deflection of the end of the cantilever is given as:

$$z = \frac{ML^2}{2EI}, \tag{6.12}$$

where L is the length of the cantilever, E is the modulus of the material of the cantilever, and I is the moment of inertia of the cantilever given by $I = wt^3/12$.

6.3.4 Tests for Other Mechanical Material Properties

The technique of nano-indentation is a commonly used method for measuring a number of important material properties. This approach uses a tip made from a very hard material, such as diamond, having material properties that are known and pressing the tip into the material being analyzed. The force applied to the tip and into the sample is increased over a predetermined range of values. The area of the residual indentation is measured to determine the hardness, H, according to the following relationship:

$$H = \frac{P_{\max}}{A}, \tag{6.13}$$

where P_{max} is the maximum load applied and A is the indentation area after the load has been removed. The resultant indentation areas can be relatively small, such as a few microns or nanometers, thereby making measurement of the residual indentation area difficult. Therefore, it is common to use sophisticated methods such as SEM imaging or AFM for these measurements.

Additionally, it is common in this technique to use a tip having a known geometry called a Berkovich tip, having a three-sided geometry. Using such a tip configuration and measurement of the depth of the indentation allows a relatively simple method to determine the residual indentation areas. Nano-indentation can be used to create a load versus displacement curve for the material being examined. This is equivalent to a stress versus strain curve. Generation of such a load-displacement curve allows the determination of the maximum load or stress relationship, the modulus, and the hardness of the material. It can also be used to determine the strain-rate sensitivity [10].

6.4 Electrical Test Structures

It is common practice in microsystems manufacturing to perform electrical measurements on specially designed test structures. These test structures are usually fabricated using the same manufacturing process as the actual devices in production. The types of commonly used electrical test structures include resistors; diodes; transistors; MOS capacitors; gate conductors; metal interconnects; dielectrics between different metal interconnect layers; simple electrical circuits; ring oscillators; and others. A few of these will be reviewed in this section.

Capacitance-voltage (CV) measurements can be used to determine the dopant profile in substrates. The measurement relies on the relationship between the capacitance variation with voltage of a reversed-biased n+/p or p+/n junction. The capacitance of a junction is given by:

$$C(V) = \left[\frac{q\varepsilon_s C_B}{2}\right]^{1/2} \left[V_{bi} \pm V_R - \left(\frac{2kT}{q}\right)\right]^{1/2}, \tag{6.14}$$

where ε_s is the permittivity of the semiconductor, C_B is the substrate dopant concentration, V_{bi} is the built-in potential of the p-n junction, and V_R is the reverse bias voltage [11]. In this technique the reverse-biased voltage across the junction is varied while the capacitance is measured. For each value of V_R, using Eq. (6.14), concentration C_B can be determined. The depth of each value of C_B is determined by the value of the reverse-biased voltage for that measurement point. The limitation of using CV measurements to determine doping profiles is that the calculated values of C_B include the zero-bias depletion width, and this technique is not accurate for abrupt profiles.

CV measurements can also be employed to measure the oxide thickness, oxide charges, contamination from mobile ions, and interface trap density in wafer processes in metal oxide semiconductor (MOS) devices. The approach involved applying a DC bias potential with a small AC high-frequency signal across the gate dielectric and measuring the capacitance as shown in Fig. 6.11. The electrical field lines created by the voltage potential across the gate dielectric pass through the insulating material and into the semiconductor substrate to impact the charge carriers. There are three regions of behavior that have been identified in this test method as follows: accumulation (Fig. 6.11a), depletion (Fig. 6.11b), and inversion (Fig. 6.11c).

During CV measurements, high- and low-frequency voltages are applied to the gate electrode in the inversion region, and the gate voltage is scanned over the range from accumulation to inversion, and from the measurement data, the oxide thickness, substrate doping profile, and oxide interface charges, including Q_f the oxide fixed charge, Q_I the inversion layer charge, Q_m the mobile charge, and Q_{ot} the oxide trapped charge, can all be determined [12]. The precision and accuracy of CV measurements depend on the electrical testing equipment that is used to perform the tests. Nevertheless, most equipment for CV measurements has accuracy in the range of 0.1% or better.

Among the most simple electrical test structures are electrical interconnects to determine the occurrence of short circuits and open circuits in electrical interconnect wiring across substrates, especially when multiple layers of electrical interconnects are involved. Often these test structures are meandering electrical lines of each conductive layer that use the smallest line widths and line-to-line spacing involved in the microsystems actual design for these test structures (see Figs. 6.12 and 6.13).

For example, the test structure in Fig. 6.12 is a metal electrical interconnect line where the resistance between the two electrical test pads is measured. If the measured resistance is as expected (expected to be a low but measureable value of resistance), then the fabrication of these lines is deemed as acceptable. If the measured resistance indicates an open circuit (i.e., $R = \infty$), then it is possible that the metal lines were over etched. If, on the other hand, the resistance is lower than expected, it may indicate stringers (small pieces of metal left after the etch process joining two metal lines where they are intended to be separated) associated with an incomplete etch or an unacceptable lithography process. This kind of test structure can also be placed over areas on the die where the topography has significant variations in order to determine if the metal patterning process is acceptable in these regions.

Another type of electrical test structure is shown in Fig. 6.13 and is composed of two metal lines that are to be electrically separated. The resistance between electrical test pads 1 and 2 can be measured, and if the measured resistance does not indicate an open circuit (i.e., $R = \infty$) between the two metal lines, then something in the processing has caused an electrical short between them. If an electrical short is measured, it could be due to stringers left over from an incomplete etch process that connects the two lines. Additionally, different electrical interconnect layers that are to be electrically separated may be stacked on top of one another with an insulating layer to separate the lines. The measured resistance across the insulating layer would be measured to determine if it contained any shorts that would be problematic.

A structure that is used to measure the sheet resistance of a conductive thin-film layer is shown in Fig. 6.14 and is referred to as the 4-point probe structure (also known as the Kelvin probe). This technique is simpler that the 4-point probe described above where the current profile was not uniformly distributed in the conductive material layer.

A current of I (in units of amps) is forced through the outer two electrodes, and the voltage (in units of volts) of the two inner electrodes is measured. The resistance (in units of ohms) is simply the ratio of $R = V/I$. This structure is not sensitive to any contact resistances that may exist since the measurement of the voltage does not result in a significant current flow or voltage drop. For the structure shown in Fig. 6.14, the resistance and the sheet resistance are related by the following:

$$R = R_S\left(\frac{L}{W}\right). \tag{6.15}$$

Conveniently, from Eq. (6.15) the resistance is simply the product of the sheet and the ratio of length over width or as sometimes written the number of squares (i.e., the ratio of L/W). Often it is useful to perform a sweep of the current and measure the resultant voltages as well.

Another commonly used test structure is the van der Pauw test structure, shown in Fig. 6.15. This structure can be used to measure the resistivity and the Hall coefficient of a material sample. The van der Pauw structure requires that the thickness of the material layer is uniform and much smaller than the width or length of the structure. The structure should also be symmetrical. The structure in Fig. 6.15 is also called the "Greek cross," and there are other types of layout patterns that can be used, although this one is the most common.

The technique is as follows: a current is forced between two electrodes along one edge of the structure, say between electrodes numbered 1 and 2, given by I_{12}, and the voltage between the two electrodes along the other edge of the structure is measured, and this is given by V_{23}. The resistance is given by:

$$R_{12,\,34} = \frac{V_{34}}{I_{12}}. \tag{6.16}$$

This measurement is repeated by forcing current between electrodes 4 and 1, given by I_{41}, and measuring the voltage across the electrodes 2 and 3, given by V_{23}, to find the resistance $R_{23,\,41}$. The sheet resistance, given by R_s, is related to these measured resistances by the equation given as:

$$e^{\frac{-\pi R_{12,34}}{R_s}} + e^{\frac{-\pi R_{23,41}}{R_s}} = 1 \tag{6.17}$$

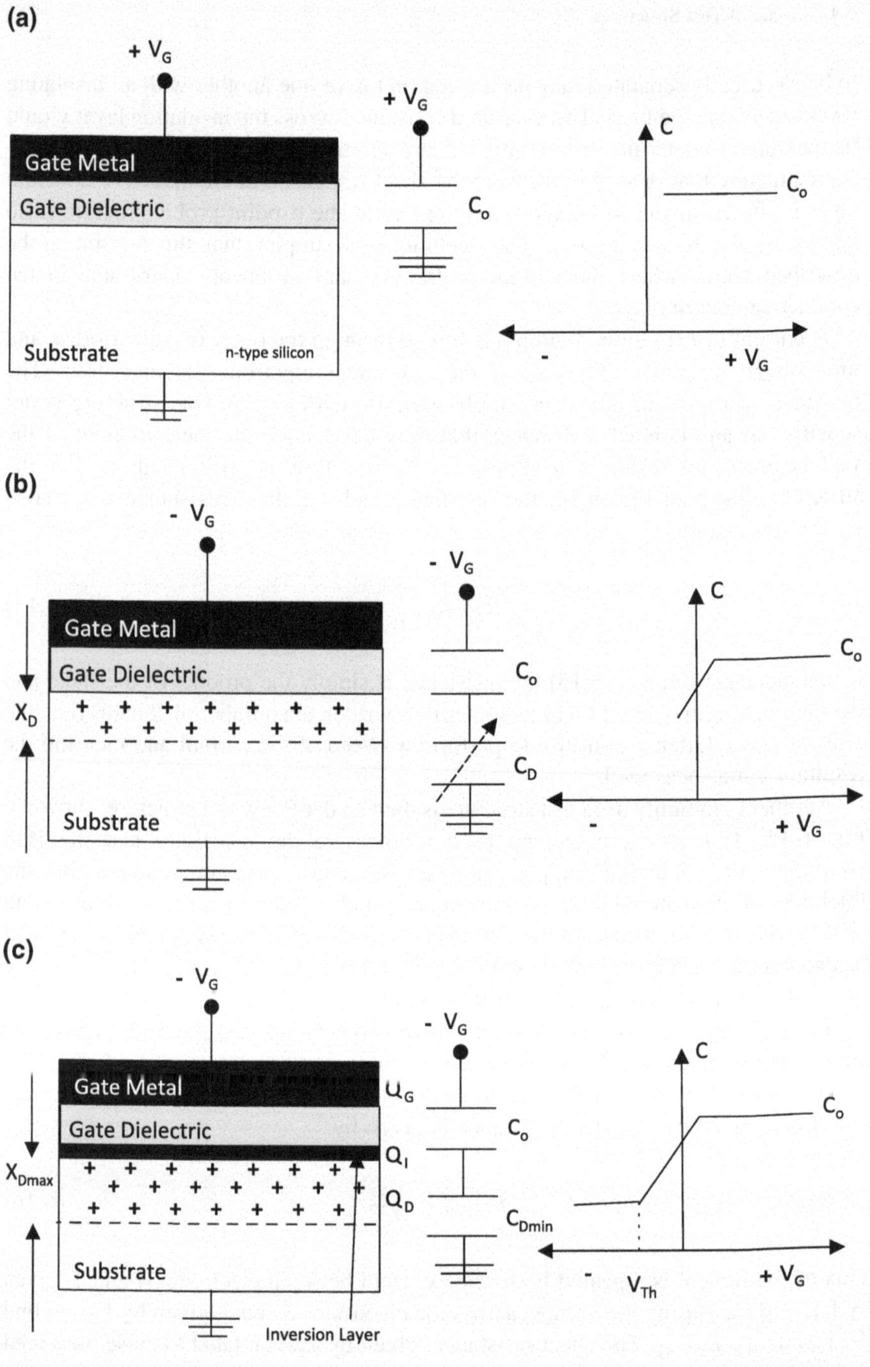

Fig. 6.11 Illustration of the CV measurement on a MOS gate structure. On the right is shown a cross-sectional model of the gate structure, in the middle the electrical circuit model, and on the right the CV curves. A voltage V_G is applied to the gate metal electrode of the MOS device composed of a DC potential with a small signal AC. As shown in (**a**), when V_G is positive, the

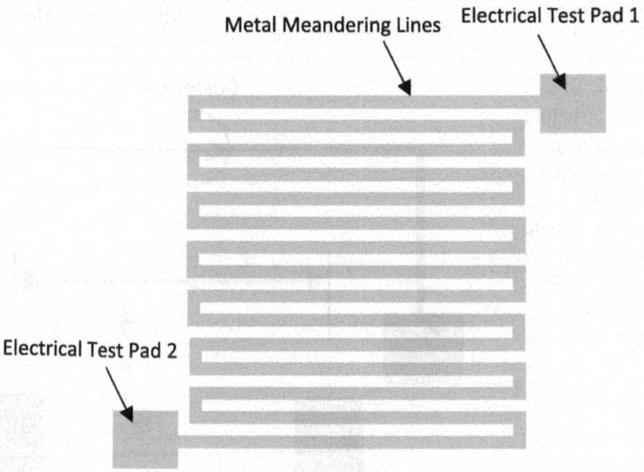

Fig. 6.12 Example of a simple test structure composed of a meandering line with contact points, labeled as testing pads, on the ends. The resistance of the metal line is measured between the testing pads and can indicate important information about the processes used to make the metal interconnect lines including whether there are any opens or shorts and whether the interconnect lines can maintain electrical continuity over aggressive topology

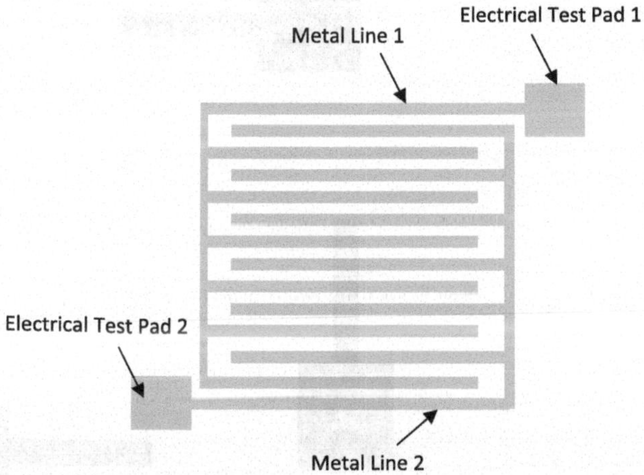

Fig. 6.13 Example of a simple test structure composed of two metal lines that are electrical separated. The resistance between the two electrical test pads is measured and should indicate an open circuit. If a finite resistance is measured, it indicates that something in the lithography or etch process is unacceptable

---

Fig. 6.11 (continued) effect is a constant capacitance of value C_O. In (**b**), when the applied voltage decreases, a depletion region is created near the substrate and gate dielectric interface that has a varying capacitance given by C_D that is in series with C_O, and therefore the total combined capacitance decreases. In (**c**), an inversion layer composed of holes is created at the interface, and the overall capacitance is the series combination of C_O and C_{Dmin}. (**a**) Accumulation. (**b**) Depletion. (**c**) Inversion

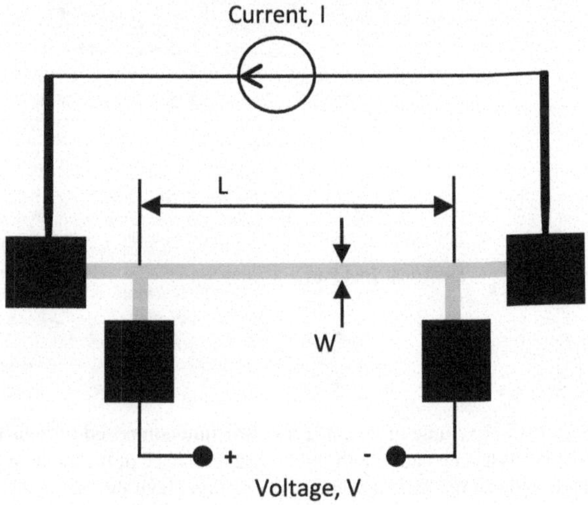

Fig. 6.14 Example of a 4-point probe test structure for measuring the sheet resistance of a conductive thin-film layer on a substrate. Current is forced on the outer two electrodes, and the voltage across the inner two electrodes is measured. This structure is not sensitive to the contact resistances

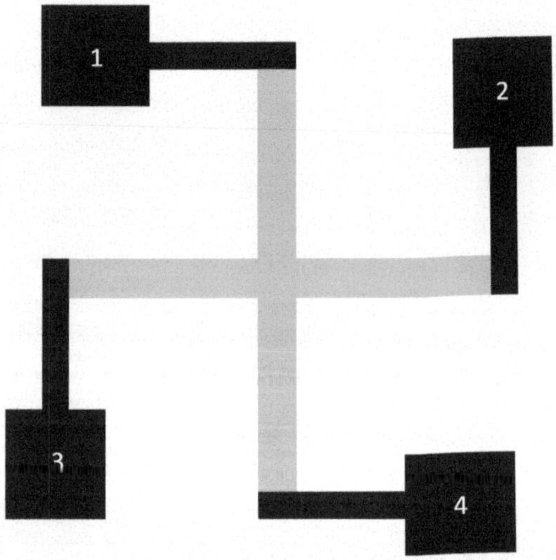

Fig. 6.15 Illustration of the van der Pauw test structure for determining the resistivity

Most of the measurements using van der Pauw structures are for resistances that are very small, typically a few ohms. Consequently, the measured voltages will also be small, typically less than 10 mV. Therefore, a digital multimeter with 7 or more digits will be used.

Additionally, in order to improve the accuracy of this technique, multiple measurements can be made. It should be recognized that the reciprocity theorem for this structure indicates that:

$$R_{12,34} = R_{34,12}. \tag{6.18}$$

Therefore, to obtain a more accurate measurement value of the resistances $R_{12,34}$ and $R_{23,41}$, two additional measurements can be taken to obtain the reciprocal resistance values of $R_{34,12}$ and $R_{41,23}$, and averaging the results as follows:

$$R_{\text{Vertical}} = \frac{R_{12,34} + R_{34,12}}{2}, \text{ and} \tag{6.19}$$

$$R_{\text{Horizontal}} = \frac{R_{23,41} + R_{41,23}}{2}. \tag{6.20}$$

The van der Pauw equation can now be written as:

$$e^{\frac{-\pi R_{Vertical}}{R_s}} + e^{\frac{-\pi R_{Horizontal}}{R_s}} = 1 \tag{6.21}$$

Further improvements in accuracy can be made by reversing the polarities of the currents and voltages and repeating the measurements of the resistances. In this case, Eqs. (6.19) and (6.20) above can be written as:

$$R_{\text{Vertical}} = \frac{R_{12,34} + R_{34,12} + R_{21,43} + R_{43,21}}{2}, \text{ and} \tag{6.22}$$

$$R_{\text{Horizontal}} = \frac{R_{23,41} + R_{41,23} + R_{32,14} + R_{14,32}}{2}. \tag{6.23}$$

These expressions for R_{Vertical} and $R_{\text{Horizontal}}$ can be substituted into Eq. (6.17) above to find the value of R_s.

If the different interconnect layers have connections, these can be tested using via chains or other appropriate structures. For example, shown in Fig. 6.16 is a test structure, in plan and cross-sectional views, for a via chain with a two-layer metal scheme. The metal layer 1 has a meandering pattern with breaks in the electrical continuity at various points. Metal layer 2 makes electrical contact to the first layer metal through the various vias that penetrate through an insulating layer (not shown) between metal layers 1 and 2. If there are any discontinuities in the electrical pathway, an open circuit would be exhibited. While this structure allows the electrical continuity of the lines and vias to be measured, it is not a good structure for measuring the contact resistances.

Likewise, test structure made of one or more diodes or transistors or some combination thereof can be inserted into the mask sets to enable the determination of the functionality of single active devices. Additionally, simple test circuits can be developed and inserted into the mask sets to give more complex information. For example, the insertion of ring oscillator circuits is very common and is used to provide quantitative performance data about how the processing conditions may be impacting the speed and capacitive loading effects. Shown in Fig. 6.17 is a three-stage ring oscillator composed of three inverter gates in series with the output connected to the input. The voltage in these inverters oscillates between two voltage levels, with the output of the ring oscillator alternating at a frequency (6 times the inverter delay)$^{-1}$.

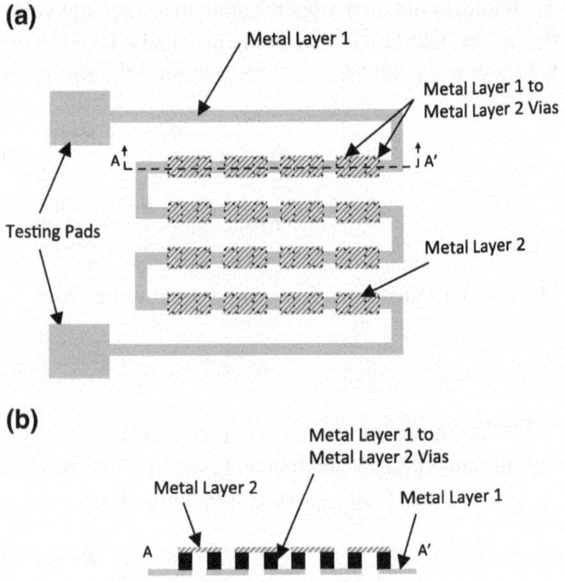

Fig. 6.16 Example of a two-metal layer test structure to monitor the electrical continuity of these metal layers through the connecting vias

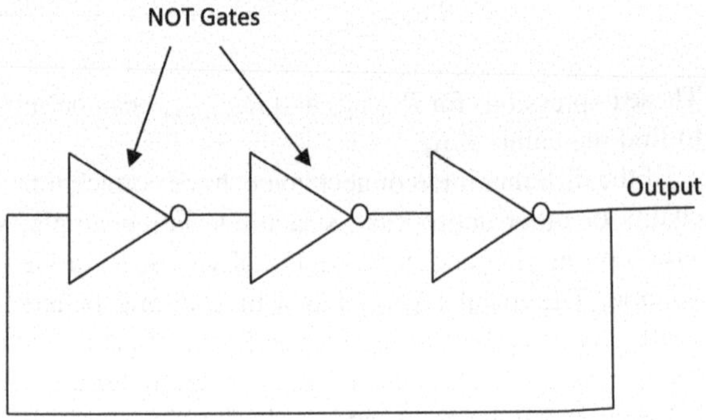

Fig. 6.17 Schematic of a simple three-stage ring oscillator that can be used to monitor the operational frequency of electronic devices used in the circuitry

6.5 Thin-Film Material Properties

In this section we review some of the reported values of the material properties of the most commonly used thin-film material layers in microsystems manufacturing. These are taken from the literature and are presented with the processing conditions.

6.5.1 Thermal SiO_2

The important material properties of thermally grown silicon dioxide are presented in Table 6.1 below [13]. Unlike many of the other fabrication processes for thin films used in microsystems manufacturing, thermally grown silicon dioxide layers do not exhibit significant variations in most of the material properties depending on the processing conditions. This is due to the fact that it is a direct thermal conversion process rather than a chemical or physical vapor process. A notable exception is the residual stress as can be clearly seen in Table 6.1.

Part of the reason why the residual stress of thermal oxide can vary is because the thermal expansion coefficient of this material is significantly different from that of the silicon substrate that it is grown on, and therefore thermally grown silicon dioxide will result in a significant residual stress in the oxide layer. Some of the reported measured values of the residual stresses in oxide layers on silicon are shown in Table 6.2. The variation in the residual stress appears to correlate with the oxide growth temperature. The values of Young's modulus also exhibit some amount of variation depending on whether the oxide is grown wet or dry, but in general does not vary significantly otherwise with processing conditions.

6.5.2 LPCVD Polysilicon

Polysilicon is the most commonly used material as a structural layer in surface micromachining process sequences in the manufacturing of MEMS devices. Many of the material properties of LPCVD polysilicon thin-film layers are highly dependent on the processing conditions [16]. There is a strong relationship between the processing conditions used during deposition, the microstructure of the deposited thin-film layers, and the material properties.

Table 6.1 Thermally grown silicon dioxide layer material properties [13]

Property	Value
Crystal structure	Amorphous
Melting point temperature	1732 °C
Density	2200 kg/m^3
Young's modulus	70 GPa (wet) and 57 GPa (dry)
Poisson ratio	0.17
Tensile strength	110 MPa (tension) and 690–1380 MPa (compression)
Residual stress	−225 to −331 MPa
Thermal expansion coefficient	5×10^{-10} /°C
Specific heat	1.0 J/gK
Thermal conductivity	1.4 W/mK
Dielectric constant	3.9
Index of refraction	1.46
Electrical resistivity	>10^{14} ohm-cm
Electrical breakdown strength	>5 MV/cm (although may be much lower for very thin films)
Bandgap	9 eV
Tensile/fracture strength	110 MPa (tensile strength); 690–1380 (compressive strength)

Table 6.2 Residual stresses measured in silicon dioxide layers grown using wet thermal oxidation

Deposition temperature (°C)	Process gases	Thickness (μm)	Residual stress (MPa)	Fracture stress (GPa)	Modulus (GPa)	References
1000	O_2/H_2	0.430	331 (compressive)	–	–	[14]
950–1050	O_2/H_2	0.5–1	258 (compressive)	0.89	49	[15]

With respect to microstructure and processing conditions, it is known that the transition temperature from amorphous to polycrystalline occurs at approximately 570 °C, and films deposited above this temperature will have a polycrystalline microstructure. At deposition temperatures of 600 °C, the grains in the deposited layers tend to be small and equiaxed, while they tend to be large and columnar at depositions at 625 °C [16]. The crystal orientation of the deposited thin-film material layers is largely (110) silicon for deposition temperatures ranging from 600 to 650 °C. At deposition temperatures from 650 to 700 °C, the films mostly have (100) grain orientation. Since polysilicon is used as an electromechanical material, both electrical and mechanical material properties are of interest and will be reviewed in this section. The material properties of most interest in polysilicon thin films are usually the electrical resistivity and the residual stress.

The electrical resistivity of LPCVD polysilicon can be modified by in situ doping during deposition, source diffusion, or ion implantation on previously deposited thin films. Table 6.3 provides some of the reported electrical property data for LPCVD polysilicon thin-film layers, including resistivity, for films in situ doped. As can be seen, the electrical resistivity tends to increase as the deposition temperature increases. Shown in Table 6.4 are the measured electrical material properties for a LPCVD polysilicon thin film that was diffusion doped.

The in situ doping of polysilicon thin films can be performed by including a dopant gas, a dopant source gas in the furnace during deposition. Common source gases are diborane (B_2H_6) or phosphine (PH_3). In situ doping of boron increases, while in situ doping of phosphorus reduces, the deposition rate of polysilicon compared to the deposition of undoped films [23]. The use of in situ dopants in LPCVD polysilicon results in electrically conductive thin films having very uniform profiles of dopants that are electrically active without the need for a high-temperature process such as required for dopant diffusion or anneal after ion implantation.

The amount and type of residual stress in deposited polysilicon thin films are highly dependent on the deposition conditions, especially temperature. As shown in Fig. 6.18, the residual stresses in as-deposited polysilicon layers at a constant silane flow rate and differing pressures and temperatures are compressive for deposition temperatures below 580 °C irrespective of the pressure, but at a deposition temperature of 605 °C, the stress transitions to tensile [24]. Moreover, at a deposition temperature of 620 °C, the stress changes back to being compressive [24]. This finding correlates with the observed microstructure of the polysilicon grains.

Table 6.3 Electrical properties of in situ doped LPCVD polysilicon

Temperature (°C)	Source gases	Flow rates (sccm or ratios)	Pressure (mtorr)	Deposition rates (Å/min)	Electrical resistivity (mΩ-cm)	Anneal conditions	References
555	SiH_4 BCl_3 N_2	200–400 0–180 3%	350	24–105	0.002–0.03	N/A	[17]
560–610	SiH_4/PH_3	–	375–800	22–33	0.46–5.3	RTA at 900 °C for 10–120 s	[18]
580	SiH_4 PH_3/SiH_4	500 0.01	1000	45	0.5	980 °C for 30 min	[19]
625	SiH_4 PH_3 N_2	60 30 300	750	46.3	1	900 °C for 30 min	[20]
650	SiH_4/PH_3	100	320	26.7	1.7	N/A	[21]

Table 6.4 Electrical properties of diffusion doped LPCVD polysilicon

Temperature (°C)	Source gases	Flow rates (sccm or ratios)	Pressure (mtorr)	Diffusion temperature (°C)	Sheet resistance (Ω/sq)	Anneal conditions	Reference
585	SiH_4 O_2 N_2 $POCl_3$	50 100 2000 100	200	850–950	12	650 °C for 3 h at 500 mtorr	[22]

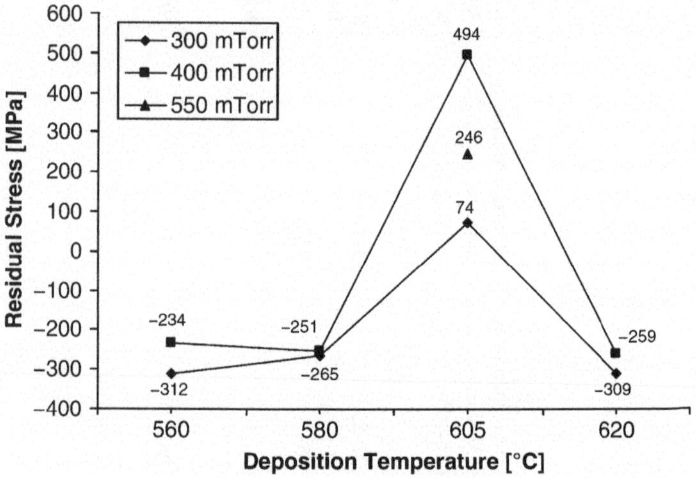

Fig. 6.18 The residual stress in as-deposited polysilicon thin films with varying deposition pressures and temperatures at a silane flow rate of 30 sccm [24]

Table 6.5 is a compilation of a number of different reported values of residual stress in LPCVD undoped polysilicon thin-film material layers as a function of the processing conditions. As can be seen, there is a huge amount of variability in these values of residual stress.

Due to the variability of the residual stress in polysilicon material layers deposited by LPCVD, many microsystems manufacturers use post-deposition anneals as a more controllable method to modify the resultant residual stresses in these materials. It has been shown that the residual stress in polysilicon layers can be reduced from 500 MPa compressive to less than about 10 MPa compressive after an anneal at 1000 °C [31, 32]. Usually the annealing is performed in a nonoxidizing ambient such as nitrogen. While these anneals can be performed in conventional furnaces, it is usually preferable to perform them in a rapid thermal anneal (RTA) system since this will reduce the thermal impact on dopants and other materials that may be present on the substrate [33].

The annealing of polysilicon thin films at temperatures of 1000 °C or higher will usually result in a residual stress level of near zero. More specifically, for polysilicon

Table 6.5 Residual stresses in as-deposited, undoped LPCVD polysilicon

Deposition temperature (°C)	Silane flow rate (sccm)	Pressure (mtorr)	Thickness (μm)	Residual stress (MPa)	References
560–630	30	300–550	2	−340 to 1750	[24]
570	80	150	1.3	82	[25]
570	100	300	2	270	[26]
600	125	550	0.1	12	[27, 28]
615	100	300	2	−200	[26]
570–615	100	300	2.72	<10	[26]
620	70	100	0.46	−350 ± 12	[29]
630	20	400	0.2	−180	[30]

films that were deposited at or near the amorphous-to-crystalline transition (i.e., deposition performed at around 570 °C), an anneal at 1000 °C will reduce the residual stress to near zero, whereas for polysilicon films that are highly textured (i.e., deposition performed at around 625 °C), an anneal at 1100 °C will be required to reduce the residual stress to near zero [26]. High-temperature anneals reduce the stress in polysilicon thin films for undoped and doped layers as seen in Tables 6.6 and 6.7.

With regard to other material properties of LPCVD deposited polysilicon, Table 6.8 shows some of the reported measured values of some of the mechanical properties of LPCVD polysilicon thin-film layers as a function of deposition temperature and thicknesses. As can be seen, these other material properties do not vary nearly to the extent as seen in the residual stress. Additionally, it has also been reported that the density of LPCVD deposited polysilicon ranges in value from 2.25 to 2.33 g/cm^3, and the refractive index ranges from 3.22 to 3.4 depending on the processing conditions [41]. The thermal conductivity for fine-grained material LPCVD polysilicon has been reported to be approximately 25% of the value of single-crystal silicon. For large-grain films, the thermal conductivity ranges between 50% and 85% of the single-crystal value [16].

6.5.3 LPCVD Silicon Dioxide (SiO_2)

Like other CVD thin films, the residual stress in LPCVD oxides, specifically LTO and PSG, is process dependent. The processing conditions and resultant residual stresses in deposited LPCVD oxide layers are shown in Table 6.9 for as-deposited thin films and Table 6.10 for layers that were annealed after deposition.

The dielectric constant of LTO has been reported to be 4.3, and the dielectric strength of LTO is about 80% of that for thermal oxide [44]. Many of the other material properties of LPCVD SiO_2 also have values of about 80% of the same material property for thermal oxides depending on the exact processing conditions.

Table 6.6 Residual stress in undoped LPCVD polysilicon thin-film material layers after high-temperature annealing

Deposition temperature (°C)	Silane flow rate (sccm)	Pressure (mtorr)	Thickness (μm)	Anneal conditions	Residual stress (MPa)	References
565	–	–	1	1050 °C for 10 s using RTA	142	[34]
570	80	150	1.3	1200 °C for 6 h	17	[25]
570	100	300	2	1100 °C for 30 min	30	[26]
580	–	–	3.5	1000 °C for 1 h	12 ± 5	[35]
615	100	300	2	1100 °C for 30 min	−20	[26]
620	70	300	0.5	900–1150 °C for 1–10 s, RTA	−340 to 90	[33]
620	70	100	0.46	1100 °C for 2 h	Low stress	[29]
625	80	180	3	1200 °C for 6 h	−205	[25]
630	–	–	4	1000 °C for 90 min	42	[36]

Table 6.7 Resultant residual stresses in doped polysilicon thin films deposited using LPCVD

Deposition temperature (oC)	Silane flow rate (sccm)	Deposition pressure (mtorr)	Thickness (μm)	Doping and anneal conditions	Resultant residual stress (MPa)	References
560–610	100	375–800	2	PH_3, anneal 900 °C, 10–120 s	−195 to 310	[18]
580	–	–	2	P implant, anneal 950 °C, 1–10 h	40.3–83.9	[37]
580	–	350	2	P implant or diff, variable anneal	26–72	[38]
585	50	200	0.5	$POCl_3$ diff, 850–950 °C	−110	[22]
625	80	180	3	$POCl_3$ diff, anneal 1200 °C for 6 h	−98 to 11	[25]

Table 6.8 The measured mechanical properties of undoped polysilicon thin films deposited using LPCVD

Deposition temperature (°C)	Thickness (μm)	Young's modulus (GPa)	Tensile strength (GPa)	Fracture toughness (MPa$\sqrt{m}$)	References
565	1	–	2.84 ± 0.09	–	[34]
580	2.1	175 ± 21	–	–	[39]
580	3.5	–	–	1.2 ± 0.2	[35]
620	1–1.14	175 ± 21	2.7–3.4	–	[40]
620	0.46	151 ± 6	–	–	[29]
620	0.46	162 ± 8	–	–	[29]
630	0.2	160	–	–	[30]
630	4	190	–		[36]

Table 6.9 Resultant residual stresses in as-deposited SIO_2 thin films deposited using LPCVD

Deposition temperature (°C)	Source gases	Flow rate ratio	Pressure (mtorr)	Deposition rate (nm/min)	Residual stress (MPa)	References
425	SiH_4/O_2	0.5	200	20	−10	[42]
700	TEOS/O_2	0.81	–	70	200	[43]
	TEOS/PH_3	5.3				

Table 6.10 Resultant residual stresses in annealed SiO_2 thin films deposited using LPCVD

Material type	Source gases	Deposition temperature (°C)	Anneal conditions	Residual stress (MPa)	References
PSG	SiH_4/O_2/PH_3	425	600 °C for 30 min	0	[40]
LTO	SiH_4/O_2	425	600 °C for 30 min	20	
PSG	TEOS/O_2/PH_3	700	950 °C for 10 min	100	[42]

6.5.4 LPCVD Silicon Nitride (SiN)

The material properties of SiN allow its use as an electrical insulator, an etch masking material, a mechanical structural material, and a mechanical biasing material. The mechanical biasing is based on its high tensile residual stress that can be used to make microstructures taut. SiN has a resistivity of about 10^{14} ohm-cm, a bandgap of about 5 eV, and a dielectric field strength of 10^7 V/cm. The mass density is about 3.1 g/cm^3. It is also relatively unaffected by chemical attack, and therefore it is can be used as an etch mask material, particularly for HF etching of sacrificial

Table 6.11 Material property values of as-deposited LPCVD SiN thin-film layers

Temperature (°C)	Gases	Ratio	Pressure (mtorr)	Thickness (μm)	Young's modulus (GPa)	Residual stress (MPa)	References
785	SiH_2Cl_2/ NH_3	0.33	368	–	330	1020	[52]
785	SiH_2Cl_2/ NH_3	6	368	–	230	430	[52]
–	SiH_2Cl_2/ NH_3	4	–	0.3	295	322	[53]
850	SiH_2Cl_2/ NH_3	0.33	150	–	320	967	[51]
850	SiH_2Cl_2/ NH_3	5	–	1	186	108	[54]
850	SiH_2Cl_2/ NH_3	5.7	150	–	230	430	[51]
880	SiH_2Cl_2/ NH_3	4	600	–	–	1 ± 10	[55]

layers and silicon etching using anisotropic etchant solutions. For example, the etch rate in HF is about 1 nm/min. One of the more critical material properties of SiN thin films in microsystems manufacturing is the residual stress.

The residual stress in stoichiometric SiN (e.g., Si_3N_4) has been reported to have a tensile value of approximately 1 GPa [45]. As a practical matter, this very high value of residual stress means that films of this material are susceptible to cracking if they are more than a few hundred nanometers in thickness. The high-residual stress state of stoichiometric SiN severely limits the maximum thickness of material that can be deposited and the uses of this material. As a result, researchers have developed alternative recipes for LPCVD SiN that lower the residual stresses in these types of material layers.

It has been shown that modifying the ratio of dichlorosilane to ammonia to about 6 parts dichlorosilane to 1 part ammonia can result in almost stress-free SiN thin-film layers deposited using LPCVD [46–48]. The resulting material layers are silicon-rich compared to the stoichiometric SiN layers, with a chemical makeup of about $Si_1N_{1.1}$ [49], although other chemistries are possible depending on the processing conditions. Studies of the ratio of silicon and nitrogen and the effect on the residual stress in the deposited thin-film layers have been reported wherein it was shown that increasing the silicon content results in a lowering of the residual stress [47, 48]. A significant lowering of the residual stress allows SiN layers of 2 μm to be deposited without the cracking seen in the stoichiometric materials. Additionally, the presence of the additional silicon content in these layers also results in films that have lower etch rates in HF. The ability to deposit nearly stress-free SiN layers has been an important advancement particularly for MEMS microsystems [50, 51].

Table 6.11 shows some of the reported values of the mechanical properties of SiN films as a function of the processing conditions, including the ratio of dichlorosilane to ammonia. As can be seen from the data in this table, the residual stress can vary over a very large range depending on the exact process recipe used for the deposition. More importantly, this table illustrates the relationship between the ratio of dichlorosilane to ammonia and residual stress (as well as the measured Young's modulus) in the deposited layers. As can be seen, while the residual stress in the layers tends to decrease with increasing flow rates of ammonia to dichlorosilane, other processing conditions also have an effect on the residual stress.

Table 6.12 is a compilation of reported values of mechanical properties of post-deposited annealed SiN films as a function of processing conditions. The takeaway from this table is that performing a high-temperature anneal has the effect of resulting in a lower state of residual stress.

In addition to modifying the residual stresses in deposited layers, the other material properties of silicon-rich SiN layers are also different from those of stoichiometric material layers. For example, the tensile strength of the silicon-rich material was found to be 5.5 GPa compared to 6.4 GPa for Si_3N_4, and a similar decrease in the fracture toughness has been reported [53, 59]. Low-stress SiN thin-film layers were reported to have a measured mass density of 3.0 g/cm^3, a thermal conductivity of 3.2×10^{-2} W/cm °K, and a heat capacity of 0.7 J/g °K [49].

6.5.5 *PECVD Silicon Dioxide (SiO_2)*

PECVD SiO_2 is heavily used in MEMS microsystems manufacturing due to the low deposition temperatures. Additionally, another advantage of PECVD SiO_2 is that the mechanical properties can be modified over a range of values to suit the requirements for many device applications.

The material properties of PECVD deposited layers are even more strongly sensitive to the processing conditions than LPCVD due to the complicated interaction of process gases and plasma. The low deposition temperatures often result in significant hydrogen incorporation into the deposited layers that can also affect the material properties. The residual stress of PECVD SiO_2 can vary over a large range of values for different processing condition as exhibited in Table 6.13.

A post-deposition anneal of PECVD SiO_2 is often performed to modify the residual stress as well as to densify the material layer. Densification is a particularly important consideration for this material due to the incorporation of hydrogen in the layers that usually results in lower mechanical and electrical strength levels. Additionally, if another material layer is deposited on top of the PECVD SiO_2, densification may be a prerequisite before deposition of a new material since the PECVD SiO_2 may outgas and shrink significantly. This can result in the top material layer cracking or delaminating. The formation of bubbles in the top layer is also a possibility. Table 6.14 shows some of the measured mechanical properties of PECVD SiO_2 thin-film layers after annealing was performed. As can be seen, the residual stress in PECVD SiO_2 layers can be reduced to low values after annealing at or above 350 °C.

Table 6.12 Material property values of post-annealed LPCVD SiN thin-film layers

Temperature (°C)	Gases	Flow rate (sccm) or ratio	Pressure (mtorr)	Thickness (μm)	Anneal conditions	Young's modulus (GPa)	Residual stress (GPa)	References
840	SiH_2Cl_2 NH_3	6–1	170	0.76	1100 °C, in N_2 for 2 h	202	291	[56]
840	SiH_2Cl_2 NH_3 N_2	1000 16 100	200	0.3–0.7	1100 °C, in forming gas for 2 h	248	–	[57]
850	SiH_2Cl_2/ NH_3	4	–	–	1100 °C, in O_2 for 3 h	–	10	[58]

Table 6.13 Processing conditions for PECVD SiO_2 layers

Temperature (°C)	Gases	Gas flow rate (sccm)	Deposition rate (μm/min)	Thickness (μm)	Residual stress (MPa)	References
300	SiH_4 N_2O N_2	– – –	– – –	0.3	−397	[14]
350	TEOS O_2	2.3 mL/min 9500	0.25	12.5	−45	[60]
400	SiH_4 N2O N_2	300 9500 1500	1	20	−80	[60]

Table 6.14 Mechanical properties as a function of processing conditions for PECVD SiO_2 after an anneal was performed

Temperature (°C)	Gases	Gas flow rate (sccm)	Thickness (μm)	As-deposited residual stress (MPa)	Anneal process	Post anneal residual stress (MPa)	References
300	SiH_4 N_2O N_2	– – –	0.3	−397	800 °C for 30 min, O_2	−172	[14]
350	TEOS O_2	2.3 mL/min 9500	12.5	−45	500 °C	<10	[60]
400	SiH_4 N_2O N_2	300 9500 1500	20	−80	500 °C	20	[60]

6.5.6 *PECVD Silicon Nitride (SiN)*

PECVD SiN has application in MEMS microsystems as a structural material layer or as a passivation layer. The mechanical properties of PECVD SiN layers reported that Young's modulus increased with the temperature used during deposition. Specifically, in one study the measured values of Young's modulus were found to be 79 GPa for films deposited at 125 °C and 151 GPa for films deposited at 205 °C [61]. Additionally, the residual stress in the deposited layers was found to be less than 30 MPa at both deposition temperatures. However, other studies have reported measured values of residual stress ranging from moderately compressive to moderately tensile depending on the process recipe used.

It should be cautioned that PECVD SiN has been reported to have high etch rates in exposure to hydrofluoric acid (HF) depending on the processing conditions. Table 6.15 is a compilation of some of the reported values of mechanical properties of PECVD SiN layers as a function of the processing conditions.

Table 6.15 Mechanical properties for PECVD SiN thin-film layers for various process recipes

Temperature (°C)	Gases	Gas flow rate (sccm)	Pressure (torr)	Power (W)	Thickness (μm)	Young's modulus (GPa)	Residual stress (MPa)	References
50–300	SiH_4 NH_3 N_2	5 45 100	0.45	100	0.5	50–150	−225 to 300	[62]
55–330	SiH_4 NH_3 N_2	5 45 100	0.88	75	0.7	150	−75 to 375	[62]
125–300	SiH_4 NH_3 N_2	– 10 2020	0.2–0.6	40–200	0.6–1.2	106–198	−250 to 250	[63]
300	SiH_4 NH_3 N_2	600 55 1960	0.9	100	0.7	133	178	[64]
300	SiH_4 N_2	15 535	0.5	–	0.5	210	110	[30]

6.5.7 PECVD Polycrystalline Silicon

PECVD is not as commonly used as LPCVD for the deposition of polysilicon thin-film material layers. However, PECVD does offer one significant advantage compared to LPCVD, and that is the ability to deposit at lower temperatures. As expected, material layers of silicon deposited using PECVD are amorphous in microstructure. Deposition of polysilicon at temperatures as low as 100 °C has been reported using PECVD [65]; however in most of the literature, the deposition temperatures are in the range of 250–300 °C, which is still much lower than for LPCVD polysilicon deposition processes. PECVD silicon films can also be in situ doped using the typical dopant gases such as PH_3 for n-type material and B_2H_6 for p-type material to modify the films' conductivity [66].

Like all CVD deposited materials, the material properties of PECVD silicon are heavily influenced by the exact processing conditions. Obviously, the electrical properties of PECVD silicon are not comparable to those of single-crystal silicon. The mechanical properties include a fairly high compressive residual stress (i.e., −130 to −575 MPa) depending on the deposition temperature. It has been reported that the compressive residual stress can be modified to a tensile residual stress with an anneal of the material at 367 °C for 100 min for a film that was deposited at 300 °C [67]. This indicates a remarkable amount of sensitivity of the residual stress to the temperatures that these materials are exposed to.

Due to the amorphous nature of PECVD silicon, which does not have favorable electrical properties, this material is not used for IC microsystems manufacturing but has found application in MEMS microsystems as a device or structural layer [68, 69]. Table 6.16 provides some of the reported processing conditions used and the measured material properties for PECVD silicon layers.

6.5.8 Evaporative Physical Vapor Deposition

The material properties of evaporative deposited thin films depend on a number of factors including processing conditions during deposition; material type being deposited; substrate type; material surface onto which the material is deposited; thickness of the deposited layer; and more. While each situation should be examined and evaluated individually, some general trends are discerned from published results of material deposited using evaporation PVD as follows [3]:

- Metal thin films deposited using evaporation usually exhibit tensile residual stress in the range from 10 MPa to 1 GPa.
- Dielectric films exhibit both tensile and compressive residual stresses.
- There appears to be no strong dependence between the residual stress in a thin film deposited using evaporation and the type of substrate.
- The magnitude of the residual stress exhibited in nonmetallic thin-film layers tends to be small.

Table 6.16 Mechanical and electrical material properties of PECVD silicon as a function of processing conditions

Temperature (°C)	Gases	Gas flow rate (sccm)	Press (torr)	Power or power density	Thick (μm)	Young's modulus (GPa)	Residual stress (MPa)	Conductivity (1/ohm cm)	References
150	SiH_4 Ar	10 7	0.5–0.75	77–114 mW/cm^2	0.5–3	–	−370	Dark 10^{-10} Light 10^{-6}	[66]
250	SiH_4/ H_2 PH_3	10 –	0.1	50 mW/cm^2	0.6–0.8	146	−130	10^{-3}	[70]
300	SiH_4 Ar	0.125–0.25 –	0.8	100–300 W	–	–	−360 to −380	–	[71]
300	SiH_4 Ar	40 20	0.33–0.53	50–215 W	1	–	−520 to −575	–	[67]

Table 6.17 Residual stresses and Young's moduli for a number of different materials deposited using evaporation physical vapor deposition [3]

Material type	Residual stress (MPa)	Young's modulus (GPa)
Metals		
Ag (silver)	20	76
Al (aluminum)	−74	69
Au (gold)	260	80
Cu (copper)	60	117
Cr (chromium)	850	248
In (indium)	0	11
Mo (molybdenum)	1080	324
Pd (palladium)	60	112
Ti (titanium)	0	115
Nonmetals		
C (carbon)	−400	40
Ge (germanium)	230	158
Si (silicon)	300	200
ZnS (zinc sulfide)	−190	54
MgF_2 (magnesium fluoride)	300–700	117
SiO (silicon oxide)	10	–

A simplistic rationale for the difference seen in the behaviors of metals and nonmetals deposited using evaporation has been offered that is based on the fact that in general metals are strong in tension, but not in compression, whereas dielectrics and semiconductors are strong in compression but tend to be weak in tension. Therefore, the metals tend to be in a state of tension, while nonmetals tend to be in the state of compression [3]. Table 6.17 provides some of the measured values of residual stress and Young's modulus for metals and nonmetal deposited using evaporation from various published references.

The residual stress has also been found to depend on the thickness of the evaporated layer. Metals with high melting temperature and hard refractory metals both tend to exhibit higher residual stresses than softer metals and metals with lower melting temperatures. It has also been reported that the residual stress in thin film deposited using evaporation only rises to appreciable levels after the thickness has reached 100 Å thick of material. The residual stress increases to large values up to a thickness of about 600 Å, after which the residual stress does not change significantly [3].

The temperature during deposition also has a considerable impact on the residual stress in thin films deposited using evaporation. Specifically, heating of the substrate during deposition modifies the residual stresses by increasing the rates of defect annealing, recrystallization, and growth of grains. The stresses associated with the growth of a thin-film material layer decrease rapidly with increasing temperature. Additionally, the diffusion of impurities in and out of the material layer substantially increases with temperature and also has an impact on the residual stress. As a result of all these temperature-related effects, the residual stress may reach a minimum value or exhibit a reversal in sign [3].

6.5.9 *Sputter Physical Vapor Deposition*

Due to the interactions of the plasma and the gases involved, there is far more complexity associated with sputtering processes compared to evaporation. Therefore, it is far more challenging to develop useful general conclusions about residual stresses in layers deposited using sputtering processes. Nevertheless, sputtering also allows much more freedom in obtaining desired material properties.

Some general tendencies about sputtering: when the substrate temperature during deposition is not elevated, the residual stress tends to be compressive in nature regardless of the material type [3]. It is also known that the amount of gas trapped in the thin film during deposition directly relates to the compressive stress exhibited in the layers, and it is suspected that this may partially explain stress seen in film deposited at lower relative temperatures.

One of the most comprehensive studies of effects of processing conditions on the residual stresses in sputter-deposited thin-film layers was performed by Hoffman and Thornton [72]. Their study was performed using magnetron sputtering which made it possible to deposit the layers over a wide range of pressures and deposition rates. This work reported two distinct regions with an almost discontinuous change in the material properties. Specifically, at low sputtering pressures, using lighter-mass sputtering gases, targets with higher masses, and lower deposition rates, the deposited layers exhibited compressive intrinsic stresses and also exhibited values of electrical resistivity and optical reflectance that were very similar to the bulk values of these properties. These films also tended to entrap more gases species into the films during deposition. Conversely, tensile stresses were observed in sputtered thin-film layers deposited at higher pressures, using heavier sputtering gases, lighter-mass target materials, and oblique angle of incidences. These films also tended to incorporate less trapped gases into the layers during deposition.

It is known that elevated working pressures induce columnar grain growth having intercrystalline voids, which is the so-called zone 1 developed by Hoffman and Thornton, and such layers tend to exhibit tensile stress combined with higher values of electrical resistivity and lower values of optical reflectance. At lower operating pressures, the zone 1 structure is suppressed, and the energetic particle bombardment of the sputtered atoms leads to films having a compressive residual stress that is thought to be due to an atomic peening mechanism during deposition.

6.5.9.1 Sputter-Deposited Silicon

The advantage of sputtering deposition processes for the deposition of silicon layers is of course the low temperatures that these processes can be performed at compared to other methods such as LPCVD and PECVD. Sputter-deposited silicon thin-film layers on silicon dioxide surfaces have been demonstrated with low residual compressive stresses at practical deposition rates and smooth surface roughness [73]. Silicon layers deposited using sputtering are amorphous unless the substrate

Table 6.18 Mechanical material properties of sputter-deposited silicon layers on various substrates and processing conditions [75]

Substrate	Power (kW)	Pressure (mtorr)	Deposition rate (nm/min)	Residual stress (MPa)
Silicon	1.5	8	23	34
Silicon	1.5	14	19	141
Phosphosilicate glass	1.5	8	23	97
Phosphosilicate glass	1.5	14	19	106
Aluminum on silicon	1.5	8	23	31
Aluminum on silicon	1.5	14	19	109
Silicon	2.5	8	37	−22
Silicon	2.5	14	30	164
Phosphosilicate glass	2.5	8	37	27

is heated substantially during deposition or the substrate is annealed after the deposition. It has been reported that annealing temperatures of at least 800 °C are needed in order to induce crystallization of the deposited silicon layers [74]. The silicon can be doped or undoped based on the dopant levels present in the target materials. Since the sputtering efficiency varies with element atomic mass, most silicon layers are sputtered using undoped source targets. Table 6.18 provides some of the reported mechanical properties of sputter-deposited silicon layers onto different substrates and processing conditions at room temperature.

6.5.10 Electrochemical Deposition

With respect to material properties of electroplated materials, as with other processes covered for microsystems manufacturing, electroplating is also very dependent on the processing conditions. For example, it has been reported that electroplated nickel exhibited a residual stress that varied from −110 to 150 MPa over a range of plating current densities from 0 to 30 mA/cm^2 at a temperature of 60 °C. Therefore, by adjusting the process settings, it is possible to plate a nickel film with a moderate to zero state of stress. Moreover, it was also reported that Young's modulus of the plated nickel varied from about 200 to 100 GPa as plating temperature was varied from 20 to 80 °C. Higher plating temperatures resulted in reduced Young's modulus, and this has been attributed to inclusion of organics in the plated thin films [76]. It should also be noted that not all of the material properties were observed to vary significantly with the processing conditions. For example, the resistivity of the plated nickel was measured as a function of plating current density, and the variation in

Table 6.19 Material properties for various electroplated materials as a function of processing conditions

Temperature (°C)	Material plated	Substance	Plating mixture	Plating current density (mA/cm^2)	Plating rate (nm/min)	Resistivity (μΩ-cm)	Residual stress (MPa)	Young's modulus (GPa)	References
20–80	Ni	5 nm Cr and 60 nm Cu on Si wafer	$Ni(SO_3NH_2)$ $4H_2O$ – 300 g/L $NiCl_2 6H_2O$ – 10 g/L H_3BO_3 – 40 g/L	0–30	0–550	17–22	−110 to 150 at 60 °C	200–100	[76]
–	Cu	Cu seed layer on top of Ta or TiW layer on Si wafer	$CuSO_4 \cdot 5H_2O$ with proprietary additives	–	–	–	0–200 as deposited depending on thickness, increased by 200 MPa after anneal at 350 °C	–	[77]
–	Cu	SiNx on Si wafer	–	–	–	–	100	117–137	[78]
–	Au	50 nm Ti and 100 nm Au on Si wafer	Sulfite gold chemistry	1–9 pulsed	20–185	–	−90 to 90	–	[79]
50	Au	Si wafer	Sulfite gold chemistry with agitation	1–5	–	–	−105 to −10	–	[80]

resistivity was relatively modest. Nevertheless, the resistivity values of the plated nickel were about three times that level seen in bulk nickel material [76]. Table 6.19 provides some of the reported material properties of electroplated materials including nickel, copper, and gold.

6.6 Summary

This chapter reviewed some of the important properties of the most commonly used materials in MEMS microsystems manufacturing. Most attention was given to two specific mechanical material properties, namely, Young's modulus and residual stress, due to the fact that these properties usually have an important impact on the behavior of MEMS devices and the fact that these properties can vary quite significantly depending on the exact processing conditions. As stated in the introduction to this chapter, the purpose of providing information about reported values of Young's modulus and residual stress in deposited thin-film layers is not to give values that can be used by MEMS designers, but instead to give an appreciation of the amount that these properties can vary with processing conditions and to give some guidance about the ranges that these properties may span. In all circumstances, the MEMS designers and developers should perform experiments on measuring the actual properties of all the important materials used in the implementation of MEMS devices. We also reviewed a number of test structures that can be used to determine some material properties so that MEMS designers will have the resources to perform measurements themselves.

References

1. R. Ghodsi, P. Lin (eds.), *MEMS Materials and Process Handbook* (Springer, New York, 2011)
2. E.P. Popov, *Introduction to the Mechanics of Solids* (Prentice-Hall, Englewood Cliffs, 1968)
3. M. Ohring, *The Materials Science of Thin Films* (Academic Press, London, 1992)
4. M. Huff, A thermally isolated microstructure suitable for gas sensing applications, S.M. Thesis, MIT, 1988
5. S. Senturia, Can we design microrobotic devices without knowing the mechanical properties of the materials? Micro robots and teleoperators workshop, Hyannis, 1987
6. W.C.K. Tang, Ph.D. thesis, University of California at Berkeley, 1990
7. L. Lin, Ph.D. thesis, University of California at Berkeley, 1993
8. H. Guckel, T. Randazzo, D.W. Burns, A simple technique for the determination of mechanical strain in thin films with application to polysilicon. J. Appl. Phys. **57**, 1671 (1985)
9. W.H. Chu, M. Mehregany, X. Ning, P. Pirouz, Measurement of residual stress-induced bending moment of p+ silicon films. Mater. Res. Soc. Symp. **239**, 169 (1992)
10. W.C. Oliver, G.M. Pharr, Measurement of hardness and elastic modulus by instrumented indentation: Advances in understanding and refinements to methodology. J. Mater. Res. **19**, 3 (2004)
11. S. Wolf, R.N. Tauber, *Silicon Processing for the VLSI Era, Volume 1-Process Technology* (Lattice Press, Sunset Beach, 1986)

12. E.H. Nicollian, J.R. Brews, *MOS Physics and Technology* (Wiley, New York, 2002)
13. MEMSNet Material Property Information, See: https://www.memsnet.org/material/silicondioxidesio2film/. Accessed 24 Sept. 2018
14. J. Leconte, F. Iker, S. Jorez, N. Andre, J. Proost, T. Pardoen, D. Flandre, J.P. Raskin, Thin films stress extraction using micromachined structures and wafer curvature measurements. Microelectron. Eng. **76**, 219–226 (2004)
15. J. Yang, J. Gaspar, O. Paul, Fracture properties of LPCVD silicon nitride and thermally grown silicon oxide thin films from the load-deflection of long Si3N4 and SiO2/Si3N4 diaphragms. J. Microelectromech. Syst. **17**, 1120–1134 (2008)
16. T. Kamins, *Polycrystalline Silicon for Integrated Circuits and Displays*, 2nd edn. (Kluwer, Boston, 1998)
17. G. Fresquet, C. Azzaro, J.P. Couderc, Analysis and modeling of in situ boron-doped polysilicon deposition by LPCVD. J. Electrochem. Soc. **142**, 538 (1995)
18. M. Biebl, G.T. Mulhern, R.T. Howe, In situ phosphorus-doped polysilicon for integrated MEMS, technical digest, in 8th International Conference on Solid-State Sensors and Actuators, Eurosensors IX, Stockholm, Sweden, pp. 198–201, 1995
19. J.G.M. Mulder, P. Eppenga, M. Hendriks, J.E. Tong, An industrial LPCVD process for in situ phosphorus-doped polysilicon. J. Electrochem. Soc. **137**, 273 (1990)
20. S. Bouwstra, E.L. de Weerd, M.C. Elwenspoek, In situ phosphorus-doped polysilicon for excitation and detection in micromechanical resonators. Sens. Actuators A **24**, 227–235 (1990)
21. R.C. Anderson, R.S. Muller, C.W. Tobias, Porous polycrystalline silicon: A new material for MEMS. J. Microelectromech. Syst. **3**, 10–18 (1994)
22. L. Elbrecht, R. Catanescu, J. Zacheja, J. Binder, Highly phosphorus-doped polysilicon films with low tensile stress for surface micromachining using POCl3 diffusion doping. Sens. Actuators A **61**, 374–378 (1997)
23. J.J. McMahon, J.M. Melzak, C.A. Zorman, J. Chung, M. Mehregany, Deposition and characterization of in-situ boron doped polycrystalline silicon films for microelectromechanical systems applications. Mater. Res. Symp. Proc. **605**, 31–36 (2000)
24. L. Chen, J. Miao, L. Guo, R. Lin, Control of stress in highly doped polysilicon multi-layer diaphragm structure. Surf. Coat. Technol. **141**, 96–102 (2001)
25. Y.B. Gianchandani, M. Shinn, K. Najafi, Impact of high-thermal budget anneals on polysilicon as a micromechanical material. J. Microelectromech. Syst. **7**, 102–105 (1998)
26. J. Yang, H. Kahn, A.Q. He, S.M. Phillips, A.H. Heuer, A new technique for producing large-area as-deposited zero-stress LPCVD polysilicon films: The multipoly process. J. Microelectromech. Syst. **9**, 485–494 (2000)
27. G.M. Dougherty, A.P. Pisano, T.D. Sands, Processing and morphology of permeable polycrystalline silicon thin films. J. Mater. Res. **17**, 2235–2242 (2002)
28. G.M. Dougherty, T.D. Sands, A.P. Pisano, Microfabrication using one-step LPCVD porous polysilicon films. J. Microelectromech. Syst. **12**, 418–424 (2003)
29. D. Maier-Schneider, J. Maibach, E. Obermeier, D. Schneider, Variations in Young's modulus and intrinsic stress of LPCVD-polysilicon due to high-temperature annealing. J. Micromech. Microeng. **5**, 121–124 (1995)
30. O. Tabata, K. Kawahata, S. Sugiyama, I. Igarashi, Mechanical property measurements of thin films using load-deflection of composite rectangular membrane, in Proceedings of Micro Electro Mechanical Systems, Salt Lake City, pp. 152–156, 1989
31. H. Guckel, T. Randazzo, D.W. Burns, A simple technique for the determination of mechanical strain in thin films with application to polysilicon. J. Appl. Phys. **57**, 1671–1675 (1983)
32. R.T. Howe, R.S. Muller, Stress in polysilicon and amorphous silicon thin films. J. Appl. Phys. **54**, 4674–4675 (1983)
33. X. Zhang, T.Y. Zhang, M. Wong, Y. Zohar, Rapid thermal annealing of polysilicon thin films. J. Microelectromech. Syst. **7**, 356–364 (1998)

34. M. Biebl, H. von Philipsborn, Fracture strength of doped and undoped polysilicon, technical digest, in 8th International Conference on Solid-State Sensors and Actuators, Eurosensors IX, Stockholm, Sweden, pp. 72–75, 1995
35. H. Kahn, N. Tayebi, R. Ballarini, R.L. Mullen, A.H. Heuer, Fracture toughness of polysilicon MEMS devices. Sens. Actuators A **82**, 274–280 (2000)
36. J.A. Walker, K.J. Gabriel, M. Mehregany, Mechanical integrity of polysilicon films exposed to hydrofluoric acid solutions. J. Electron. Mater. **20**, 665–670 (1991)
37. F. Ericson, S. Greek, J. Soderkvist, J. Schweitz, High sensitive internal film stress measurement by an improved micromachined indicator structure, technical digest, in 8th International Conference on Solid-State Sensors and Actuators, Eurosensors IX, Stockholm, Sweden, pp. 84–87, 1995
38. M.A. Benitez, L. Fonseca, J. Esteve, M.S. Benrakkad, J.R. Morante, J. Samitier, J.A. Schweitz, Stress-profile characterization and test-structure analysis of single and double ion-implanted LPCVD polycrystalline silicon. Sens. Actuators A **54**, 718–723 (1996)
39. H. Guckel, D.W. Burns, H.A.C. Tilmans, D.W. DeRoo, C.R. Rutigliano, Mechanical properties of fine grained polysilicon-the repeatability issue, technical digest, Solid-State Sensor Actuator Workshop, Hilton Head, pp. 96–99, 1988
40. J. Koskinen, J.E. Steinwall, R. Soave, H.H. Johnson, Microtensile testing of free-standing polysilicon fibers of various grain sizes. J. Micromech. Microeng. **3**, 13–17 (1993)
41. D. Maier-Schneider, A. Kprll, S.B. Holm, E. Obermeier, Elastic properties and microstructure of LPCVD polysilicon films. J. Micromech. Microeng. **6**, 436–446 (1996)
42. Y. Shioya, M. Mamoru, Comparison of phosphosilicate glass films deposited by three different chemical vapor deposition methods. J. Electrochem. Soc. **133**, 1943–1950 (1986)
43. R.M. Levin, A.C. Adams, Low pressure deposition of phosphosilicate glass films. J. Electrochem. Soc. **129**, 1588–1592 (1982)
44. S.K. Ghandhi, *VLSI Fabrication Principles – Silicon and Gallium Arsenide* (Wiley, New York, 1983)
45. J. Yang, O. Paul, Fracture properties of LPCVD silicon nitride thin films from the load deflection of long membranes. Sens. Actuators A **97–98**, 520–526 (2002)
46. M. Sekimoto, H. Yoshihara, T. Ohkubo, Silicon nitride single-layer x-ray mask. J. Vac. Sci. Technol. **21**, 1017–1021 (1982)
47. J.G.E. Gardeniers, H.A.C. Tilmans, C.C.G. Visser, LPCVD silicon-rich silicon nitride films for applications in micromechanics studied with statistical experimental design. J. Vac. Sci. Technol. A **14**, 2879–2892 (1996)
48. P. Temple-Boyer, C. Rossi, E. Saint-Etienne, E. Scheid, Residual stress in low pressure chemical vapor deposition SiNx films deposited from silane and ammonia. J. Vac. Sci. Technol. A **16**, 2003–2007 (1998)
49. C. Mastrangelo, Y.-C. Tai, R. Muller, Thermophysical properties of low-residual stress, silicon-rich, LPCVD silicon nitride films. Sens. Actuators **856–860**, A21–A23 (1990)
50. P.P. Tsai, I.-C. Chen, C.J. Ho, Ultralow power carbon monoxide microsensor by micromachining techniques. Sens. Actuators B **76**, 380–387 (2001)
51. P.J. French, P.M. Sarro, R. Mallee, E.J.M. Fakkeldij, R.F. Wolffenbuttel, Optimization of a low-stress silicon nitride process for surface micromachining applications. Sens. Actuators A **58**, 149–157 (1997)
52. S. Hong, T.P. Weihs, J.C. Bravman, W.D. Nix, Measuring stiffness and residual stresses of silicon nitride thin films. J. Electron. Mater. **19**, 903–909 (1990)
53. A. Kaushik, H. Kahn, A.H. Heuer, Wafer-level mechanical characterization of silicon nitride MEMS. J. Microelectromech. Syst. **14**, 359–367 (2005)
54. E.I. Bromley, J.N. Randall, D.C. Flanders, R.W. Mountain, A technique for the determination of stress in thin films. J. Vac. Sci. Technol. B **1**, 1364–1366 (1983)
55. J.M. Olson, Analysis of LPCVD process conditions for the deposition of low stress silicon nitride. Part 1: Preliminary LPCVD experiments. Mater. Sci. Semicond. Process. **5**, 51–60 (2002)

56. T.Y. Zhang, Y.J. Su, C.F. Qian, M.H. Zhao, L.Q. Chen, Microbridge testing of silicon nitride thin films deposited on silicon wafers. Acta Mater. **48**, 2843–2857 (2000)
57. Y. Ren, D.C.C. Lam, Characterization of elastic behaviors of silicon nitride films with varying thicknesses. Mater. Sci. Eng. A **467**, 93–96 (2007)
58. B.C.S. Chou, J.S. Shie, C.N. Chen, Fabrication of low stress dielectric thin film for microsensor applications. IEEE Electron Device Lett. **18**, 599–601 (1997)
59. L.S. Fan, R.T. Howe, R.S. Muller, Fracture toughness of brittle thin films. Sens. Actuators **872–874**, A21–A23 (1990)
60. X. Zhang, K.-S. Chen, R. Ghodssi, A.A. Ayon, S.M. Spearing, Residual stress and fracture in thick tetraethylorthosilicate (TEOS) and silane-based PECVD oxide films. Sens. Actuators A **91**, 373–380 (2001)
61. B.A. Walmsley, Y.L. Liu, X.Z. Hu, M.B. Bush, J.M. Dell, L. Faraone, Poisson's ratio of low-temperature PECVD silicon nitride thin films. J. Microelectromech. Syst. **16**, 622–627 (2007)
62. M. Martyniuk, J. Antoszewski, C.A. Musca, J.M. Dell, L. Faraone, Dielectric thin films for MEMS-based optical sensors. Microelectron. Reliab. **47**, 733–738 (2007)
63. H. Huang, K.J. Winchester, A. Suvorova, B.R. Lawne, Y. Liud, X.Z. Hud, J.M. Dell, L. Faraone, Effect of deposition conditions on mechanical properties of low-temperature PECVD silicon nitride films. Mater. Sci. Eng. A **435–436**, 453–459 (2006)
64. W. Zhou, J. Yang, Y. Li, A. Ji, F. Yang, Y. Yu, Bulge testing and fracture properties of plasma-enhanced chemical vapor deposited silicon nitride thin films. Thin Solid Films **517**, 1989–1994 (2009)
65. J. Gaspar, T. Adrega, V. Chu, J.P. Conde, Thin-film paddle microresonators with high quality factors fabricated at temperatures below 110°C, in Proceedings of the 18th International Conference on Microelectromechanical Systems, Miami, pp. 125–128, 2005
66. S. Chang, S. Sivoththaman, Development of a low temperature MEMS process with a PECVD amorphous silicon structural layer. J. Micromech. Microeng. **16**, 1307 1313 (2006)
67. S. Chang, W. Eaton, J. Fulmer, C. Gonzalez, B. Underwood, Micromechanical structures in amorphous silicon, technical digest, in International Conference on Solid State Sensors and Actuators, San Francisco, pp. 751–754, 1991
68. S.B. Patil, V. Chu, J.P. Conde, Surface micromachining of a thin film microresonator using dry decomposition of a polymer sacrificial layer. J. Vac. Sci. Technol. B **25**, 455–458 (2007)
69. S. Chang, S. Sivoththaman, A tunable RFMEMS inductor on silicon incorporating an amorphous silicon bimorph in a low-temperature process. IEEE Electron Device Lett. **27**, 905–907 (2006)
70. P. Alpuim, V. Chu, J.P. Conde, Amorphous and microcrystalline silicon films grown at low temperatures by radio-frequency and hot-wire chemical vapor deposition. J. Appl. Phys. **86**, 3812–3821 (1999)
71. C.-K. Chung, M.-Q. Tsai, P.-H. Tsai, C. Lee, Fabrication and characterization of amorphous Si films by PECVD for MEMS. J. Micromech. Microeng. **15**, 136–142 (2005)
72. D.W. Hoffman, J.A. Thornton, J. Vac. Sci. Technol. **20**, 355 (1982)
73. T. Abe, M.L. Reed, Low strain sputtered polysilicon for micromechanical structures, in Proceedings of the 9th International Workshop on Micro Electro Mechanical Systems, San Diego, pp. 258–262, 1996
74. P. Pal, S. Chandra, RF sputtered silicon for MEMS. J. Micromech. Microeng. **15**, 1536–1546 (2005)
75. K.A. Honer, G.T.A. Kovacs, Integration of sputtered silicon microstructures with prefabricated CMOS circuitry. Sens. Actuators A **91**, 392–403 (2001)
76. J.K. Luo, M. Pritschow, A.J. Flewitt, S.M. Spearing, N.A. Fleck, W.I. Milne, Effects of process conditions on properties of electroplated Ni thin films for microsystem applications. J. Electrochem. Soc. **153**(10), D155–D161 (2006)
77. A.A. Volinsky, M. Hauschildt, J.B. Vella, N.V. Edwards, R. Gregory, W.W. Gerberich, Residual stress and microstructure of electroplated Cu films on different barrier films. Mater. Res. Soc. Symp. **695**., Materials Research Soc., L1.11.1 (2002)

78. Y. Xiang, X. Chen, J.J. Vlassak, The mechanical properties of electroplated Cu thin films measured by means of the bulge test technique. Mater. Res. Soc. Symp. Proc. **695**., Materials Research Soc., L4.9.1 (2002)
79. P.H. Lawyer, C.H. Fields, Film stress versus plating rate for pulse-plated gold, HRL Laboratories Report, 2001
80. S.H. Pu, A.S. Holmes, E.M. Yeatman, Stress in electroplated gold on silicon substrates and its dependence on cathode agitation. Microelectron. Eng. **112**, 21 (2013)

Chapter 7
Microsystems Process Integration, Testing, and Packaging

7.1 Introduction

The most commonly used processing steps (and process modules) for the manufacturing of microsystems were reviewed in Chaps. 3 and 4. These individual processing steps should be considered as the elemental building blocks for making microsystems devices. However, it is only when a collection of selected processing steps are placed into an ordered and sequentially performed array of processing steps can actual microsystems devices be implemented. That is, an individual processing step is not very useful unless it can be used with other processing steps to form a process sequence. Depending on what is being produced, the number of processing steps in the process sequence for the manufacturing of a microsystem can vary greatly. Process sequences for some advanced microsystems may involve hundreds of individual processing steps. Importantly, each of these processing steps must work together in order for the resultant device to function properly, or function even at all.

This chapter is directed at the extremely important topic of process integration. Process integration *is the technique by which processing steps are combined into an ordered sequence with the goal of creating a methodology that allows the production of functional microsystems devices with acceptable yields and cost levels.* Process integration involves the selection of individual processing steps to be used in the device fabrication. But it also entails the understanding, characterizing, and optimizing of these steps and their interrelationships with the goal of having them effectively work together as a production method to meet a specific acceptable performance, yield, and cost level. Process integration is one of the most important topics in MEMS microsystems manufacturing because the choices made about how to produce a MEMS device have a large impact on the success or failure of any MEMS-related product development activity or business venture.

M. Huff, *Process Variations in Microsystems Manufacturing*, Microsystems and Nanosystems, https://doi.org/10.1007/978-3-030-40560-1_7

This chapter will describe why process integration is important in MEMS and how it is performed. It will also review two notable examples of successfully developed microsystems process technologies that have transitioned into manufacturing production.

This chapter will also briefly review testing and packaging of MEMS microsystems. Both of these topics are both challenging and important for the manufacturing of MEMS devices. Testing is important since it is the methods by which manufactured devices are deemed to meet the application/product requirements. Packaging is important since it is the enclosure that protects the device while simultaneously allowing it to interact with the environment. Most MEMS testing equipment setups and packages are customized in MEMS.

7.2 What Is Process Integration?

Definitions of processing step, process module, process sequence, and process technology were given in Chap. 2. An example of a very simple process sequence for fabricating a MEMS polysilicon cantilever is shown in Fig. 7.1. The cantilever is anchored to the substrate at one end and free on the other. The method for making this structure would be classified as a "polysilicon surface micromachining process

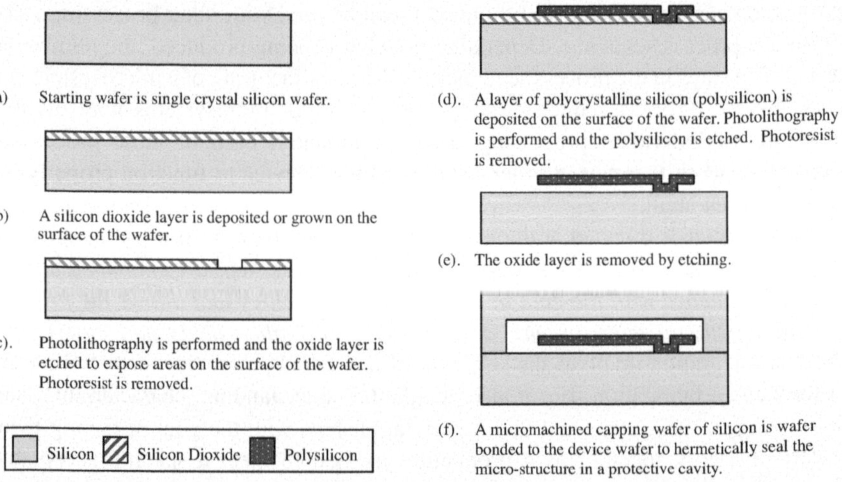

Fig. 7.1 Example of a very simple surface micromachined process sequence surface with a wafer-bonded capping layer. (**a**) Starting wafer is single crystal silicon wafer. (**b**) A silicon dioxide layer is deposited or grown on the surface of the wafer. (**c**) Photolithography is performed and the oxide layer is etched to expose areas on the surface of the wafer. Photoresist is removed. (**d**) A layer of polycrystalline silicon (polysilicon) is deposited on the surface of the wafer. Photolithography is performed and the polysilicon is etched. Photoresist is removed. (**e**) The oxide layer is removed by etching. (**f**) A micromachined capping wafer of silicon is wafer bonded to the device wafer to hermetically seal the micro-structure in a protective cavity

sequence." The illustration in Fig. 7.1 shows a cross section of the wafer at various important points in the process sequence. These types of drawings are extremely useful for designing a fabrication process for any MEMS device and are commonly used by MEMS designers, particularly in the early stages of a development effort. These types of illustrations will also be frequently used to describe many of the example process technologies that will be reviewed later in this chapter.

As can be seen, the first step (Fig. 7.1a) describes the starting wafer as a single-crystal silicon wafer. Subsequently, a thermal oxide thin-film material layer is grown on the surface of the wafer (Fig. 7.1b). This oxide layer is commonly termed the *sacrificial layer* in surface micromachining process sequences. Next, the thermal oxide layer is photolithographically patterned and etched (Fig. 7.1c) to open up areas in the oxide to expose the underlying silicon wafer surface. A thin-film material layer of polycrystalline silicon (polysilicon) is then deposited on the surface of the wafer, and this layer is patterned and etched in the shape of the cantilever (Fig. 7.1d). The polysilicon thin-film material layer is called the *structural layer* in surface micromachining process sequences. The oxide is then removed from the wafer (Fig. 7.1e). The process of removing the oxide sacrificial layer is called a *release*. Subsequently, a micromachined capping wafer is wafer bonded to the device wafer. The capping wafer has a cavity made into its surface that will encapsulate the polysilicon cantilevers (Fig. 7.1f).

The description of the above process is highly simplified and leaves out many of the important details, including thickness of the deposited layers; depth of the etches; type of photolithography process performed to pattern the layers; what type of mask was used; as well as the critically important process step parameters. Additionally, for each processing step in a process sequence, there will be one or more process step parameters including the exact details of the equipment that is used to perform the process step; the temperature and chemistries used in the process step; etc. Obviously, the fabrication of a MEMS device has many levels of detailed information that must be specified before implementation can begin.

Process sequences (and technologies) are commonly defined in what are called *process runcards* or *process travelers*. These are documents that define the processing steps and process modules for the fabrication of a device as a set of sequentially performed processing steps. The level of detail of the information on runcards is sufficient to represent the process sequence and includes the processing steps employed (including possibly an identifier of the process recipe on the tool the processing step is performed); thickness of the layers; equipment that the processing step is performed on; specification of the masks and wafers to be used; etc. Generally, the process runcard would not include extremely detailed information about the processing steps such as the tool settings of the process recipes, as these would usually be documented elsewhere. In a production level facility, this hierarchy of documentation would be kept on a computerized manufacturing support system that may be networked to the production equipment.

Figure 7.2 shows an example of a process runcard for the polysilicon surface micromachining process sequence to implement the cantilever illustrated in Fig. 7.1. As can be seen, it starts with a specification of the masks to be used in the process

Run: Simple Surface Micromachining

Masks

Polysilicon Layer

Equipment type contact	Substrate material soda lime	Coating chromium	
Magnification 1.0	Field bright	Source customer	
Length 5 inch	Width 5 inch	Thickness 0.09 inch	

Oxide Layer

Equipment type contact	Substrate material soda lime	Coating chromium	
Magnification 1.0	Field dark	Source customer	
Length 5 inch	Width 5 inch	Thickness 0.09 inch	

Wafer Descriptions

Group 1

Number of wafers 10	Material silicon	Diameter 150 mm
Surface finish single side polish	Thickness 625 μm	Orientation 100
Doping type none	Quality prime	Resistivity 10 Ω*cm

Process Sequence

Step 1: Wet oxidation

Material silicon dioxide Thickness 1 μm ±10%

1 Piranha Clean
2 HF 10:1 Batch Etch
3 Wet oxidation
4 Spectrophotometric film thickness measurement

Instructions 1) Wafer loaded into furnace with flats up. 2) 20 point metrology

Step 2: Contact photolithography (Automated)

1 HMDS Vapor Prime
2 Photoresist Spin Coat ACS200 (AZ 5214E)
3 Photoresist Softbake ACS200
4 Edge bead removal
5 Contact flat alignment and exposure
6 Photoresist Develop (5214E)
7 Linewidth Microscope Measurement
8 Microscope inspection
9 Stylus profilometer 1-D step measurement
10 Manual Photoresist Hardbake (hotplate)

Step 3: BOE Etch

Depth 1 μm ±10% Etch type wet isotropic Material silicon dioxide

Step 4: Resist strip

Depth 1.5 μm ±10% Material AZ 5214e

Step 5: Undoped polysilicon LPCVD

Excluded materials gold (category), copper Material polysilicon Thickness 1 μm ±10%

1 RCA clean with HF Dip (Pre-furnace clean)
2 Undoped polysilicon LPCVD
3 Spectrophotometric film thickness measurement

Step 6: Contact photolithography (Automated)

1 HMDS Vapor Prime
2 Photoresist Spin Coat ACS200 (AZ 5214E)
3 Photoresist Softbake ACS200
4 Edge bead removal
5 Contact flat alignment and exposure
6 Photoresist Develop (5214E)
7 Linewidth Microscope Measurement
8 Microscope inspection
9 Stylus profilometer 1-D step measurement
10 Manual Photoresist Hardbake (hotplate)

Step 7: Silicon ICP Etch

Batch size 1 Etch rate 1.1 μm/min Material silicon
Depth 1 μm ±1% Gas SF6, Ar Temperature 25 °C

Step 8: Resist strip

Depth 1.5 μm ±10% Material AZ 5214e

Step 9: BOE Etch

Depth 1 μm ±10% Etch type wet isotropic Material silicon dioxide

Instructions Release Etch.

Step 10: Anodic bonding (vacuum, with alignment)

Alignment type optical Bonded materials silicon, Pyrex (Corning 7740), glass (Hoya)
Batch size 2 Pressure 0.1 mTorr

1 Megasonic clean
2 Anodic bonding (vacuum, with alignment)

Step 11: Wafer dicing

Batch size 1 Diced wafer material silicon (single crystal) Ship wafer on dicing frame yes

Fig. 7.2 Example of process runcard or process traveler for the simple example of the polysilicon surface micromachined cantilever surface

sequence, then details the wafer descriptions to be used, and then lists each of the processing steps in the process sequence, numbered Step 1 through Step 11. Some of the processing steps, such as Steps numbered 1, 2, 5, 6, and 10, are actually groups of processing steps, such as processing steps 2.1 through 2.10 included in step 2. A document such as this would be conveyed from the MEMS designer to the foundry to specify the fabrication of a MEMS device.

7.3 How Is Process Integration Performed?

The development of a microsystems manufacturing process sequence is a very challenging endeavor and for it to be successful requires the careful balance of creative ideas tempered with objective and judicious decision-making. The technologies employed as well as the people involved must all be carefully selected and may vary considerably in specific skill sets from one type of development project to another. It is highly advisable to include domain experts in the specific application area who can provide important guidance on the product requirements and MEMS fabrication experts who can design and develop production-worthy manufacturing methods.

A successful MEMS development effort requires the determination to overcome the constant stream of problems and disappointments that often occur. The development team requires experienced technologists led by people with both strong leadership skills and technical understanding, who can direct and focus the team on the most practical and timely solutions to problems. It is important that the process designers and developers be brought into the development effort very early and work closely with the device and product designers to provide their guidance on the implementation of the device and product. It is also important that the decision maker(s) for the process approaches be "process agnostics." It is also important that testing and packaging experts be brought onto the team very early since their input on the needs, requirements, and constraints on both testing and packaging is critically important.

It would be incorrect to assume that there is a "one size fits all," "foolproof," or even "well-proven and demonstrated" strategy or set of principles for successfully developing a production worthy manufacturing process sequence for a microsystems device. Nevertheless, some of the core competencies for successful process development include: very creative instincts; an encyclopedic knowledge of what has been previously demonstrated in the art; a good level of engineering intuition; good device design capability; the ability to correlate facts from different domains; the ability to handle multivariable spaces with ease; and years or decades of hands-on experience. In short, these are not skills that can be easily conveyed or taught in a class or book. It should be understood that even for the most gifted process developer, process development entails a considerable amount of iteration and perfecting, based on the ability to quickly learn from previous experiments, including unsuccessful outcomes, and developing strategies for advancing the process development to a successful conclusion.

Figure 7.3 is a simple high-level illustration of the general approach taken for developing a MEMS microsystems process sequence. In the first phase, called the "Concept Phase," the device design and process sequence for implementing the design are created. This work will involve several key stakeholders including device designers and fabrication experts as well as business and marketing people. Many compromises will be discussed and made by these stakeholders in order to obtain an outcome that will be thought to meet the performance, cost, and time-to-market goals. It is prudent to have several potential approaches for the implementation of the microsystems device at the conclusion of this stage. The reason is that some of the implementation approaches may be discovered to have deficiencies when the actual fabrication development begins. Importantly, the design models, whether analytical or numerical, may not be based on completely accurate values for the material properties since these are process dependent. Therefore, the designers may need to

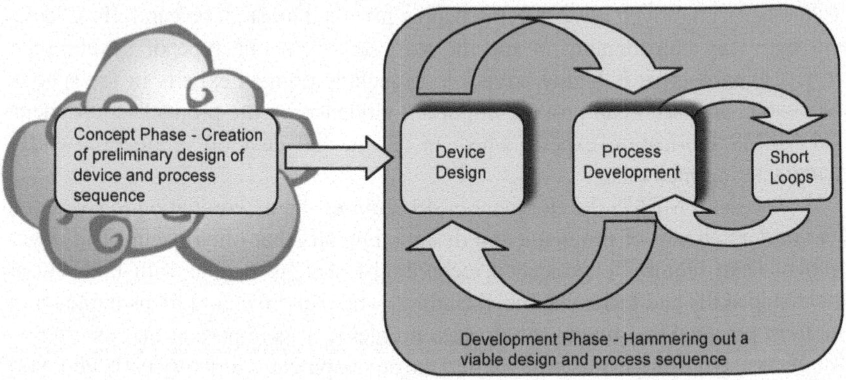

Fig. 7.3 The strategy for microsystems process development

make reasonable estimates of these values in their design, with the goal of "bounding the problem." The output of the designer's efforts will be a set of models, one or more preliminary process sequence(s), and a prototype mask layout set.

After conceptualizing one or more draft process sequences, the fabricators will identify any individual processing steps that may need development as well as portions of the process sequence that have higher risk. With regard to processing steps that need development, this includes both processing steps that may need to be slightly modified as well as processing steps that may be completely new.

Importantly, at this point in development, the fabrication experts will not be able to predict the outcome of either the process sequence(s) or even each processing step with certainty. Instead, they will use their experience and judgment to point out the processing steps and portions of the processing sequence(s) that will likely need the most development work.

The fabricators will formulate strategies, based on the principles of design of experiments (DOE), to develop the individual processing steps that will need to be generated. A DOE is a technique wherein selected process parameters are varied in a methodical way and the results of the process are quantified. The data from the DOE is then fed into a sophisticated statistical computer algorithm that determines the optimal tool settings for a specific desired outcome. It should be noted that performing a DOE is a costly and time-consuming enterprise and requires considerable experience and skill to perform efficiently and effectively. For the modification of existing processing steps, performance of more limited DOEs may also be used.

The fabricators will also design short-loop runs (i.e., sets of processing steps for a segment of the process sequence) also based on statistical methods such as DOE to develop portions of the process sequence (i.e., process modules) as well as the entire sequence. These short-loop process runs are typically performed on the sections of the process sequence that are deemed to present higher risk and/or uncertainty.

Additionally, the fabricators will work with the designers to create the designs for a set of test structures that will be useful for measuring the properties of the materials used in the process sequence as well diagnostic structures and devices to help with the process development. The output at this stage will be a set of planned experiments for all process steps that need development and short-loop process runs.

As a method to partly deal with the variations in the device parameters as a result of the manufacturing processes, it is common practice to develop a mask design wherein the critical lateral dimensions of the microsystems devices are purposefully varied over a range of values. Specifically, the lateral dimensions of devices will be varied over a range of values. The purpose of doing this is to collect as much information as possible about the effects of these manufacturing variations on device performance and use this data to develop more accurate models. Additionally, an added benefit of this exercise is that it increases the probability of getting some devices that work or even meet the performance goals in the first run of the process sequence. Normally, the dimensions are varied above and below a central value wherein the central value is equal to the desired lateral dimensions of the element. The maximum amount of variation in the parameter, both positive and negative, is usually set at more than the expected manufacturing variations. The vertical dimensions of the critical features can also be varied, but this requires different recipe times, such as shorter or longer depositions.

With the designs of the devices and the fabrication process in hand, along with a set of planned process experiments, the work in the fabrication laboratory begins to develop the processing steps, modules, and ultimately the sequence. Much of this work is iterative at many levels. The development of process steps varies the processing parameters over a set of reasonable values, followed by measurement of the outcomes, which are then documented and statistically analyzed. Similarly the efforts to develop modules and sequences will rely on the outcome of a number of short-loop experiments, which will also be documented and statistically analyzed. The benefit of well-thought-out, statistically significant experiments for such work cannot be overemphasized. Also, it is extremely important to plan what measurement data is required, what measurement tools are needed, methods to document and track each experiment, and the methods to be used for analysis before the development effort begins.

Moreover, it should be recognized that even with excellent preparation, fabrication development work often does not go as planned. Wafers break, process equipment sometimes fails during a run with the wafers in the tool, and accidents and mistakes can be made. Not to mention that the processing steps may not result in a successful outcome. Even the best planned process sequences and/or device designs will be found to have some issues. Therefore, the fabricators should attempt to mitigate for unexpected problems to the extent possible, such as processing more wafers that are needed and breaking wafer lots into smaller lots for batch processes. Similarly, fabricators will often use one wafer in each new previously unperformed processing step in the process sequence. This wafer is sometimes called a *pioneer* or *scout wafer*. The reason is that each new processing step that has not be previously performed represents some amount of risk and uncertainty, and therefore only one wafer is lost if the results of that processing step are unacceptable.

Once the process sequence is beginning to provide good results (again usually after many iterations and hard work), test structures will be fabricated and measurements taken to determine accurate resultant device dimensions and the material property values. These values will be fed back into the design models, and new more accurate models will be created. From these more accurate models, new device designs will be created, and the device designs may be revised. Even when working devices are beginning to be yielded from the runs, the design and models will be further refined, new mask sets will be created, and the iterations of process development will continue. It is reasonable to expect that the first two or more cycles of a new and customized process sequence may not initially yield any working devices. It is also reasonable to expect that once working devices are beginning to be yielded on the process runs, the yields will be below 25%, although higher yields are possible.

During subsequent process runs, various quality improvement methods (e.g., Pareto, histograms, etc.) can be employed to increase the yield to above 50%. At some point, the process of volume yield learning begins whereby a large enough volume of wafers has been run that the most causes of yield degradation can be reduced or eliminated. Obviously, the cost and development time are heavily influenced by the complexity of the process sequence for any device type. It is not uncommon to see very large development costs (e.g., in excess of $10 M or even in excess of $100) for the development of MEMS microsystems process technologies.

Given the difficulty, cost, and risk associated with the approach of complete customization for the development of a new process sequence, it should be obvious that the first step of any microsystems development activity should be to determine if an existing process technology can be used or leveraged (by leveraging it is meant to make relatively simple modifications or to use portions of an existing process technology) since this will often be the most efficient and effective approach available. The leveraging of existing process technologies is extremely common in IC microsystems, but has not been widely used in MEMS.

The leveraging of existing process technologies in commercial MEMS foundries has frequently not been the preferred option for designers. There are several reasons for this, including the following: (1) most MEMS process technologies in existence were developed by captive foundries that are not interested in making their capabilities available to outside entities; (2) MEMS is a relatively new field and has an enormous range of applications, and therefore the number of existing process technologies available is relatively limited; and (3) most MEMS process technologies in existence were not designed with the intent to be *generic* for a range of applications but instead were developed and optimized for a single device and application area.

Historically, the leveraging of an existing process sequence requires that the intended application be closely aligned to the original or else the development of a customized process sequence will be warranted. Most of the MEMS process technologies that are in existence were developed for the intended original application and not for the widespread general production of MEMS devices.

7.4 What Is an Integrated MEMS Process Sequence?

An *integrated MEMS process* is one wherein the MEMS device(s) is fabricated on the same substrate as microelectronics. Many MEMS devices require some type of electronics in order to be functionally useful. The electronics can be located on the same substrate as the MEMS or can be located on a separate substrate (as well as a printed circuit board). Therefore, the natural question is: when does it make sense to produce a MEMS device and microelectronics on the same wafer? This is a difficult question to answer and usually depends on whether such integration provides the best solution in terms of performance and cost. There are other issues as well, such as the magnitude of the development cost and the length of time associated with developing an integrated MEMS process technology that provides adequate manufacturing yields and cost levels.

In cases where the system-level trade-offs do point to the need for merging MEMS and electronics on the same substrate, an integrated MEMS process must be designed and developed. This may seem relatively straightforward; however, the complexities of such integration are quite significant and often drive the development cost (and time to market) up very significantly relative to MEMS not integrated with electronics.

In addition to the significant challenges associated with process integration for the development of a process sequence for making any MEMS device, there are additional process integration challenges with attempting to merge integrated circuits with MEMS onto the same semiconductor substrate. Some of these relate to the fact that IC foundries will not allow previously processed wafers into their facilities due to contamination risks, and therefore an integrated MEMS process sequence will most likely involve performing IC fabrication first, followed by MEMS fabrication last. We will review one integrated MEMS process technology in the examples given below.

7.5 Examples of MEMS Microsystems Process Technologies

This section reviews two process technologies for microsystems in the commercial sector. These examples were selected to represent a range of different microsystem technologies including a generic surface micromachined process technology that can be used for a number of MEMS applications and a surface micromachined process technology for the implementation of an integrated MEMS microsystems used in a digital display application.

7.5.1 PolyMUMPS™ Process Technology

The PolyMUMPS™ process technology is part of the MUMPS® (Multi-User MEMS Processes) prototyping program offered by MEMSCAP Inc. The PolyMUMPS™ process is a MEMS process technology available on a periodic basis in a multi-user environment, wherein designs from different customers are pooled together onto a single mask set. PolyMUMPS™ is a three-layer polysilicon surface micromachining process technology that is widely known for implementing MEMS and has been offered continuously since 1992. Importantly, this is a rare example of a relatively generic process technology that can be used to implement a number of MEMS devices for different applications. The process sequence begins with 150 mm n-type (100) silicon wafers of 1–2 ohm-cm resistivity [1]. The surfaces of the wafers are first heavily doped with phosphorus in a standard diffusion furnace using a phosphosilicate glass (PSG) thin-film material layer as the dopant source.

The doping of the substrate surface helps to reduce or prevent charge feedthrough from the substrate to electrostatic MEMS devices made on the surface. Next, after removal of the PSG film, a 600-nm-thick low-stress LPCVD (low-pressure chemical vapor deposition) silicon nitride layer is deposited on the wafers as an electrical isolation layer. This is followed directly by the deposition of a 500-nm-thick LPCVD polysilicon film, referred to as Poly0 (Fig. 7.4a).

Poly0 is then patterned by photolithography (mask level 1 using mask name POLY0), a process that includes the coating of the wafers with photoresist, exposure of the photoresist with the appropriate mask, and developing the exposed photoresist to create the desired etch mask for subsequent pattern transfer into the underlying layer. After patterning the photoresist, the Poly0 layer is then etched in a plasma etch system, and the photoresist is stripped (Fig. 7.4b).

A 2.0-μm-thick phosphosilicate glass (PSG) sacrificial layer is then deposited by LPCVD and annealed at 1050 °C for 1 hour in argon as shown in Fig. 7.4c. This layer of PSG, known as Oxide1, is a sacrificial material layer and will be removed at the end of the process to free the mechanical structural layer(s) composed of polysilicon. The Oxide1 layer is lithographically patterned (mask level 2 using mask name DIMPLE), and the dimple features are transferred into the sacrificial PSG layer using a RIE (reactive ion etch) system. After this etch, the photoresist is stripped as shown in Fig. 7.4d. The nominal depth of the dimples is 750 nm.

The wafers then undergo another photolithography step to pattern the Oxide1 layer (mask level 3 using mask name ANCHOR1), and subsequently a reactive ion etch is performed through Oxide1 layer. This etch provides anchor holes that subsequently will be filled by the Poly1 layer. After the resist is stripped, the cross section of the wafer is seen in Fig. 7.4e. After etching Oxide1 using the ANCHOR1 mask pattern, the first structural layer of polysilicon (Poly1) is deposited at a thickness of 2.0 μm. Then a thin (200 nm) layer of PSG is deposited over the polysilicon and the wafer is annealed at 1050 °C for 1 hour (Fig. 7.4f). The anneal drives dopants into the polysilicon layer with phosphorus from the PSG layers both above and below it. The anneal also serves to significantly reduce the residual stresses in the Poly 1 layer.

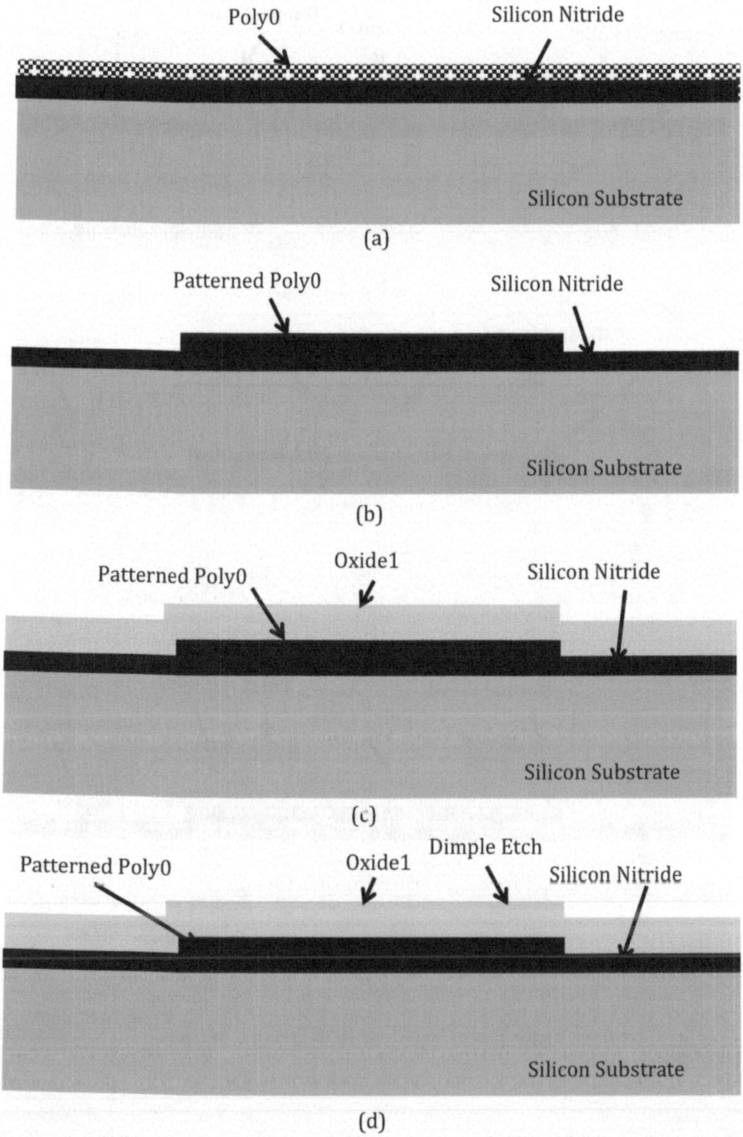

Fig. 7.4 **(a–d)** First four major processing steps in the process sequence for manufacturing the PolyMUMPS™ process

The PSG layer is photolithographically patterned (mask level 4 using mask name POLY1) and then etched so as to form a PSG hard mask on the polysilicon Poly1 layer. The hard mask is more resistant to the polysilicon etch chemistry than the photoresist and ensures better transfer of the pattern into the polysilicon. After etching the polysilicon using reactive ion etching (Fig. 7.4g), the PSG hard mask layer is stripped.

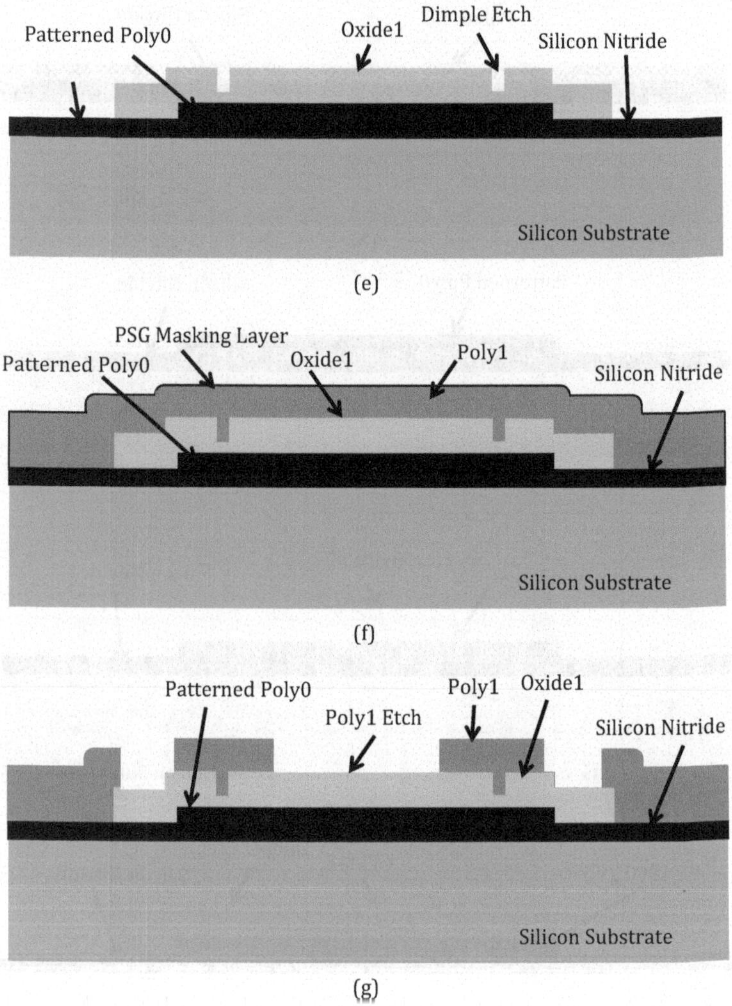

Fig. 7.4 **(e–g)** Next three major processing steps in the process sequence for manufacturing the PolyMUMPS™ process

After Poly1 is etched, a second sacrificial PSG layer (Oxide2 having a thickness of 750 nm) is deposited and annealed (Fig. 7.4h). The Oxide2 layer is photolithography patterned using two different etch masks (mask level 5 using mask name POLY1_POLY2_VIA and mask level 6 using mask name ANCHOR2) with two different objectives. The POLY1_POLY2_VIA level provides for etch holes in the Oxide2 layer down to the Poly1 layer. This provides a mechanical and electrical connection between the Poly 1 and Poly 2 layers. The POLY1_POLY2_VIA layer is lithographically patterned and etched by RIE as shown in Fig. 7.4i.

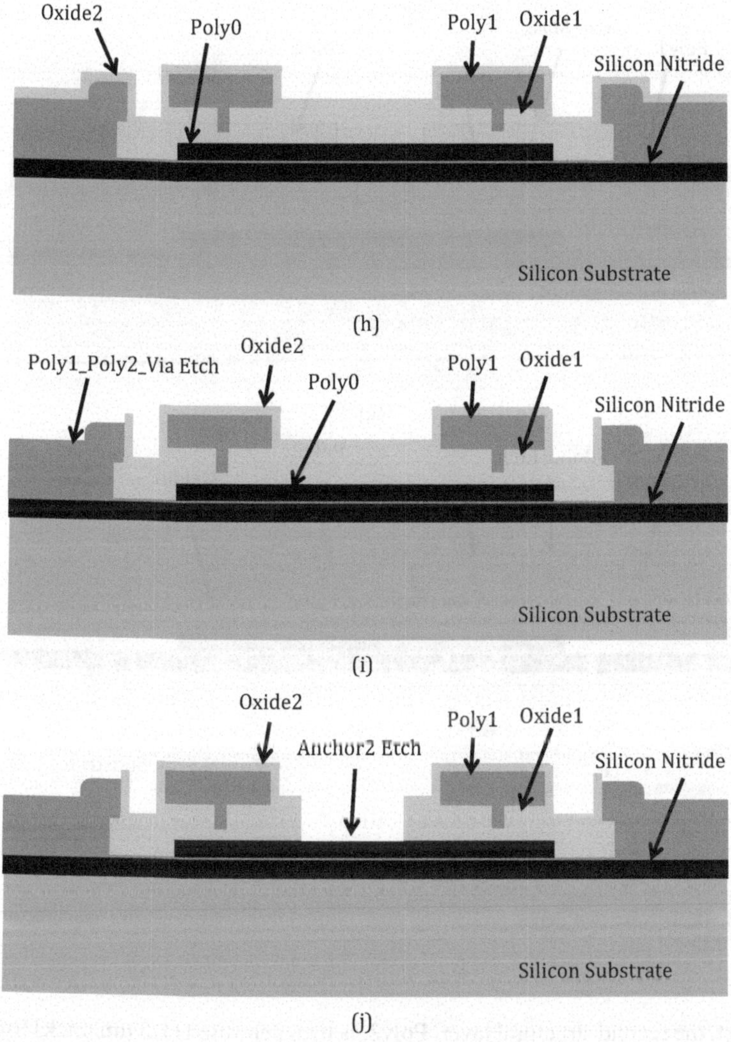

Fig. 7.4 (**h–j**) Next three major processing steps in the process sequence for manufacturing the PolyMUMPS™ process

The ANCHOR2 level is provided to etch both the Oxide1 and Oxide2 layers in one step, thereby eliminating any misalignment between separately etched holes. More importantly, the ANCHOR2 etch eliminates the need to make a cut in Oxide1 layer unrelated to anchoring a Poly1 structure, which needlessly exposes the substrate to subsequent processing that can damage either the Poly0 or silicon nitride layers. The ANCHOR2 layer is lithographically patterned and etched by RIE in the same way as POLY1_POLY2_VIA. After this etch is performed and the resist is stripped, the cross section of the wafer is shown in Fig. 7.4j.

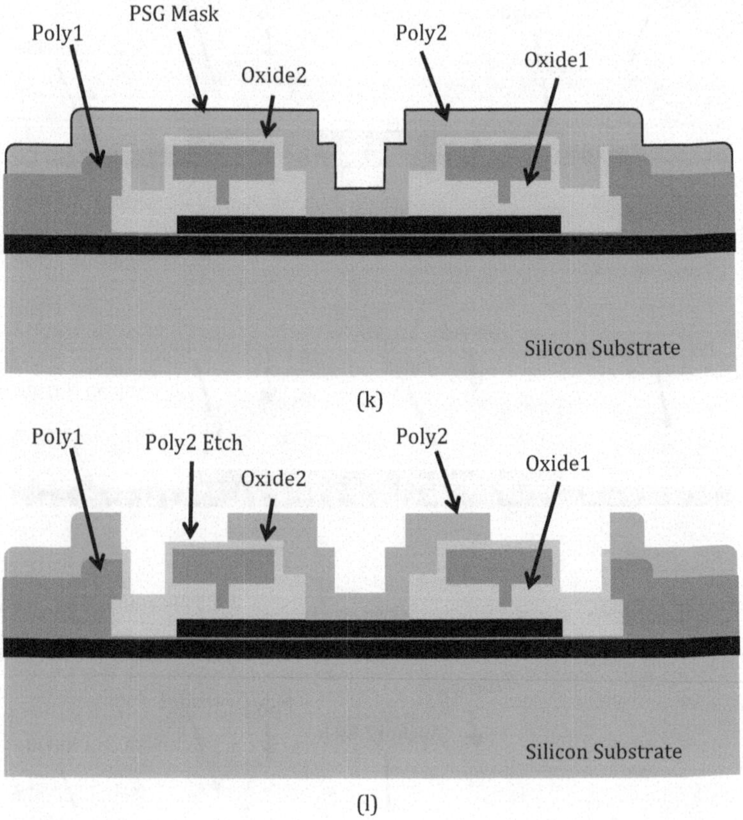

Fig. 7.4 (**k**, **l**) Next two major processing steps in the process sequence for manufacturing the PolyMUMPS™ process

Next, the second structural layer, Poly2, is then deposited (1.5 μm thick) followed by the deposition of 200 nm of PSG. As with Poly1, the thin PSG layer acts as both an etch mask and dopant source for the Poly2 layer (Fig. 7.4k). The wafer is annealed to dope the polysilicon and reduce the residual film stress in the polysilicon material layer.

The Poly2 layer is photolithographically patterned (mask level 7 using mask name POLY2)). The PSG is etched using the POLY2 mask pattern to form a hard mask for the subsequently polysilicon etch. The polysilicon layer Poly2 is etched using reactive ion etching, similar to that used for Poly1. The photoresist is then stripped and the masking oxide is removed (Fig. 7.4l).

The final deposited layer in the PolyMUMPS™ process is a 0.5 μm metal (Au with Cr adhesion layer) layer that provides for probing, bonding, electrical routing, and highly reflective mirror surfaces. The wafer is patterned lithographically

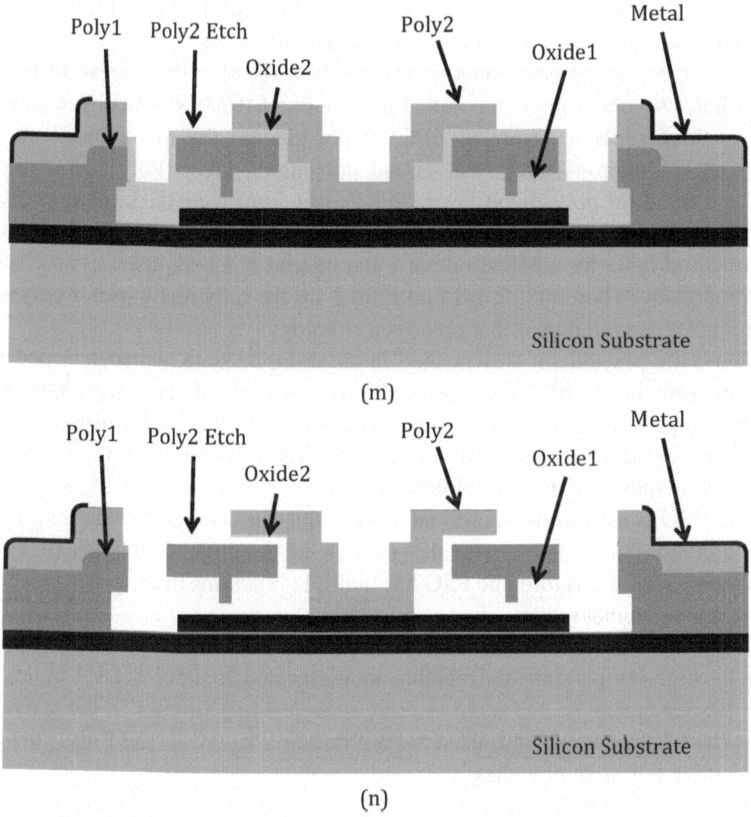

Fig. 7.4 (**m, n**) Last two major processing steps in the process sequence for manufacturing the PolyMUMPS™ process

(mask level 8 using mask name METAL), and the metal is deposited and patterned using lift-off. The final, unreleased structure is shown in Fig. 7.4m. Figure 7.4n shows the device after sacrificial oxides are removed using hydrofluoric acid (HF).

7.5.1.1 Some Important Elements About PolyMUMPS™

The PolyMUMPS™ process technology was established and made available as a fabrication service in the United States in the early 1990s. The PolyMUMPS™ process technology was intended to be as generic as possible so that it could be used to implement a variety of different MEMS device types. On that account, this process technology has been successful since a large number of different MEMS devices and device types have been demonstrated using this process technology.

Nevertheless, the PolyMUMPS™ process technology has had very limited viability in the commercial production of any specific MEMS device.

With respect to process integration, the PolyMUMPS™ process technology demonstrates several important points. First, the PolyMUMPS™ process technology employs three separate layers of LPCVD polysilicon: the first polysilicon layer (Poly0) is 0.5 microns thick, the second polysilicon layer (Poly1) is 2 microns thick, and the third polysilicon layer (Poly2) is 1.5 microns thick. Due to the first polysilicon layer being deposited onto the silicon nitride layer, it really is not useful as a structural layer for a MEMS device and instead is mostly used as an electrical grounding plane. Therefore, Poly1 and Poly 2 are the only really useful polysilicon structural layers available in this process technology.

Second, the polysilicon is deposited in an undoped state and then doped using diffusion from layers of PSG deposited between each of the polysilicon layers. During a high-temperature step, the phosphorus in the PSG diffuses into the polysilicon to make it electrically conductive. Using the PSG layer to dope the polysilicon avoids an extra diffusion step.

Third, PSG is used as the sacrificial layer in this surface micromachining process as opposed to other glass material type due to the high etch rates of PSG in hydrofluoric acid. Therefore, the PSG has multiple functions in the process sequence since it is a sacrificial material layer and a dopant source.

Fourth, the diffusion of the phosphorus from the PSG into the polysilicon occurs during the high-temperature anneals that are performed at 1050 °C for 1 hour. These anneal processing steps also serve to reduce the residual stresses in the polysilicon layers. The PSG layers are also used as hard masking layers for the Poly1 and Poly2 polysilicon material layer etches.

Fifth, the dimple etch that is performed creates small but important standoffs on the bottom of the first polysilicon structural layer (Poly1) that help to prevent stiction effects from rendering any surface micromachined devices using this process technology as being nonfunctional. Without these dimples, Poly1 would contact Poly0 over a large area, and since these layers will have smooth surfaces, they are prone to stiction effects.

Sixth, the alternating layers of PSG and polysilicon help on the etching processes of these material layers since the selectivity of etching each of these material layers is usually quite good. That is, performing a plasma etch of a polysilicon layer and stopping on the underlying PSG layer can be achieved, and, vice versa, performing a plasma etch of a PSG layer and stopping on the underlying polysilicon layer can also be achieved.

Seventh, the sixth patterning step (using Mask level 6) whereby both the first and second PSG layers are etched at once enables the features in each oxide layer to be perfectly aligned. If these were done as separate photolithographic and etch steps, there would invariably be a small amount of misalignment. This is important for performance of certain MEMS devices implemented with this process technology such as micromotors.

Eighth, gold is used as the metal layer in the process sequence and is patterned using liftoff. Gold is highly conductive and also holds up reasonably well when immersed in HF for removing the sacrificial PSG layers. Liftoff is chosen as the patterning method of the gold layer due to the fact that the dimensional requirements are not very strict and liftoff only requires a solvent immersion, whereas if the gold where to be etched, it would expose the MEMS structures to more aggressive chemical mixtures.

Ninth, the Poly0 etch is expected to terminate on the silicon nitride layer. The problem is that the selectivity of polysilicon etching is not high with respect to silicon nitride plasma etching. Therefore, the silicon nitride layer will be attacked when exposed and during the over etch. This is a deficiency in this process technology. However, since the Poly0 thickness is relatively small, the likelihood of piercing the silicon nitride is low. Having said that, attempting to stack and pattern Poly1 and Poly2 on Poly0 would be very problematic and would result in the silicon nitride layer being removed in those areas.

7.5.2 Digital Light Processor (DLP®) Technology

Digital Light Processor (DLP®) technology developed by Texas Instruments Corporation is one of the more well-known MEMS development endeavors to date with market sales of chip sets that reached $100 M annually and is used in various large volume commercial markets such as projection televisions and projection display systems [2]. DLP® technology is a disruptive technology for movie theaters since it replaces the 100 year-old celluloid film with a completely digital format. DLP® is currently used in most theaters, with many using the advanced 3-D systems.

The key component of DLP® technology is the micromirror array chip (called the Digital Micromirror Device or DMD for short [3]). Essentially the DMD chip is a large array of individually and digitally controlled MEMS micromirrors. There are up to 2 M mirrors in each chip. Each mirror in the array measures just over 10 microns by 10 microns and is electrostatically actuated by the microelectronics physically located underneath the mirror array [4]. This enables the fill factor of the DMD to reach levels of over 90%, thereby allowing very high optical efficiency and contrast.

The DLP® technology is an example of an integrated MEMS process technology. Integrated microelectronics is a necessity in DLP® technology since the number, density, and size of the mirrors combined with individual addressability would preclude having all the electronics off-chip. That is, there is no practical method using off-chip electronics to individually address of each mirror in the array. The DMD fabrication is made using a "MEMS last" integrated process technology wherein the microelectronics are fabricated in the base substrate first and then the MEMS mirrors are fabricated on the surface of the microelectronics substrate [5].

Figure 7.5 is a SEM of a portion of the DMD device showing an array of individual pixels, with some in the front of the image in the actuated state and the surrounding pixels in the unactuated state. As can be seen, the mirror rotates by approximately 12 degrees in the actuated state, and without actuation the mirrors

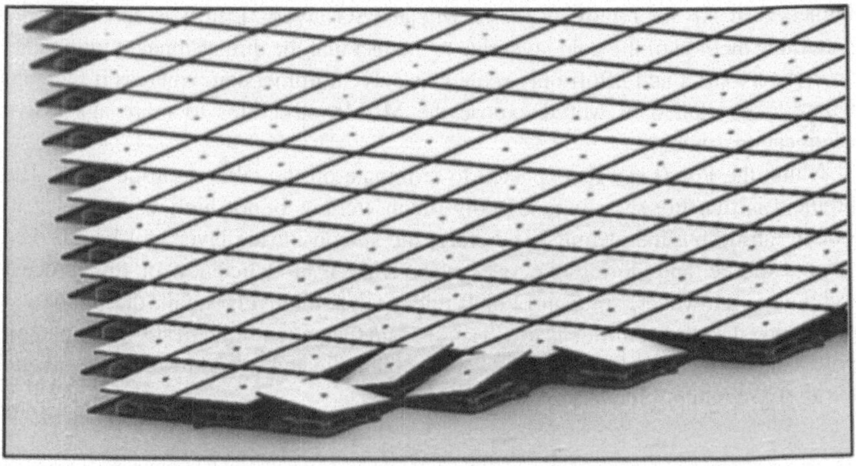

Fig. 7.5 SEM image of DMD chip shown with the pixels near the leading edge of the array in the actuated (i.e., non-flat) state. (Reprinted with permission, Copyright Texas Instruments, Inc.)

(shown around the center pixel) are nominally flat. For a description of how DMDs are used in projection optical systems, see [3, 5]. The mirrors have a complex hinge mechanism to allow them to rotate by a sufficient angular displacement while keeping the strain in the hinges to an acceptable level. Remarkably, individual mirrors have been operated over 10^{12} full mechanical movements without failure and thereby have demonstrated an extremely high level of reliability [6].

The DMD pixel consists of a CMOS microcircuit fabricated in the silicon substrate with a MEMS superstructure implemented on top. The MEMS last process sequence requires that all post-CMOS processing steps used to fabricate the MEMS superstructure be compatible, both thermally and chemically, with the CMOS. Consequently, the thermal budget for these processing steps is quite restricted, and no processing temperatures can exceed 400 C. This is achieved in the DMD process sequence by fabricating the MEMS superstructure utilizing a series of relatively low-temperature deposition and etch processes.

The DMD process sequence begins (see Fig. 7.6) with a relatively conventional CMOS process technology having a two-layer metallization (M1 and M2) [5]. After the CMOS is completed, a relatively thick oxide layer is deposited on the wafers and subsequently planarized using chemical-mechanical polishing (CMP). The planarization of the surface of the CMOS wafer is very important to ensure that the wafer surface is completely flat prior to beginning the MEMS fabrication in order not to degrade the brightness uniformity or contrast ratio of the DMDs. The CMP oxide layer is patterned, etched, and filled with metal to form plugged vias to electrically connect the CMOS electronics to the actual mirror bus lines as shown in Fig. 7.6a.

Next, a so-called "dark" metal layer (M3) is deposited, patterned, and etched to form the lower address electrodes (Fig. 7.6b). M3 is actually a stack composed of a

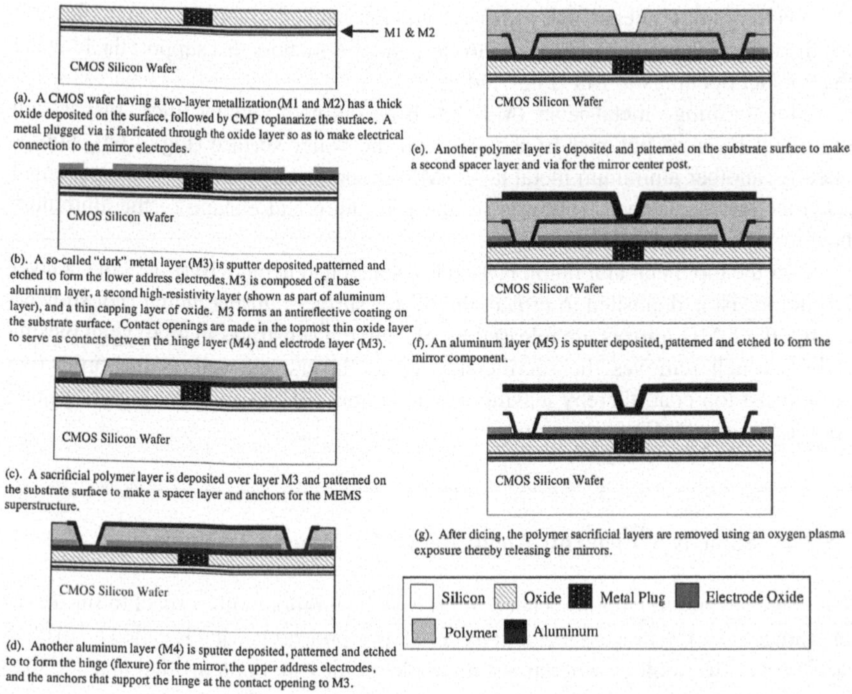

Fig. 7.6 Texas Instruments' DMD process technology. (**a**) A CMOS wager having a two-layer metallization (M1 and M2) has a thick oxide deposited on the surface, followed by CMP to planarize the surface. A metal plugged via is fabricated through the oxide layer so as to make electrical connection to the mirror electrodes. (**b**) A so-called "dark" metal layer (M3) is sputter deposited, patterned, and etched to form the lower address electrodes. M3 is composed of a base aluminum layer, a second high-resistivity layer (shown as part of aluminum layer), and a thin capping layer of oxide. M3 forms an antireflective coating on the substrate surface. Contact openings are made in the topmost thin oxide layer to serve as contacts between the hinge layer (M4) and electrode layer (M3). (**c**) A sacrificial polymer layer is deposited over layer M3 and patterned on the substrate surface to make a spacer layer and anchors for the MEMS superstructure. (**d**) Another aluminum layer (M4) is sputter deposited, patterned, and etched to form the hinge (flexure) for the mirror, the upper address electrodes, and the anchors that support the hinge at the contact opening to M3. (**e**) Another polymer layer is deposited and patterned on the substrate surface to make a second spacer layer and via for the mirror center post. (**f**) An aluminum layer (M5) is sputter deposited, patterned, and etched to form the mirror component. (**g**) After dicing, the polymer sacrificial layers are removed using an oxygen plasma exposure thereby releasing the mirrors

base aluminum layer, a second high-resistivity film, and a capping layer of thin oxide. Together these three layers form an antireflective coating to reduce stray light reflections from the substrate. Contact openings are then patterned and etched in the thin cap oxide layer to serve as electrical contacts between the hinge layer (M4) and electrode layer M3. Next, a planarizing, sacrificial polymer layer is deposited over the M3 and subsequently patterned (Fig. 7.6c). Another aluminum metal layer

(M4) is then sputter deposited, patterned, and RIE etched to form the hinge (flexure) for the mirror, the upper address electrodes, and the anchors that support the hinge at the contact openings to M3 (Fig. 7.6d).

After the hinge metal layer (M4) has been defined, another thick, planarizing sacrificial layer of polymer is deposited on the wafer surface (Fig. 7.6e). Subsequently, another aluminum metal layer (M5) is sputter deposited onto the patterned polymer sacrificial layer, followed by the patterning and etching of the aluminum layer to define the DMD mirrors.

After the top metal aluminum layer has been etched using RIE (Fig. 7.6f), a layer of photoresist is deposited to protect the mirror surfaces during dicing (not shown). Lastly, the DMD mirrors are released by placing the substrates in an oxygen plasma etcher, which removes the sacrificial polymer layers, as well as the protective photoresist top coat, thereby leaving the functional MEMS mirrors (Fig. 7.6g) and completing the DMD fabrication.

7.5.2.1 Some Key Elements About the DLP® Process Technology

There are several key points related to process integration with respect to the Texas Instruments DLP® process technology. First, this process is what is called a "MEMS last" process technology wherein the microelectronics are implemented first and then the MEMS are fabricated on top of the microelectronics. This is important since it allows the manufacturing process to be separated into two parts: the microelectronics fabrication followed by the MEMS fabrication. This approach enables the use of any suitable IC foundry for the microelectronics portion of the process sequence. Most IC foundries will not allow wafers to come into the facility from outside foundries, particularly wafers having MEMS fabrication previously of them due to the potential risk of contamination. However, the DLP® process avoids this problem.

Second, since the wafers have prefabricated IC microelectronics on them, includ ing multiple layers of metallization, the wafers are sensitive to processing steps involving elevated temperatures. To overcome this limitation, the DLP® process technology uses all low-temperature processing steps for implementation of the MEMS.

Third, the sacrificial layers in the DLP® process are composed of photoresist. Moreover, the photoresist layer can be easily removed without any wet chemical immersion that could affect the microelectronics. Instead, it is removed using a simple oxygen plasma. This avoids the harshness of a chemical wet etch and any negative effects of stiction.

Fourth, the structural layer is composed of thin-film aluminum material. This metal is highly reflective which is extremely important in an optical display technology. Also, the deposited aluminum metal layers do not require a high-temperature anneal to reduce the residual stress. Instead, the stress in these layers can be adequately controlled using physical vapor deposition.

7.6 Process Integration and Manufacturing Variations

The individual processing steps reviewed in Chaps. 3 and 4 indicated that these processes result in variations in the device dimensional and material property parameters. The variations mostly relate to non-uniformities of deposition, etch, and growth processes, variations due to pattern transfer processes, and variations due to lateral etching of features. It is important to note that when a group of processing steps are combined to form a process sequence, the interrelationships between the different processing steps will also result in additional variations in the device dimensional and material property parameters. This section reviews some of these multiple step variations and their causes.

7.6.1 Causes of Device Parameter Variations in Process Sequences

When a number of processing steps are performed sequentially in a process sequence, there are often interdependencies between the processing steps that occur and can have an impact on the device dimensional and material property parameters. We review some of the more common types of these interdependencies in this section.

As multiple layers are sequentially deposited and patterned on the substrate, the topography of the surface of the substrate will increase.[1] Specifically, as the underlying layers are etched and new layers are deposited, the result will be features that are partly filled by subsequent layer depositions. Since the subsequent layer cannot be both uniform in thickness and fill the existing features, the result will be an increase in surface topography. The amount of topography directly relates to the number of layers deposited and their thicknesses and the number of photolithography and etching steps that have been completed on these layers.

Topography of the substrate surface causes photolithographic constraints on the device's planar dimensions and limits the layer-to-layer alignment accuracy that can be achieved. As noted in the section of Chap. 3 on photolithographic technologies, as the resolution of the optical system allows smaller and smaller features to be transferred, the depth of focus decreases. Therefore, if there is sufficient topography on the surface, it is possible that some areas of the surface that will be in focus while other areas will not.

This is demonstrated in the illustration below (Fig. 7.7), which shows a projection optical system being used for photolithography on a substrate having some amount of topography. As can be seen in the insert to the right, even though the mask image is in focus in certain locations, other locations on the wafer surface are out of focus

---

[1]This is true unless some processing step is performed to planarize the substrate surface such as chemical-mechanical polishing (CMP).

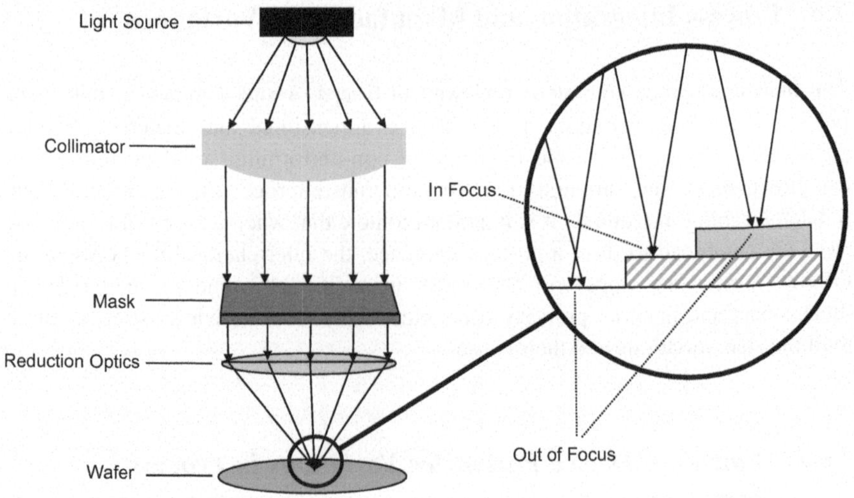

Fig. 7.7 Illustration of how various regions of the photolithographic pattern can be in or out of focus based on the topography of the substrate surface

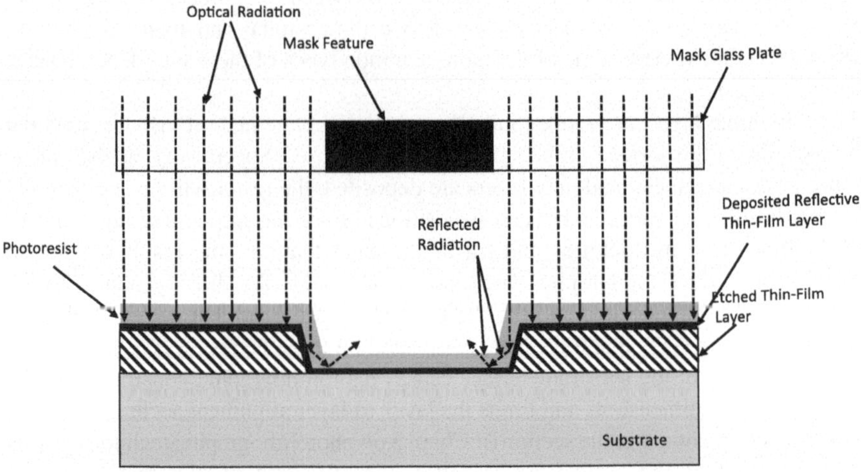

Fig. 7.8 Illustration of the problem in attempting to perform photolithography on substrate surface having topography from previous processing steps

due to the topography of the surface. If the image projected onto the substrate surface is out of focus, then the photolithography process will be inherently limited in the resolution that can be obtained.

Another issue related to fabrication problems caused by topography is that it can result in scattering of the radiation in optical lithography thereby diminishing the resolution of the features being printed. For example, as illustrated in Fig. 7.8, a substrate is undergoing an exposure process in an optical photolithography step.

Prior to performing this photolithography, the wafer had a thin-film material layer that was patterned, followed by the deposition of a reflective thin-film material layer. As can be seen, the reflective material layer conforms over the surface of the features. A photoresist was deposited before performing the photolithography, and it too conforms over the surface features. The optical radiation that strikes the sloped sidewalls of the etched feature will be reflected from the reflective material layer at angles different from normal to the substrate surface. That is, the optical radiation will be reflected at an angle equal to the incident angle. Therefore, the reflected optical radiation will propagate through the photoresist so as to expose photoresist that should have been unexposed if there was no off-angle reflections. The result will be that the photoresist will be exposed over some of the area under the mask thereby making the printed features either smaller or larger depending on the polarity of the resist. Obviously, this can limit the resolution that can be obtained. It should be noted that this effect can occur even when the material layers under the resist are not very reflective.

Existing surface topography can also negatively degrade the layer-to-layer alignment accuracy that can be achieved. This is illustrated in Fig. 7.8 wherein the sidewalls of the etched features have a slope. This slope is present due to the fact that the etch performed on the thin-film layer is not perfectly anisotropic and would be present in the etched thin-film layer features everywhere over the wafer surface, including alignment marks. The problem arises in attempting to align a mask during a photolithography processing step to features having slope sidewalls is that the region to distinguish the edge of the features is blurred by the sidewall slope. This will result in some uncertainty on the alignment registration accuracy.

Topography on the substrate surface can also cause problems with continuity of layers in subsequent thin film layer depositions. At the edge of high-topography structures where there is a drop-off, the thin-film material layer after thin-film deposition may be thinner than in the field regions (Fig. 7.9a). At worst case, there may even be a discontinuity in the layer across large variations of topography. A discontinuity in an insulating layer could result in an electrical shorting, and a discontinuity in an electrically conducting layer could result in an open circuit

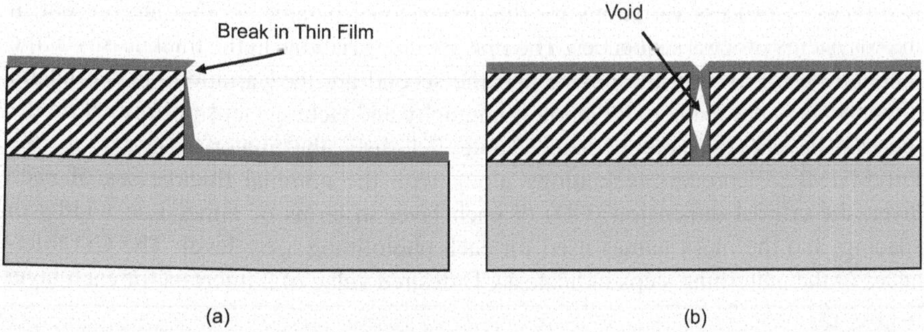

Fig. 7.9 Illustration of the problems caused by attempting to perform thin-film layer depositions on a substrate having topography

Table 7.1 PolyMUMPS™ material layers, nominal thicknesses, CD tolerances, and mask level names

Material layer	Nominal thickness (microns)	CD tolerances[a] (microns)	Lithography level mask names
Silicon nitride	0.6	–	–
Poly0	0.5	1.800–2.200 (lines)	POLY0 (HOLE0)
Oxide1	2.0	1.800–2.200 (spaces) 1.700–2.300 (spaces)	DIMPLE ANCHOR1
Poly1	2.0	1.750–2.250 (lines)	POLY1 (HOLE1)
Oxide2	0.75	1.750–2.250 (spaces) 1.750–2.250 (spaces)	POLY1_POLY2_VIA ANCHOR2
Poly2	1.5	1.700–2.300 (lines)	POLY2 (HOLE2)
Metal	0.5	2.500–3.500 (lines)	METAL (HOLEM)

[a]It is assumed that these CD variations are for ± 3-sigma. The meaning of 3-sigma will be covered in a later chapter

condition. Additionally, it can be difficult to obtain conformal and continuous coverage thin-film deposition in the deeper high-aspect ratio features that are frequently seen contact vias (Fig. 7.9b).

These are only a few examples to demonstrate how a multiplicity of processing steps in a process sequence can impact the dimensional variations in the device parameters. Additionally, the material properties of the materials used in the manufacturing process are also dependent on the process sequence as well.

7.6.2 An Example of Parameter Variations for a Process Technology: The PolyMUMPS™ Process

The PolyMUMPS™ is one of the few MEMS process technologies where the manufacturer publishes relatively detailed information about the design rules and discloses dimensional parameter variations[2] [1]. At the most basic level, there are two types of process variations for the dimensional parameters that are present in microsystems process sequences. The first are the variations in the thicknesses of the material layers used in the device and the second are the variations in the lateral dimensions that are due to the photolithography and etching steps performed on the material layers to pattern them. Table 7.1 lists the material layers in the PolyMUMPS™ process technology along with the nominal thicknesses of each layer, the critical dimension (CD) of each layer in terms of either line widths or spacing, and the mask names used for each photolithography level. The CD tolerances of the patterning steps indicate a CD desired value of 2 microns for each layer

[2]Most foundries consider this type of information highly proprietary and require the execution of a nondisclosure agreement (NDA) before disclosing this type of process technology information.

Table 7.2 PolyMUMPS™ layer thickness statistics [1]

Layer	Silicon nitride	PolySi 0	PSG 1	PolySi 1	PSG 2	PolySi 2
Nominal (Å)	6000	5000	20,000	20,000	7500	15,000
Sample mean (Å)	5952	5001	19,804	19,576	7590	15,038
Σ (Å)	224	60	792	512	371	320
No. of samples	99	99	126	72	126	89

Table 7.3 PolyMUMPS™ statistics for line widths for each layer [7]

Layer	Minimum feature (microns)	Mean feature (microns)	Σ (microns)	Number of samples
PolySi 0	2.0	2.153	0.060	59
Dimple	2.0	1.963	0.052	67
Anchor 1	2.0	2.02	0.105	74
PolySi 1	2.0	1.956	0.070	60
P1 to P2 Vias	2.0	1.977	0.068	75
Anchor 2	2.0	2.050	0.074	75
PolySi 2	2.0	1.974	0.080	60

except the metal layer, which has a desired CD of 3 microns. It can be seen that as the layers stack on top of one another, it results in an increase of the CD tolerances values.

Table 7.2 shows the thicknesses of the deposited layers used in the PolyMUMPs™ process sequence along with the nominal (i.e., desired value) of the thickness, the sample mean, the sample standard deviation for each layer,[3] and the number of samples taken to derive these statistical parameters. The nominal value of the thickness for each layer in a standardized process technology, such as the PolyMUMPs™ process, is not something within the control of the device designer. That is, the foundry selects the values of the thickness for each layer.

The manufacturing variations in the lateral feature dimensions are mostly due to the lithography and etch processes used, as well as the amount of topography present. Measured values for the minimum feature sizes of the layers used in the PolyMUMPs™ process sequence are given in Table 7.3. These measurements were made for each layer implemented by performing a photolithography and etch of 2 micron-wide features, which is the minimum feature size for each layer.

In addition to the variations in the dimensional parameters in the PolyMUMPs™ process technology, the material properties will also exhibit variations as well. The PolyMUMPs™ design handbook reports some of the material properties, which are listed in Table 7.4. Additionally, the PolyMUMPs™ design handbook also reports a value of Young's modulus for the polysilicon used in the process technology of 158 ± 10 GPa, a Poisson's rate of 0.22 ± 0.01 and a fracture strength of 1.21 ± 0.08

[3]The sample standard deviation is represented by the variable "S" since it is taken for a sample rather than the population.

Table 7.4 Reported values of residual stresses and electrical resistances for material layers using in the PolyMUMPS™ [1]

Material layer	Nominal thickness (Angstroms)	Residual stress (MPa) Minimum	Residual stress (MPa) Typical	Residual stress (MPa) Maximum	Resistance (ohm/sq) Minimum	Resistance (ohm/sq) Minimum	Resistance (ohm/sq) Minimum
Silicon nitride	6000	0	90	180	N/A	N/A	N/A
Poly0	5000	0	−25	−50	15	30	45
Oxide1	20,000	N/A	N/A	N/A	N/A	N/A	N/A
Poly1	20,000	0	−10	−20	1	10	20
Oxide2	7500	N/A	N/A	N/A	N/A	N/A	N/A
Poly2	15,000	0	−10	−20	10	20	30
Metal	5200	0	50	100	0.05	0.06	0.07

Table 7.5 Some of the important material property values of the polysilicon layers used in the PolyMUMPS™ process sequence reported in literature

Thickness (microns)	Young's modulus (GPa)	Poisson's ratio	Tensile strength (GPa)	Residual stress (MPa)	Sheet resistance (ohm/square)	Refs.
3.5	16 ± 6.15	0.22 ± 0.01	120 ± 0.15	–	–	[8]
2.0	149 ± 10	–	–	-3.5 ± 0.5	–	[9]
0.5–2.0	158 ± 10	0.22 ± 0.01	121 ± 1.65	−50	1–45	[1]
2.0	162 ± 4	0.20 ± 0.03	–	–	–	[10]

to 1.65 ± 0.28 GPa [1]. The mechanical properties for the polysilicon structural material layers used in the PolyMUMPs™ process sequence have also been reported in the literature and are shown in Table 7.5. Note that the number of samples in these references was not provided; however, the uncertainties in the values of the material properties are given.

7.7 Microsystems Design Rules

Most IC manufacturing process technologies will provide the circuit designers with the number and function of the individual design mask layers (e.g., gate polysilicon layer; metal layer 1; etc.) along with a set of associated design rules. The design rules incorporate information about the minimal resolution of each layer, the manufacturing dimensional variations, and the spacing between the masking layers, to enable designers to develop designs that work and within the smallest area possible. In general, there are two types of layout design rules: resolution constraints and alignment/overlap constraints. These are shown in Fig. 7.10. The resolution constraints typically specify the smallest size feature dimension that can be used for each layer (Fig. 7.10a) and the smallest dimensional spacing between electrically conducting layers so as ensure that no electrical shorts occur. The alignment constraints specify the alignment between different layers in the process (Figs. 7.10b, c) and the overlap constraints specify the minimum overlaps or overhangs of one layer to another layer (Fig. 7.10d). Design rules also incorporate information about the manufacturing variations that are present in each layer as well as layers that are interrelated.

7.7.1 MEMS Microsystems Design Rules

Design rules are also used for MEMS process technologies even though there are far fewer MEMS compared to IC process technologies available commercially. Similar to IC design rules, MEMS design rules account for the manufacturing dimensional

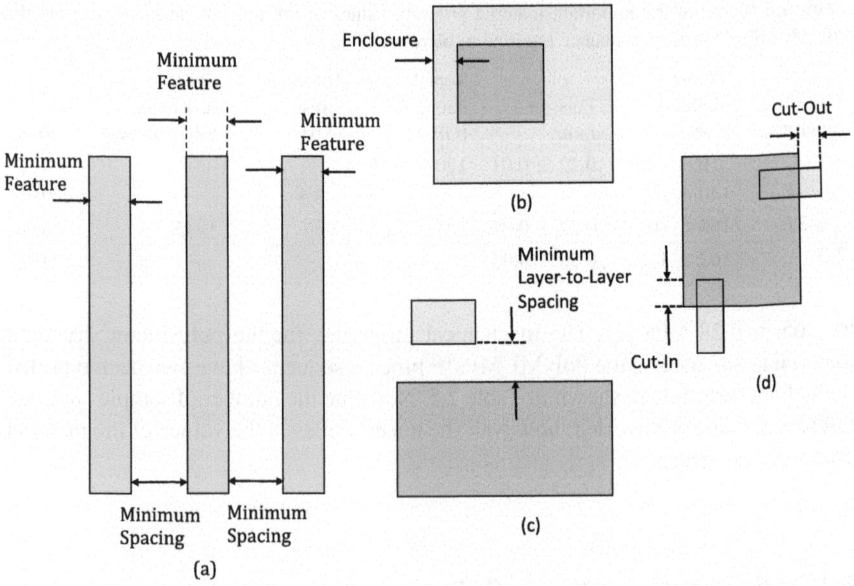

Fig. 7.10 Illustration of some common design rules. In (**a**), both the minimum width and the spacing between features on a single layer are shown. In (**b**), the amount of spacing between one layer and the next that are connected, which is called the "enclosure," is shown. In (**c**), the minimum spacing between two separate layers is shown. And, in (**d**) the cut-in which is the minimum penetration of one layer into the next layer is shown along with the cutout which is the minimum protrusion of one layer from the next layer is also shown

Table 7.6 Design rules for the PolyMUMPS™ for layer Poly2 [1]

Design rule	Rule label	Figure no.	Minimum value (microns)
Poly2 enclose Anchor2	1	7.11	5.0
Poly2 enclose Poly1_Poly2_Via	L 2	N/A	4.0
Poly2 cut in Poly1	P 3	N/A	5.0
Poly2 cut out Poly1	Q 4	N/A	4.0
Poly2 enclose metal	M 5	N/A	3.0
Poly2 space Poly1	I 6	N/A	3.0
Hole2 enclose Hole1	T 7	N/A	2.0
HoleM enclose Hole2	U 8	N/A	2.0
Lateral etch holes space Poly2	S 9	N/A	=30 (max. value)

variations to enable designers to develop designs that work. Table 7.6 shows the design rules of the Poly2 layer for the PolyMUMPS™ process technology [1]. Figure 7.11 illustrates one of the design rules from Table 7.6. This rule is intended such that polysilicon in Poly2 layer overlaps sufficiently the Anchor2 opening that the opening is ensured to be completely covered by polysilicon Poly2 layer material.

There are similar tables of design rules for each layer, and therefore there are a few dozen design rules total for the PolyMUMPs™ process technology having eight

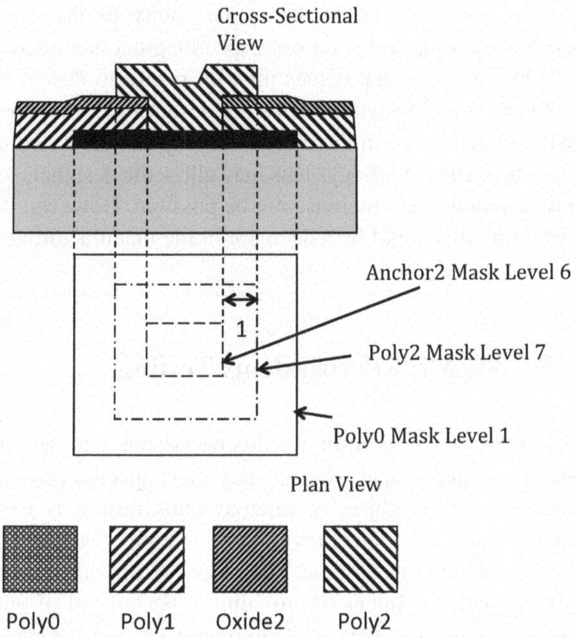

Fig. 7.11 Illustration of Rule 1 from Table 7.6 that is a design rule that ensures that Poly2 layer expends beyond Anchor2 opening to ensure that the opening is completing covered with polysilicon from Layer Poly2

masking levels. For simplicity, only Rule 1 in Table 7.6 is shown. As noted from Table 7.6, Rule 1 has a minimum value of enclosure of 5 microns. This dimensional rule takes into account the layer-to-layer registration accuracy of patterning the polysilicon material layer Poly2 with respect to the opening made in polysilicon layer Poly1 that was made using the Anchor2 mask level, as well as the topography of the pre-existing layers of Poly0, Poly1, Poly2, and Oxide2. Notice that while the general guidance of the PolyMUMPs™ process technology is a layer-to-layer registration alignment accuracy of 2 microns, that the existing topography in Fig. 7.11 requires that the design rule, namely Rule 1, have a minimum value of 5 microns. As this example demonstrates, the topography has a large impact on the alignment accuracy of material layers. Additionally, the organization running the PolyMUMPs™ process technology states that designers can use a uniform value of 5 microns for all of the design rules for simplicity.

7.7.2 Design Rule Checking

Design rule checking (DRC) in software is extensively used in integrated circuit (IC) microsystems design to automatically determine whether the mask layout design adheres to the specified design rules of a process technology. Design rule checking is also used in MEMS manufacturing. However, as noted previously, many MEMS process sequences are custom, and these cases there may not be any existing design rules. It will be the responsibility of the designers to develop a set of estimated design rules for each specific process sequence.

One question that comes up with regard to design rules is the following: what happens if the designer breaks the design rules? Microsystems developers can and will sometimes push the design envelope to break the manufacturer's design rules for a process technology. This may allow the designers to obtain higher performance levels that would otherwise not be possible. However, the potential price to be paid for doing this could be a decrease in the manufacturing yields.

7.8 MEMS Microsystems Testing

The purpose of the testing of microsystems is to determine whether the fabricated devices function and whether they meet specific predetermined performance specifications. The subject of microsystems testing is very complicated, broad, and diverse, and a complete review of this topic would require far more space than can be allocated in this volume. Therefore, we provide a brief overview of microsystems testing with the intent of providing a basic level of understanding of some of the issues involved as they relate to manufacturing parameter variations.

There is both an importance and cost associated with the testing of microsystems. The importance of testing and detecting failures as early as possible, preferably immediately at the completion of wafer-level fabrication and before the device is placed in use, is illustrated by the "rule of ten." This rule asserts that the cost of using a faulty device for every major step in subsequent manufacturing stages increases by a factor of ten. There are also significant benefits associated with identifying the cause(s) of the failure(s) as well. If the reason why the microsystems device is defective can be determined, it may allow the device engineers to develop methods that can potentially remedy the cause(s) and thereby increase manufacturing yields.

At the most basic level, device testing involves applying input signals to the device under test and then measuring the output signals in response to these inputs (Fig. 7.12). Often electrical power must be applied to the microsystems device during testing in order for it to be placed into an operational state. The input signals

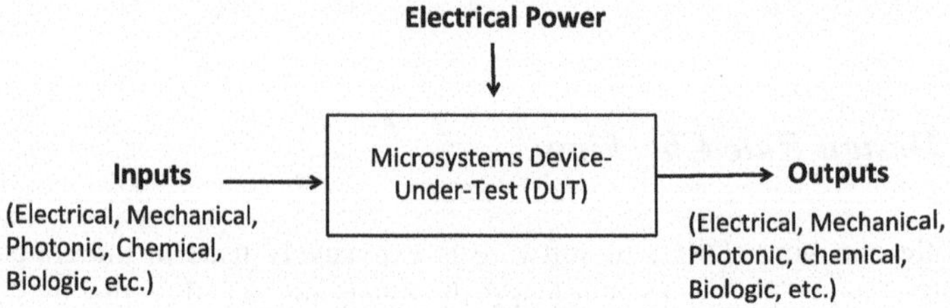

Fig. 7.12 Illustration of microsystems device testing wherein various inputs are applied to the device under test (DUT) and the various types of outputs are measured. Often the device requires electrical power to turn the device on

applied to the microsystems device can be electrical, mechanical, photonic, etc. Likewise, the output signals are usually electrical but can also be mechanical, photonic, etc.

The level of comprehensiveness of the testing procedure varies depending on the circumstances and device type. For some devices, performing a test that exercises every possible operational state may not be practical (i.e., it would take too long or would not even be theoretically possible). Therefore, in these situations the test and design engineers must devise an appropriate testing procedure composed of a more limited set of testing exercises that will allow a specified level of statistical confidence that the devices passing the testing have the required performance level. For other microsystems, the testing is comparatively exhaustive. For MEMS, the testing is often fairly extensive and also may even involve additional operations such as a calibration and/or trimming procedures performed on the devices to place them at the desired operational state.

Some basic definitions are in order. Functional yield testing is often called the "hard" or "catastrophic" yield since it is meant to determine whether the devices work or not. Often functional yield testing is performed on whole wafers at the completion of the wafer-level fabrication steps and before the devices are diced and packaged. The intent of functional testing is to identify those dies on the wafer that are not suitable for further manufacturing steps.

Parametric testing is usually performed on devices that have passed the functional tests and is comparatively a much more exhaustive set of tests. It is not meant to determine whether the device works (since this was already determined in the functional tests performed) but how well it works. Moreover, how well the device works is based on the ability of the device to meet or exceed certain pre-defined performance requirements. Parametric testing may also involve calibration, trimming, or even repair. The measurement data from parametric testing can also be very valuable in enabling the diagnoses for the causes of the device not meeting the performance requirements. Variations in the device parameters will cause a range of output values from one device to another, and if these output variations are too large, it can result in devices not meeting the performance requirements, thereby rendering them unacceptable for the intended application. Importantly, the information derived and analyzed from parametric testing can be used to modify the device designs and/or modify the manufacturing processes to improve the manufacturing yield.

Both functional and parametric testing may be performed at multiple stages in the manufacturing process of microsystems. Additionally, there may be other forms of testing performed such as burn-in testing, environmental testing, lifetime testing, and reliability testing. More information about these types of testing methods can be found elsewhere [11].

7.8.1 Example of MEMS Microsystems Testing

The testing of MEMS microsystems devices varies greatly depending on the MEMS device type and application. Given that there often are no existing markets for specific types of equipment for MEMS device tests, frequently they must be custom designed and developed.

For instance, the parametric testing of a MEMS piezoresistive pressure transducer would likely entail applying an appropriate biasing voltage to the Wheatstone bridge electrical circuit of the piezoresistive elements, combined with application of a range of pressure loadings to the mechanically compliant sensor membrane, and measuring the output voltage signal as a function of the pressure loads applied to the sensor. Obviously, this testing procedure is radially different from that of an IC given that the test system needs to apply controlled pressures to the sensor devices. That is, the testing of MEMS most often involves electrical and mechanical inputs and electrical and perhaps mechanical outputs. Moreover, in order to perform accurate testing, it is important that the pressure loads are sealed around the perimeter of the sensor membrane. This makes the test system complicated compared to an IC-based tester that only has to apply input and measure output electrical signals.

Additionally, a different type of MEMS sensor will almost always require a completely different testing configuration. For example, an inertial MEMS sensor would also need to have an input mechanical signal to operate the devices and measure the output electrical signals during testing. But, in this case, the mechanical input signal would be a controlled range of applied inertial signals, such as acceleration along different coordinate axes, rate of rotation along different axes, or both. Therefore, the testing of the MEMS inertial device is much different from the MEMS pressure transducer.

7.8.2 MEMS Microsystems Device Trimming

The process of trimming involves methods to alter the input–output relationship of microsystems device usually by modifying specific elements within the microsystems. The purpose of trimming is to reduce the output range of the microsystems device due to variations in the device manufacturing parameters (e.g., dimensional and material properties). Importantly, trimming is performed after the microsystem has been manufactured. The trimming of microsystems devices is often based on using some method to remove selected and controlled amounts of material from elements within the microsystems device active area. The amount of removal material can be to adjust the input/output property of the element or alternatively to render the element completely nonfunctional. Frequently, the elements to be trimmed on the microsystems device die are one or more thin-film resistors made from some metal, metal alloy, or metal silicide thin-film material layer. Trimmable capacitors are also possible and are usually adjusted in value by ablating the top electrode of the capacitor using a laser.

The trimming resistors are commonly connected as thin-film circuit elements into the microsystems device layout design. The resistors are adjusted in their resistance values by removing controlled amounts of the material layer of the resistor or completely opening up the resistor, making the resistive element an open circuit.

Methods to adjust the resistor values include using a laser to cut material from the area of the resistor or cut to completely open the resistor or applying a sufficient current through the resistor to cause it to fuse out. In most situations, laser trimming is used since it affords a higher level of control and accuracy. Laser resistive trimming techniques can only be used to increase the value of the resistors being trimmed since material can only be removed.

Passive trim is the adjustment of a resistor's (or capacitor's) value to a desired value when the microsystems device is inactive (i.e., no power is applied to turn "on" the microsystems device). Since this method does not involve an active measurement to obtain the desired value, it has a serious disadvantage. Alternatively, a much more desirable technique of trimming is called "active trimming." In this technique, the microsystems device output (e.g., output voltage, frequency, etc.) is monitored continuously and compared to the desired value while the laser is removing material to adjust the element(s) value(s). Once the desired output level is reached, the laser is turned "off." Figure 7.13 illustrates some of the resistor configurations used in trimming.

Laser trimming of microsystems devices has the advantages of being an automated, high-speed, low-cost, and relatively good accuracy method for tuning elements so as to adjust the output of the device. Typically, a single laser-trimmed resistor can reach a tolerance level of 0.1%. Higher overall precision is possible using more complicated trimming circuit configurations composed of a number of trimmable resistors. Of course, the trade-off on more precision on the trimming resistors is that it will consume increased valuable die area.

7.8.3 MEMS Microsystems Calibration

Calibration is roughly defined as the comparison of the output values of a device with a calibration standard of known accuracy, followed by an adjustment of the device so that it provides an output signal equal to that of the standard. Some MEMS microsystems devices, such as pressure and inertial sensors, often require the device to be calibrated as part of the manufacturing process. In many circumstances, the calibration process of a device includes a trimming operation to adjust the device output, based on comparison to a standard. As might be inferred, the process of calibration is a method to mitigate some of the variation in microsystems device output signals due to the random and bias variations of device parameters from the fabrication processes used to implement the device [12].

7.9 MEMS Microsystems Packaging

Microsystems packaging has a number of very important functions. First, a major objective of the package is to protect the encapsulated microsystems device from the environment. Second, the package also provides pathways for various input and

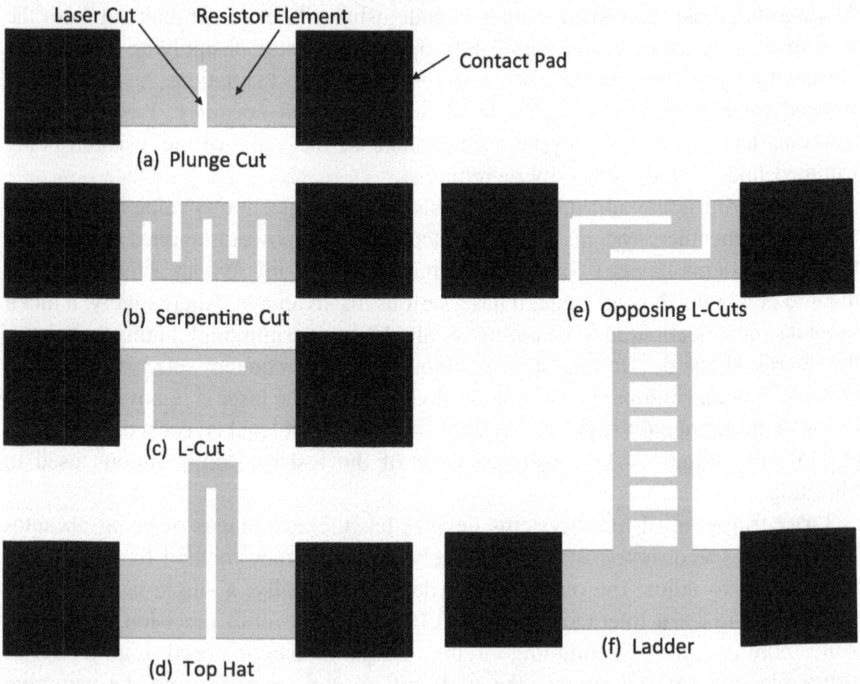

Fig. 7.13 An illustration of some of the configurations used in trimming resistors. The resistors are of the thin-film type and have two contact pads on either side. In most instances, the resistors would be configured into an electrical circuit of a MEMS microsystems device. The trimming configurations have varying degrees of precision with cuts shown in (**a**) having minimum level of precision. The serpentine cut pattern in (**b**) allows large changes in the value of resistance. The L cut in (**c**) and opposing L cuts in (**e**) and the ladder in (**f**) each enable higher levels of precision in trimming the resistance. The ladder in (**f**) is adjusted by cutting through the desired number of branches

output signals, as well as power, if needed, between the inside and outside of the package. These pathways can be electrical, mechanical, chemical, optical, etc. Therefore, the package is an interface between the device die and the outside. Third, the package allows the microsystems die to be more easily handled for assembly onto boards and into systems. Lastly, many microsystems devices can heat up substantially during operation and can cause drift and/or reliability issues. The package is often designed as a mechanism to transport away the heat generated by the die, allowing the die temperature to be maintained at a desired level.

MEMS packaging is more complicated and expensive compared to packaging for ICs. While MEMS packaging shares many of the same objectives as IC packaging, there are important differences that make packaging for MEMS difficult. One difference is that with IC packaging, input and output signals are electrical, while the inputs to MEMS devices can be a wide variety of physical stimuli. Consequently, the packaging for MEMS devices not only needs to encapsulate the device and protect it from the environment; it must also allow the device to interact with the

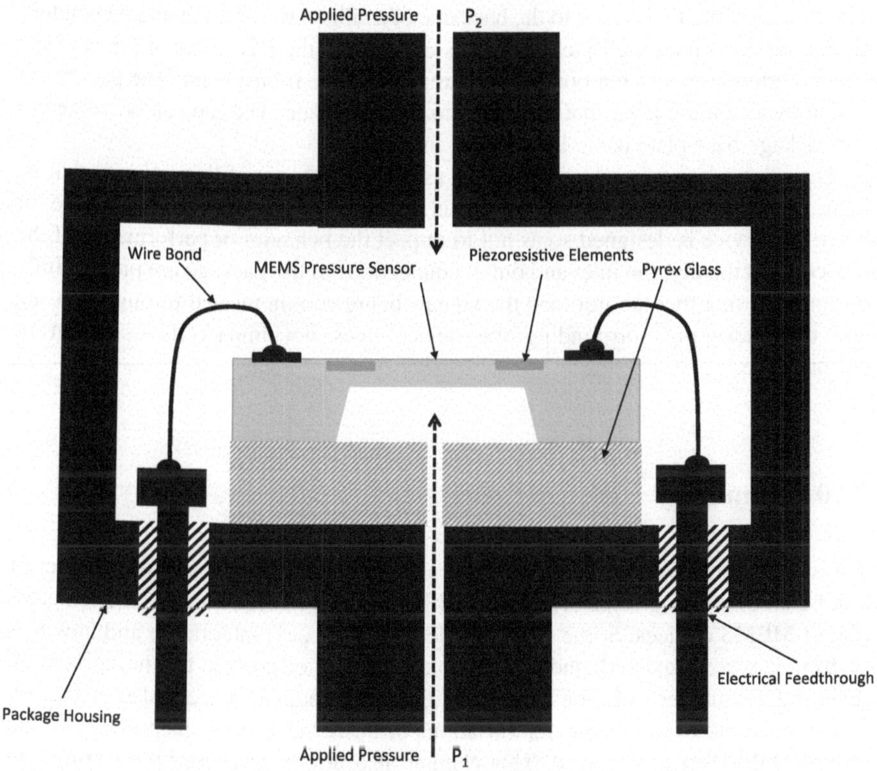

Fig. 7.14 An illustration of a MEMS packaging approach for a pressure sensor. The sensor is housed in a metal can enclosure having a base plate that has pressure ports and electrical feedthroughs

environment in a very limited and controlled fashion. Another difference relates to the market sizes. In the case of IC microsystems packaging, it is very common that a packaging solution can be used for a relatively large number of different IC device types thereby enabling large economies of scale and lower costs. MEMS often require customized and costly packaging solutions and thereby higher costs.

An example of a MEMS packaging approach for a differential pressure sensor is shown in Fig. 7.14. There are two pressure ports shown, and the sensor measures the differential pressure between the two ports. It should be noted that this is only one example and there are many different types of packaging solutions that can be employed for pressure sensors. Nevertheless, this example demonstrates some of the unique issues involved with packaging of MEMS microsystems. Importantly, the package provides for pressure ports that extend through the housing so that the pressures to be sensed can reach the two sides of the mechanically compliant membrane that has piezoresistors fabricated on the top surface. The pressure sensor is attached to a relatively thick layer of Pyrex glass to better isolate the sensor from the attachment mechanical stresses associated with attaching the die to the package.

The Pyrex is directly bonded to the backside of a silicon wafer using anodic bonding techniques, and then the Pyrex surface is attached to the base plate of the package using thermo-compression bonding to form a leak-free robust bond. The package is sealed by a capping layer that is bonded to the base plate. The cap can be affixed to the package base plate using a variety of methods.

The importance of packaging related to the major topic of this volume, that is, manufacturing parameter variations, is that in nearly every instance, the package for a MEMS device is designed so as not to impact the behavior or performance of the device. Specifically, the inlet and outlet connections to the package are purposefully designed so that they do not load the signals being communicated through housing and the packaging surrounding the device does not impact its operation or performance.

7.10 Summary

This chapter reviewed the important topic of process integration where a number of individual processing steps are combined into a process sequence for implementation of MEMS devices. Some of the challenges of process integration and how it is performed were discussed, and two notable examples of process technologies were reviewed. Parameter variations that arise in process sequences were also discussed, and an example of the parameter variations of a MEMS process technology called PolyMUMPS™ was reviewed. This chapter also briefly discussed the testing and packaging of MEMS, including device trimming and calibration.

References

1. A. Cowen, B. Hardy, R. Mahadevan, S. Wilcenski, POlyMUMPs Design Handbook, a MUMPs® process. Revision 13.0, copyright 2011, MEMSCap
2. Wicht Technologie Consulting, "MEMS 30" Top 30 MEMS manufacturers in 2006 by revenue, www.wtc-consult.de
3. L.J. Hornbeck, From Cathode Rays to Digital Micromirrors: A History of Electronic Projection Display Technology. Texas Inst. Tech. J. **15**(3), 7–46 (1998). Also at http://www.dlp.com
4. J. Grimmett, J. Huffman, Advancements in DLP® technology – the new 10.8 μm pixel and beyond. IDW/AD'05, Proc 12th Intl. Display Workshops, (in conjunction with Asia Display 2005) **2**, 1879–1882 (2005)
5. L.J. Hornbeck, *Combining Digital Optical MEMS, CMOS, and Algorithms for Unique Display Solutions* (IEEE International Electron Devices Meeting Technical Digest, Plenary Session, 2007), pp. 17–24
6. M.R. Douglass, MEMS reliability – coming of age. Proc. SPIE **6884**, 688402 (2008)
7. D. Koester, B. Hardy, MCNC PolyMUMPS staff engineering communication
8. W.N. Sharpe Jr., B. Yuan, R. Vaidyanathan, R.L. Edwards, Measurements of young's modulus, poisson's ratio, and tensile strength of polysilicon, in *Proceedings of the 10th Annual International Workshop on Micro Electro Mechanical Systems, Nagoya, Japan*, (1997), pp. 424–429

9. R.K. Gupta, P.M. Osterberg, S.D. Senturia, Material property measurements of micromechanical polysilicon beams. Proc. SPIE **2880**, 39 (1996)
10. S. Jayaraman, R.L. Edwards, K.J. Hemke, Relating mechanical testing and microstructural features of polysilicon thin films. J. Mater. Res. **14**, 688–697 (1999)
11. A.L. Hartzell, M.G. da Silva, H.R. Shea, *MEMS Reliability* (Springer, New York, 2011)
12. S. Senturia, *Microsystems Design* (Springer, New York, 2001)

Chapter 8
Device Parameter Variations in Microsystems Manufacturing

8.1 Introduction

This chapter discusses the analysis of device parameter variations resulting from the manufacturing of microsystems. There is considerable confusion, sometimes even from experts in the field about this topic. It is often proclaimed that microsystems manufacturing methods provide for high levels of dimensional *precision* compared to other production methods commonly employed in industry. This is not the case in most circumstances. While it is true that microsystems manufacturing allows the dimensions of devices to be *very small*, this is not equivalent to being able to make things with high levels of precision. Additionally, declaring a manufacturing technology as "precise" is necessary, but not always sufficient to ensure that the variations due to fabrication processes are optimized since accuracy is also very important. Precision and accuracy are not equivalent as will be discussed later in this chapter.

The topic of manufacturing variation is germane to the dimensional parameters of microsystems, but it is also relevant to other device parameters involved in the implementation of microsystems. Specifically, the material properties can also have a significant impact on the resultant microsystem's output behavior and performance. Therefore, the resultant variation of both the dimension and material properties needs to be considered since each has an impact on whether the microsystems device performs as it was designed for the intended application(s).

As will be shown later, the physics involved in the microsystems device's operation must be considered in the analysis of parameter variation effects since some of the equations describing the device's behavior may involve one or more device parameters that are raised to higher powers (e.g., squared, cubed, etc.). This can dramatically amplify the impact of the variations of these device parameters on the device's output performance. Given the relatively large variances in both dimensions and material properties involved in MEMS microsystems manufacturing, this is an extremely important topic since it directly relates to the amount of variability in

M. Huff, *Process Variations in Microsystems Manufacturing*, Microsystems and Nanosystems, https://doi.org/10.1007/978-3-030-40560-1_8

the manufactured device output behavior and thereby the resultant yield of the manufacturing processes (or even whether there is any yield at all!) and consequently the manufacturing cost.

For many microsystems devices, there are a number of elements involved in the device's composition each having multiple dimensions (i.e., thickness, width, and length) and composed of different material types that can influence the microsystems device's resultant behavior. Consequently, the analysis of the effects of the parameter variations on the device output behavior can be complex.

8.2 Manufacturing Variations

Anything that is manufactured will exhibit some amount of variation from one item to the next, both in the dimensions as well as material properties. That is to say that no manufacturing process is perfectly repeatable wherein every item made is an exact copy of the other elements made or of the specified parameter values in the original design. The magnitude of these device parameter variations has an impact how the item performs.

The resultant variation of any manufacturing method will be described by quantifying the difference between the *desired value* of a device parameter in a microsystem and the *actual* or *resultant value* of this same parameter after the manufacturing has been completed. The desired value of the device parameter would normally be the value that resulted from engineering calculations used in the design process. The resultant value is the measured, either directly or indirectly, value of the parameter after the device has been manufactured.

In general, the manufacturing variation, given by ε, of some device parameter, such as the dimension of a device feature, can be expressed as the absolute value as follows:

$$\varepsilon = |x_r - x_a|, \tag{8.1}$$

where x_r is the resultant (or actual) measured value of a parameter from a manufacturing process and x_a is the desired design parameter value. Variations are also sometimes defined on the basis of the maximum variation, ε_{max}, or:

$$\text{Maximum variation} = \varepsilon_{max} = |x_{r\,max} - x_a|, \tag{8.2}$$

where x_{rmax} is the maximum variation in the actual measured value of the parameter from the designed desired value.

When analyzing variations in device parameters, there is often the need to estimate the range that the variation may have with some level of confidence. This can be expressed as follows:

$$-u \leq \varepsilon \leq +u \ \ (n:1), \tag{8.3}$$

where u is the limit of the range in the variation value of the parameter with a probability of n to 1. As a general rule, the probability of the variation of a parameter being within the range of values is greater when the range is larger and vice versa.

A related term called "relative variation" is the device parameter variation value normalized with respect to the desired design value and given in percentage as follows:

$$\text{Relative variation (or error)} = \left(\frac{1}{x_\text{a}}\right)(x_\text{r} - x_\text{a}) \times 100\%. \tag{8.4}$$

A relative variation can be very useful since it allows different manufacturing methods even at different dimensional size scales to be easily compared to one another to determine how well they are able to produce elements meeting the desired designed parameter values.

There are at least three other characteristics that are desired in any manufacturing process[1]: repeatability; reproducibility; and stability. Repeatability of a manufacturing process refers to the ability to produce products that are identical to one another. Reproducibility refers to the ability to be able to reproduce the manufacturing process. This could refer to running a production lot for some period of time, switching to the production line to the making of another product using a different manufacturing process, and then restarting the manufacturing of the original product again. A reproducible process would be one where the manufacturing process has not changed or drifted even though the product has not been made for a period of time. Stability refers to a manufacturing process that does not change over time. The period of time over which stability is measured depends on the type of manufacturing process employed, kind of manufacturing equipment used, and many other factors.

8.3 Measurement of Manufacturing Variations

In order to determine that there is a manufacturing variation, it must be measured using some technique. In Chap. 5, we reviewed a number of methods of metrology used in microsystems manufacturing to make various measurements and also provided some broad guidance on their associated levels of resolution, precision, and accuracy. As a general rule, it is desirable that the precision of the method used to measure a device parameter is significantly less than the expected smallest variation of the parameter being measured.

[1] There may be additional desirable characteristics, such as availability, low cost, ability to scale with demand, etc.

This is commonly expressed as the ratio of P/V, where P is the precision of the metrology measurement system and V is the variation [1]. In this context variation refers to the magnitude of the variation of the parameter being measured. A P/V ratio of 0.1 or less is considered good, while a P/V ratio of less than 0.3 is considered adequate depending on the circumstances. This is commonly known as the so-called "watch makers" rule. In most circumstances presented in this volume, the P/V ratio will be assumed to be 0.1 or less. However, in actual practice the measurement P/V ratio should always be determined to ensure that the measurement technique is adequate.

8.4 Bias and Random Variations

In manufacturing processes there are generally two types of variations that are of concern[2]: bias variations and random (or statistical) variations. In general, device parameters can exhibit both of these types of variations in microsystems manufacturing.

A bias variation, also sometimes called a systematic variation or error, is an offset of a fixed value occurring consistently in a multiplicity of measured values of the parameter.[3] This offset may be due to something in the manufacturing process that results in a consistent and repeatable variation in the parameter from the desired designed value. Bias variations are repetitive, meaning that the same amount of offset occurs in every device made and is observed over and over again in the manufacturing process. Therefore, by definition this type of variation is not disposed to statistical analysis techniques.

An example to illustrate bias variation is shown in Fig. 8.1. This set of processing steps is composed of the deposition of a thin-film material layer as shown in Fig. 8.1a, followed by the deposition of a photoresist onto this material layer that is exposed and patterned using photolithography as shown in Fig. 8.1b and ending with the reactive ion etching (RIE) of the material layer as shown in Fig. 8.1c. As shown in Fig. 8.1b, the "as drawn" or "desired" dimension of the patterned thin-film material layer has a dimension given by "L." The desired value of "L" is derived from the engineering design activity performed so as to select a value of L to allow the device to exhibit the desired output behavior for the intended application. However, after etching of this material layer, as shown in Fig. 8.1c, there is some amount of lateral etch of the material layer whereby the photoresist layer is undercut by the etching process used. While it is desired that an etch process used for this sort

[2]There can also be other types of variations (or errors) due to mistakes made in the manufacturing processes, but these types of variations and errors will be ignored in this discussion.

[3]It should be noted that the bias variation could be due to a systematic variation in the parameter or a systematic variation in the measurement of the parameter or both. However, we are assuming in this discussion that the systematic variation is not due to the measurement instrument or method used.

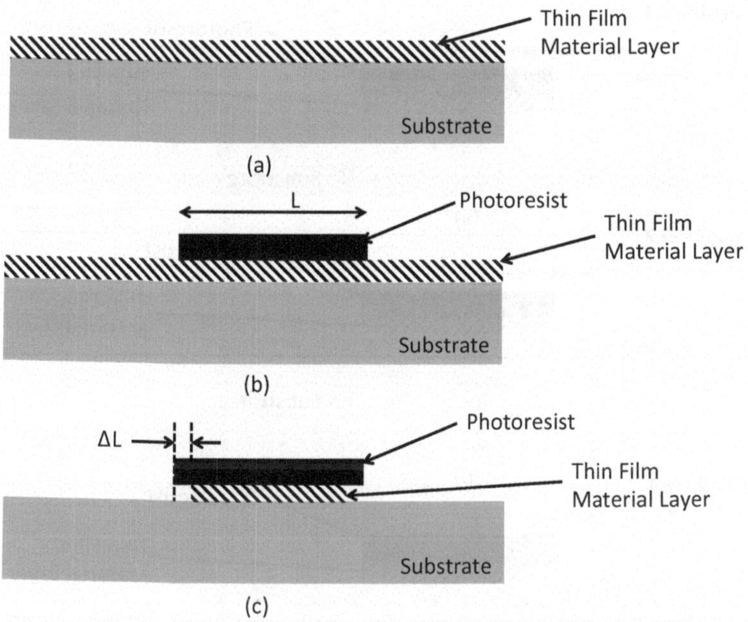

Fig. 8.1 Commonly used in microsystems manufacturing process composed of deposition of a layer, shown in (**a**), followed by photolithography, shown in (**b**), and ending with etching of the material layer as shown in (**c**). The desired or design value of the parameter of interest is given by "*L*." Due to the fact that no etch is perfectly anisotropic, there will be some amount of lateral etch. Due to the lateral etching, the resultant dimension of "*L*" is reduced to "$L - 2\Delta L$"

of patterning be as anisotropic as possible, in practice there will always be some amount of lateral etching in any etch process. The result of this lateral etching is a resultant manufacturing value of "*L*" that is smaller on each edge by an amount shown as "ΔL" in Fig. 8.1c from the desired design value. The parameter of interest in Fig. 8.1 is "*L*," and the bias variation in this example is given by "ΔL" on each side of the etched thin-film material layer. Since this amount of lateral etching would be present on all of the devices fabricated using a specific etch method, it would be considered as a systematic variation that biases the value of *L* by an amount that is directly related to the manufacturing process employed.

Random variations, unlike bias variations, are by definition inconsistent. That is, they exhibit varying values from one sample to the next. These types of variations result from random deviations in the manufacturing processing conditions across a substrate, from one substrate to the next and from one production lot to the next. These variations are inconsistent and are always present to some extent. As a general rule, random variations will exhibit a range of values. Most microsystems manufacturing processing steps are dependent on a number of different process parameters such as temperature, gas flow rates, RF power levels, stirring or mixing, power level, light intensity and/or wavelength of light, as well as others. None of

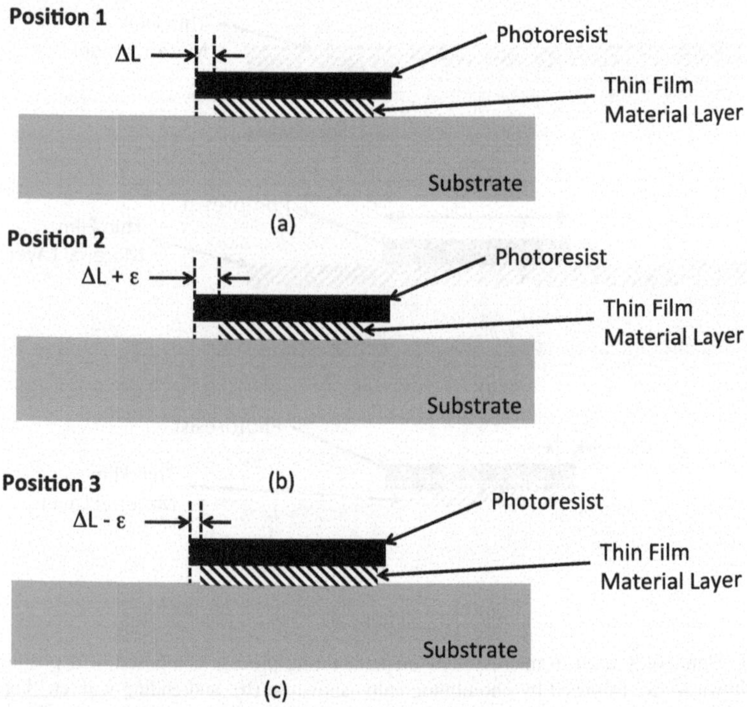

Fig. 8.2 An illustration of presence of both bias and random variations whereby the thin-film material layer is etched. A photoresist layer that has been patterned is used a masking layer for a subsequent etch of the thin-film material layer. The consistent amount of lateral etch is given by the bias variation ΔL labeled Position 1, and the inconsistent amount is given by a range of values across the substrate with two positions being shown above as in (**b**) and (**c**) with an amount of lateral etch given by $\Delta L + \varepsilon$ and $\Delta L - \varepsilon$, respectively. These are labeled Positions 2 and 3, respectively

these parameters will be controlled with perfection either spatially or temporally across a wafer, a batch of wafers, or from one lot of wafers to the next.

Random and bias variations are illustrated in Fig. 8.2. As can be seen, in Fig. 8.2a labeled Position 1, the lateral etch or undercut of the photoresist masking layer is given by ΔL. This is the bias variation, and it is consistently present from one location to the next. At a different position on the substrate shown in Fig. 8.2b labeled Position 2, the lateral etch or undercut of the photoresist masking layers is slightly different and given by $\Delta L + \varepsilon$, and at yet a different position on the substrate shown in Fig. 8.2c labeled Position 3, the lateral etch is given by another slightly different value of $\Delta L - \varepsilon$. The ΔL is the bias variation and the ε is the random variation.

If one were to take measurements over and over again different locations across the substrate the amount of lateral etch of the thin-film material layer, etch sample will likely differ from ΔL by a small (but changing) amount in either the positive or negative direction depending on the exact location where the measurement is taken. It is these inconsistent differences in widths that are an example of random variations. If enough measurements are taken, statistical techniques may be applied to analyze various important attributes about these random variations. In many circumstances, both bias and random variations may be present, and therefore both need to be properly analyzed, estimated, and managed.

8.5 Resolution, Precision, and Accuracy

Engineering disciplines, including microsystems manufacturing, have created a set of important terms that are used when describing attributes of production processes.

Resolution The resolution in a manufacturing process is the smallest value parameter that the equipment can implement or measure. For example, a contact photolithography aligner will not be able to transfer feature dimensions of 10 nm. Resolution strongly depends on the equipment used to perform the manufacturing process.

Precision Precision refers to the resultant spread in the parameter values over a collection of values. For example, precision can be used to define a range of values of some device's dimensional features as measured over a sample set of these features. Repeatability and reproducibility are related to precision. Repeatability can be defined as the lack of variation in the parameter values when the manufacturing process is repeated some number of times under the same exact conditions (e.g., the same device made on the same tool, by the same operator, using the same measurement equipment, etc.). Reproducibility refers to being able to get the same parameter values on different manufacturing tools, operators, production facilities, etc.

Accuracy The accuracy of the manufacturing process is the difference in a device's resultant parameter values from the desired (sometimes also called designed or true) parameter values. Accuracy refers to the offset between the resultant and desired value of a parameter rather than to the spread in the values of the parameter.

Accuracy and precision are sometimes confused with one another. A diagram that helps to visualize the difference between precision and accuracy is illustrated in Fig. 8.3 wherein a target is shown that has a number of arrows shot at it attempting to hit the center. Figure 8.3a is an example of high precision, but low accuracy. The arrows in Fig. 8.3a are considered to have precise locations on the target because they do not deviate very far from one another, and therefore the spread in the arrows is relatively small. However, since the average of this group of arrows is far from the center of the target, this outcome does not exhibit good accuracy. Figure 8.3b shows a scenario where the arrows have both low precision and low accuracy. As seen, there is a considerable amount of scatter in the locations of the arrows indicating low precision, and averaging over the arrow locations results in a position far from the center of the target also indicating inaccuracy. Figure 8.3c is an example where the arrows have low precision and high accuracy. There is a considerable amount of

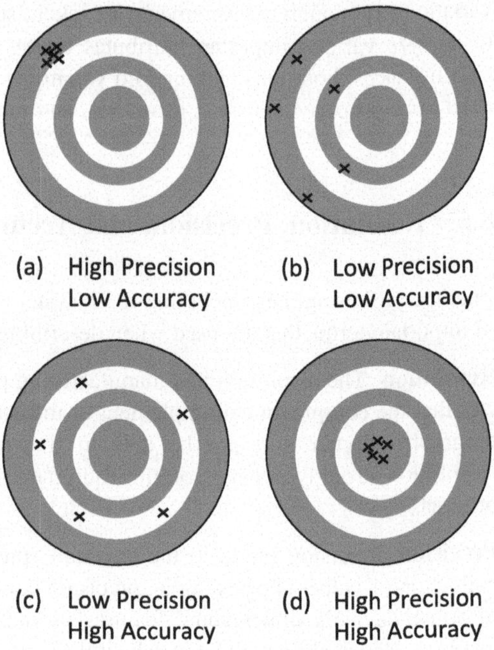

Fig. 8.3 Illustration of the concept of precision and accuracy

scatter in the arrows over the target indicating low precision, but the average of the arrows indicates accuracy. Finally, in Fig. 8.3d, an example of a scenario of high precision and high accuracy is shown. The arrows are grouped together near the center, and the average of the arrows is also in the center.

This exercise illustrates a few important points. First, precision is not meaningful unless there is a collection of data points. Second, since precision has meaning for collections of data points, it therefore becomes a form of data that can be subjected to statistical analysis. Third, precision is not the same as accuracy. As shown in Fig. 8.3a, it is possible to have high precision but still have a large inaccuracy. Fourth and most importantly, both precision and accuracy of resultant parameter values are important for a well-controlled manufacturing process that produces products that will meet the application requirements. Obviously, this applies to device parameters such as the dimensions and material properties of manufactured microsystems devices. In short, the accuracy of a manufacturing process relates to the ability to make devices with parameter values that are equal to or nearly equal to the desired design value of the design value of the parameter. The precision relates to how much spread there is in the actual device's parameters.

It should be obvious that precision and accuracy have a direct linkage to the bias and random parameter variations described above. Specifically, the magnitude of the bias variations relates to the inaccuracy of a manufacturing process, while the amount of random variations relate to the precision of a manufacturing process.

In order to obtain a precise and accurate manufacturing process, that is, a manufacturing process with a low amount of total variation in the device parameters, both the bias and random variations of the parameters should be minimized.

Another term related to manufacturing is *tolerance*. In general, the tolerance refers to the allowable ranges of the values of the device parameters to be considered as acceptable. Allowable tolerances are usually established during the design phase. An acceptable device is one that meets the performance requirements, and parameter tolerances can be specified so that any device made having parameter values falling with the tolerances should also meet the performance requirements. A device having one or more dimensions outside of the tolerance range would be rejected since the device would likely not be able to meet the performance requirements and therefore would not be useable for the intended application.

The specification of the tolerances is part of the design process, and not directly the result of the manufacturing process. Nevertheless, tolerances have a relationship to precision and accuracy, which are inherently dependent on the manufacturing processes. Ideally, the engineer performing the design of a device would know the precision and accuracy of the possible manufacturing processes that could be used in the fabrication of that device, as well as their associated costs, before initiating the design effort. It should be apparent that setting tolerances that cannot be manufactured is not a fruitful exercise, and therefore the designer should have a reasonable knowledge of what tolerances are possible before starting the design process.

As a general rule, manufacturing processes that provide for higher levels of accuracy and precision are more expensive than those having lower levels of accuracy and precision. The manufacturing costs will always be a strong factor in determining whether any microsystems device goes into production. Therefore, the engineering team performing development will typically not specify tolerances on the microsystems device that are more aggressive than required for the device to meet its functional design intent since this would increase production costs needlessly. Consequently, an important element of the design process is not only conceiving of the device design but also determining and selecting (and sometimes developing) manufacturing processes that allow the device to made and function as intended and can be produced at the cost levels required in order to derive a profit.

8.6 Comparison of the Dimensional Parameter Variations in Manufacturing Technologies

It was stated at the beginning of this chapter that the device dimensional parameter variations obtained in most microsystems manufacturing processes can be large in comparison to other manufacturing technologies. This section provides a basis for this assertion. Two forms of commonly used manufacturing techniques at different

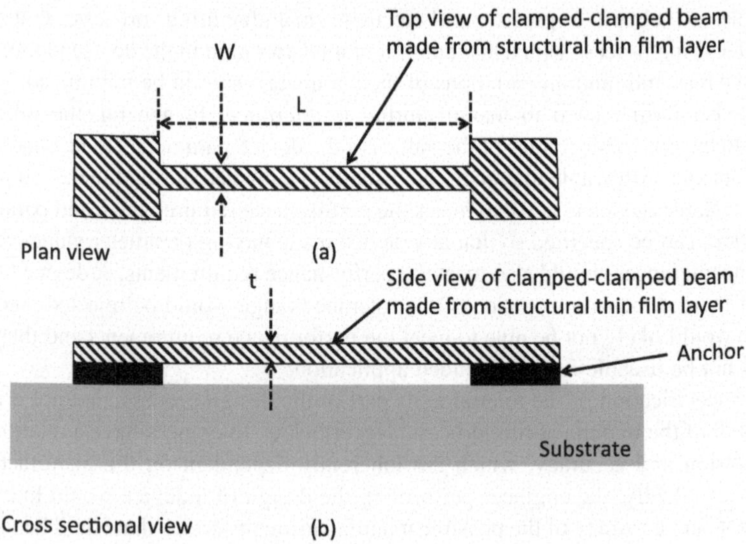

Fig. 8.4 An illustration of a clamped beam microscale beam that could be used in a number of different important MEMS applications having width "*W*," length "*L*," and thickness "*t*"

dimensional scales will be compared: one on the micro- and the other on the macro-dimensional scale.[4]

Consider a representative example in the microsystems dimensional domain, specifically a freestanding micro-beam made from polysilicon that is clamped on both sides as shown in Fig. 8.4. The microscale beam shown in Fig. 8.4 has dimensions given by a width of "*W*," a length of "*L*," and a thickness of "*t*."

It is assumed that the desired width of the micro-beam shown in Fig. 8.4 is 5.0 microns. A typical resultant random variation of the micro-beam width made using the process shown in Fig. 8.4 and having the above dimensions likely will be at least ±0.50 micron.[5] In terms of relative variation, this equates to the following:

$$\text{Relative variation} = \frac{\pm(0.5\ \text{micron})}{(5.0\ \text{micron})} \times 100\% = \pm 10\%. \tag{8.5}$$

What is immediately evident to those experienced in manufacturing domains is the large magnitude of the relative variation of the beam width parameter resulting

[4]The terms "micro-"and "macro-"as they relate to dimensional scales are usually meant to mean the following: "Micro-dimensional" scale refers to devices that have dimensions in the range from microns to tens of microns, sometimes even to the level of hundreds of microns. "Macro-dimensional" scale usually refers to devices that are visible and can be handled without aid. Macro-dimensional most often refers to devices that are several millimeters or larger.

[5]It is assumed that contact photolithography is used. The variation in the width was given in Chap. 4.

Fig. 8.5 Illustration of a commonly used machine tool for making items at the macroscale. This machine tool is a lathe and rotates the workpiece around an axis while performing various machining operations

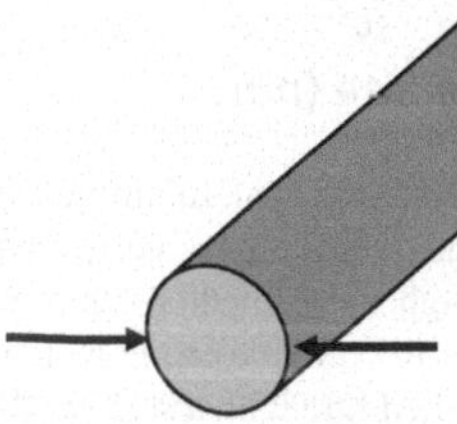

Fig. 8.6 Illustration of a commonly manufactured item at the macroscale dimensions consisting of a 100 mm nominal diameter cylindrical shaft made from steel and having an variation of ±1 mil on the diameter. This type of item could be manufactured on a standard machine tool such as a lathe

from using a microfabrication process. It is also important to note that this example is not atypical of many microfabrication technologies commonly used in microsystems manufacturing.

To compare this example to that of a more conventional manufacturing process, let us calculate the resultant relative parameter variation of a commonly used method of making mechanical elements at the macro-dimensional scale. Let us assume that we employ a conventional mechanical machine shop tool such as a lathe (see Fig. 8.5) and also assume that we desire to fabricate a cylindrical shaft from steel having a nominal dimensional diameter of 100.00 millimeters (see Fig. 8.6) using a lathe.

An experienced machinist operating a commonly available lathe can obtain fairly small dimensional parameter variation using this standard machine tool and would easily be able to make this shaft with a variation of the diameter of the steel shaft to within ±1 mil (±25.4 microns) or better. For this macroscale machining example, the relative error equates to:

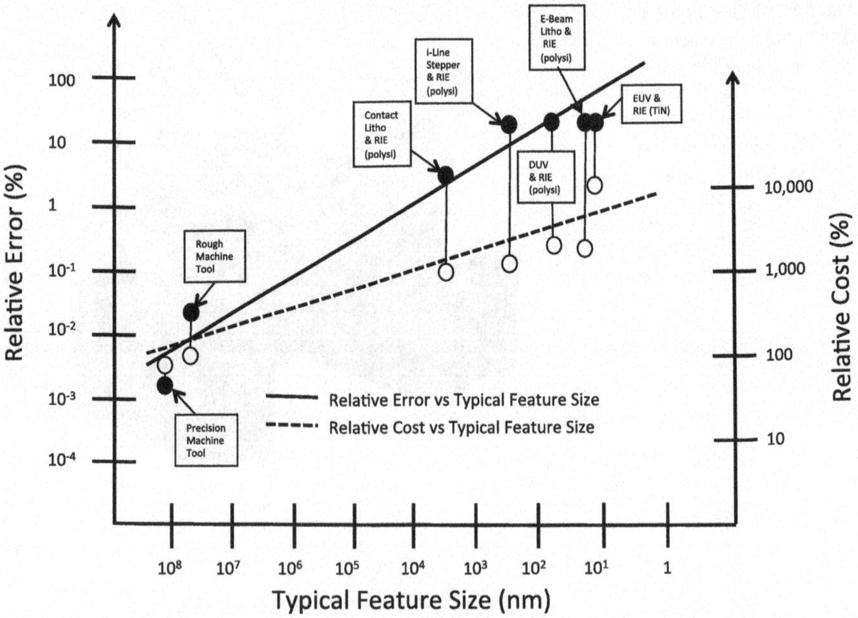

Fig. 8.7 A chart plotting relative variations versus typical minimum feature size for various methods of manufacturing (on left side axis) and relative cost versus typical minimum feature size (on right side axis). A solid black line through these methods is drawn showing that as the dimensions of the items being made reduce in size, the relative errors increase. The relative costs of these fabrication methods are shown along the dashed line with a scale at the right. This shows that as the dimensions of the items being machined decrease, the cost increases fairly dramatically

$$\text{Relative variation} = \frac{\pm(25.4\ \text{micron})}{(10^5\ \text{micron})} \times 100\% = \pm 0.0254\%. \tag{8.6}$$

In fact, some of the newer and more advanced ultraprecision computer numerical control (CNC) machine tools have demonstrated dimensional variations down to a few microns or less, thereby translating into a relative variation of around 0.002% for the sized shaft of this example.

If we take the ratio of the relative variation of the manufacturing of the shaft using conventional macroscale machining techniques to the relative variation of the width of the beam made using microfabrication techniques, we find that the macroscale machining has nearly 400 times lower variation and as much as 5000 times lower variation for an ultraprecision machine tool. This example clearly demonstrates that the relative variations of the dimensions using microfabrication methods are significantly larger compared to those obtained using macroscale machining methods.

Additionally, the relative variations of dimensions made with microfabrication methods tend to degrade (i.e., get larger) as the dimensions are reduced. The chart shown in Fig. 8.7 has relative variations (in percentages) along the y-axis plotted against the typical minimum feature sizes (in nm) along the x-axis. On this chart are shown a number of methods of manufacturing starting at more conventional macroscale machining techniques on the left and moving to progressively more advanced microsystems fabrication techniques from left to right.

The fabrication processes shown in this chart include from left to right a precision macroscale machine tool; a rough macroscale machine tool; a fairly low-resolution microfabrication process to make a line using contact photolithography followed by a reactive ion etch (RIE) of a polysilicon layer; a higher-resolution microfabrication process using an i-line stepper photolithography followed by an reactive ion etch (RIE) of a polysilicon layer to define the width of a line; the use of a higher-resolution microsystem process involving a deep-ultraviolet (DUV) photolithography followed by an reactive ion etch (RIE) of a polysilicon layer to define the width of a line; a high-resolution process involving the use of an e-beam direct-write lithography followed by an reactive ion etch (RIE) of the polysilicon layer to define the width of a line; and lastly the currently highest known resolution process involving the use of an extreme ultraviolet (EUV) lithography followed by an reactive ion etch (RIE) of a titanium-nitride layer to define the width of a line.

A solid black line is drawn approximately through the points representing these manufacturing processes and shows the general tendency that the relative parameter variations increase as the dimensions of the features being manufactured decrease. Figure 8.7 also plots along the right-side y-axis the relative cost in percentage of each of these fabrication processes against the typical feature size. A dashed black line is drawn through the dots representing each of the fabrication processes. This chart shows that as the typical feature sizes of the dimensions of the features being fabricated decrease, the costs increase fairly dramatically. This is despite the fact that the relative variations in the dimensions are getting larger in magnitude.

8.7 The Nature of Random Parameter Variations

If a number of the manufactured devices have their parameter values measured to form a set of data points, such as measurement of the dimensions of a critical feature of the device, there will be a distribution in the values of these dimensions. The totality of a group of measurements of the parameters, or data points, from manufactured devices is commonly termed a "population." A collection of a number of data points that is less than the population is called a "sample."

As a starting point for the discussion of populations and samples, the example of the micro-beam fabrication (Fig. 8.4) from above will be used. Specifically, it will be assumed that a manufacturing process has been performed to make a large number of the micro-beams, 415 micro-beams to be exact, with a desired design width of 5.00 ± 0.05 microns.[6] The measurement system used to measure the micro-beam

[6]It is noted that the variation of the manufacturing process used to make these micro-beams is better than that obtained with contact photolithograph and could be obtained with a higher-resolution form of lithography.

Table 8.1 Table of the class intervals, midpoints, and observed frequency of the measured widths of the 415 micro-beams manufactured

Class interval	Midpoint	Observed frequency
4.35–4.45	4.40	3
4.45–4.55	4.50	4
4.55–4.65	4.60	9
4.65–4.75	4.70	29
4.75–4.85	4.80	49
4.85–4.95	4.90	68
4.95–5.05	5.00	80
5.05–5.15	5.10	71
5.15–5.25	5.20	52
5.25–5.35	5.30	31
5.35–5.45	5.40	11
5.45–5.55	5.50	5
5.55–5.65	5.60	3

widths has a resolution better than 0.01 micron. Further, it is assumed that there are no bias variations present. After these micro-beams have been made and their widths measured, a tabulation of the measured widths of the micro-beams correlated to the number of micro-beams having each specific width dimension can be created as shown in Table 8.1.

A term "class intervals" is used in Table 8.1, and this refers to the range of values over which a specific bin has been determined, and the midpoint is the center of this class interval. In general, the endpoints of the class intervals are decided somewhat arbitrarily, but it is desired that these endpoints have equal size ranges and are selected so as to ensure that there are at least five or more class intervals and no individual class interval has less than a few data points. The frequency listed in the far right column in Table 8.1 is the number of micro-beams that have a width that has been measured to be within the specific class interval.

The data from this tabulation can be displayed graphically as shown in Fig. 8.8 wherein the widths of the bins of the manufactured micro-beams in this production lot are plotted along the x-axis with dimensions in microns and the number of micro-beams having the width within the defined bin is plotted along the y-axis. A bin size of 0.1 micron has been chosen.

This chart is called a histogram, which is a graphical representation of the distribution of the measured micro-beam widths and can be used to make various estimates about the likelihood of that parameter. From Fig. 8.8 it can be seen that a number of the micro-beams, 80 micro-beams to be exact, have widths centered at the desired value of 5.00 ± 0.05 microns, but it is also important to note that most of the micro-beams in this production lot have widths that are either larger or smaller than 5.00 ± 0.05 microns. The amount of beams varying from the desired width is an important characteristic of this manufacturing process.

We can also use Fig. 8.8 to calculate certain probabilities related to our production lot. For example, the probability that a micro-beam from this batch has a width of 5.00 ± 0.05 microns is given by the number of micro-beams having this width,

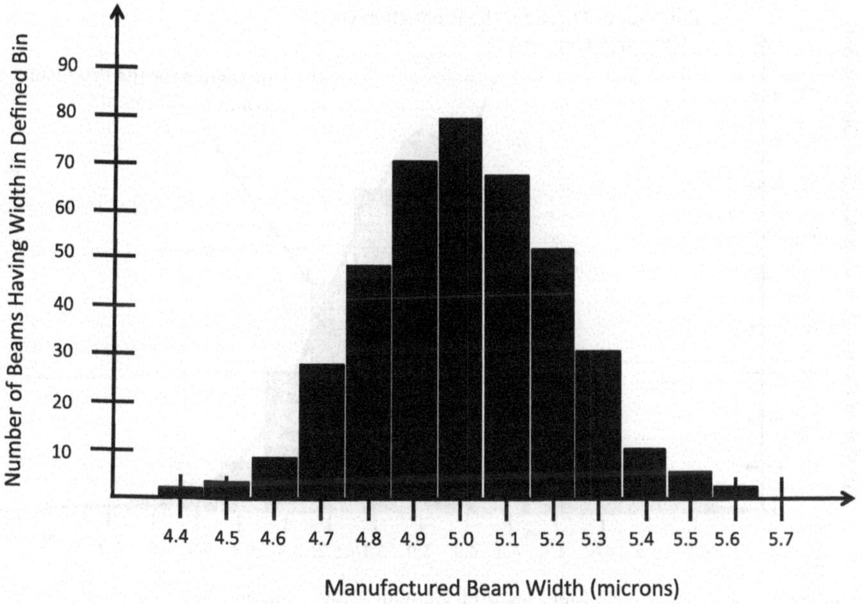

Fig. 8.8 A histogram of the measured width of a production lot of clamped-clamped freestanding polysilicon micro-beams

divided by the total number of micro-beams made in the production lot. Therefore, the probability of any of the micro-beams in the batch having a width of 5.00 ± 0.05 microns is 80/415 or 19.3%. This also means that 100% − 19.3% = 80.7% of the micro-beams have widths different from our desired design width of 5.00 ± 0.05 microns. Since the probability of events can be calculated from a histogram, it is also referred to as a *probability distribution*. Furthermore, since deviations from the desired width represent the random variation of our manufacturing process of the micro-beams, the histogram can be thought of as a probability distribution of the random variations of the manufacturing process employed to manufacture this lot of micro-beams.

If the manufacturing process from above is resumed and the resultant widths of these micro-beams are measured and plotted, the distribution shown in Fig. 8.8 will lose its discrete nature and start to increasingly resemble a continuous curved line as the number of measured micro-beams increases. This is shown as an overlay of the histogram as seen in Fig. 8.9. The distribution given by the solid curve shown in Fig. 8.9 is a type of distribution called a "normal," "bell-shaped," or "Gaussian distribution." A normal distribution is one wherein there are a large number of samples, the samples of all independent of one another, and the samples each have equal probabilities of occurring. A normal distribution is the most common type of probability distribution.

A continuous normal probability distribution curve is shown in Fig. 8.10 wherein each measurement taken of a parameter, such as the width, is given by x. The

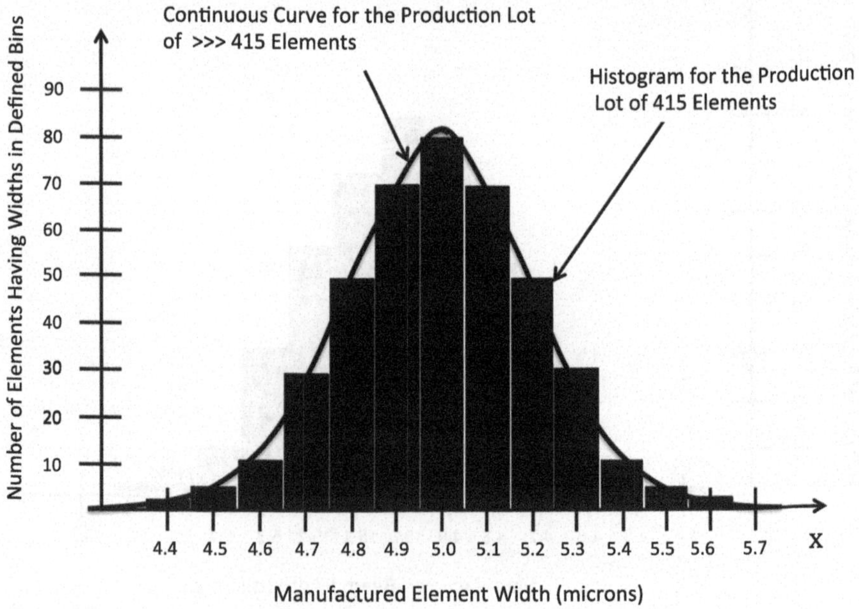

Fig. 8.9 Figure 8.8 from above with a continuous curve of the measured resultant widths of a very large number of the manufactured micro-beams superimposed onto the histogram of beams made in the production lot consisting of 415 micro-beams

probability of obtaining a value of x is dependent on the value of x, and the probability distribution function of x is given by $f(x)$.

The ordinate of Fig. 8.10 represents the occurrence per unit change of x. That is, the probability of measuring a specific value of x is not given by $f(x)$ but instead by the area of the function $f(x)$ in the interval Δx which is given as follows [2]:

$$\text{Probability}(x_1 \rightarrow x_2) = \int_{x_1}^{x_2} f(x)dx, \tag{8.7}$$

The integral of the total area under the probability distribution function $f(x)$ is given by unity. That is to say, the probability of measuring some value of x under the function $f(x)$ is 100%. The probability distribution function $f(x)$ for a normal distribution of a parameter x is given by the Gaussian probability distribution as follows [2]:

$$f(x) = \frac{1}{\sigma\sqrt{2\pi}} \exp\left[-\frac{(x-\mu)^2}{2\sigma^2}\right], \tag{8.8}$$

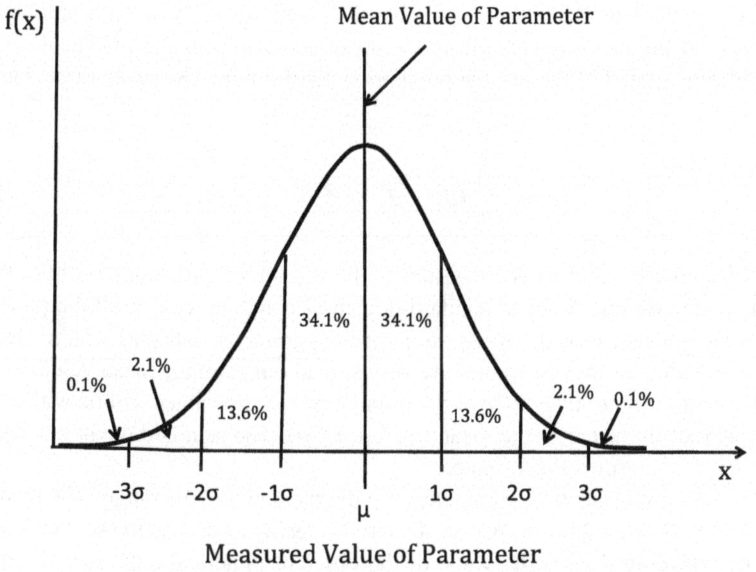

Fig. 8.10 A normal probability distribution function, $f(x)$, whereby the frequency of the specific occurrences is plotted on the y-axis and values of the parameters are plotted on the x-axis. The average or mean of the population is given by the μ, and the standard deviation (or spread) of the error values is given by σ. The areas under the normal distribution as they relate to the standard deviations are listed in percentages of the total area

where x is the value of a particular measured parameter, μ is the mean or average value of the entire population of the parameter, and σ is the standard deviation of the entire population of the parameter.

The average value or mean, μ, of the population of a parameter x is given by:

$$\mu = \frac{1}{n}\left(x_1 + x_2 + x_3 \rightleftharpoons + x_n\right) = \frac{1}{n}\sum_{i=1}^{n} x_i. \tag{8.9}$$

The mean value is the most probable value of the population of measured values and coincides with the peak or maximum of the distribution function as shown in Fig. 8.10.

The manner that the measured values of a parameter, such as the variations of a dimension of a device feature, are dispersed around the center value of the population is also important. The amount in which a single measured value of a parameter, given by x, differs from the mean value of that parameter, μ, of the population is the deviation, d, and is given by:

$$d = (x - \mu). \tag{8.10}$$

The standard deviation of a population, given by σ, is an indication of the deviation of the measurements from the mean and also provides information about the width or spread of the normal population distribution. The standard deviation is given by:

$$\sigma \approx \sqrt{\frac{1}{n}\left(d_1^2 + d_2^2 + d_3^2 + \ldots + d_n^2\right)}. \tag{8.11}$$

The standard deviation is a measure of the amount of variation present in the set of parameter values. A small standard deviation is indicative that all the parameter values are situated near the mean value of the parameter, whereas a large standard deviation indicates that the values are very spread out from the mean value. Most of the population (more than 99% of the population) of the values will be within 3σ of either side of the value of the mean (e.g., $\mu \pm 3\sigma$). The standard deviation squared, given by σ^2, is termed the variance.

Two less used statistical parameters are the median and the range. The median is obtained by arranging the numerical data in increasing or decreasing sequential order and then locating the middle value of the ordered numerical data. It is the middle value of the sequential ordering of numbers if there are an odd number of values of data and the arithmetic mean of the two middle numbers if there is an even number of values of data. Both the mean and median provide information about the middle values of a sample or population, but the mean is usually considered more useful. The one disadvantage of the mean is that it can be skewed by extreme values for very small sample sizes, whereas the median is less likely to be significantly shifted by one outlier data point in smaller data sets.

The range is another parameter that is defined as the difference between the largest and smallest numerical values in a population or sample. The range is easy to compute but is usually not considered a good measure of the amount of variation in a data set since it considers only the extreme values. Specifically, it does not reveal anything about how the data is distributed between these extreme values. Therefore, the range is usually considered as a poor measure of the variation compared to the standard deviation, particularly for large data sets. Having said that, the range is sometimes used in manufacturing quality control operations. For example, a manufacturing processes may be set up to reject devices coming off the production line that have parameters that fall outside what is considered a predetermined "acceptable range" of those parameters.

It is a common practice to transform the normal distribution using the variable z, which is given by [3]:

$$z = \frac{1}{\sigma}(x - \mu). \tag{8.12}$$

Replacing z into Eq. (8.8) above results in the following standard normal equation for the probability distribution function in terms of the variable z as follows:

$$f(z) = \frac{1}{\sigma\sqrt{2\pi}} e^{-\frac{z^2}{2}}. \tag{8.13}$$

The reason for performing this operation is to simply calculate. Specifically, use of the z variable allows one to avoid performing the integrals of Eqs. (8.7) and (8.8) for each circumstance and instead use a simple look-up table of pre-calculated z values. Tabulated values of the areas under the standard normal curve for various values of z are shown in Table 8.2. The tabulated data in Table 8.2 represents the area under the curve for various values of z. The $f(z)$ distribution curve is symmetric around zero, and therefore the tabulated values represent half of the total area. Also, similar to the total area under the probability distribution function $f(x)$ from above, the total area under the $f(z)$ distribution is also equal to 1. The distribution based on the variable z is shown in Fig. 8.11 wherein the probability for a given value of z is given by the area under the curve multiplied by 2.

It is often not practical or cost-effective to take detailed measurements on every microsystem manufactured, much less to take measurements of all the dimensions on each and every element on every microsystem produced. Fortunately, the measured parameter data, such as the micro-beam widths, can form a sampling of the population that can be used to predict the characteristics of the population from which the sample was derived under certain circumstances.

The principal method is to use the mean value from the samples of the population to estimate the mean and standard deviation of the population. Consequently, we calculate the mean of a sample from the population as an approximation of the population mean as follows:

$$\overline{x} = \frac{1}{n}\sum_{i=1}^{n} x_i = \frac{1}{n}\left(x_1 + x_2 + \rightleftharpoons + x_n\right). \tag{8.14}$$

Additionally, the standard deviation of the sample population that can be taken as an approximation to the population standard deviation σ is given by:

$$S_x = \sqrt{\frac{(x_1 - \overline{x})^2 + (x_2 \quad \overline{x})^2 + \ldots + (x_n - \overline{x})^2}{(n-1)}} = \sqrt{\frac{\left(\sum_{i=1}^{n} x_i^2 - n\overline{x}^2\right)}{(n-1)}}. \tag{8.15}$$

The number n is the number of data in the sample and $n - 1$ in the denominator is the number of degrees of freedom, which is the number of independent discrete data points being evaluated. Since the sample average uses all data that is independent, and the standard deviation uses the sample average, which is not independent of the remaining data, the number of degrees of freedom must be reduced by one as shown by the denominator in Eq. (8.15).

It is useful to be able to estimate with a level of confidence that the sample mean and standard deviation are reasonably accurate estimates of the actual mean and

Table 8.2 Tabulated z-values for half of the area under a normal curve

	Second decimal place in z									
z	0.00	0.01	0.02	0.03	0.04	0.05	0.06	0.07	0.08	0.09
0.0	0.0000	0.0040	0.0080	0.0120	0.0160	0.0199	0.0239	0.0279	0.0319	0.0359
0.1	0.0398	0.0438	0.0478	0.0517	0.0557	0.0596	0.0636	0.0675	0.0714	0.0753
0.2	0.0793	0.0832	0.0871	0.0910	0.0948	0.0987	0.1026	0.1064	0.1103	0.1141
0.3	0.1179	0.1217	0.1255	0.1293	0.1331	0.1368	0.1406	0.1443	0.1480	0.1517
0.4	0.1554	0.1591	0.1628	0.1664	0.1700	0.1736	0.1772	0.1808	0.1844	0.1879
0.5	0.1915	0.1950	0.1985	0.2019	0.2054	0.2088	0.2123	0.2157	0.2190	0.2224
0.6	0.2257	0.2291	0.2324	0.2357	0.2389	0.2422	0.2454	0.2486	0.2517	0.2549
0.7	0.2580	0.2611	0.2642	0.2673	0.2704	0.2734	0.2764	0.2794	0.2823	0.2852
0.8	0.2881	0.2910	0.2939	0.2967	0.2995	0.3023	0.3051	0.3078	0.3106	0.3133
0.9	0.3159	0.3186	0.3212	0.3238	0.3264	0.3289	0.3315	0.3340	0.3365	0.3389
1.0	0.3413	0.3438	0.3461	0.3485	0.3508	0.3531	0.3154	0.3577	0.3599	0.3621
1.1	0.3643	0.3665	0.3686	0.3708	0.3729	0.3749	0.3770	0.3790	0.3810	0.3830
1.2	0.3849	0.3869	0.3888	0.3907	0.3925	0.3944	0.3962	0.3980	0.3997	0.4015
1.3	0.4032	0.4049	0.4066	0.4082	0.4099	0.4115	0.4131	0.4147	0.4162	0.4177
1.4	0.4192	0.4207	0.4222	0.4236	0.4251	0.4265	0.4279	0.4292	0.4306	0.4319
1.5	0.4332	0.4345	0.4357	0.4370	0.4382	0.4394	0.4406	0.4418	0.4429	0.4441
1.6	0.4452	0.4463	0.4474	0.4484	0.4495	0.4505	0.4515	0.4525	0.4535	0.4545
1.7	0.4554	0.4564	0.4573	0.4582	0.4591	0.4599	0.4608	0.4616	0.4625	0.4633
1.8	0.4641	0.4649	0.4656	0.4664	0.4671	0.4678	0.4686	0.4693	0.4699	0.4706
1.9	0.4713	0.4719	0.4726	0.4732	0.4738	0.4744	0.4750	0.4756	0.4761	0.4767
2.0	0.4772	0.4778	0.4783	0.4788	0.4793	0.4798	0.4803	0.4808	0.4812	0.4817
2.1	0.4821	0.4826	0.4830	0.4834	0.4838	0.4842	0.4846	0.4850	0.4854	0.4857
2.2	0.4861	0.4864	0.4868	0.4871	0.4875	0.4878	0.4881	0.4884	0.4887	0.4890
2.3	0.4893	0.4896	0.4898	0.4901	0.4904	0.4906	0.4909	0.4911	0.4913	0.4916
2.4	0.4918	0.4920	0.4922	0.4925	0.4927	0.4929	0.4931	0.4932	0.4934	0.4936

2.5	0.4938	0.4940	0.4941	0.4943	0.4945	0.4946	0.4948	0.4949	0.4951	0.4952
2.6	0.4953	0.4955	0.4956	0.4957	0.4959	0.4960	0.4961	0.4962	0.4963	0.4964
2.7	0.4965	0.4966	0.4967	0.4968	0.4969	0.4970	0.4971	0.4972	0.4973	0.4974
2.8	0.4974	0.4975	0.4976	0.4977	0.4977	0.4978	0.4979	0.4979	0.4980	0.4981
2.9	0.4981	0.4982	0.4982	0.4983	0.4984	0.4984	0.4985	0.4985	0.4986	0.4986
3.0	0.4987	0.4987	0.4987	0.4988	0.4988	0.4989	0.4989	0.4989	0.4990	0.4990
3.1	0.4990	0.4991	0.4991	0.4991	0.4992	0.4992	0.4992	0.4992	0.4993	0.4993
3.2	0.4993	0.4993	0.4994	0.4994	0.4994	0.4994	0.4994	0.4995	0.4995	0.4995
3.3	0.4995	0.4995	0.4995	0.4996	0.4996	0.4996	0.4996	0.4996	0.4996	0.4997
3.4	0.4997	0.4997	0.4997	0.4998	0.4997	0.4997	0.4997	0.4997	0.4997	0.4998
3.5	0.4998	0.4998	0.4998	0.4999	0.4998	0.4998	0.4998	0.4998	0.4998	0.4998
3.6	0.4998	0.4998	0.4999	0.4999	0.4999	0.4999	0.4999	0.4999	0.4999	0.4999
3.7	0.4999	0.4999	0.4999	0.4999	0.4999	0.4999	0.4999	0.4999	0.4999	0.4999
3.8	0.4999	0.4999	0.4999	0.4999	0.4999	0.4999	0.4999	0.4999	0.4999	0.4999
3.9	0.5000	0.5000	0.5000	0.5000	0.5000	0.5000	0.5000	0.5000	0.5000	0.5000

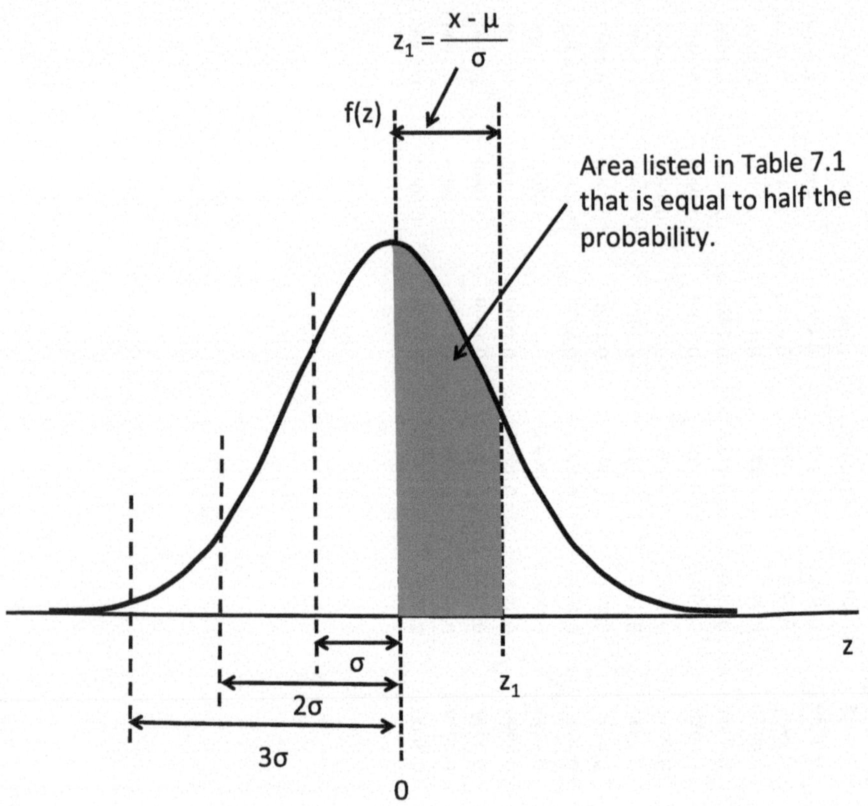

Fig. 8.11 The curve of distribution function $f(z)$ for calculating areas (and probabilities) under curve for various values of z

standard deviation of the population. Additionally, it is also useful to estimate characteristics about the probability distribution of the population using a sample. These types of analyses on samples depend on the number of the data points included in the sample. In the case of a sample with a large number of data points, it will be shown that the sample mean and sample standard deviation can be used as a good estimate of the population mean and population standard deviation, and the characteristics of the probability distribution can be estimated with a low level of uncertainty.

For large sample sizes (typically more than 30 samples), estimating the uncertainty of the sample mean as an approximation for μ depends on the central limit theorem. Specifically, if we take some number of measurements to form a sample, we can find the mean and standard deviation of that sample. Suppose we repeat this measurement process several times to create a mean and standard deviation for a set of sample measurements. It would be found that the average of these individual averages of sample measurements would display a dispersion around a central value that essentially would be equivalent to calculating the mean of the means. The central limit theorem states that if the number of measurements n for each sample is very large, the distribution of the mean values will be a normal distribution with a mean given by the sample mean and a standard deviation given by:

$$\sigma_X = \frac{\sigma}{\sqrt{n}}. \tag{8.16}$$

This equation provides a statement of how the standard deviation of the means of the samples decreases as n increases. Additionally, it can be shown that this equation is valid for large samples, that is, for large values of n, even if the distribution of the underlying population does not have a normal distribution.

Equation (8.16) allows us to estimate the uncertainty in using the sample mean as an approximation for μ. Using Eq. (8.12) from above, coupled with the knowledge that the sample mean has a normal distribution with a standard deviation given by σ_X, we can derive the following result:

$$z = \frac{\overline{x} - \mu}{\sigma_X} = \frac{\overline{x} - \mu}{\frac{\sigma}{\sqrt{n}}}. \tag{8.17}$$

Using Eq. (8.17), we can conclude that with a confidence level of $b\%$, the actual value of the population mean that falls within an interval around the value of the sample mean is given by:

$$\overline{x} - z_{\frac{b}{2}}\left(\frac{\sigma}{\sqrt{n}}\right) < \mu < \overline{x} + z_{\frac{b}{2}}\left(\frac{\sigma}{\sqrt{n}}\right). \tag{8.18}$$

However, one issue with Eq. (8.18) is that we may not know the population standard deviation σ. Nevertheless, if n is large, then we can reasonably assume that σ is approximately equal to the sample standard deviation S_x, thereby allowing the following to be written as:

$$\overline{x} - z_{\frac{b}{2}}\left(\frac{S_x}{\sqrt{n}}\right) < \mu < \overline{x} + z_{\frac{b}{2}}\left(\frac{S_x}{\sqrt{n}}\right). \tag{8.19}$$

Equation (8.19) enables the estimation of the uncertainty in using the sample mean as a representation of the population average, μ. Additionally, the standard deviation of the sample mean can be expressed by:

$$S_{\overline{x}} = \frac{S_x}{\sqrt{n}}. \tag{8.20}$$

The normal distribution is the most useful type of distribution, but it is important to note that there are other types of distributions that are possible depending on the nature of the relationship or dependences, if any, of the samples in the distribution.

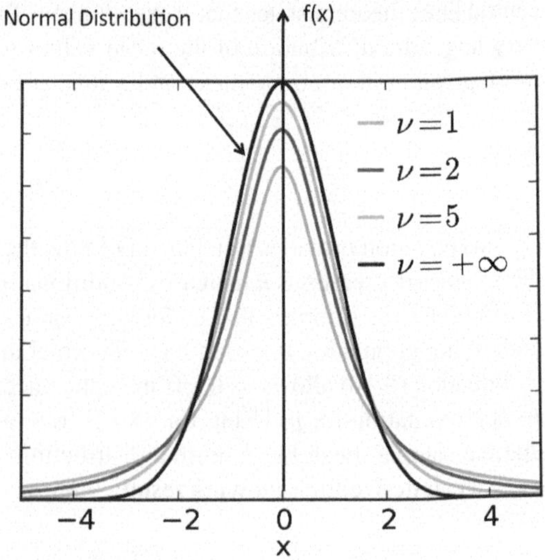

Fig. 8.12 The probability distribution function of the t-statistic for various values of ν

Among the more important types of other continuous distributions, we will only mention a few that will have usefulness to important practical situations for microsystems manufacturing.

One of the other important continuous distributions is the *Student's t-distribution*, which is useful in estimating the mean value of a normal population when the sample size is small. For small samples, usually defined as sample sizes less than 30 samples, the confidence levels can be estimated using the Student's t-distribution, given by [3]:

$$t = \frac{(\bar{x} - \mu)}{\left(\frac{s}{\sqrt{n}}\right)}. \tag{8.21}$$

Equation (8.21) assumes that the small sample size follows a normal distribution, which may or may not actually be the case.

The Student's t-distribution is a probability distribution similar to the normal distribution discussed above (see Fig. 8.12). Similar to the normal distribution, the probabilities using the t-distribution are calculated based on the area under the curve wherein the total area is equal to one. Like the z-distribution, the t-distribution is also symmetric about zero. Moreover, for sample sizes larger than about 30 samples, the t-distribution essentially becomes a normal distribution as can be seen in Fig. 8.12 as ν approaches ∞. The Student's t-distribution directly relies on the number of samples chosen through the degrees of freedom that is given by:

$$\nu = \nu - 1. \tag{8.22}$$

Table 8.3 Tabulated values for Student's t-distributions

ν	$t0.10,\nu$	$t0.05,\nu$	$t0.025,\nu$	$t0.01,\nu$	$t0.005,\nu$
1	3.078	6.314	12.706	31.821	63.656
2	1.886	2.920	4.303	6.965	9.925
3	1.638	2.353	3.182	4.541	5.841
4	1.533	2.132	2.776	3.747	4.604
5	1.476	2.015	2.571	3.365	4.032
6	1.440	1.943	2.447	3.143	3.707
7	1.415	1.895	2.365	2.998	3.499
8	1.397	1.860	2.306	2.896	3.355
9	1.383	1.833	2.262	2.821	3.250
10	1.372	1.812	2.228	2.764	3.169
11	1.363	1.796	2.201	2.718	3.106
12	1.356	1.782	2.179	2.681	3.055
13	1.350	1.771	2.160	2.650	3.012
14	1.345	1.761	2.145	2.624	2.977
15	1.341	1.753	2.131	2.602	2.947
16	1.337	1.746	2.120	2.583	2.921
17	1.333	1.740	2.110	2.567	2.898
18	1.330	1.734	2.101	2.552	2.878
19	1.328	1.729	2.093	2.539	2.861
20	1.325	1.725	2.086	2.528	2.845
21	1.323	1.721	2.080	2.518	2.831
22	1.321	1.717	2.074	2.508	2.819
23	1.319	1.714	2.069	2.500	2.807
24	1.318	1.711	2.064	2.492	2.797
25	1.316	1.708	2.060	2.485	2.787
26	1.315	1.706	2.056	2.479	2.779
27	1.314	1.703	2.052	2.473	2.771
28	1.313	1.701	2.048	2.467	2.763
29	1.311	1.699	2.045	2.462	2.756
30	1.31	1.673	2.042	2.457	2.75
∞	1.282	1.645	1.96	2.326	2.576

The t-statistic distribution has some similarity to the z-distribution discussed above, with areas under the distribution tabulated for various values of t and ν seen in Table 8.3. However, the t-distribution is used differently since the area, as given by α, is calculated by taking the integral from t to ∞ and represents the probability that t will have a value larger than that given for the value from the table for a sample size of $n = \nu + 1$. Since the t-distribution is symmetric about zero, α is also the probability that t has a value less than the negative of the value from the table. Using values of the area, α, and the level of confidence, c, in percent that the actual value of t does not reside in the area α is given by:

$$\text{Confidence } (c\%) = (1 - \alpha). \tag{8.23}$$

The level of confidence, $c\%$, that a small sample mean, $\bar{x}$, lies within the interval of the true population mean is given by:

$$\bar{x} - t_{\frac{\alpha}{2},\nu}\frac{S_x}{\sqrt{n}} < \mu < \bar{x} + t_{\frac{\alpha}{2},\nu}\frac{S_x}{\sqrt{n}}, \tag{8.24}$$

where $\alpha = (1 - c)$ and $\nu = (n - 1)$. Therefore, the uncertainty at a confidence level c % in the value of the small sample mean is given by:

$$P_x = t_{\frac{\alpha}{2},\nu}\frac{S_x}{\sqrt{n}}. \tag{8.25}$$

Another important probability distribution is the χ^2 (chi-squared) distribution. If a sample size of n is taken from a normal distribution that has variance of σ^2, the variance of the sample can be computed to be s^2 and can be used an estimate of σ^2. The interval estimate of σ^2 is computed as:

$$\chi^2 = \frac{s^2(n-1)}{\sigma^2}, \tag{8.26}$$

where χ^2 is the chi-squared statistic having a distribution with $\nu = n - 1$ degrees of freedom and ν is the divisor in the calculation of s^2. The curve of χ^2 is more complicated than the normal or Student's t-distributions but can be constructed by making a series of random samples of size n from a normal population, calculating the values of χ^2 for each sample, and plotting the results by drawing a line of the plotted χ^2 values. A series of curves of χ^2 for different values of ν is shown in Fig. 8.13. This distribution is skewed to the right as shown. As with other important distributions, the total area under this distribution is one, and as the number of degrees of freedom, ν, increases, the distribution approaches the shape of a normal distribution.

The probability of a random sample having a χ^2 value larger than a specified value is equivalent to the area to the right of the computed χ^2 value. It is usual practice to have χ_α^2 represent a value of χ^2 above the area of α as shown in Fig. 8.14. Table 8.4 provides the values of the chi-squared, χ^2, for various values of α and ν.

The area $1 - \alpha$ is shown in Fig. 8.15, wherein the following relationship can be written as:

$$P\left(\chi^2_{1-\alpha/2} < \chi^2 < \chi^2_{\alpha/2}\right) = 1 - \alpha, \tag{8.27}$$

wherein $\chi^2_{1-\alpha/2}$ and $\chi^2_{\alpha/2}$ are the values of the chi-squared distribution having $n - 1$ degrees of freedom and demarcating the boundaries of the area under the curve equal to $1 - \alpha$.

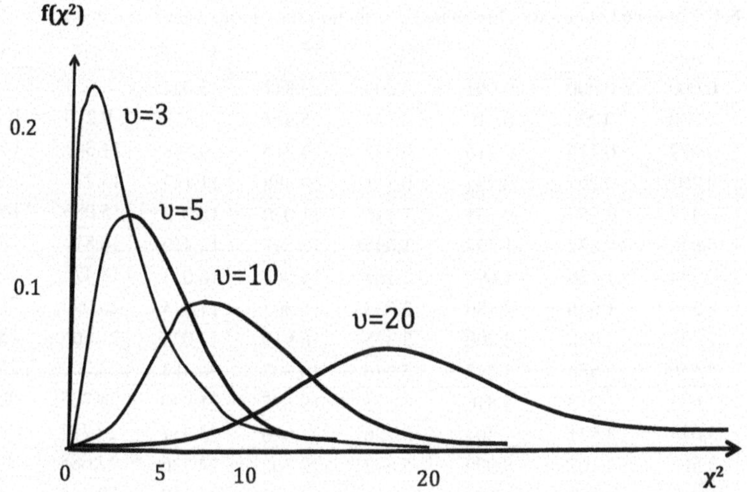

Fig. 8.13 The chi-squared distribution for four values of the degrees of freedom, ν. Note that as the degrees of freedom increases, the distribution more resembles a normal distribution

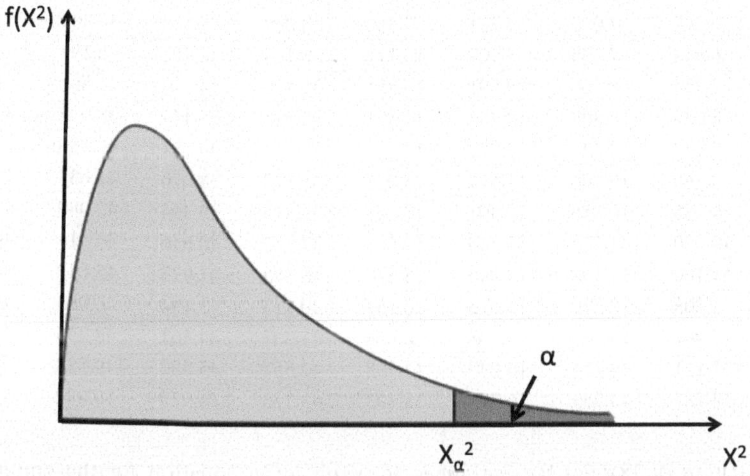

Fig. 8.14 The chi-squared distribution showing the area of α that represents the area of the curve above the value of χ_α^{2}

Equation (8.27) can be rewritten by substituting Eq. (8.26) from above to write the following:

$$P\left(\chi^2_{1-\alpha/2} < \frac{s^2(n-1)}{\sigma^2} < \chi^2_{\alpha/2}\right) = 1 - \alpha. \tag{8.28}$$

Table 8.4 Tabulated values for chi-squared distribution for various values of α and ν

ν	$X^2_{0.995}$	$X^2_{0.99}$	$X^2_{0.975}$	$X^2_{0.95}$	$X^2_{0.05}$	$X^2_{0.025}$	$X^2_{0.01}$	$X^2_{0.005}$
1	0.000	0.000	0.001	0.004	3.841	5.024	6.635	7.879
2	0.010	0.020	0.510	0.103	5.991	7.378	9.210	10.597
3	0.072	0.115	0.216	0.352	7.815	9.348	11.345	12.838
4	0.207	0.297	0.484	0.711	9.488	11.143	13.277	14.860
5	0.412	0.554	0.831	1.145	11.070	12.832	15.086	16.750
6	0.676	0.872	1.237	1.635	12.592	14.449	16.812	18.548
7	0.989	1.239	1.690	2.167	14.067	16.013	18.475	20.278
8	1.344	1.646	2.180	2.733	15.507	17.535	20.090	21.955
9	1.735	2.088	2.700	3.325	16.919	19.023	21.666	23.589
10	2.156	2.588	3.247	3.940	18.307	20.483	23.209	25.188
11	2.603	3.053	3.816	4.575	19.675	21.920	24.725	26.757
12	3.074	3.571	4.404	5.226	21.026	23.337	26.217	28.300
13	3.565	4.107	5.009	5.892	22.362	24.736	27.688	29.819
14	4.075	4.660	5.629	6.571	23.685	26.119	29.141	31.319
15	4.601	5.229	6.262	7.261	24.996	27.488	30.578	32.801
16	5.142	5.812	6.908	7.962	26.296	28.845	32.000	34.267
17	5.697	6.408	7.564	8.672	27.587	30.191	33.409	35.718
18	6.265	7.015	8.231	9.380	28.869	31.526	34.805	37.158
19	6.844	7.633	8.907	10.117	30.144	32.852	36.191	38.582
20	7.434	8.260	9.591	10.851	31.410	34.170	37.566	39.997
21	8.034	8.897	10.283	11.591	32.671	35.479	38.932	41.401
22	8.643	9.542	10.982	12.338	33.924	36.781	40.289	42.796
23	9.260	10.196	11.689	13.091	35.172	38.076	41.638	44.181
24	9.886	10.856	12.401	13.848	36.415	39.364	42.980	45.558
25	10.520	11.524	13.120	14.611	37.652	40.646	44.314	46.928
26	11.160	12.198	13.844	15.379	38.885	41.923	45.642	48.290
27	11.808	12.879	14.573	16.151	40.113	43.194	46.963	49.645
28	12.461	13.565	15.308	16.928	41.337	44.461	48.278	50.993
29	13.121	14.256	16.047	17.708	42.557	45.722	49.588	52.336
30	13.787	14.953	16.791	18.493	43.773	46.979	50.892	53.672

Equation (8.28) can be rearranged to result in an equation for the confidence interval of σ^2 as:

$$\frac{s^2(n-1)}{\chi^2_{\alpha/2}} < \sigma^2 < \frac{s^2(n-1)}{\chi^2_{1-\alpha/2}}), \tag{8.29}$$

where s^2 is the variance of a random sample of size n and $\chi^2_{\alpha/2}$ and $\chi^2_{1-\alpha/2}$ are values of $\chi^2$with $n - 1$ degrees of freedom and $\alpha/2$ and $1 - \alpha/2$ are the boundaries of the areas represented under the curve. The $(1 - \alpha)$ 100% confidence interval for the standard deviation can be found by taking the square root of σ^2. Consequently, with a sample standard having been calculated, Eq. (8.29) enables the determination of the

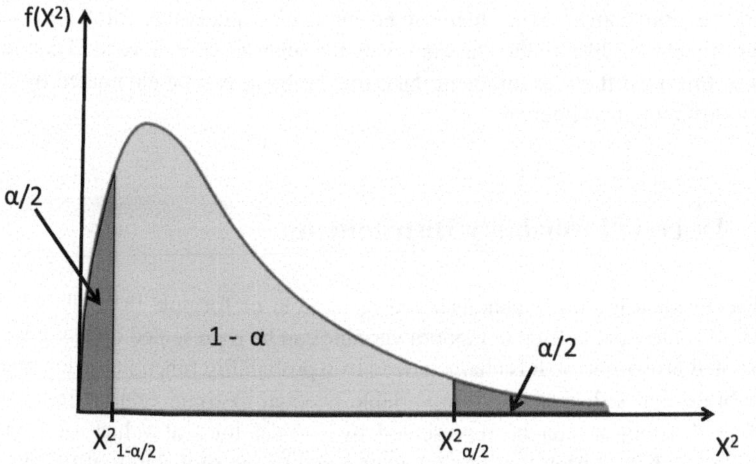

Fig. 8.15 The chi-squared distribution showing the area of $1 - \alpha$ that represents the area of the curve between $\chi^2_{\alpha/2}$ and $\chi^2_{1-\alpha/2}$

confidence interval for the population standard deviation using the standard deviation of the sample.

The chi-squared function is also used as a method to test the "goodness of fit" of experimental data. Specifically, it is used to determine whether a distribution can be considered to be a normal distribution within a specific confidence level. The chi-squared goodness of fit is defined as follows:

$$\chi^2 = \sum_{i=1}^{k} \frac{(O_i - E_i)^2}{E_i}, \tag{8.30}$$

where O_i is the observed frequency, E_i is the expected frequency, k is the total number of variables being analyzed, and $\nu = k - 1$. When the observed frequency values approach the expected frequency values, the value of the chi-squared distribution approaches zero, and therefore it can be stated that there is a good fit of the data to a normal distribution.

While the chi-squared distribution is often used for goodness of fit hypothesis testing, it should be noted that this method does not give a definitive answer about whether a distribution of data can be properly classified as a normal distribution. Instead, it can only provide a certain level of confidence that the data may be normally distributed. Importantly, this technique requires the exercise of prudent judgment in order to use most effectively. Also, the method is not well suited when the sample sizes are small.

A few other important points about using the chi-squared testing are as follows: First, the experimental values of O_i must be integer frequencies. That is, no events can occur as a fraction. Second, the frequency values of O_i for each class interval should be more than zero (i.e., there are no empty class intervals). Third, the use of χ^2 is questionable if 20% of the class intervals for either O_i or E_i have total counts less than 5. Provided there is sufficient data, this problem can be eliminated by developing different class intervals.

8.8 Discrete Probability Distributions

A discrete distribution is characterized by a table or formula that lists all of the possible values that a discrete random variable can be represented by along with the associated probabilities. It is characterized by a probability function, which gives the probability that a discrete random variable is equal to some value. Graphically, a discrete distribution can be represented by vertical lines at values of a random variable x_i wherein each line has a height equal to the probability at that value of x_i. Therefore, the example of manufacturing a production lot of 415 micro-beams began as a discrete distribution, and as the number of samples increased, the distribution started to resemble a continuous probability distribution. While the continuous normal distribution is useful for analyzing data, there are also circumstances where discrete distributions have utility, particularly concerning yield analysis. This section reviews some important discrete distributions [3].

One of the most useful discrete probability distributions is called the binominal distribution. The properties of a binominal experiment that generates this distribution are the following:

- There are n repeated trials in the experiment.
- Each trial has only two possible outcomes, sometimes referred to as a "success" or a "failure.[7]"
- The probability of each outcome remains constant.
- Repeated trials are independent of one another.

The binomial distribution is commonly used in manufacturing yield analysis wherein the dies are either accepted (i.e., a success) or rejected (i.e., a failure). The binomial experiment is one wherein each trial has a probability of success of p and a failure of $1 - p$; then the probability of obtaining x number of successes in n independent trails can be written as:

$$P(x) = \binom{n}{x} p^x (1-p)^{n-x}, \quad \text{for } x = 0, 1, 2, \ldots, n, \tag{8.31}$$

where the term $\binom{n}{x}$ is given by:

[7]These are called "Bernoulli" trials.

$$\binom{n}{x} = \frac{n!}{x!(n-x)!}. \tag{8.32}$$

The mean and variance of the binomial distribution are given by:

$$\mu = np \quad \text{and} \quad \sigma^2 = np(1-p). \tag{8.33}$$

A sometimes useful relationship is the sample fraction that is a failure or non-conforming given by:

$$\widehat{p} = \frac{x}{n}. \tag{8.34}$$

The probability distribution for $\widehat{p}$ based on the binomial distribution is given as:

$$P(\widehat{p} \leq a) = \sum_{x=0}^{na} \binom{n}{x} p^x (1-p)^{n-x}, \tag{8.35}$$

where $[na]$ is the greatest integer value less than or equal to na. The mean and variance of this distribution are as follows:

$$\mu(\widehat{p}) = p \quad \text{and} \quad \sigma^2(\widehat{p}) = \frac{p(1-p)}{n}. \tag{8.36}$$

A distribution related to the binomial distribution is called the Pascal distribution. It is also based on repeated trials with only two outcomes, either success or failure, wherein for a series of independent trials, each with a probability of success given by p, x is the number of trials required in order to obtain the rth occurrence of success. The Pascal probability distribution is given by:

$$P(x) = \binom{x-1}{r-1} p^r (1-p)^{x-r}, \tag{8.37}$$

where r is an integer, with $r \geq 1$ and $x \geq r$. The mean and variance of the Pascal distribution are:

$$\mu = \frac{r}{p} \quad \text{and} \quad \sigma^2 = \frac{e(1-p)}{p^2}. \tag{8.38}$$

The Pascal distribution has two special cases that are useful for yield analysis. The first is when $r = 1$ and results in what is called the geometric distribution. In this case, repeated trials are performed that can result in a success with a probability of p and failure with a probability of $1 - p$; therefore the probability distribution of the random variable x, representing the number of the trials when the first success happens, is given by:

$$P(x) = p(1-p)^{x-1} \quad \text{for } x = 1, 2, 3 \ldots. \tag{8.39}$$

The other special case is the negative binomial experiment that is one wherein repeated trials are conducted that can result in a success with a probability p and a failure with a probability of $q = 1 - p$. The probability distribution of the random variable x, which is the number of trials conducted before a success on the kth trial, is given by:

$$P(x) = \binom{x-1}{k-1} p^k (1-p)^{x-k} \quad \text{for } x = k, k+1, \ldots \tag{8.40}$$

The hypergeometric distribution is one where a population of size N contains k successes and $N - k$ failures. The probability distribution of a random variable x wherein the number of successes of a sample size of n is given by:

$$P(x) = \frac{\binom{k}{x}\binom{N-k}{n-x}}{\binom{N}{n}} \quad \text{for } x = 0.1, 2, 3 \ldots, n. \tag{8.41}$$

The mean and variance of the hypergeometric distribution are written as:

$$\mu = \frac{nk}{N} \quad \text{and} \quad \sigma^2 = \frac{(N-n)}{(N-1)} * n * \frac{k}{N}\left(1 - \frac{k}{N}\right). \tag{8.42}$$

Lastly, the Poisson distribution is useful in experiments that result in numerical values of the random variable x or the number of outcomes during a specific time interval or within a specified area. A Poisson experiment has the following properties:

- The number of outcomes occurring in a time interval or specified area is independent of the number of outcomes in other time intervals or areas.
- The probability that an outcome occurs during a short time interval or a small area is proportional to the time interval or size of area.
- The probability that an outcome occurs during a short time interval or a small area does not depend on the number of outcomes outside this time interval or area.
- The probability that more than one outcome can occur in a short time period or area is negligible.

The probability distribution for a random variable obeying the Poisson experimental criteria above wherein the number of outcomes occurs in a specified time interval or area is given as:

$$P(x) = \frac{e^{-\mu}\mu^x}{x!} \quad \text{for } x = 0, 1, 2, \ldots \tag{8.43}$$

where n is the mean number of outcomes in a specified time interval or area. The mean and variance of the Poisson distribution are:

$$\mu = \lambda \quad \text{and} \quad \sigma^2 = \lambda. \tag{8.44}$$

It can be shown using the binomial distribution with n approaching infinity and p approaching 0, where $np = \lambda$ is constant, that the binomial distribution approaches the Poisson distribution.

8.9 Some Examples of Statistical Analysis of Variations

Using the statistical methods described above, we can demonstrate some examples of how to employ these methods to make interferences about device parameter data collected during MEMS microsystems manufacturing.

8.9.1 *Confidence Interval for Manufacturing Large Samples (N > 30)*

As a first example, consider a scenario involving a microsystems manufacturing process whereby measurements are taken of the width of micro-beams fabricated using a surface micromachining process, such as discussed above. It is assumed that the production lot of devices is 415 and each device has one micro-beam. Therefore, the sample size is 415 micro-beams and thereby represents a large sample (i.e., >30 samples). The calculated mean value, $\bar{x}$, of the sample can be found as 5.01 microns, and the calculated standard deviation, S_x, is 0.21 microns as follows:

$$\bar{x} = 5.01 \text{ microns}, \quad \text{and} \quad S_x = 0.21 \text{ microns}. \tag{8.45}$$

Suppose we are interested in knowing the 95% confidence interval that the sample mean from the sample of 415 micro-beams represents the actual population mean.

The first step is to note that since the sample size is well above 30, we can use the z-statistics for this analysis. Next, we calculate the desired value of the confidence interval divided by two or 0.95/2 which equals 0.475 since the z-statistic is two-sided. The value of $z_{0.475}$ is determined from Table 8.2 as $z_{0.475} = 1.96$. We assume that the sample standard deviation is equal to the population standard deviation, that is, $S_x = \sigma_x$. Therefore, using Eq. (8.19), it can be stated that the population mean will fall in the following range with a confidence interval of 95%:

$$\mu = 5.01 \pm 1.96 \frac{S_x}{\sqrt{n}} \text{microns} = 5.01 \pm 1.96 \frac{0.21}{\sqrt{415}} \text{microns}, \quad \text{or} \tag{8.46}$$

$$\mu = 5.01 \pm 0.02 \text{ microns (95\% confidence interval).} \tag{8.47}$$

This is a useful result since it shows that we can estimate the level of uncertainty based on a number of sample measurements of the manufactured beam widths. Moreover, this assertion can be made based on the sample and not the population.

Suppose that instead of a 95% confidence interval, a 99% confidence interval was desired for the same micro-beam data from above. Proceeding as before and dividing 0.99 by 2 results in 0.495. We can determine the value of $z_{0.495}$ from Table 8.2 as $z_{0.495} = 2.575$. The population mean will fall in the following range with a confidence interval of 99%:

$$\mu = 5.01 \pm 2.575 \frac{S_x}{\sqrt{n}} \text{microns} = 5.01 \pm 2.575 \frac{0.21}{\sqrt{415}} \text{microns}, \quad \text{or} \tag{8.48}$$

$$\mu = 5.01 \pm 0.03 \text{ microns (99\% confidence interval).} \tag{8.49}$$

This is a larger range than what was found for the 95% probability based on the same sample size. This is expected since the larger confidence level necessitates a larger range of possible values of the population mean.

Now suppose that the sample size is 31 micro-beams (i.e., $n = 31$) instead of 415 micro-beams, and repeat the calculation to determine the 95% confidence interval that the sample mean represents the population mean. As before, the value of $z_{0.475}$ is determined from Table 8.2 as $z_{0.475} = 1.96$. Using Eq. (8.19) the population mean will fall in the following range with a confidence interval of 95%:

$$\mu = 5.01 \pm 1.96 \frac{S_x}{\sqrt{n}} \text{microns} = 5.01 \pm 1.96 \frac{0.21}{\sqrt{31}} \text{microns}, \quad \text{or} \tag{8.50}$$

$$\mu = 5.01 \pm 0.07 \text{ microns (95\% confidemce interval).} \tag{8.51}$$

Comparing this result to Eq. (8.47) above, it can be seen that the 95% confidence interval results in a larger range compared with a much larger sample size of 415 beams. This is expected since the number of samples is much smaller, and therefore the amount of uncertainty in the sample mean being equal to the population mean would be expected to be greater.

8.9.2 Confidence Interval for Small Samples (N < 30)

The examples above all employed large sample sizes ($n > 30$ samples). The next example assumes that the sample size is small and is composed of less than 30 samples. In the case of small samples, the Students t-distribution is used. The underlying assumption when using the Student's t-distribution is that the underlying population has a normal distribution.

Consider the following set of data composed of nine samples of the measured micro-beam widths shown in Table 8.5 wherein the micro-beams have been fabricated.

The 95% confidence interval of the population mean is calculated. First, the mean and standard deviation of the sample with $n = 9$ are calculated as follows:

$$\bar{x} = 5.03 \text{ microns}, \quad \text{and} \quad S_x = 0.39 \text{ microns}. \tag{8.52}$$

It is recalled that the confidence interval, represented by c, is given by $(1 - \alpha)$. Therefore, in our example α is equal to $(c - 1)/2$ or 0.025, wherein $(c - 1)$ is divided by 2 since the t-statistic is two-sided. Using the Student's t-distribution and Table 8.3, wherein $\nu = n - 1 = 8$, we determine that $t_{\alpha/2,\nu} = t_{0.025,8} = 2.306$. Therefore, using Eq. (8.24), we can write:

$$\bar{x} - (2.306)\frac{(0.39)}{\sqrt{9}} < \mu < \bar{x} + (2.306)\frac{(0.39)}{\sqrt{9}}, \quad \text{or} \tag{8.53}$$

$$\mu = 5.03 \pm 0.30 \text{ microns (95\% confidence interval)}. \tag{8.54}$$

Comparing Eq. (8.54) to the confidence intervals calculated for large sample sizes, it can be seen that when the sample size is smaller, the range over which the true population mean is expected to reside with a 95% confidence interval has gotten considerably larger. This is expected since the level of uncertainty about the value of the population mean for a small sample is higher than for a larger one.

8.9.3 Hypothesis Testing for Small Sample Sizes (N < 30)

Table 8.5 Measured micro-beam widths

5.35 microns	5.04 microns	4.86 microns
4.72 microns	5.14 microns	4.69 microns
4.44 microns	5.58 microns	5.46 microns

Hypothesis testing is used in manufacturing to make decisions about device parameter sample data sets. Typically, hypothesis testing is performed by analyzing a first hypothesis called the null hypothesis, followed by a second hypothesis called the alternative hypothesis. Hypothesis testing can be used for many statistical parameters. For example, the null hypothesis for the mean can be written as follows:

$$H_o : \mu = \mu_o, \tag{8.55}$$

where μ_o is some specified value of the mean. The alternative hypothesis employs development of a set of hypotheses to attempt to demonstrate a useful answer or outcome. In general, the alternative hypothesis can involve three possible outcomes:

A. Two-tailed hypothesis test is used to determine if the population mean, μ, is different from some specific value of mean, μ_o, to be tested:

$$\mathrm{Ha} : \mu \neq \mu_o. \tag{8.56}$$

The rejection and acceptance ranges of the two-tailed hypothesis testing are illustrated in Fig. 8.16. It should be noted that Fig. 8.16 is based on the Student's t-distribution wherein the degrees of freedom is given by $\nu = n - 1$ and the confidence level is equal to $1 - \alpha$.[8]

B. The right-tailed hypothesis test is used to determine if the population mean, μ, is greater than some specific value of mean, μ_o, to be tested:

$$\mathrm{Ha} : \mu > \mu_o. \tag{8.57}$$

The rejection and acceptance ranges of the right-tailed hypothesis testing are illustrated in Fig. 8.17.

C. The left-tailed hypothesis test is used to determine if the population mean, μ, is less than some specific value of mean, μ_o, to be tested:

$$\mathrm{Ha} : \mu < \mu_o. \tag{8.58}$$

The rejection and acceptance ranges of the left-tailed hypothesis testing are illustrated in Fig. 8.18.

As an example of the use of hypothesis testing, let us use the beam width measurement data shown in Table 8.5 that has a sample size of 9. Suppose it is desired to determine if the sample of beam widths comes from a population whose true mean is greater than 5.00 microns with a confidence level of 95%. This would be

[8]Hypothesis testing can also be used on large samples wherein the procedure is the same as for small samples except that z_α is used to replace $t_{\alpha,\nu}$.

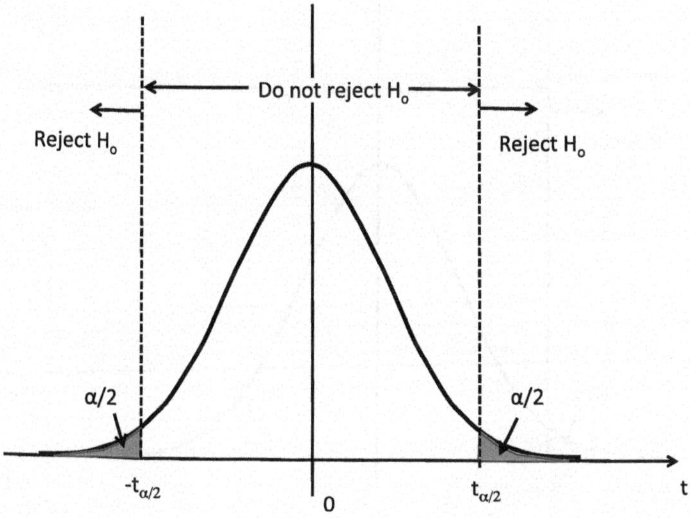

Fig. 8.16 Two-tailed hypothesis test

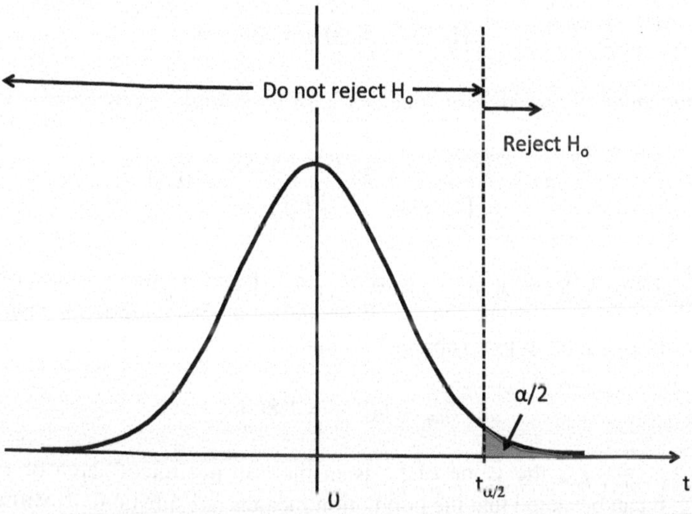

Fig. 8.17 Right-tailed hypothesis test

a hypothesis test using the null hypothesis and right-tailed hypothesis tests with μ_o replaced by 5.00 microns, as follows:

$$H_o : \mu = 500 \text{ microns, and} \tag{8.59}$$

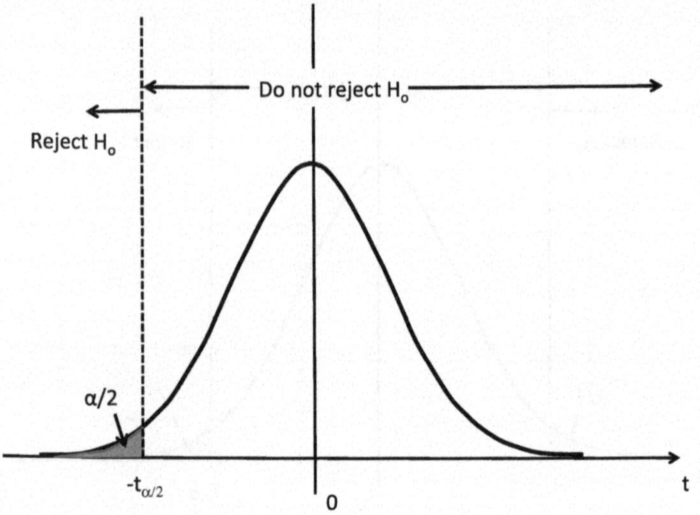

Fig. 8.18 Left-tailed hypothesis test

$$\text{Ha}: \mu > 5.00 \text{ microns}. \tag{8.60}$$

First the value of t_{ds} ("ds" for "data set") for the sample is calculated, as follows:

$$t_{ds} = \frac{(\overline{x} - \mu_o)}{\left(\frac{s}{\sqrt{n}}\right)} = \frac{(5.03 - 5.00)}{\left(\frac{0.39}{\sqrt{9}}\right)} = 0.23. \tag{8.61}$$

The degrees of freedom is given by $\nu = n - 1$ and in this example is equal to 8. This allows us to calculate the $t_{\alpha,\nu}$. If we desire a 95% confidence interval, then $\alpha = 0.05$, allowing us to use Table 8.2 to find:

$$t_{\alpha,\nu} = t_{0.05,8} = 1.860. \tag{8.62}$$

Since $t_{0.05,8} > t_{ds}$, the value of t_{ds} is in the "do not reject" area of Fig. 8.18. Therefore, it can be stated that the population mean is not substantially different than 5.00 microns within a 95% confidence interval.

Hypothesis testing is also useful for comparing certain statistical parameters for different sets of samples. For example, it can be used to compare two data sets based on the means of the two samples. If the sample sizes are both small (i.e., <30 samples), the value of t for the means of the two samples is written as follows:

$$t = \frac{(\bar{x}_1 - \bar{x}_2)}{\sqrt{\left(\frac{S_1^2}{n_1}\right) + \left(\frac{S_2^2}{n_2}\right)}}, \tag{8.63}$$

where $\bar{x}_1$, S_1, and n_1 are the mean, standard deviation, and number of data in the respective sample 1 and $\bar{x}_2$, S_2, and n_2 are the mean, standard deviation, and number of data in the respective sample 2. The calculated value of t for the two samples from Eq. (8.63) is compared to the interval $\pm\, t_{\alpha/2}$ using Table 8.3 for a confidence interval given by $(1 - \alpha)$. The degrees of freedom, ν, is found using the following:

$$\nu = \frac{\left[\left(\frac{S_1^2}{n_1}\right) + \left(\frac{S_2^2}{n_2}\right)\right]^2}{\frac{\left(\frac{S_1^2}{n_1}\right)^2}{n_1 - 1} + \frac{\left(\frac{S_2^2}{n_2}\right)^2}{n_2 - 1}}. \tag{8.64}$$

When calculating ν using Eq. (8.64), it is important to remember to round the number down to the nearest integer value.

8.9.4 Hypothesis Testing of Goodness of Fit

The chi-squared distribution can be used on measured manufacturing parameter data to determine if it can be considered to be normally distributed. We again examine the parameter sample of 415 surface micromachined micro-beams. The bins on either end of the distribution in Table 8.1 have less than five entries and therefore are combined with the neighboring bins as shown in Table 8.6. The chi-squared distribution is calculated using Eq. (8.30). However, the expected values of frequency, E_i, must be determined first.

The expected values of frequency are found using Eq. (8.17) to find the values of z for the boundaries of the class intervals and then finding the area between adjacent boundaries using these calculated z values. For example, in calculating the area in the second class interval (4.55–4.65 microns), the z values of the boundaries on either side of the class interval are determined as follows:

$$z_1 = \frac{(x_{4.55} - \bar{x})}{S_x} = \frac{(4.55 - 5.01)}{.21} = -2.19, \text{ and} \tag{8.65}$$

$$z_2 = \frac{(x_{4.65} - \bar{x})}{S_x} = \frac{(4.65 - 5.01)}{.21} = -1.71. \tag{8.66}$$

The area between these two z boundary points is found using Table 8.2 as follows:

Table 8.6 Class intervals, midpoints, observed frequencies, and expected frequencies for the measured surface micromachined micro-beam widths

Class interval	Midpoint	Observed frequency	Expected frequency
4.45–4.55	4.50	7	5.81
4.55–4.65	4.60	9	12.16
4.65–4.75	4.70	29	26.52
4.75–4.85	4.80	49	48.18
4.85–4.95	4.90	68	67.35
4.95–5.05	5.00	80	78.60
5.05–5.15	5.10	71	71.92
5.15–5.25	5.20	52	51.59
5.25–5.35	5.30	31	30.92
5.35–5.45	5.40	11	14.20
5.45–5.55	5.50	8	7

$$\begin{aligned}\text{Area} &= P(-2.19 < z < -1.71) \\ &= P(z < -1.71) - P(z < -2.19) \\ &= 0.0436 - 0.0143 = 0.0293.\end{aligned} \tag{8.67}$$

The expected frequency is then found by multiplying the result from Eq. (8.67) by the total number of samples, 415, to find:

$$E_2 = (0.293)(415) = 12.16. \tag{8.68}$$

This calculation for the expected frequencies, E_i, is repeated for each of the class intervals, and the results are shown in Table 8.6.

Using the calculated values of expected frequencies, we can determine the chi-squared value to be:

$$\chi^2 = \sum_{i=1}^{k} \frac{(O_i - E_i)^2}{E_i} = \frac{(7 - 5.81)^2}{5.81} + \frac{(9 - 12.16)^2}{12.16} + \frac{(8 - 7)^2}{7} = 2.30. \tag{8.69}$$

The degrees of freedom for this example is the number of intervals (11 as shown in Table 8.6) minus three since the total frequency, sample mean, and sample deviation were all employed to find the expected frequencies. Therefore, the degrees of freedom is $\nu = n - 1 = 11\text{–}3 = 8$. We need to determine the chi-squared value from Table 8.4 when $\nu = 8$. We assume that we are determining the goodness fit of the beam width data to a confidence interval of 95%. Therefore, $\alpha = 0.05$ and ${\chi_{0.05}}^2 = 2.733$ when $\nu = 8$. Since the computed value of χ^2 from Eq. (8.69) above is less than ${\chi_{0.05}}^2$, it can be concluded that the null hypothesis is not to be rejected, and a normal distribution is a good fit (within a confidence interval of 95%) of distribution of measured beam widths.

8.9.5 *Sample Size Required to Estimate Population Mean*

The examples of using statistical analysis on manufacturing samples of micro-beams reviewed above allowed the parameters of the population to be estimated within a certain specified confidence interval. However, in each case the assumption was that the sampling for the parameter data had already taken place. An important question is how large does a sample need to be in order for the confidence interval to have a desired margin of error? This is a more realistic scenario since often in a microsystems manufacturing production situation, a method to be able to determine an adequate sample size to be able to estimate statistical parameters within some level of confidence is needed. This section shows how the sample size can be determined [4].

Since the standard deviation of the population is unknown, the Student's t-distribution is used. The standard deviation of the population is estimated using the standard deviation of a sample that must be obtained using either a prior analysis or from performing a small sampling study. Equation (8.25) can be rearranged to find the sample size as follows:

$$n = \left(t_{\frac{\alpha}{2},\nu} \left(\frac{S_x}{P_x} \right) \right)^2. \tag{8.70}$$

The problem with Eq. (8.70) is that since n is unknown, the degrees of freedom, ν, is also unknown and therefore the value of t cannot be determined.

The solution to this problem is to use what is called the iterative method wherein the t-statistic is replaced with the z-statistic to develop an estimated sample size from the z-statistic. Then the sample size is recomputed using the t-statistic with the degrees of freedom estimated by equating it to the sample size minus 1. The number of samples computed using the z- and t-statistics should be relatively close in value. However, if these numbers are different, then use the calculated number from the t-statistic to compute the number of samples again. Additionally, if the calculated sample size is less than 30, the sample size should be increased to at least 30 samples unless it is known that the population has a normal distribution.

For example, suppose we need to determine the number of samples that would be needed to estimate the population mean of a microsystems dimension to within ±0.1 microns with a 90% level of confidence. The first step would be to perform a small study to determine the standard deviation of a sample. Assume that this sample contained 31 samples and had a calculated standard deviation of 0.35 microns. Using the z-statistic, the number of samples is found by:

$$n = \left(z_{\frac{b}{2}} \frac{S_x}{P_x} \right)^2, \tag{8.71}$$

where P_x is the uncertainty. For a 90% confidence, we can write $1 - b = 0.90$, or $b = 0.10$, and $b/2 = 0.05$. Therefore, using Table 8.2, the $z_{0.05}$ is found to be 1.645. Substituting values for $z_{0.05}$, S_x, and P_x into Eq. (8.71) allows us to estimate that the number of samples is $n = 33.15 \approx 34$.

This preliminary approximation of the sample size, n, is used to estimate the degrees of freedom ν which is set equal to $n - 1 = 33$. Therefore, using a t-statistic table, the value of $t_{0.05,33}$ is determined as 1.692. The number n can be calculated using Eq. (8.70) and found to be equal to 35.07, which is rounded up to 36 samples. Therefore, the minimum number of samples required in order to achieve a confidence level of 90% with an uncertainty of ±0.1 microns is 36 samples.

8.9.6 *Example of Use of the Hypergeometric Distribution*

Suppose it is assumed that there is a pilot production run of 10 wafers, each containing 50 dies, for a total number of 500 dies in the run. Suppose it is known that there are 2 dies per wafer that are defective. Based on this, if 10 dies are selected at random from the production run, what is the probability of finding at least 1 defective die?

This example uses the hypergeometric distribution wherein $N = 500$ and $n = 10$. The total number of defective die per production run is 20, and therefore, $k = 20$. To determine the probability of finding at least 1 die that is defective, Eq. (8.41) is written as:

$$P(x \le 1) = P(0) + P(1) = \frac{\binom{50}{0}\binom{450}{50}}{\binom{500}{10}} + \frac{\binom{50}{1}\binom{450}{49}}{\binom{500}{10}} = 0.281. \tag{8.72}$$

That is, there is a 28.1% probability that we will select 1 defective die in the sample of 10 dies selected.

8.9.7 *Example of Poisson Distribution*

It has been determined on a production line that there are on average ten defects per substrate. What is the probability that a randomly selected substrate will have three or fewer defects?

Applying the Poisson equation from above as:

$$P(x \le 3) = \sum_{x=0}^{3} \frac{e^{-10}10^x}{x!} = 0.0103 = 1.03\% \tag{8.73}$$

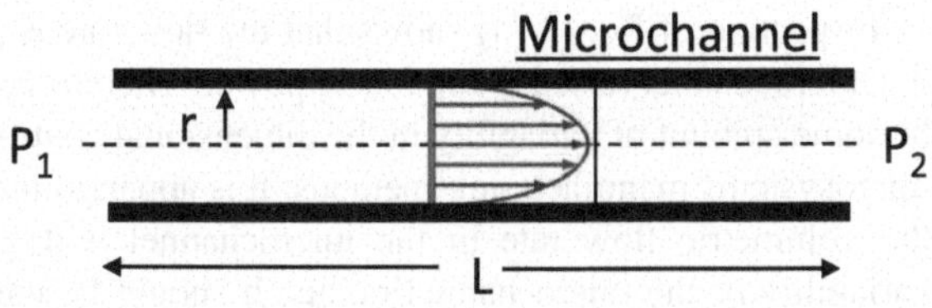

Fig. 8.19 Cross section of a fluid microchannel with differential pressure $\Delta P = P_1 - P_2$, radius r, and length L

Therefore, there is a small probability, namely, 1.03%, that any substrate randomly selected will have three or fewer defects.

8.10 Impact of Physics and Random Parameter Variations

Examination of the equations involving the physical laws that are used to describe, model, and predict a microsystems device behavior indicates that various device input parameters are often raised to higher powers. This is especially the case for many MEMS microsystems. This section explores this impact in more detail.

As an example, consider the volumetric fluid flow rate in a relatively simple microsystems microfluidic device, namely, a microchannel flow restrictor. In microfluidic devices, the Hagen-Poiseuille equation is commonly employed to relate the volumetric flow rate of a fluid in a conduit to the pressure differential, conduit dimensions, and fluid properties.[9] This equation assumes that the flow path is long and cylindrically shaped[10] and has a constant cross section (see Fig. 8.19). It is also further assumed that the fluid is incompressible and Newtonian and has laminar flow characteristics. These assumptions are usually applicable for microfluidic devices handling liquids. The Hagen-Poiseuille for a fluid transport in a microchannel is commonly written as [5]:

$$Q = \frac{\pi r^4 \Delta P}{8\mu L}, \tag{8.74}$$

where ΔP is the differential pressure across the microchannel (equal to $P_1 - P_2$ in Eq. (8.74), μ is the fluid dynamic viscosity, L is the microchannel length, Q is the volumetric flow rate, and r is the radius of the microchannel.

[9]The Hagen-Poiseuille equation is derived from the more fundamental Navier-Stokes equations that are used to describe the motion of viscous fluids. The Navier-Stokes are based on application of Newton's second law to fluid motion combined with an assumption that the stress in the fluid is the summation of a diffusing viscous term that is proportional to the fluid velocity gradient and the fluid pressure.

[10]It may seem that the condition of a cylindrical conduit would exclude this law for use on modeling microfluidic conduits since fabrication of a cylindrical conduit would be difficult. However, it should be noted that correction factors to take into account the conduit cross-sectional shape are available. For simplicity we will ignore these correction factors in the present discussion.

Inspection of Eq. (8.74) shows that the flow rate is proportional to the radius of the microchannel raised to the fourth power. Therefore, given the fact that there will be some amount of variability in the dimensional control of device parameters using microsystems manufacturing methods, it is apparent that the amount of variation of the volumetric flow rate in the microchannel will be highly dependent on the variability of the microchannel radius. It should be noted that the flow rate is also dependent on the inverse of the length of the microchannel and the fluid viscosity to the first powers. Consequently, depending on the relative magnitudes of variations of these parameters, they may also have an impact on the overall variability of the microchannel flow rate.

A simplified manufacturing process for implementing fluid microchannels is shown in cross-sectional form in Fig. 8.20. The process begins with a blank substrate that could be single-crystal silicon. The substrate has a masking thin-film material layer deposited as shown in cross-sectional form in Fig. 8.20a. This layer could be composed of silicon dioxide, silicon nitride, or any other suitable material layer that can be used as a masking layer. A layer of photoresist is deposited and patterned using optical photolithography as shown in Fig. 8.20b. Next, the thin-film material layer is etched to expose the silicon substrate surface underneath. This etching could be performed using either a dry plasma reactive ion etching or wet etching. The photoresist layer is then stripped. Next, the exposed silicon substrate is etched using an appropriate etch method, such as wet anisotropic etching or deep, reactive ion etching (DRIE), as seen in Fig. 8.20c. The thin-film material layer used as a masking layer is then removed, Fig. 8.20d. Finally, a glass substrate, with previously patterned or machined through holes to act as a fluidic inlet and outlet ports, is anodic bonded to the top surface of the silicon substrate to complete the fabrication as shown in Fig. 8.20e.

It is important to note that the manufacturing process outlined in Fig. 8.20 provides for different options for some of the fabrication steps. Specifically, the thin-film material layer that acts as a masking layer could be etched using either dry or wet etching. Additionally, the silicon substrate could be etched using a dry plasma DRIE technique or using anisotropic wet etching such as KOH. As noted in the previous chapters on fabrication processes, dry plasma etching almost always allows much better dimensional control of the features than does wet etching. Therefore, it is expected that there will be differences in the magnitudes of the variations of the microchannel dimensional parameters using these two fabrication methods.

The impact of the device physics and microchannel dimensional variations on the device behavior will now be examined. It will be assumed that the manufacturing process outlined in Fig. 8.20 is used to make a quantity of five (5) microchannel devices. Further, as part of the manufacturing process, measurements of the radii of the fabricated conduits that have been produced are shown in Table 8.7. The features reported in this table were based on using dry plasma etch processes rather than wet etching techniques, and therefore the dimensional accuracy and precision are expected to be relatively good. Although the sample size is small, $n = 5$, this is to keep the complexity of the calculations minimized for demonstration purposes.

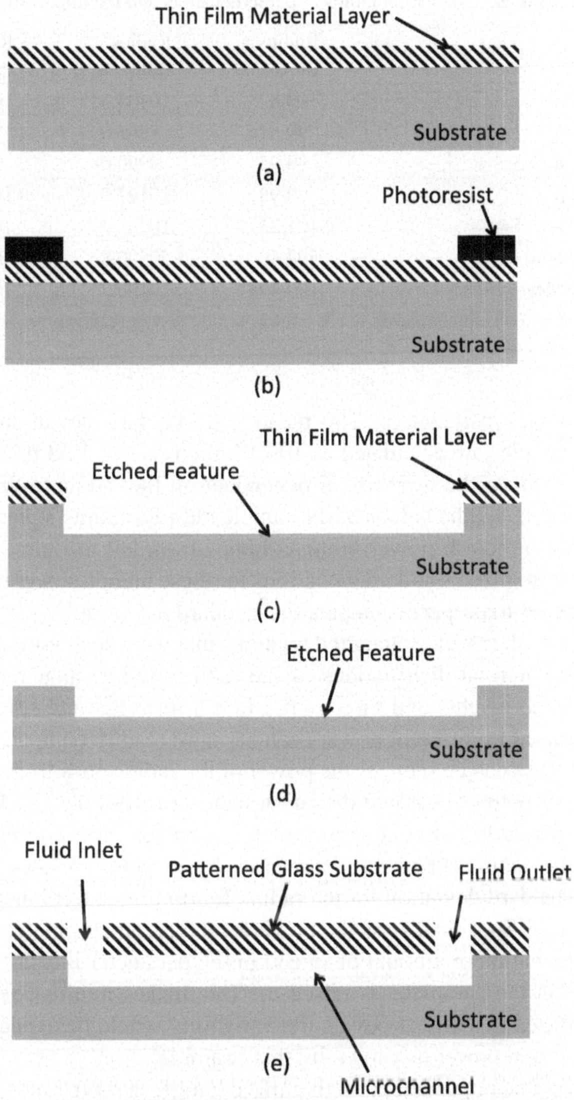

Fig. 8.20 Cross section of substrates illustrating the manufacturing process to implement the microchannel devices

Based on the engineering device design, the desired radius of the conduit is 100.00 microns,[11] and the five samples show some amount of variation from the desired value. We assume that statistical methods can be applied to the measured

[11]The desired microchannel radius would likely be the result of a desired flow resistance value of the device, and this flow resistance value would be derived from the application specification. In other words, the 100.00 micron microchannel radius would not be arbitrarily selected but instead derived from the device output response.

Table 8.7 Measured radius of a microchannel and the impact of raising the radii to higher powers

	Radius (microns)	Radius^2 (microns^2)	Radius^3 (microns^3)	Radius^4 (microns^4)
$n = 1$	100.52	10104.3	1015681.3	102096280.3
$n = 2$	99.81	9962.0	994310.8	99242163.26
$n = 3$	99.95	9990.0	998500.7	99800149.95
$n = 4$	100.98	10197.0	1029689.1	103978001.4
$n = 5$	101.23	10247.5	1037355.7	105011520.6
Sum=	502.49	50500.8	5075537.6	510128115.6
Mean=	100.498	10100.2	1015107.5	102025623.1
Normalized standard deviation	0.62	1.25	1.88	2.52

device parameters. The mean and standard deviation of the radius for the five samples are calculated as 100.14 microns and 1.63 microns, respectively.

Since the flow rate is dependent on the radius to the fourth power, we show in Table 8.7 the values of the sample radii for radius-squared, radius-cubed, and radius to the fourth power in the columns from left to right. Additionally, the means and normalized standard deviations for these higher-power radii terms are also shown. In order to properly compare the standard deviations on an equivalent basis, the bottom row shows the standard deviations that have been normalized. This results in each of the normal distributions of the radii raised to their respective powers to have the same heights and thereby put into a form that the spreads in the radius parameter raised to different powers can be more readily compared to one another.

As can be seen, as the power of the radius goes from radius to radius-squared, the normalized standard deviation increases from 0.62 to 1.25. That is, the normalized standard deviation approximately doubles. The normalized standard deviation for the radius-cubed term increases by a factor of about three, and the normalized standard deviation for the radius-fourth power term increases by a factor of roughly four. Specifically, as the power of the parameter increases, the normalized standard deviation or amount of spread in the parameter increases in proportion to the power that the parameter is raised to. This makes intuitive sense that as the parameter is raised to a higher power, the variability would be expected to increase by a multiple of that power as shown by this example.

The important point that this example demonstrates is that the random variations that are present in all manufactured elements can have an outsized impact on the resultant variation depending on the physics of the equations used to describe the device behavior. It is important to remember that the equation describing the device behavior usually contains other parameters that also have an impact on the resultant device behavior. For example, the flow rate of the microchannel device is also proportional to one over the length of the channel, L, and one over the fluid viscosity, μ.

The example of the microfluidic flow channel where the flow rate is proportional to the radius to the fourth power is not unusual in MEMS microsystems. Many

MEMS devices have physical equations describing their behavior where one of more parameters is raised to higher powers. For example, MEMS pressure sensor devices employ a square-shaped diaphragm as a mechanically compliant structural element that is important to the device performance. Under uniform pressure loading, the maximum deflection, w, of a square membrane, with an edge length of 2a, is at the center and is approximately given by [6]:

$$w \approx \frac{Pa^4}{47D}, \tag{8.75}$$

where P is the applied uniform pressure loading and D is the flexural rigidity of the membrane which is given by:

$$D = \frac{Et^3}{12(1 - \nu^2)}. \tag{8.76}$$

In Eq. (8.76), E is the Young's modulus of the material from which the diaphragm is made, t is the thickness of the diaphragm, and ν is the Poisson's ratio of the diaphragm material. As can be seen from these equations, the deflection of the diaphragm is proportional to the diaphragm edge length to the fourth power and inversely proportional to the thickness of the diaphragm cubed. Consequently, it would be expected that the behavior of the diaphragm would be highly dependent on the amount of manufacturing variations present in both of these dimensions. Many MEMS microsystems have these types of parameter raised to higher power relationships, and consequently the behavior of MEMS devices is strongly impacted by the resultant manufacturing parameter variations.

8.11 Combination of Both Bias and Random Manufacturing Parameter Variations

As noted above, in microsystems manufacturing circumstances, both bias and random variations may be present. These variations can be represented as shown in Fig. 8.21.

The bell-shaped curve is shown representing the distribution of random variations related in some parameter of the microsystems device (e.g., width or length of a micro-beam, radius of a microfluidic flow channel, etc.). Additionally, due to the presence of a bias variation, the normal distribution in Fig. 8.21 is shifted to the right from the "desired" design value of the parameter. For a specific sample point measurement of a parameter represented by the vertical line at the right, the random variation is 2σ, and the total variation of a given sample point is the sum of both the bias and random variation.

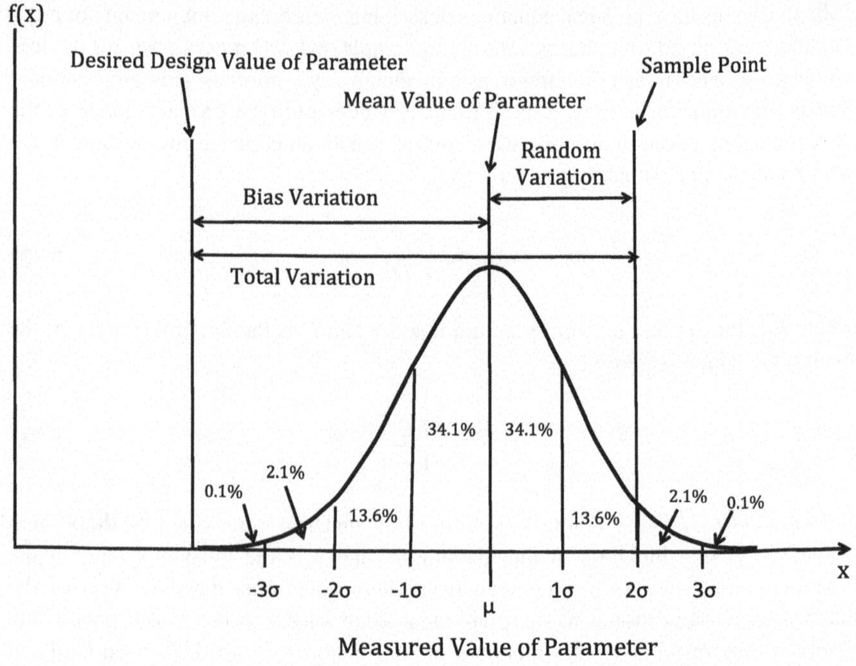

Fig. 8.21 An illustration of the two types of variations that are typically found in manufacturing processes. The random variations are statistical in nature and can be presented as a normal bell-shaped distribution, and the bias variation causes the resultant shift of the normal distribution from the true or desired value. The total variation is the sum of the bias and random variations

One issue is how to combine the effects of random and bias variation to determine the total variation in a parameter. As it turns out, simply adding the bias and random variations together to estimate the total variation usually calculates a total variation that is too large. More often these two sources of variation are combined to find the total variation, E_x, using a root mean square function as follows:

$$E_x = \sqrt{\left(B_x^2 + P_x^2\right)}, \tag{8.77}$$

where B_x is the bias variation and P_x is the random variation. This equation assumes that the random and bias variations are independent of one another. If the bias and random variations are not independent, then the variations are added to estimate the total variation present.

8.12 Device Output Behavior Variation Due to Parameter Variations

In this section, a method to estimate the variations in the output behavior of a device based on the combined variations in the device's parameters is developed. This analysis directly relates to the discussion above regarding the impact of variables that are raised to higher powers in the physical equations employed to describe microsystems device behavior.

Suppose there is a physical relationship that describes the output behavior of a MEMS microsystems device that is a function of n independent variables, $x_1, x_2, x_3, \ldots x_n$, that can be written as:

$$y = f(x_1, x_2, x_3 \ldots x_n). \tag{8.78}$$

The desire is to calculate the total standard deviation of the function $y = f(x_1, x_2, x_3 \ldots x_n)$., given by σ_y, when the only information available is about the standard deviations of the independent variables, $x_1, x_2, x_3, \ldots x_n$,. The equation for σ_y can be written as:

$$\sigma_y^2 = \frac{\sum_{i=1}^{r} (y_i - u_y)^2}{r}, \tag{8.79}$$

where u_y is the mean of y and r is the number of measurements in the population y. Let $\Delta_y = y_i - u_y$. Assuming Δ_y is small, then Δ_y can be approximated by:

$$\Delta_y \approx dy = \frac{\delta f}{\delta x_1} dx_1 + \frac{\delta f}{\delta x_2} dx_2 + \ldots + \frac{\delta f}{\delta x_n} dx_n. \tag{8.80}$$

This allows Eq. (8.79) to be rewritten as $\sigma_y^2 = \frac{\sum_{i=1}^{r} dy^2}{r}$. And using Eq. (8.80), the following can be written as:

$$\begin{aligned} dy^2 &= \left(\frac{\delta f}{\delta x_1} dx_1 + \frac{\delta f}{\delta x_2} dx_2 + \ldots + \frac{\delta f}{\delta x_n} dx_n \right)^2 \\ &= \left(\frac{\delta f}{\delta x_1} dx_1 \right)^2 (dx_1)^2 + \left(\frac{\delta f}{\delta x_2} dx_2 \right)^2 (dx_2)^2 + \ldots + \left(\frac{\delta f}{\delta x_n} dx_n \right)^2 (dx_n)^2 \\ &\quad + \sum_{j=1}^{n} \sum_{k=1}^{n} \left[\left(\frac{\delta f}{\delta x_j} \right) \left(\frac{\delta f}{\delta x_k} \right) (dx_j)(dx_k) \right]_{j \neq k}. \end{aligned} \tag{8.81}$$

If the variables x_i are independent, then:

$$\sum_{j=1}^{n}\sum_{k=1}^{n}\left[\left(\frac{\delta f}{\delta x_j}\right)\left(\frac{\delta f}{\delta x_k}\right)(dx_j)(dx_k)\right]_{j\neq k}=0. \tag{8.82}$$

This same result can be shown to be true for similar terms, allowing the following:

$$\sum_{i=1}^{r}(dy_1)^2=\sum_{i=1}^{r}\left[\left(\frac{\delta f}{\delta x_1}dx_1\right)^2(dx_1)^2+\left(\frac{\delta f}{\delta x_2}dx_2\right)^2(dx_2)^2+\ldots\left(\frac{\delta f}{\delta x_n}dx_n\right)^2(dx_n)^2\right]_i. \tag{8.83}$$

Each of the partial derivatives can be evaluated at its respective mean value, allowing:

$$\frac{\delta f}{\delta x_i}=C_i, \tag{8.84}$$

where C_i is a constant for each of the x_i. Therefore, Eq. (8.83) can be rewritten as:

$$\sum_{i=1}^{r}(dy_i)^2=\left(\frac{\delta f}{\delta x_1}\right)^2\sum_{i=1}^{r}(dx_1)_i^2+\left(\frac{\delta f}{\delta x_2}\right)^2\sum_{i=1}^{r}(dx_2)_i^2+\ldots+\left(\frac{\delta f}{\delta x_2}\right)^2\sum_{i=1}^{r}(dx_n)_i^2. \tag{8.85}$$

Using Eq. (8.85) and inserting into Eq. (8.79) result in the following for the variance of function y:

$$\sigma_y^2=\frac{\left(\frac{\delta f}{\delta x_1}\right)^2\sum_{i=1}^{r}(dx_1)_i^2+\left(\frac{\delta f}{\delta x_2}\right)^2\sum_{i=1}^{r}(dx_2)_i^2+\ldots+\left(\frac{\delta f}{\delta x_2}\right)^2\sum_{i=1}^{r}(dx_n)_i^2}{r}, \tag{8.86}$$

which can be rewritten as:

$$\sigma_y^2=\left(\frac{\delta f}{\delta x_1}\right)^2\frac{\sum_{i=1}^{r}(dx_1)_i^2}{r}+\left(\frac{\delta f}{\delta x_2}\right)^2\frac{\sum_{i=1}^{r}(dx_2)_i^2}{r}+\ldots+\left(\frac{\delta f}{\delta x_n}\right)^2\frac{\sum_{i=1}^{r}(dx_n)_i^2}{r}. \tag{8.87}$$

Recognizing that the $\frac{\sum_{i=1}^{r}(dx_n)_i^2}{r}$ terms are variances of each of the x_i's, the following can be written for the variance of function y:

$$\sigma_y^2 = \left(\frac{\delta f}{\delta x_1}\right)^2 \sigma_{x_1}^2 + \left(\frac{\delta f}{\delta x_2}\right)^2 \sigma_{x_2}^2 + \ldots + \left(\frac{\delta f}{\delta x_n}\right)^2 \sigma_{x_n}^2. \tag{8.88}$$

This can be rewritten in terms of the standard deviation as follows:

$$\sigma_y = \sqrt{\left(\frac{\partial y}{\partial x_1}\sigma_1\right)^2 + \left(\frac{\partial y}{\partial x_2}\sigma_2\right)^2 + \ldots + \left(\frac{\partial y}{\partial x_n}\sigma_n\right)^2}. \tag{8.89}$$

The same technique can be used to develop a similar relationship for the total amount of systematic offset (i.e., bias variations) in the output behavior of a MEMS device given by e_y. Again, it is assumed that the output function, given by $y = f(x_1, x_2, x_3, \ldots. x_n)$, is a function of n independent variables $x_1, x_2, x_3, \ldots .x_n$ and the total offset is due to the individual offsets in each parameter [6, 7]:

$$e_y = \sqrt{\left(\frac{\partial y}{\partial x_1}e_1\right)^2 + \left(\frac{\partial y}{\partial x_2}e_2\right)^2 + \ldots + \left(\frac{\partial y}{\partial x_n}e_n\right)^2 (n:1)}, \tag{8.90}$$

Equation (8.90) can be used to estimate the total bias variations, and Eq. (8.89) is used to estimate the total standard deviation of the random variations. The variables in Eqs. (8.89) and (8.90) are the device parameters used in the physical equations to describe the output behavior of the device. In many situations the variables representing the parameters involved in a device output function are independent of one another, and these equations are good estimates of the total bias variation and total standard deviation.

8.13 Example of Device Output Behavior Variation Analysis

As an example, the manufacturing variation analysis of the microfluidic microchannel discussed previously will be examined. It is desired to find the variation in the volumetric flow rate due to the manufacturing variations in the parameter dimensions of the microchannel. Using the Hagen-Poiseuille given in Eq. (8.74) that relates the volumetric flow rate of a fluid in a channel to the pressure differential, channel dimensions, and fluid properties (see Fig. 8.20), where it is assumed that the fluid is incompressible, is Newtonian, and has laminar flow characteristics, can be rearranged as follows:

$$R = \frac{\Delta P}{Q} = \frac{8\mu L}{\pi r^4}, \tag{8.91}$$

where R is the fluid flow resistance of the microchannel, ΔP is the differential pressure across (equal to $P_1 - P_2$ in Fig. 8.20), Q is the volumetric flow rate, μ is the fluid viscosity, L is the conduit length, and r is the radius of the conduit.

The reason for putting Eq. (8.74) into the form of Eq. (8.91) is that it allows the input and output variables to be placed on one side of the equation and the dimensional and material property parameters to be placed on the other side, thereby putting the relationship into a form of a transfer function that can be readily used in Eqs. (8.79) and (8.80). Using these equations, the total normalized bias variation in the microchannel device flow resistance can be written as follows:

$$\frac{e_R}{R} = \sqrt{\left(\frac{e_\mu}{\mu}\right)^2 + \left(\frac{e_L}{L}\right)^2 + \left(-4\frac{e_r}{r}\right)^2}, \tag{8.92}$$

where the individual bias variations in the independent variables, μ, L, and r are given by e_μ, e_L, and e_r, respectively. Additionally, the following can be written for the total normalized standard deviation of the random variations in the microchannel device flow resistance as:

$$\frac{\sigma_R}{R} = \sqrt{\left(\frac{\sigma_\mu}{\mu}\right)^2 + \left(\frac{\sigma_L}{L}\right)^2 + \left(-4\frac{\sigma_r}{r}\right)^2}, \tag{8.93}$$

where σ_μ, σ_L, and σ_r are the standard deviations of the random variations of the fluid viscosity, microchannel length, and microchannel radius, respectively. Each of the variations and standard deviations of the random variables must be determined, with the bias and random variations estimated separately.

The starting place is to determine the desired nominal value of the microchannel flow resistance. It will be assumed that the flow resistance has a desired value of 4.54 × 10^9 Pa•s/m based on the application that this device is to be used in. Using Eq. (8.91) from above, we can determine the design values of the dimensions and material properties of the microchannel device as follows:

$$R = \frac{\Delta P}{Q} = \frac{8\mu L}{\pi r^4} = \frac{8\left(8.9 \times 10^{-4}\,\mathrm{Pa \cdot s}\right)\left(2{,}000 \times 10^{-6}\,\mathrm{m}\right)}{\pi\left(100 \times 10^{-6}\mathrm{m}\right)^4}$$
$$= 4.54 \times 10^9\,\frac{\mathrm{Pa \cdot s}}{\mathrm{m}}. \tag{8.94}$$

Specifically, the chosen desired design dimensions of the microchannel are as follows:

$$\text{Desired Design Dimensions:} \quad L = 2{,}000.00 \text{ microns; and} \quad r = 100.00 \text{ microns.} \tag{8.95}$$

Additionally, the desired design material property related to the dynamic viscosity of the fluid (water at room temperature) used in the microchannel is given as:

$$\text{Desired design fluid property}: \mu = 8.90 \times 10^{-4}\ \text{Pa sec}. \tag{8.96}$$

Note that these desired design dimensional parameter values are derived from the engineering design activities and other values of the microchannel radius and length could be selected to obtain the same flow resistance value. That is, the device designer has freedom to select these values based on engineering judgment, device area constraints, manufacturing constraints, cost, etc. In any case, these dimensions will be used in this example.

It is assumed that a production quantity of 50 microchannel devices has been fabricated and measured. The measured dimensional data from these microchannels includes the microchannel radius and length. It is assumed that the microchannel dimensions exhibit a normal distribution, which can be verified by performing goodness of fit tests on the sample data. It is noted that these means and standard deviations are higher than the values displayed in Table 8.7, and this is due to the fact that the microchannels in the present case were manufactured using wet etching techniques to make the microchannel in the silicon substrate. Therefore, it would be expected that the values of the mean and standard deviation for the wet etched microchannels would be larger.

The calculated mean value of the sample of microchannel widths is found to be 102.08 microns, and the calculated standard deviation is 2.44 microns as follows:

$$\bar{x}_r = 102.08\ \text{microns}, \text{ and } S_x = 2.44\ \text{microns}. \tag{8.97}$$

Similarly, the calculated mean value of the sample of microchannel lengths is found to be 2002.08 microns, and the calculated standard deviation is 2.52 microns as follows:

$$\bar{x}_L = 2002.08\ \text{microns}, \text{ and } S_x = 2.52\ \text{microns}. \tag{8.98}$$

Since there are 50 measurements of the microchannel device dimensions, this represents a sufficiently large sample to use z-statistics. It is assumed that we are interested in knowing the mean with a 95% confidence interval. This allows us to write for the microchannel width the following:

$$\boldsymbol{\mu}_r = 102.08 \pm 1.96 \frac{S_x}{\sqrt{n}}\ \text{microns} = 102.08 \pm 1.96 \frac{2.44}{\sqrt{50}}\ \text{microns, or} \tag{8.99}$$

$$\boldsymbol{\mu}_r = 102.08 \pm 0.68\ \text{microns (95\% confidemce interval)}. \tag{8.100}$$

And, similarly for the microchannel length:

$$\mu_L = 2002.08 \pm 1.96\frac{S_x}{\sqrt{n}} \text{ microns} = 2002.08 \pm 1.96\frac{2.52}{\sqrt{50}} \text{ microns, or} \quad (8.101)$$

$$\mu_L = 2002.08 \pm 0.70 \text{ microns (95\% confidence interval)}. \quad (8.102)$$

The first item seen from examination of these uncertainties is that the range of the mean values of the radii and length does not overlap with the desired design value of these device parameters. Therefore, there appears to be an offset in these means that can likely be attributed to some amount of bias variation present in the processing steps used to make the microchannel. We are fortunate since we know the desired design values for the microchannel radii and lengths as given above. Therefore, we can estimate the normalized bias variations in the microchannel radius and length as follows:

$$\text{Bias variation microchannel radius} = 2.08 \text{ microns; and} \quad (8.103)$$

$$\text{Bias variation microchannel length} = 2.08 \text{ microns}. \quad (8.104)$$

While these bias variation estimates are useful, it should be noted that the mean values of the microchannel dimensional parameters each had a range of uncertainty associated with them. The bias variation estimates shown above used the central value of the range of possible mean values. Therefore, these bias variation estimates maybe somewhat limited in their accuracy.

The only option available to reduce the uncertainty in these bias variation estimates would be to manufacture a larger sample size, measure the microchannel dimensions, and re-perform the analysis on this larger sample set. In order to reduce the uncertainty by a significant amount, a significantly large sample set will be required. Of course, a larger sample size will escalate the cost of the bias variation analysis accordingly. As a matter of practice, in most situations a reasonably sized sample size would be created first in order to get a meaningful notion of the magnitude of the bias variations in the manufacturing processes. This process would take place in development and/or early pre-production. However, at a later stage of pre-production, a larger sample size would be used so as to get a more precise and accurate estimate of the bias variation, and this would be used to fine-tune the microsystems design and process parameters so as to remove the remaining bias variations. We will assume in the present example that a larger sample size of microchannels was also fabricated and that the mean and standard deviations shown above did not appreciably change, and the bias variations given in Eqs. (8.103) and (8.104) are accurate.

In addition to the dimensional parameters of the microchannel, the total bias variation is also dependent on any variation associated with the dynamic viscosity of the fluid used in the device. For the sake of simplicity, it will be assumed that we

Table 8.8 Normalized bias variations for microchannel device

Parameter	Normalized bias variations (95% confidence interval)
Microchannel radius	0.02
Microchannel length	0.001
Fluid viscosity	0

have previously performed measurements in order to determine the mean value of the fluid viscosity at a 95% confidence interval, and it was found to be:

$$\text{Fluid dynamic viscosity}: \mu = 8.90 \times 10^{-4} \pm 0.10 \times 10^{-4}\ \text{Pa sec}. \quad (8.105)$$

Since the calculated mean of the fluid viscosity is centered at the value of the nominal design value, the bias variation in the viscosity is negligible. Moreover, since this is a material property that would not be expected to have any bias, we can safely ignore the viscosity's contribution to the total bias variation. Therefore, the normalized bias variations for the parameters of the microfluidic microchannel device can be summarized in Table 8.8 as follows:

Substituting these values of bias variations into Eq. (8.92) allows the following to be written for the estimate of the total normalized bias variation in the microchannel device flow resistance:

$$\frac{e_R}{R}(\text{bias}) = \sqrt{(0.001)^2 + (-4 \times 0.02)^2} = 0.08. \quad (8.106)$$

The standard deviations of the microchannel radius and length dimensions given above can be used to calculate the normalized standard deviations for each of these microchannel parameters as shown in Table 8.9. Substituting these values of standard deviations into Eq. (8.93) allows the estimated normalized standard deviation of the microchannel flow resistance to be written as follows:

$$\frac{\sigma_R}{R}(\text{random}) = \sqrt{(0.001)^2 + (0.0013)^2 + (-4 \times 0.024)^2} = 0.10. \quad (8.107)$$

Equation (8.107) gives an estimate of the normalized standard deviation of the random variation of the microchannel flow resistance. What is of interest to us in the present analysis is an interval that covers a specified proportion of the population distribution with a certain level of confidence. This is often called a tolerance interval.

A tolerance interval is used to make a reasonable estimate of the manufacturing function location depending on the sample size. Specifically, the purpose of the tolerance interval is to increase the number of sample standard deviations that are used to define the possible region covering the proportion of the population distribution desired with a given level of confidence.

Table 8.9 Normalized standard deviations for variations in microchannel parameters

Parameter	Normalized standard deviations
Microchannel radius	0.024
Microchannel length	0.0013
Fluid viscosity	0.011

A tolerance interval overlaid onto a manufacturing function is shown in Fig. 8.22. The first important point illustrated in this figure is that the manufacturing function is shown as having a possible range of locations. The central (solid black line) normal distribution manufacturing function has identically shaped normal distribution manufacturing to the right and left (dotted gray lines). The range of the mean is the confidence interval of the mean value. Therefore, it can be observed that in order to capture a certain percentage of the parameter x within the tolerance interval, the boundaries of that interval must accommodate for the possible range of the values of the mean.

The two-sided tolerance interval is calculated using the following equations [8]:

$$\begin{aligned} \text{Tolerance interval}_L &= \overline{x} - k_2 S_x; \text{ and} \\ \text{Tolerance interval}_U &= \overline{x} + k_2 S_x. \end{aligned} \tag{8.108}$$

In Eq. (8.108), $\overline{x}$ is the sample mean of some device parameter, S_x is the sample standard deviation, and k_2 is a factor that is determined so that the interval includes at least a proportion p of the population with confidence γ. The tolerance interval allows the calculation of the estimated boundary locations for the interval that will contain a desired percentage of the population based on the sample mean and standard deviation.

The k_2 factors are tabularized in ISO 16269-6 [9, 10], since the calculation is difficult. A reasonably good approximation of the factor k_2 can be calculated using a method reported by Howe [11] and is given by:

$$k_2 = \sqrt{\frac{\nu\left(1 + \frac{1}{N}\right) z^2_{(1-p)/2}}{\chi^2_{1-\gamma,\nu}}}, \tag{8.109}$$

where $\chi^2_{1-\gamma,\nu}$ is the critical value of the chi-square distribution with υ degrees of freedom that is exceeded with probability γ and $z^2_{(1-p)/2}$ is the cumulative probability $(1-p)/2$. The quantity υ represents the degrees of freedom that was used to estimate the standard deviation. In most instances, the same sample is used to estimate both the mean and standard deviation so υ is equal to $N - 1$. Values of k_2 can also be obtained in look-up tables.

We can now go back to our example of the microchannel device to determine the interval of the random variations that contains 95% of the population of the microchannel flow resistance with 99% level of confidence, wherein the k_2 value is found to be equal to 2.576. Using that result the tolerance interval can be

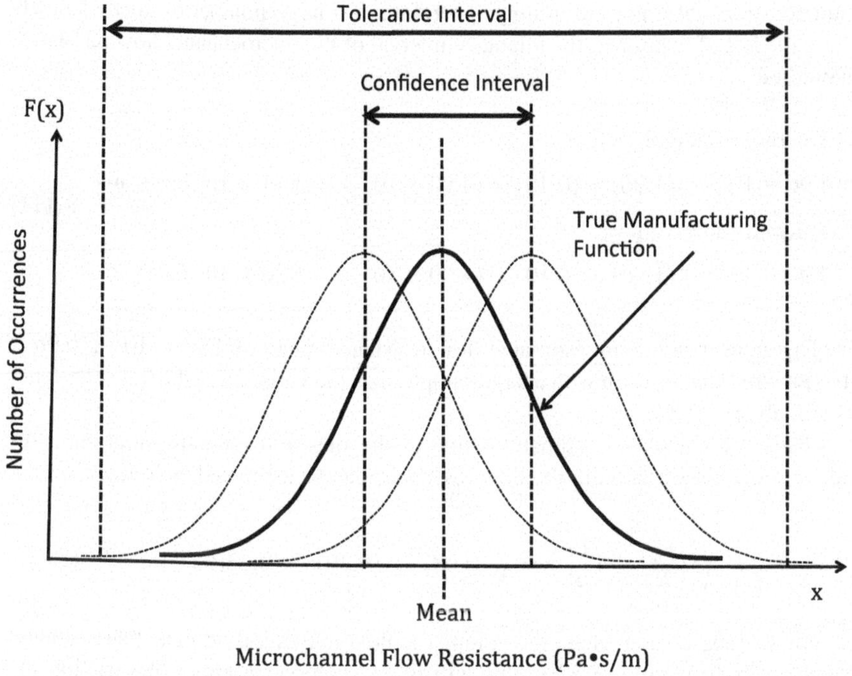

Fig. 8.22 Illustration of the tolerance internal wherein the manufacturing function for the population is represented by the solid line, and the dotted line distributions on either side represent the uncertainty of the location of the actual mean of the population lean. The difference of the centers of these dotted line distributions is the confidence interval of the mean

calculated. The estimated range of values of the microchannel flow resistance due to random variations in the device parameters can be estimated as follows:

$$\begin{aligned}
&\text{Tolerance interval}_L = \\
&4.90 \times 10^9 - \left[(2.576) \times (0.10) \times (4.90 \times 10)^9\right] = 3.64 \times 10^9\ \text{Pa} \cdot \text{s/m} \\
&\text{Tolerance interval}_U = \\
&4.90 \times 10^9 + \left[(2.576) \times (0.10) \times (4.90 \times 10)^9\right] = 6.16 \times 10^9\ \text{Pa} \cdot \text{s/m}
\end{aligned} \tag{8.110}$$

It can now be stated that 95% of the population of microchannels have a flow resistance within a range of $4.90 \times 10^9 \pm 1.26 \times 10^9$ P-s/m with a level of confidence of 99% due to the random variations in the microchannel parameters. This represents a fairly large estimated variation in the flow resistance. In fact, the range of variation is over ±26%.

If the number of samples is increased significantly, the range of variation will decrease. Specifically, the k_2 factor approaches the limiting value of 1.96 as the number of samples goes to infinity. Therefore, if a large number of microchannels were made and measured, the tolerance interval of the microchannel flow resistance would be:

$$\begin{aligned}
&\text{Tolerance interval}_L = \\
&4.90 \times 10^9 - \left[(1.96) \times (0.10) \times (4.90 \times 10)^9\right] = 3.94 \times 10^9 \text{ Pa} \cdot \text{s/m} \\
&\text{Tolerance interval}_U = \\
&4.90 \times 10^9 + \left[(1.96) \times (0.10) \times (4.90 \times 10)^9\right] = 5.86 \times 10^9 \text{ Pa} \cdot \text{s/m}
\end{aligned} \tag{8.111}$$

This translates into an estimated flow resistance range of $4.90 \times 10^9 \pm 0.96 \times 10^9$ Pa-s/m. That is, even with a large sample size, the variation in the flow resistance is still about ±20%.

Using the calculated estimated value of the bias and random variations from above, the total variation in the flow resistance can be estimated as follows:

$$\frac{E_R}{R} = \sqrt{\left(\frac{e_R}{R}\right) + \left(\frac{P_R}{R}\right)} = \sqrt{(0.08)^2 + (0.20)^2} = 0.22(95\%). \tag{8.112}$$

For any application where the volumetric flow rate of a fluid must be reasonably precisely controlled, a ± 22%, this will likely represent a range of variation that would be too large.

The manufacturing function, for the microchannel device, which is the probability distribution function of the flow resistance, is illustrated in Fig. 8.23.[12] Importantly, it is seen that the bias variation has resulted in an offset in the flow resistance, and standard deviation in the random variations in the microchannel dimensions results in the amount of spread in values of the distribution.

8.14 Simplified Variation Analysis

It is very helpful for the microsystems engineer to be equipped with a simple method to be able to quickly develop rough estimates of the variations in the microsystems device output response. This method can be useful before performing more detailed analysis, certainly before expending the time and resources for fabricating some number of devices. This is especially the case for MEMS development activities that often involve a phase of brainstorming in the attempt to come up with potential device ideas to suit a particular application. This section provides guidance on how

[12] The reader should recall that the tolerance interval was for 95% of the population and therefore represents 2σ.

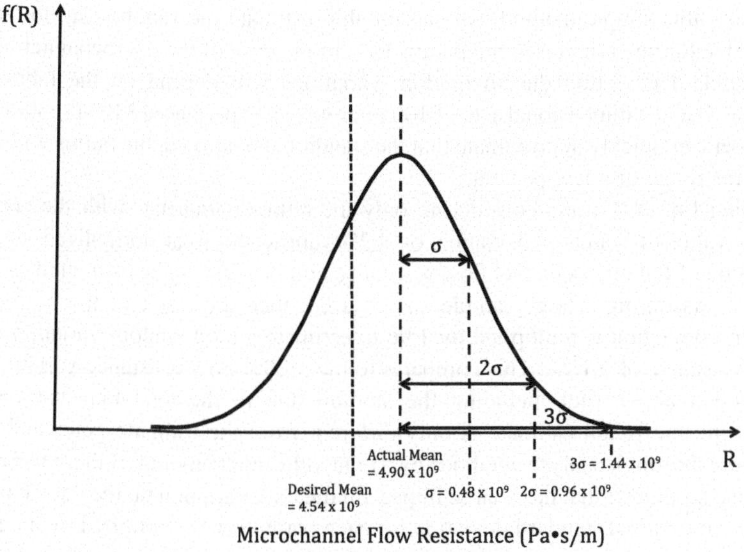

Fig. 8.23 The probability distribution function for the microchannel flow resistance showing the effects on the bias and random variations on the manufacturing function

to quickly perform rough simple estimates of the expected variation in the microsystems device behavior.

The basic approach is to examine the physical equation(s) used to describe the device behavior output response, identify those terms that are raised to higher powers, likely have the largest relative standard deviations, and calculate the result.

In the case of the microchannel device example above, the physical equation used for calculating the flow resistance was Eq. (8.91), and from that relationship we can immediately identify that the radius has by far the most potential for effecting the flow resistance since it is raised to the fourth power. We also see that the other parameters, namely, length and viscosity, are not raised to a higher power. We then examine the desired parameter values for the intended application. We know that the desired design value of the radius is 100.0 microns. We can compare the value of the radius to the desired design value of the microchannel length, which is 2000.0 microns. This allows us to immediately conclude that there is likely to be much larger variation in the manufactured dimensions of the radius compared to the microchannel length. This is due to the knowledge that the parameters with smaller dimensions typically have larger relative variations compared to the parameters with larger sizes. Therefore, the microchannel length can be ignored in the rough estimate.[13]

---

[13]It should be obvious to the reader that if it were desired to include the variation of the length of the microchannel, this would be a simple matter.

Using this simple method, we can roughly estimate the variation in flow resistance due to the variation in the parameters. In the case of the microchannel radius, the standard deviation due to random variations will depend on the fabrication method, but at a dimensional size of 100 microns, an experienced MEMS fabrication engineer can quickly approximate that the standard deviation of the radius will likely be in the range of a few percent.

Using Eq. (8.93) and considering only the radius parameter, with the approximated value of standard deviation of $\pm 2\%$, allow the total normalized standard deviation of the microchannel flow resistance and flow rate to be estimated as about $\pm 8.0\%$. Assuming a large sample size is used, then the result of the normalized standard deviation is multiplied by 1.96 to estimate a total random variation of the flow resistance of $\pm 16\%$. This compares to the $\pm 20\%$ flow resistance variation that was found above (not including the amount due to the total bias variation[14]). Therefore, the rough estimate is only different from the estimate generated using our more thorough analysis methods by a relatively small amount. If there were more than one parameter that has a large impact on the total variation on the device output, then the individual standard deviation for each parameter is multiplied by the power that they are raised to and combined using the square root of the sum of the squares.

The real usefulness of this rough estimate is that if the intended application requires a precision of the flow rate of the microchannel device to be considerably better than 15–20%, then the device design engineer quickly knows that there may be a problem with meeting the application specifications before starting an expensive development effort. The engineer can explore how to potentially improve the situation, such as using a different fabrication method to make the device, changing the design, etc. However, if the intended application requires a precision on the output response that is unachievable and no alternatives to improving the situation are available, then using this quick method can allow the engineer to quickly determine that the microchannel device may not be able to meet the requirements of the application before starting a development activity.

8.15 Important Generalizations

In many situations, such as when considering the probability distribution of the output of a device, the output would be a function of multiple device parameters. In these situations, the probability distribution function is modified so as to incorporate these other parameters and given by a multiparameter joint probability distribution function expression as follows:

[14]In Chap. 10 methods for removal of the bias variations will be examined.

$$\Phi(x_1, x_2, x_3, \ldots, x_n) \text{ where } n \text{ is the number of parameters involved.} \tag{8.113}$$

This probability distribution function would still satisfy the relationship given by:

$$\int_{-\infty}^{+\infty} \cdots \int_{-\infty}^{+\infty} \Phi(x_1, x_2, x_3, \ldots x_n) dx_1, dx_2, dx_3, \ldots dx_n = 1. \tag{8.114}$$

For a given elemental region δ of size dx_1, dx_2, dx_3,. . .dx_n, in x_1, x_2, . . .,x_n space, the probability of a combination of finding a combination of x_1, x_2, . . ., x_n is given by the multidimensional space under the PDF curve bounded by d and given by:

$$\int_{\delta} \Phi(x_1, x_2, \ldots x_n) dx_1, dx_2, \ldots dx_n = \text{Probability}(x_1, x_2, \ldots x_n \in \delta) \tag{8.115}$$

For a probability distribution function with two parameters, Eqs. (8.113) and (8.114) would reduce to:

$$\Phi(x_1, x_2) \text{ and } \int_{-\infty}^{+\infty} \int_{-\infty}^{+\infty} \Phi(x_1, x_2) dx_1, dx_2 = 1, \tag{8.116}$$

which can be represented by the volume under the curve in Fig. 8.24. Obviously the visualization of PDFs with more than two parameters is not possible. Nevertheless, calculations in multiple parameter space using computers are straightforward.

There is another issue that often of consideration in microfabricated devices. Specifically, it is known that some microfabricated device parameters may exhibit correlations. For example, it is well known in microfabrication that the results of certain processing steps can exhibit a correlation in the dimensions from one region of the substrate to another. Correlation between device parameters is usually more pronounced for parameters that are in close proximity on the substrate. Correlation is illustrated in Fig. 8.25 where a number of measurements of two device parameters are plotted. As can be seen, there appears to be somewhat of a linear relationship between the parameters x_1 and x_2. If there were no relationship, a plot of the parameters would be randomly scattered across the two-dimensional plot. It should be noted that correlation among device parameters can sometimes be used to an advantage in device design, with the most notable examples being differential transistor pairs and switched capacitor networks.

A correlation among device parameters necessitates an adjustment in the statistical analysis. First, it is important to be able to determine the amount of correlation present in the parameters. This is determined by calculating the covariance between the parameters. For the simple case of two parameters given by x_1 and x_2, the covariance can be found using:

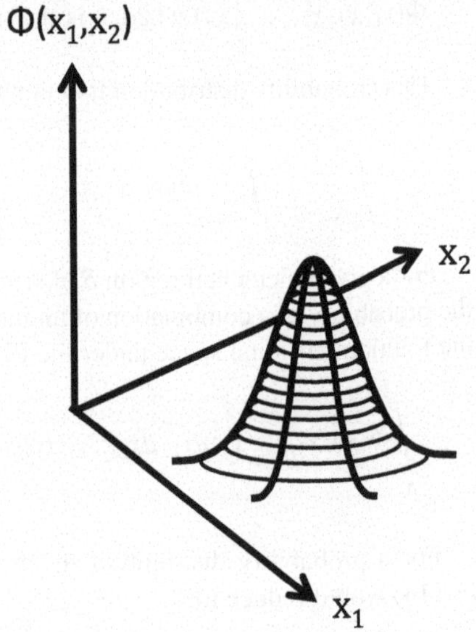

Fig. 8.24 An illustration of a joint probability distribution function of two parameters, x_1 and x_2

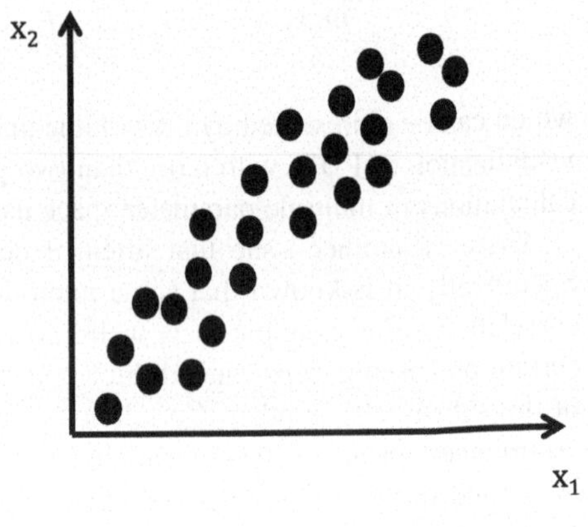

Fig. 8.25 A plotting of two device parameters, x_1 and x_2, wherein there is a somewhat linear relationship between these parameters.

$$\sigma_{1,2} = \int_{-\infty}^{+\infty}\int_{-\infty}^{+\infty}(x_1 - u_1)(x_2 - u_2)\Phi(x_1, x_{2,})dx_1dx_2, \tag{8.117}$$

where u_1 and u_2 are the mean values of parameters x_1 and x_2 and $\Phi(x_1, x_{2,})$ is the joint probability distribution function of parameters x_1 and x_2. As before, if there are more than two parameters that are potentially correlated, then the covariance Eq. (8.117) would be expanded accordingly to include these other parameters.

The measure of the correlation is given by the correlation coefficient, given by ρ, and defined for two parameters x_1 and x_2 as follows:

$$\rho_{1,2} = \frac{\sigma_{1,2}}{\sigma_1 \sigma_2}, \tag{8.118}$$

where σ_1 and σ_2 are the standard deviations of parameters x_1 and x_2.

In terms of a finite sample set of pairs of values of x_1 and x_2, the correlation coefficient can be estimated by the following:

$$\widehat{\rho}_{1,2} = \frac{\left[\frac{1}{n}\sum_{i=1}^{n}(x_{1_i} - u_1)(x_{2_i} - u_2)\right]}{\widehat{\sigma_1}\widehat{\sigma_2}}. \tag{8.119}$$

where the hat symbols over the correlation coefficient and standard deviations represents that these are estimators for these values. Regardless of the form of the correlation coefficient used, it will always have a value that ranges between -1 and 1. If the parameters x_1 and x_2 have no correlation, then the correlation coefficient is zero. If on the other hand there is complete correlation between these parameters, then the value is either -1 or 1.

In the scenario where there are more than two parameters, the expressions for covariance and correlation coefficients can be generalized for the number of parameters of relevance. Specifically, the parameters are considered two at a time and expressed as elements within an n by n matrix called the variance-covariance matrix given by [**S**]. Within the [**S**] matrix, the diagonal elements, given by S_{ii}, are the variances of the i_{th} parameter, or σ_i, and the off-diagonal elements given by S_{ij} are the covariances σ_{ij} between the i_{th} and j_{th} parameters. Including these effects, the probability distribution function can be written as follows:

$$\Phi(x_1, x_2, \ldots .x_n) = \frac{1}{\sqrt{2\pi|\mathbf{S}|}} \exp\left[\frac{-(\mathbf{X} - \mu)^T(\mathbf{X} - \mu)}{2|\mathbf{S}|}\right], \tag{8.120}$$

where |**S**| is the determinant of the variance-covariance matrix, **X** is the vector component of the parameters ($\mathbf{X} = x_1, x_2, \ldots, x_n$), and $\boldsymbol{\mu}$ is the vector of the mean parameter values ($\boldsymbol{\mu} = \mu_1, \mu_2, \mu_3, \ldots. \mu_n$). Equation (8.120) assumes that the PDF is Gaussian which is nearly always the case.

8.16 Review of Methods for Variation Analysis[15]

The method for analyzing the variations in the output behavior of a microsystems device presented in the previous section is useful in situations where the number of parameters is relatively small and when the means and standard deviations of the parameters are known. In some situations, such as the example of the microchannel,

[15] This type of analysis is also sometimes referred to as "tolerance analysis."

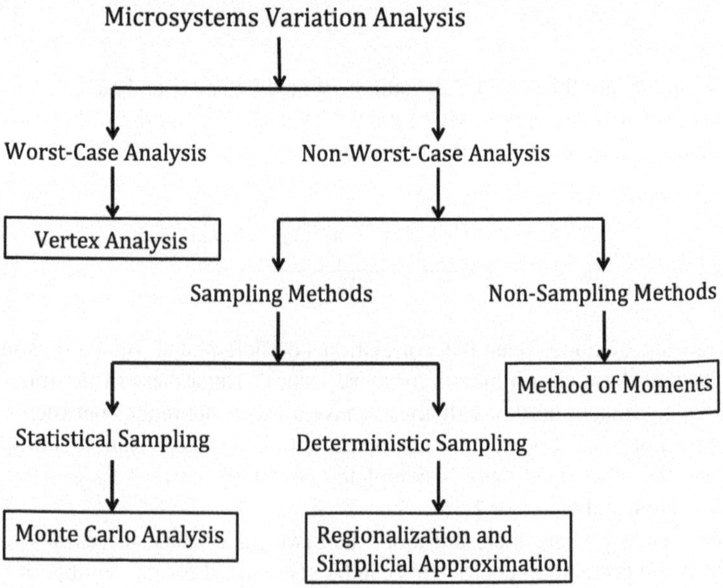

Fig. 8.26 A chart displaying the various methods of variation analysis.

where the number of parameters is relatively small, this method can be used effectively. However, there are other circumstances in microsystems analysis where the number of parameters is large and the above method may be unwieldy. Therefore, other methods of analyzing the output variations in microsystems devices are needed. Some of these methods can be performed analytically, while others require the use of computers. As a general rule, when the number of parameters is larger than about five parameters, numerical-based solutions are usually advisable. Additionally, the previous method also required knowledge of the means and standard deviations of the parameters. In some situations, these will be known, but often they are not known until they can be measured. Nevertheless, there are methods that can be used even without prior knowledge of the mean and standard deviation of the parameter. This section reviews some of the known and more widely used techniques used for analyzing the impact of device parameter variations (see Fig. 8.26).[16]

[16]The deterministic sampling methods are used to directly determine the yields and therefore will be discussed in the next chapter.

8.16.1 Worst-Case Variation Analysis

Worst-case variation analysis, also sometimes called the method of extremes, is a common method used in macroscale manufacturing, particularly related to assembled mechanical devices and systems. In worst-case analysis, the design objective is to achieve a 100% production yield level. This is sometimes expressed as 100% interchangeability of the parts in an assembly in mechanical engineering [12].

This method uses the device parameters that result in the worst (i.e., most extreme) device performance outcome. Since only the values of the parameters resulting in the worst-case output performance are used in this technique, it does not require the probability distribution functions of the device parameters to be known. The only assumption is that the parameter values are within the actual range limits of each particular parameter. The difficult part of using this method is that it requires the identification of the values of the parameters that result in the worst-case device output performance. These worst-case parameter values may exist in only certain combinations of the parameters and do not necessarily coincide with the values of the parameters at the limits of their ranges. Therefore, the values of the parameters can be hard to identify when there are many parameters involved in the device design. This method takes these combinations of parameters resulting in the worst-case performance and performs an analysis to determine the device output behavior. Each of the analyses using different identified parameter values will likely result in different device worst-case output values.

Figure 8.27 is an illustration of the use of worst-case analysis on a simple case of one element to be inserted into another element's gap during manufacturing. The two elements that are to be assembled together has element 2 inserted into a slot made in the bottom of element 1. A key assumption in this example is that the parameter means are centered within the bilateral variations of the parameter values. Element 1 has a dimension of the slot given by a nominal value of x_{P1} with a variation of $\pm t_{P1}$. Element 2 has a dimension of width given by a nominal value of x_{P2} with a variation of $\pm t_{P2}$. For it to be possible to insert element 2 into the slot of element 1, the gap must be equal to or greater than zero. The dimensions and variations of the elements are given in Table 8.10.

The mean value of the gap is calculated using the following:

$$\text{Mean gap} = \sum_{i=1}^{n} s_i x_i, \tag{8.121}$$

where n is the number of elements (or parameters involved), s_i is the sensitivity factor that defines the sign and amount in the i_{th} element, and x_i is the mean value of the dimension of the i_{th} element. Using this equation, the mean gap can be calculated as follows:

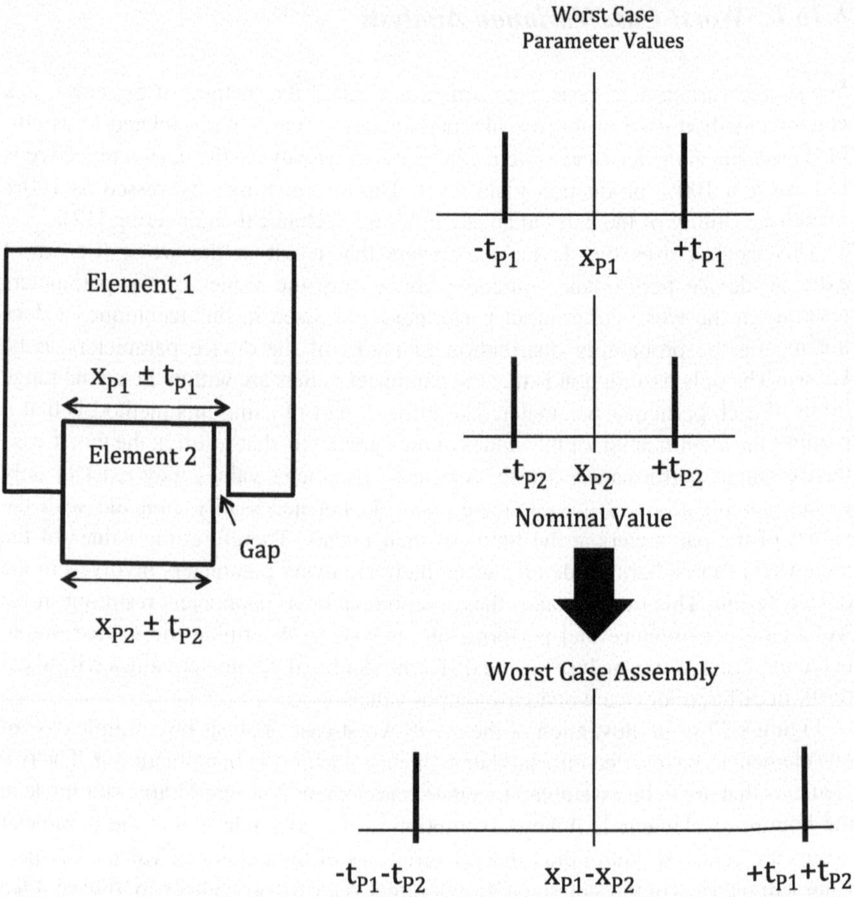

Fig. 8.27 An illustration of worst-case analysis of the two elements involved in an interlocking assembly

Table 8.10 A listing of the parts and their dimensional information

Element number	Mean width	Sensitivity	Bilateral variation
Slot in element 1	$x_{P1} = 2.00$ mm	1	$\pm t_{P1} = 0.03$ mm
Element 2	$x_{P2} = 1.98$ mm	−1	$\pm t_{P2} = 0.01$ mm

$$\text{Mean gap} = (1)(2.00) + (-1)(1.98) = 0.02. \tag{8.122}$$

The maximum expected variation in the gap can be calculated using the following equation:

$$\text{Maximum expected gap variation} = \sum_{i=1}^{n} |s_i t_i|, \tag{8.123}$$

where t_i is the bilateral variation of the i_{th} element dimension. Using this equation, we can write:

$$\begin{aligned}&\text{Maximum expected variation}\\&= |(1)(0.03\text{mm})| + |(-1)(0.01\text{mm})| = 0.04\ \text{mm}.\end{aligned} \tag{8.124}$$

Using the worst-case model, the maximum gap is equal to the mean value of the gap plus the worst-case maximum expected variation of the gap, while the minimum gap is equal to the mean value of the gap minus the worst-case maximum expected variation of the gap, or:

$$\text{Maximum gap} = 0.02\ \text{mm} + 0.04\ \text{mm} = 0.06\ \text{mm},\quad \text{and} \tag{8.125}$$

$$\text{Maximum gap} = 0.02\ \text{mm} - 0.04\ \text{mm} = 0.02\ \text{mm}. \tag{8.126}$$

According to this analysis, the minimum gap is less than zero, so therefore the design would fail. In order to satisfy the minimum gap, the slot in element 1 would need to be increased by at least 0.02 mm, or the width of element 2 would need to be reduced by 0.02 mm or some combination of these two. Alternatively, the variations in elements 1 and 2 may also be reduced to attempt to meet the minimum gap requirement. As can be appreciated by this simple example, worst-case variation analysis provides the most pessimistic prediction of the overall total device variation.

In the more general situation when the parameter space is multidimensional, it can be assumed that there are a number, m, of upper and a number, n, of lower specifications; the device output functions $f(x)$ for the microsystems can be written as follows:

$$\begin{aligned}&f_j(x) \geq L_j \text{ for } j = 1, 2, \ldots, m,\\&\text{and}\\&f_k(x) \geq U_k \text{ for } k = 1, 2, \ldots, n,\end{aligned} \tag{8.127}$$

where L_j and U_k are the lower and upper device performance limits and x is the set of parameter values. To conduct worst-case analysis, m points $x_1, x_2, \ldots x_m$ within the parameter space are to be found such that:

$$\begin{aligned} &f_j(x_j) \geq f_j(x) \text{ for all } x, \\ &j = 1, 2, \ldots .m \\ &\text{and} \\ &f_k(x_k) < f_k(x) \text{ for all } x, \\ &k = 1, 2, \ldots ., n. \end{aligned} \tag{8.128}$$

In practice the extreme values of the parameters are investigated within the known range of values. These extreme combinations of values form vertices of a variation or tolerance region that can be described by:

$$x = x_i^o \pm t_i, \tag{8.129}$$

where x_i^o, $i = 1,2,\ldots k$ are the coordinates of the nominal parameter points and t_i is the variations in these point values. However, the worst-case performance may not occur at the vertices but instead at a point within the parameter space bounded by the vertices.

A more useful approach is to predict the vertices at which the extreme performance values occur. Examining the sensitivities of the output response functions with respect to the parameters can assist in this task. For example, to find the vertex of the maximum value of the function $f(x)$, the sensitivity of $f(x)$ with respect to all values of x would be:

$$S_i = S_x^f = \left.\frac{\delta f}{\delta x_i}\right|_{x^o} \quad i = 1, 2, \ldots ., K. \tag{8.130}$$

K is the number of parameters and x^{o} is the nominal value of the parameter. A K-element vector X can be determined as follows:

$$\begin{aligned} X_i = &\ 1 \quad \text{if } S_i(x^{\mathrm{o}}) > 0, \\ &-1 \quad \text{if } S_i(x^{\mathrm{o}}) < 0. \end{aligned} \tag{8.131}$$

The worst case for the upper performance specification occurs at the vertex given by:

$$x_1^o + X_1 t_1, x_2^o + X_2 t_2, \ldots, x_K^o + X_K t_K, \tag{8.132}$$

and the worst case for the lower performance specifications occur at the vertex given by:

$$x_1^o - X_1 t_1, x_2^o - X_2 t_2, \ldots, x_K^o - X_K t_K. \tag{8.133}$$

This procedure would be repeated for each performance specification. Monotonicity must be obeyed for the above procedure to be valid. However, this can be difficult to test in practice.

Worst-case variation analysis can be performed in a number of fashions including analytically and using computers. For MEMS devices having a small number of parameters, it can be easily performed using hand calculations. Nevertheless, worst-case analysis is not often used in MEMS device design activities. The primary reason is that it has an important deficiency when applied to MEMS microsystems design since it is usually too pessimistic.

8.16.2 *Non-worst-Case Variation Analysis*

Non-worst-case tolerance analysis techniques require information of the parameter distributions and can employ non-sampling or sampling. Both are reviewed here.

8.16.2.1 Non-sampling, Non-worst-Case Variation Analysis

This method was used in Sect. 8.13 above to estimate the variation in the flow resistance of the microchannel. The premise of this method is that the device parameter values are more likely to be located in close proximity of the center of the variation range than at extremely values that result in the worst-case performance (as was assumed in the case of worst-case analysis discussed above). Specifically, Eq. (8.89) calculates the amount of variation in the output of the device based on the individual variations of each of the parameters.

If there are correlations between the device parameter values, then Eq. (8.89) is modified to the following:

$$\sigma_y = \sqrt{\sum_{i=1}^{n}\left[\frac{\delta y}{\delta x_i}\right]^2 \mathrm{var}(x_i) + 2\sum_{i=2}^{n}\sum_{j=1}^{i-1}\frac{\delta y}{\delta x_i}\frac{\delta y}{\delta x_j}\mathrm{cov}\left(x_i, x_j\right)} \tag{8.134}$$

where $\mathrm{cov}(x_i, x_j)$ is the covariance between device parameters x_i and x_j.

This method can be applied to device parameter distributions of any type. If the probability distribution functions of the device parameters are Gaussian, then knowledge of the mean and standard deviation completely defines the distributions.

This method can be performed analytically when the number of parameters is small. If there are a larger number of parameters, this method can be performed with the use of computers, but in most circumstances, a different method called Monte Carlo analysis discussed below would be used since it has additional benefits.

8.16.2.2 Monte Carlo Variation Analysis

A comparatively better method that provides for more accurate analysis of device output behavior variations (as well as other benefits) is the Monte Carlo analysis technique. Monte Carlo methods use the device parameters described by their probability density functions and performs a number of iterative calculations using randomly selected values from each of the variable's probability distributions to calculate the output behavior of the device. By using the probability distributions of the device parameters, the parameters have different probabilities of occurrence and values, thereby resulting in different outcomes. Since Monte Carlo analysis is performed on different values of the parameters, the outcome of each simulation output varies, and the result of a large number of iterations of the simulated device output behavior becomes an accurate representation of the probability distribution of the device output behavior as sufficient samples are generated. The key element is that the measured and statistically analyzed values of the parameter variances are included in the analysis, and therefore the impact of these variations is represented in the analysis result. Moreover, since the output is a probability distribution, it is relatively easy to see the expected number of devices that will meet and not meet the device performance specifications. While the normal distribution is the most common probability distribution for device parameters, Monte Carlo analysis can be performed on other probability distributions with the same level of accuracy.

Consider the scenario illustrated in Fig. 8.28, which represents an actual MEMS manufacturing process technology. The inputs to the manufacturing process technology include the MEMS device design, which will comprise a fabrication process sequence, including each processing step, and the mask layout design. The device design will employ some number of device parameters, likely dimensional, as well as material property parameters that are labeled x_1 through x_n for n number of device parameters. The device design will also employ the associated device physics (e.g., the flow resistance of a microchannel, the stiffness of a pressure sensing membrane, etc.). Due to the random variations that occur in any manufacturing process technology, the variations in the device parameter values can each be described using probability distribution functions (shown in the present case to be Gaussian PDFs but could be probability distribution functions of any form including device parameter distributions having correlations between some or all of the parameters).

The manufacturing of MEMS microsystems devices involves these inputs and associated device parameters to produce some number of devices according to the design and fabrication processes used. The manufactured microsystems devices will then be tested to measure their individual output performance as shown in the flowchart. Importantly, due to the variations in the device parameters as part of the manufacturing processes employed, the testing results of the output performance of a suitably large sample of the manufactured devices will also exhibit some amount of output performance variation from one device to another as illustrated by the PDF of the device output shown at the bottom of the flowchart in Fig. 8.28.

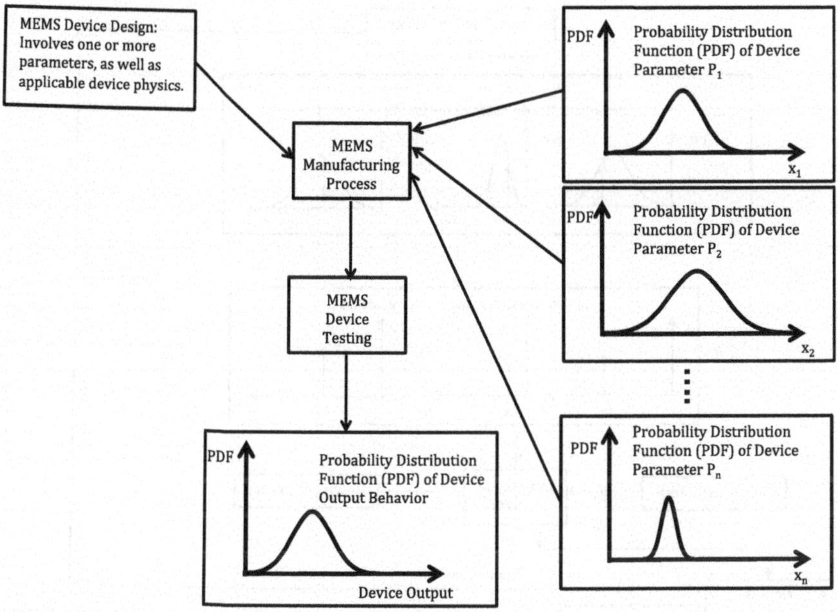

Fig. 8.28 An illustration of a MEMS manufacturing process wherein there are n number of device parameters that could be dimensions as well as material properties that each is described by their respective probability distribution functions. These randomly varying device parameters combined with the MEMS device design will result in a device output that can also be represented by its own probability distribution function

Figure 8.29 illustrates the use of Monte Carlo analysis to estimate the variation in the MEMS device output behavior. A pseudorandom number generator is used to make a single numerical sampling from the device parameter probability distribution functions for the device parameters involved in determining the device output behavior. This sampling processes results in parameter values for each parameter to be chosen (labeled as randomly selected device parameters, x_{1i}, x_{2i}, . . .x_{ni}) in Fig. 8.29. These selected device parameters combined with the MEMS device design information (including process sequence and mask layout) and MEMS device models (that include the device physics involved in determining the device output) are employed in the MEMS device analysis. The result of this device analysis is an individual-calculated device output value that is labeled "D_{1i}" in chart "analysis 1" shown in Fig. 8.29. This process of randomly selecting new device parameters and then calculating the resultant output is repeated a number of times.

The results of these individual analyses can be summed together to form a histogram of the device output as shown. Based on the principles of statistics, the accuracy of the predicted device output behavior increases as the number of iterations of the calculations increases. If a large number of iterations are performed, the histogram becomes a continuous distribution, which is the device output probability distribution function shown in the bottom right of Fig. 8.29.

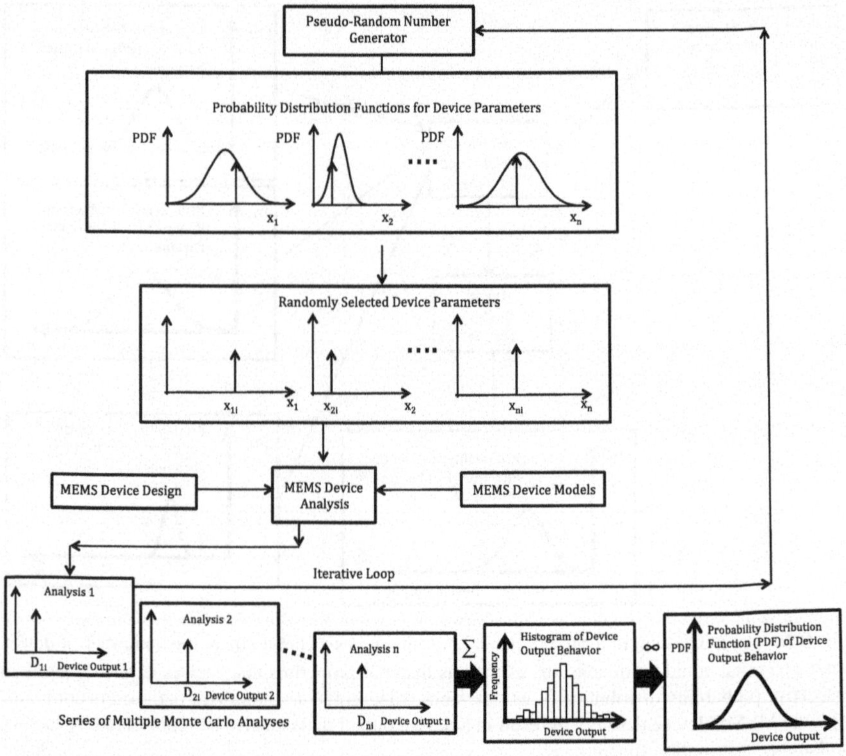

Fig. 8.29 An illustration of the use of Monte Carlo analysis to calculate the probability distribution function of the device output

Importantly, in the case of microfabricated devices, such as MEMS, where there may be correlations among some of the device parameters, these correlations can be incorporated into the Monte Carlo analysis using joint probability distribution functions of the appropriate device parameters.

It is useful to compare Figs. 8.28 and 8.29. In both cases, the device output variations are represented by probability distribution functions. Moreover, the output probability distribution functions are both based on actual data of the probability distribution functions of the device parameters. Therefore, assuming the device design and device physics and device parameter probability distribution functions are properly represented in the Monte Carlo analysis and a sufficient number of iterations of the calculations are performed, then the simulated device output behavior using the Monte Carlo method should be a good approximation of the probability distribution function of the actual output behavior of manufactured devices. Moreover, the accuracy of the results of using the Monte Carlo analysis to estimate the device output probability distribution function increases as the number of iterations of the analyses increases.

In comparing Figs. 8.28 and 8.29, it can be seen that the results of the Monte Carlo analysis in the form of a probability distribution function of the device output behavior can be treated as if it were the results of actual measurements on manufactured MEMS devices. Due to this fact, the Monte Carlo method for device output behavior variation analysis is considered as a sufficiently accurate, but a far less expensive, alternative to an actual production run of devices.

The Monte Carlo method is a very important and powerful technique for both analyzing the variations in complex MEMS microsystems and also can be an effective tool for use in yield estimation and device design as will be reviewed in later chapters. It is also useful to compare Monte Carlo to other deterministic methods of sampling for variation analysis. Unlike other methods, Monte Carlo method has dimensional independence. Other sampling methods have an accuracy that is dependent on a number of parameters; the accuracy of Monte Carlo analysis is independent of the number of parameters.

A few important points concerning Monte Carlo analysis related to microsystems manufacturing techniques are as follows. First, Monte Carlo analysis can be successfully used on any probability distribution function. The only requirement is that the distribution is known. However, in some cases for microsystems manufacturing processing steps, the probability distribution may not be known a priori. However, in those cases if the probability distribution can be assumed to be normal, which in most cases it will be, then if the mean and standard deviation are known, this can be used to completely describe the probability distribution function. Additionally, many microsystems manufacturing processing steps will result in a correlation between some of the device parameters. As noted above, if these can be properly described using joint probability distributions, then Monte Carlos analysis techniques can be used to formulate device output probability distribution functions.

In many microsystems foundry operations, "drop-in" test dies are placed onto production substrates that can be measured using a variety of metrology methods in order to generate data about each of the important device parameters. This data can be stored and used for subsequent design activities using Monte Carlo methods.

Alternatively, some number of substrates can be run through the process technology wherein the substrates have a large number of devices and/or test structures with a range of dimensions and geometries, and then perform measurements on these devices and/or test structures to generate a comprehensive database of each of the device parameter probability distribution functions. Using these device parameter PDFs, subsequent new design activities can be performed.

Another and many times better approach is to create physical models that directly relate to the materials properties of the materials used in the fabrication of devices. Specifically, material property probability distribution functions are developed that can then be used to create device models based on the physical equations that describe the device behavior. The creation of these material property PDFs and modeling equations is a considerable undertaking but can be step up as a constantly updated database and therefore allow more accurate modeling of the manufactured microsystems devices over time.

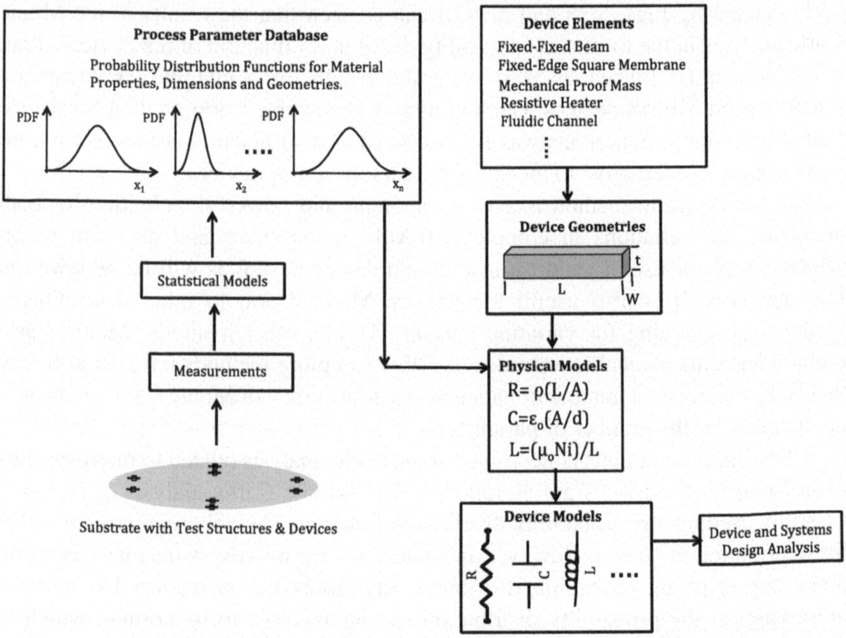

Fig. 8.30 An illustration of a method of constantly updating the material property, dimension, and geometry data from test structures and devices on production wafers to allow accurate modeling and design capability for future designs

This is illustrated in Fig. 8.30 showing a substrate with test structures and devices having a number of different dimensions, materials, and geometries and is tested after completion of the manufacturing. The data from the measurements on these test structures and devices results in considerable statistical information, specifically probability distribution functions for each of the material properties of the materials used, device dimensions, and device geometries. Microsystems developers can use this statistical data to develop accurate models of future devices to be implemented. A major benefit of this approach is that the statistical database is constantly and automatically being expanded and updated as new production lots are manufactured. Therefore, any slight changes in the processing steps that result in measureable changes in the statistics of the material properties, dimensions, and/or geometries over time can be captured in the database for future device designs.

8.17 Summary

The chapter focused on a major theme of this volume, namely, the impact of the variations in device parameters that occur in using microsystems manufacturing technologies. These variations can be in both the dimensions of the device and the material properties. These variations are important since they result in the device output behavior also exhibiting variations and thereby differing from the expected device output behavior based on the engineering design.

Two different types of variations were discussed: systematic (bias) variations and random variations. As we learned, bias variations are fixed amounts of offsets that occur in the device parameters. Random parameter variations are caused by non-systematic process variations across the wafers. These random variations are inherent in any microsystems manufacturing process. However, the magnitude of these variations can significantly vary depending on the specific details of the equipment, the process being performed, and the aggressiveness of the selected values of the device parameters.

This chapter also covered the important concepts of precision and accuracy. As discussed, both are important for a well-controlled manufacturing process. The tools of statistical analysis were also covered for both continuous and discrete probability distributions. Various examples of manufacturing were used to reinforce how these statistical methods can be effectively employed in microsystems manufacturing. Lastly, this chapter reviewed the most commonly used methods for analyzing variations in MEMS devices, with Monte Carlo analysis being the most powerful and useful.

References

1. A.C. Diebold, *Handbook of Silicon Semiconductor Metrology* (Marcel Dekker, Inc., New York, 2001)
2. B.S. Everitt, *The Cambridge Dictionary of Statistics*, 3rd edn. (Cambridge University Press, Cambridge, 2006)
3. R.E. Walpole, *Introduction to Statistics*, 3rd edn. (Macmillan Publishing, New York, 1982)
4. See: https://brownmath.com/stat/sampsiz.htm, accessed October 31, 2018
5. F.M. White, *Fluid Mechanics*, 5th edn. (McGraw-Hill, New York, 2003)
6. S. Timoshenko, S. Woinowsky-Krieger, *Theory of Plates and Shells*, 2nd edn. (McGraw-Hill, New York, 1959)
7. S.J. Kline, F.A. McClintock, Describing uncertainties in single-sample experiments. Mech. Engr. **75** (1953)
8. I. Miller, J.E. Freund, R.A. Johnson, *Miller and Freund's Probability and Statistics for Engineers*, 4th edn. (Prentice-Hall, New Jersey, 1990)
9. ISO 16269: 2005 Statistical interpretation of data – Part 6: Determination of statistical tolerance intervals
10. I. Garaj, I. Janiga, Two-sided tolerance limits of normal distribution for unknown mean and variability. Bratislava: Vydavatelstvo STU, p. 147, 2002
11. W.G. Howe, Two-sided tolerance limits for normal populations – Some improvements. J. Am. Statist. Assoc. **64**, 610 (1969)
12. P.J. Drake, *Dimensional Tolerancing Handbook* (McGraw-Hill, New York, 1999)

Chapter 9
Yield Analysis and Quality Assurance and Control Methods Used in Microsystems Manufacturing

9.1 Introduction

Manufacturing processes are designed to make products that meet the pre-defined performance specifications. It is desired that the percentage of products that are produced and meet these specifications be as high as possible. The percentage of products that work and meet the specifications is called the *yield,* and the percentage failing is called the *scape rate*. Throwing away manufactured products is expensive and directly impacts the eventual cost of all products made and also influences whether the products made can be sold at a profit.

Microsystems manufacturing is unique compared to many other types of production processes. Specifically, the manufacturing of conventional products usually entails the fabrication of components, assembling sub-systems from these components, assembling higher-level systems from these sub-systems, and so on, resulting in the complete product. Some number of components, sub-systems, and systems may be rejected for not working or meeting the specifications during manufacturing or assembly, but the processes involved frequently allow the defective elements to be identified before assembly and replaced so that they do not impact sub-systems or systems in later manufacturing stages. Additionally, there are often the capability for re-working the components, sub-systems, and systems, by swapping out bad elements or making repairs to the elements to enable them to work and meet the specifications. Such types of re-works are common in most assembled products during manufacturing such as cars, cell phones, appliances, tools, etc.

In contrast, microsystems manufacturing involves the processing of substrates using a process sequence, which is composed of an ordered array of fabrication steps. Often these fabrication steps involve chemical, electrochemical, plasma, thermal, etc. processes where the effects on the substrate and materials are not reversible. Therefore, if a processing step results in a poor outcome, the substrate (s) and all the manufacturing work-up to that point in the process sequence can be completely lost.

M. Huff, *Process Variations in Microsystems Manufacturing*, Microsystems and Nanosystems, https://doi.org/10.1007/978-3-030-40560-1_9

Indeed, most microsystems manufacturing process sequences are composed of many (e.g., dozens to hundreds) individual, sequentially performed, processing steps, and if any one step goes badly during manufacturing, it can have disastrous consequences. A batch of substrates can successfully proceed through each and every fabrication step all the way to the end of the process sequence only to have a failure at the last processing steps resulting in the complete loss of the entire batch of wafers and all the manufacturing work to get to that point. Additionally, if the devices do not function or do not meet the specifications, they usually cannot be re-worked to correct any deficiencies that may exist.

Unlike other types of manufacturing operations, microsystems devices cannot be swapped out or replaced during or after manufacturing if they don't work. Instead, if some problem occurs to render some number of device die dysfunctional, these dies continue to reside on the substrates all the way to the end of the manufacturing process. Even though these die will never function, they continue to take space on the substrate until the entire manufacturing process is completed. The result is that subsequent manufacturing steps on these nonfunctional devices are a wasted cost.

Previous chapters reviewed various important microsystems manufacturing methods, provided an understanding of the various reasons that microsystems manufacturing processing steps and process sequences result in dimensional and material property parameter variations, reviewed methods that are useful for analyzing the resultant variations in device parameters resulting from manufacturing processes, and described how the parameter variations as a result of these processes can impact device performance. The focus of this chapter is to describe the methods used in microsystems manufacturing for performing yield analysis and methods to ensure that the devices produced meet certain quality and performance standards. An important desirable attribute of microsystems manufacturing is high yield. High yield directly translates into lower manufacturing cost and is only enabled using good-quality assurance and control methods.

9.2 Importance of Manufacturing Yield

Manufacturing yield is the percentage of the products made that function and/or meet or exceed the pre-defined performance specifications relative to the total number of devices produced. The yield of any microsystems manufacturing process is dependent on a number of complex factors including the type of microsystem technology to be produced; how aggressive the critical dimensions are in relation to the lithography capability used; the overall number of fabrication steps in the manufacturing process; whether there are any processing steps that are inherently risky; etc. [1].

Regardless of the type of microsystem technology being produced, most fabrication process sequences start with low yield numbers. It is not uncommon for the first few lots at the beginning of a fabrication production run to have zero or close to zero yields. Later, as more lots are produced and the issues in processing steps, process integration, and tools are resolved and improved, the yield may reach an inflection point whereby it can rapidly increase to higher yield levels. This period of time is referred to as "yield learning." Subsequently, as the process sequence and processing steps are optimized and the process matures, the yield will eventually level out. Characteristically, a significant number of lots are required before a microsystems manufacturing process usually achieves higher yield levels. Nevertheless, it is important to point out that the actual relationship between yield and number of lots produced will depend on the specific circumstances, and each circumstance will be unique.

The economic importance of yield in microsystems manufacturing is easily demonstrated. For example, let us assume that a foundry is manufacturing microsystems devices that sell for \$10 per die, the foundry is able to produce a quantity of 10,000, 200 mm diameter substrates per month, and each substrate has 1000 die per substrate. Let us also assume that the manufacturing cost per die is \$4. Based on this information, we can calculate that the foundry produces 10M die per month, and the monthly manufacturing cost of 10M die is \$40M. Let us further assume that the production yield is currently running at 30%. At this yield level, the monthly sales revenue would be \$30M. Consequently, gross profit would be equal to \$30M minus \$40M, or equal to a negative \$10M per month, and the foundry would be losing money on each die produced.[1] However, if the yield were to be increased to 50%, the revenue would increase to \$50M per month, resulting in a gross profit of \$10M per month. If the yield were increased further to 85%, the gross profit would jump to \$45M per month. Therefore, as this example demonstrates, the yield of the manufacturing process directly relates to whether the business is not profitable, breaks even, or is profitable.

9.3 Definitions of Microsystems Manufacturing Yield

A very general definition of the yield was given above and, while helpful, is not very specific. Other more exact definitions of yield have been developed that provide a more useful guidance to determine the manufacturing yield. Substrate manufacturing yield can be defined as the ratio of the number of successfully manufactured substrates to the total number of substrates that started the manufacturing process, which can be expressed as a percentage as follows:

[1]This is a very simplistic example and assumes that indirect costs, such as marketing, sales, legal, management, etc., involved in the business operation are included in the die production cost. Additionally, there would also likely be a cost associated with the development of the manufacturing process for the microsystems that has not been included.

$$\text{Substrate manufacturing yield} = \frac{\text{\#Substrates at end of manufacturing process}}{\text{\#Substrates that began manufactured process}} \times 100\%. \tag{9.1}$$

Equation (9.1) calculates the yield before the substrates are diced and packaged. This is sometimes called the "line yield." Substrate manufacturing yield has a notable problem in that it does not say anything about how many dies on the substrates are working at the end of the manufacturing process. Specifically, each substrate has a multiplicity of die, and it is typical for some number of the die to not work. Therefore, a more useful yield is the die yield, wherein the words "working (or good) die" substituted for the word "substrate." This is commonly called the "wafer sort yield" and is given by:

$$\text{Wafer sort yield} = \frac{\text{\#Working (or good) die}}{\text{\#Die on substrate}} \times 100\%. \tag{9.2}$$

Working die is determined by each die being individually tested and marked as to whether they function or not, before the substrate is diced.

Additionally, there is a resultant yield associated with packaging of the die once they have been through the wafer sort yield and have been diced, packaged, and go through a final test, which can be expressed as follows:

$$\text{Packaging yield} = \frac{\text{\#Packaged die passed final test}}{\text{\#Good die before packaging}} \times 100\%. \tag{9.3}$$

The total yield of the manufacturing process can be expressed as the product of the individual yields of each stage of the manufacturing. For example, the yield of a microsystems production process may be defined as the product of the yields of various major manufacturing steps as follows:

$$\begin{aligned}\textbf{Microsystems manufacturing yield} = {} & (\text{Front-end substrate manufacturing yield}) \\ & \times (\text{wafer sort yield}) \\ & \times (\text{packaging yield}) \times 100\%.\end{aligned} \tag{9.4}$$

An important point about Eq. (9.4) is that since the manufacturing yield is the product of the yields of the individual manufacturing steps, the overall yield will never be higher than the lowest individual yield of any of the manufacturing steps.

It is important to point out that the yield definitions above only specified whether the die was working (i.e., functional) or not. This is one definition of yield, but there is another more detailed definition called parametric yield, which is the percentage of die that meet the product specifications. Both types of yield are explained in more detail in the following sections.

9.4 Microsystems Manufacturing Yield Monitoring and Analysis

Yield monitoring and analysis methods are dependent on yield models, which are important in microsystems manufacturing for a number of reasons. First, yield models allow the cost of the manufacturing to be more accurately estimated, which is critical for determining if the manufacturing enterprise is, will be, or can be profitable. Second, yield models allow for better planning with regard to manufacturing resource allocation. Third, yield models enable the ability to more easily identify what part(s) of the device design and manufacturing process are in need of improvement. Lastly, yield models allow different production lines and/or foundries to be compared to one another.

Yield monitoring is also called *yield testing*, of which there are generally two types. The first is *functional yield testing*, which is the percentage of the manufactured dice that are operational. Functional yield testing is not meant to be extensive but instead is a more cursory test and selection process based on whether the devices work or not. The function yield testing is often performed on substrates before dicing, packaging, and more comprehensive device parametric performance testing in order to avoid additional costs associated with later manufacturing steps and more extensive testing methods.

The second is *parametric yield testing*, which is the percentage of die that meets specific pre-defined performance requirements. Parametric yield testing is usually performed on die that have previously been determined to work either before or after the substrate has been diced. It is typified by a more extensive testing protocol to provide a sufficiently high level of confidence that the accepted dies meet or exceed the required performance specifications. Parametric yield testing may result in either acceptance (i.e., pass) or rejection (i.e., fail) of individual die depending on whether they meet or fail to meet the performance specifications. Alternatively, the tested die may be binned according to some pre-defined criteria where the working devices are categorized based on the results of the parametric testing. Higher performance devices may be sold at a higher price level than the other devices having lesser performance levels.

Since parametric testing entails conducting and collecting data on a full array of exhaustive performance tests, the results can provide very good insight into device behavior. If the device models are adequate, parametric test data can be used to enable the improvement of the device design and manufacturing processes. This can be a very important approach for increasing yields and extracting optimal performance from the devices. Additionally, the variations measured in the MEMS device outputs during parametric testing provide insight to the device parameter (both bias and random) variations. Conversely, functional yield testing does not provide much understanding of parameter variations.

9.4.1 Functional Yield

Functional yield has only two outcomes; either the device works, or it does not. Functional yield models for microsystems are usually based on the presence of defects, specifically defects that result in the device not working. These are more appropriately called "killer defects" since there is always defects on substrates, and some defects will cause devices not to work, while others do not impact the functionality of the devices.

Some examples of different types of defects commonly seen are illustrated in Fig. 9.1, which is a meandering electrical interconnection line of metal with two electrical pads on either end. Defect 1 is a point defect caused by a particle that completely overlays a portion of the metal line. This sort of defect could result in an open circuit condition, thereby rendering the circuit nonfunctional. Defect 2 is also a point defect caused by a particle located where it overlays two sections of the electrical interconnect, and this could result in an electrical shorting between the two portions of the metal interconnect. This defect would likely also render the circuit nonfunctional. Defect 3 is a point defect due to a particle on the substrate, but in this case it is located in the field region away from the meandering metal

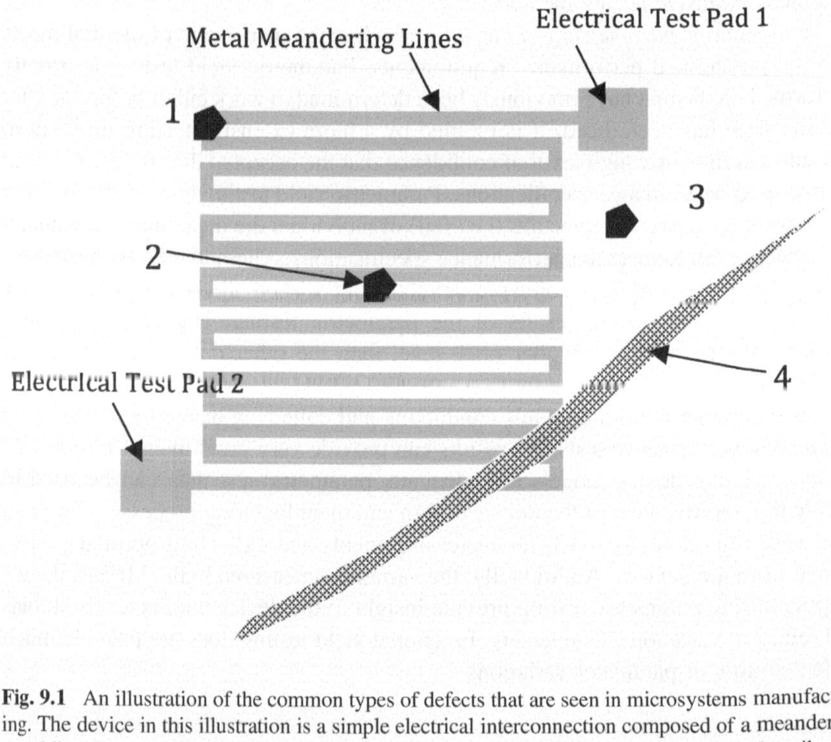

Fig. 9.1 An illustration of the common types of defects that are seen in microsystems manufacturing. The device in this illustration is a simple electrical interconnection composed of a meandering metal line that extends between two electrical testing pads. Four defects are shown as described in the text

interconnect. The presence of this defect will not likely have any direct impact on the functionality of the metal interconnect and would therefore probably not be a killer defect. Defect 4 is not a point defect. Instead this defect is a scratch on the surface of the substrate that could be caused by a wafer-handling problem. This defect overlays a section of the meandering metal interconnect and would likely result in an open circuit condition at that location. While this is not a point defect, it is a killer defect nonetheless since it renders the circuit nonfunctional.

As a general rule, the types of large area defects exhibited in defect 4 are considered outliers since if the manufacturing operations are strictly controlled, these types of defects should not happen or would happen so infrequently that they have no significant impact on the yield averages over time. The large area defects are also not amenable to analysis and therefore are usually not considered in yield models.

9.4.1.1 Functional Yield Models Based on Point Defects

The models for functional yield are based on point defects that result in die not being functional and therefore are considered as killer defects. These models were created for IC microsystems but can be applied to MEMS microsystems as well [2]. The functional yield, given by Y, can be written as a function of the average number of defects per unit area or defect density, D, and the area, A, of the substrate (or die) as follows:

$$Y = f(A, D, \alpha_i), \tag{9.5}$$

where α_i are variables representing other parameters that impact the yield depending on different yield models.

If it is assumed that the defects are uniformly distributed across the substrate surface, exist as points that are independent, and each defect results in the device not working (i.e., all defects are "killer" defects), then the calculation of yield can be based on a quantity of n defects distributed randomly over N die, where the probability that a given die contains k defects is given by the binomial distribution given by:

$$P_k = \frac{n!}{k!(n-k)!} \frac{1}{N^n} (N-1)^{n-k}. \tag{9.6}$$

If n and N are both large, but the ratio of $m = n/N$ is finite, the binomial distribution can be approximated by the Poisson distribution as follows:

$$P_k \cong e^{-m} \frac{m^k}{k!}. \tag{9.7}$$

Using Eq. (9.7), the probability that a die contains no defects is given as:

$$P_0 = e^{-m}. \tag{9.8}$$

Note that Eq. (9.8) is also the calculation of the yield.

If the area of the die is equal to A on a substrate having N die, then the total area of the substrate with die is NA. Therefore, the defect density can be written as $n/NA = D$, and the average number of defects per die can be rewritten as:

$$m = \frac{n}{N} = AD. \tag{9.9}$$

Using Eqs. (9.8) and (9.9), the calculated yield can now be expressed as follows:

$$Y = P_0 = e^{-AD.} \tag{9.10}$$

While Eq. (9.10) was one of the first equations developed for estimating yields in the early days of IC microsystems manufacturing, it was discovered that it tended to underestimate the actual resultant yield in many circumstances. Two noticeable limitations of Eq. (9.10) are that it assumes that the entire area of the die is equally impacted by the presence of defects and the defects are uniformly distributed over the substrate. These assumptions are often incorrect.

A simple modification to Eq. (9.10) to account for the fact that some areas of the die will be impacted by defects and others will not is to replace the area, A, with a new variable called the critical area, A_c, where the critical area is only that portion of the microsystems die where a defect could result in the device not working.[2]

The other limitation of the yield model above is that it assumes that the defect density is uniform across the substrate. However, in practice it is found that point defects are caused by particulates having densities and sizes that are not uniform. An improved derivation of the functional yield accounting for a non-uniform defect density assumes that the defect density, D, can be represented by a probability density function that has a shape representative of the actual defect density spatial distribution. This model is often called the Murphy yield model where the yield is then found using the following integral:

$$Y = \int_0^{\infty} e^{-A_c D} f(D) dD, \tag{9.11}$$

where $f(D)$ is a function that represents the distribution of non-uniform defects [2]. Some general probability distributions have been shown to provide reasonably good estimations of $f(D)$ [3]. Specifically, the triangular function is widely used in IC manufacturing and results in the following for the yield:

[2]Most dies have regions where the presence of a defect may not impact the device operation. This can be especially true for MEMS devices.

$$\text{Triangular Function}: Y = \left[\frac{1 - e^{-A_c D_o}}{A_c D_o}\right]^2, \tag{9.12}$$

where D_o is the average defect density.

Another yield model uses a gamma probability distribution, which is believed to more accurately represent defect density distributions seen in practice. The gamma function accounts for the defect density that may not be uniform across the die and is given as [4, 5]:

$$f(D) = \frac{1}{\Gamma(\alpha)\beta^{\alpha}} D^{\alpha-1} e^{\frac{D}{\beta}}, \tag{9.13}$$

where $\Gamma(\alpha)$ is the gamma function and α and β are two parameters used to describe the defect density distribution. In the gamma function, the average defect density is $D_o = \alpha\,\beta$, and the variance of the defect density is $\mathrm{var}(D) = \alpha\,\beta^2$. The probability of a die having k defects is calculated as:

$$P_k = \int_0^{\infty} e^{-m} \frac{m^k}{k!} f(D) dD = \frac{\Gamma(k+\alpha)}{k!\Gamma(\alpha)} \frac{(A_c\beta)^k}{(A_c\beta + 1)^{k+\alpha}}, \tag{9.14}$$

where m is the average number of defects per die. The yield using this model is given by:

$$Y = \left(1 + \frac{A_c D_o}{\alpha}\right)^{-\alpha}. \tag{9.15}$$

Using the gamma function, the yield depends on the average defect density, the die area, and the parameter, α, which is referred to as the cluster parameter. The cluster parameter increases with decreasing variance in the defect density distribution and is the factor accounting for a non-uniform defect density across the die.

If the cluster parameter, α, is large wherein the variability of the defect density is small and there is little clustering of defects, the gamma function approaches a delta function, which is the Poisson model, and the yield is given by:

$$Y = \lim_{\alpha \to \infty} \left(1 + \frac{A_c D_o}{\alpha}\right)^{-\alpha} = e^{-A_c D_o}. \tag{9.16}$$

Conversely, if the cluster parameter, α, is large where the defect density is highly non-uniform, the yield in the limit is given as:

$$Y = \lim_{\alpha \to \infty} \left(1 + \frac{A_c D_o}{\alpha}\right)^{-\alpha} = \frac{1}{(1 + A_c D_o)}. \tag{9.17}$$

In practice, the clustering parameter should be found empirically and is usually determined by counting particles [6, 7]. Nevertheless, in the absence of empirical data, it has been found that using a value of $\alpha = 2$ is often a reasonably good estimate for certain microsystems manufacturing environments.

The defect density has been shown to adhere to an inverse power-law relationship that is dependent on the defect size as follows:

$$D(x) = \frac{N}{x^p}, \tag{9.18}$$

where x is the diameter of the defect, N is a parameter that depends of the type of technology being made, and p is a parameter that must be determined experimentally. The average defect density, D_o, is found by integrating Eq. (9.18) over the substrate surface as follows:

$$D_o = N \int_{x_o}^{\infty} D(x)dx = N \int_{x_o}^{\infty} \frac{1}{x_p} dx = \frac{N}{(1-p)} \left(1 - x_o^{1-p}\right), \tag{9.19}$$

where x_o is the minimum defect size and usually taken to be equal to the minimum feature size for the technology being produced.

Another model for calculating the defect density uses a specific method to determine D_o. As an example, using the negative binomial model for a die results in the following [8]:

$$D_o = \frac{\alpha\left(\sqrt[\alpha]{\frac{1}{Y}} - 1\right)}{A_c}. \tag{9.20}$$

The critical area was mentioned above as a better representation of the area used for the computation of the defect densities and yield since it accounts for the fact that some areas of the die are more impacted by the presence of defects than others. Specifically, the critical area, A_c, is defined as the area of the die that a defect will probably result in a failure. The critical area can be found using the following equation [3]:

$$A_c = A \int_{x_o}^{\infty} P_o F(x) D(x) dx, \tag{9.21}$$

where A is the area of the entire die, x_o is the minimum defect size, $D(x)$ is the defect size distribution, and $P_oF(x)$ is the probability of a failure and is dependent on the size of the defect. It should be noted in all of the yield models that the estimated yield Y decreases with increasing values of the defect density and critical area as expected.

It is important to point out that there is an implied assumption that defects caused by particulates always result in the microsystems devices not functioning. The yield models discussed above do not provide a justification for this assumption. Nevertheless, these models are reasonably accurate predictors for advanced IC microsystems. However, it should be cautioned that attempting to apply these functional yield models to MEMS microsystems without modification may result in errors of the yield estimates.

9.4.1.2 Functional Yield Measurement Tools

There are a number of different equipment tools that can be used to help determine functional yield, many of which were briefly reviewed in the chapter on metrology. For larger defects, optical microscopy can be used to count defects and calculate the defect densities. Confocal microscopy and conventional SEMs are also used for defect density measurements, especially for defects that are too small to be observed using optical microscopy. These systems are manually operated and therefore will entail a tedious process but may be the best option for customized MEMS microsystems devices. The measurement process will often involve making maps or taking images of the areas being inspected and using this data to develop custom models for evaluation of the defects and their impact. Since the microsystems devices can be of a wide variety of different device types, the engineers will usually need to develop specialized techniques in order to properly count and quantify the defects, their size, and distribution and impact on the devices for each individual MEMS design.

The IC industry has available automated SEM defect review and classification systems that are able to quickly scan the substrate surfaces, identify impactful defects, and quantify and compare the results. However, these high-end automated systems require a substantial setup time and are very costly, and therefore they are almost exclusively used in high-volume advanced IC manufacturing environments. These systems can be used in MEMS manufacturing environments, but the production volumes would need to be sufficiently high for them to be financially justifiable.

While imaging inspection and analysis often provide a quick method for defect measurements, the obvious limitation is that it is entirely based on what can be seen or observed and not an actual functional test of the device. The conclusive functional test of any microsystems device is whether the device works or not, and this requires more sophisticated test methods including electrical, electromechanical, etc., types of tests. The electrical and mechanical testing of devices can be used for deterministic measurements of whether the microsystems devices are functional or not. Given the highly customized nature of most MEMS microsystems, the type and design of functional test performed will vary greatly and like many devices are also usually customized to each MEMS device design.

9.4.2 *Parametric Yield*

In contrast to the functional yield models and tests discussed above, parametric yield analysis is focused on determining whether the microsystems devices meet or exceed some pre-defined level of performance metrics specified by the product or application requirements. More exactly, the parametric tests are designed to determine if the output of the device exhibits a performance level that is within an acceptable range of values. This is usually conducted by performing parametric tests on each device or some representative sample set of the devices. Parametric testing can be performed at the wafer level, before or after packaging. Parametric testing reveals the variations in the device output behavior due to variations in the device parameters as a result of manufacturing variations.

9.4.2.1 Parametric Yield Model

As discussed in the previous chapter, random variations exhibited in the device parameters, such as the dimensions and material properties, can be represented by probability distribution functions. Moreover, the device output can also be represented by a probability distribution function derived from the results of device parametric testing or alternatively that has been generated using Monte Carlo analysis. The probability distribution functions of the device output behavior can be used to estimate the parametric yields. As a matter of practice to avoid confusion among the types of parameters being referred to, the characteristics used to describe the device, such as the dimensions and material properties, are called device input parameters, while the characteristics used to describe the device output behaviors are called device output performance parameters, and their distributions are called manufacturing functions. Types of microsystems device output performance parameters include the flow resistance of a microchannel; the frequency response of a MEMS resonator; and the voltage output of a piezoresistive strain sensor.

For simplicity, it is assumed that the device output performance parameters can be represented by a normal distribution function. In one dimension, x, this would be written as follows:

$$f(x) = \frac{1}{\sigma\sqrt{2\pi}} \exp\left[-\frac{(x-\mu)^2}{2\sigma^2}\right]. \tag{9.22}$$

The probability of an occurrence of a specific value of some device output performance parameter represented by the variable, x, in Eq. (9.22) is found by integrating over the area from x_1 to x_2 as follows:

$$\text{Probability}(x_1 \to x_2) = \int_{x_1}^{x_2} f(x)dx. \tag{9.23}$$

Note that Eqs. (9.22) and (9.23) are identical to what we saw in Chap. 8 except now the normal probability distribution function is being applied to the output performance of the device rather than one of the input parameters, such as microchannel radius.

An adequate sample size of the individual measurements of the variations of the device output performance parameters can be represented by a normal distribution having a mean value, $\bar{x}$, and standard deviation, S. The mean value provides information about the location (along a defined parameter axis) of the maximum probability of the device output performance parameter, and the standard deviation provides information on how much spread there is in the device output performance parameter sample data.

Since there are often several device input parameters, each having its own distribution, used to calculate the microsystems device output performance as was seen in Chap. 8, the analysis of the device output performance response becomes complicated very quickly. Therefore, the calculation of the probability distribution function of the device output performance is usually performed using methods such as Monte Carlo analysis.

In this section, it will be assumed that the device output performance response is the form of an device output probability distribution function that has been measured using either some number of device tests or calculated using variation analysis methods such as Monte Carlo, and the focus will be on how the device output performance probability distribution functions relate to the whether the device meets the application specification, that is, passes the parametric yield testing.

An example of the resultant probability distribution function of a microsystems output performance response represented by the variable x is shown in Fig. 9.2. Importantly, the distribution is not based on the device performance output values that are acceptable for the device in the intended application or product. That is, what constitutes an acceptable value or range of values of any parameter is determined separately from the probability distribution function shown in Fig. 9.2. The probability distribution function in Fig. 9.2 is based on the manufacturing process sequence used to make the device and is called the manufacturing function.

Typically, there is not one specific acceptable value of device's output performance parameter, x, that is acceptable but instead a range of values of x where the device performs adequately to meet or exceed the performance requirements for the intended application. This range is usually called the "acceptance range." For example, for a certain application of a microchannel, the acceptable output response of the device could be a certain volumetric flow rate, Q. That is, based on a criterion for the device to function adequately in the intended application, the volumetric flow rate must be within this range of values. Accordingly, there will also be a specification of the differential pressure across the microchannel as well, and therefore an acceptable range of flow resistance can be determined.

The acceptance function, $A_f(x)$, can be represented by a function that states that the microsystems device has acceptable performance if the value of the device output performance parameter, x, has a value that is between two limits of the acceptance range. For example, the acceptance range limits can be defined, as $x_o \pm \omega$, where x_o

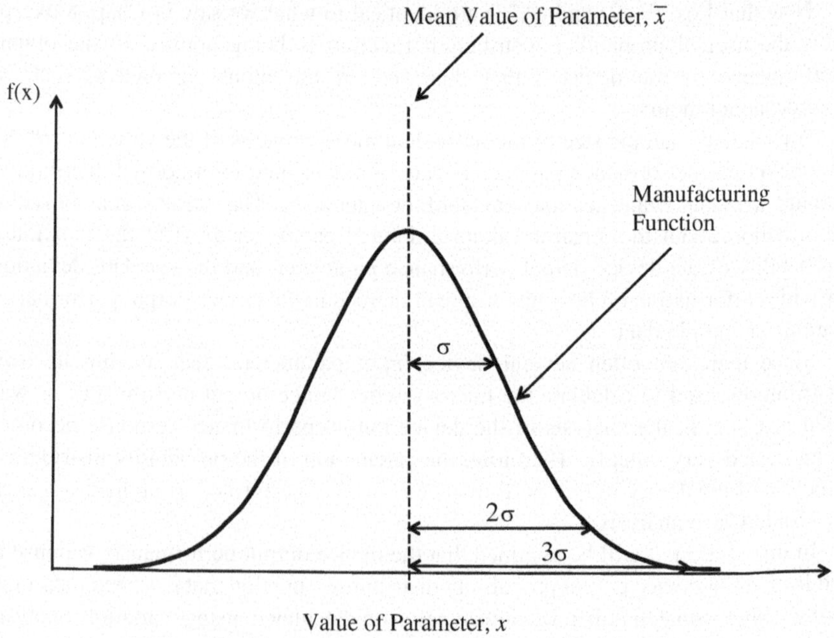

Fig. 9.2 An illustration of the probability distribution function of a device output performance parameter given by the parameter, *x*. For instance, the parameter *x* could be the flow resistance of a microchannel. This distribution is referred to as the manufacturing function

is the midpoint of the optimal range of values of x and x_o has a width equal to 2ω. The acceptance function can be represented by a relatively simple relationship as follows:

$$A_f(x) = \begin{cases} 0, |x - x_o| < \omega \\ 1, |x - x_o| \leq \omega \end{cases}. \tag{9.24}$$

The acceptance function is shown graphically in Fig. 9.3 and is equal to 1 if the value of the parameter x is between $x_o \pm \omega$ and equal to 0 outside of this range. In practice, it is desired that the value of x_o is equal to the mean value of the output performance parameter x, thereby aligning the center of the manufacturing and acceptance functions. The value of ω that is the cutoff point of either side of x is based on the values of the parameter when the device element transitions from meeting the performance requirements to not meeting the performance requirements.

Using Eq. (9.22) for the manufacturing function, where the population mean, μ, and standard deviation, σ, are replaced with the sample mean, $\overline{x}$, and standard deviation, S, and combining with Eqs. (9.23) and (9.24), allows the parametric yield to be determined as follows:

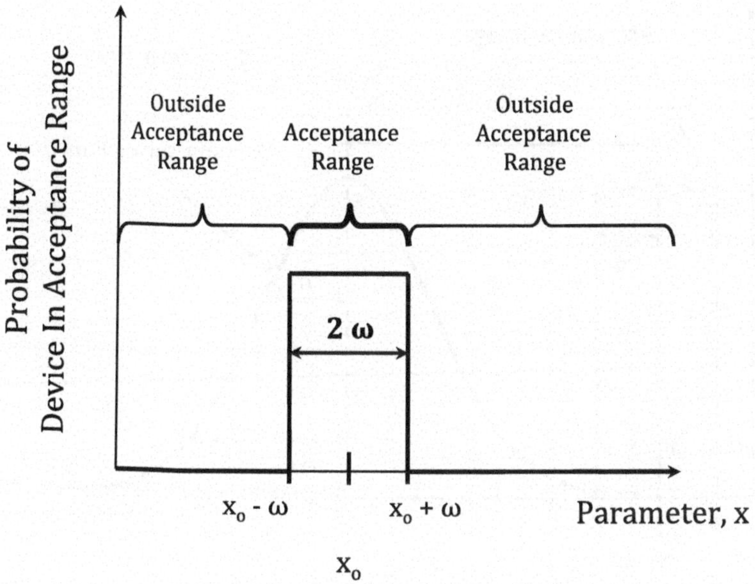

Fig. 9.3 An illustration of an acceptance function that is used in setting the boundaries for whether a microsystems device meets or exceeds the required performance specifications

$$Y(x, x_o, \omega) = \int_{x_o-\omega}^{x_o+\omega} \frac{1}{S\sqrt{2\pi}} \exp\left[-\frac{(x-\bar{x})^2}{2S^2}\right] dx. \tag{9.25}$$

Equation (9.25) can be rewritten as follows:

$$Y(x, x_o, \omega) = \frac{1}{2}\left[\text{erf}\left(\frac{x_o+\omega-x}{\sqrt{2}S}\right) - \text{erf}\left(\frac{x_o-\omega-x}{\sqrt{2}S}\right)\right]. \tag{9.26}$$

It is often useful to replace the boundaries of the acceptance function, $x_o \pm w$, with a single variable such as a and b, which allows Eq. (9.26) to be rewritten as:

$$Y(x, x_o, \omega) = \frac{1}{2}\left[\text{erf}\left(\frac{b-x}{\sqrt{2}S}\right) - \text{erf}\left(\frac{a-x}{\sqrt{2}S}\right)\right]. \tag{9.27}$$

The overlap of the manufacturing and acceptance functions is shown in Fig. 9.4. The microsystems device has a manufacturing function with a mean and standard deviation, given by $\bar{x}$ and S. The acceptance function has been superimposed over the manufacturing function and is the shaded box having width 2ω and is equal to 1 over the parameter values from $x_o \pm \omega$.

The parametric yield is given by the number of samples, or devices, that would be included within the overlap of the manufacturing function and the acceptance range. In Fig. 9.4, the left-hand tail of the normal distribution overlapping with the shaded

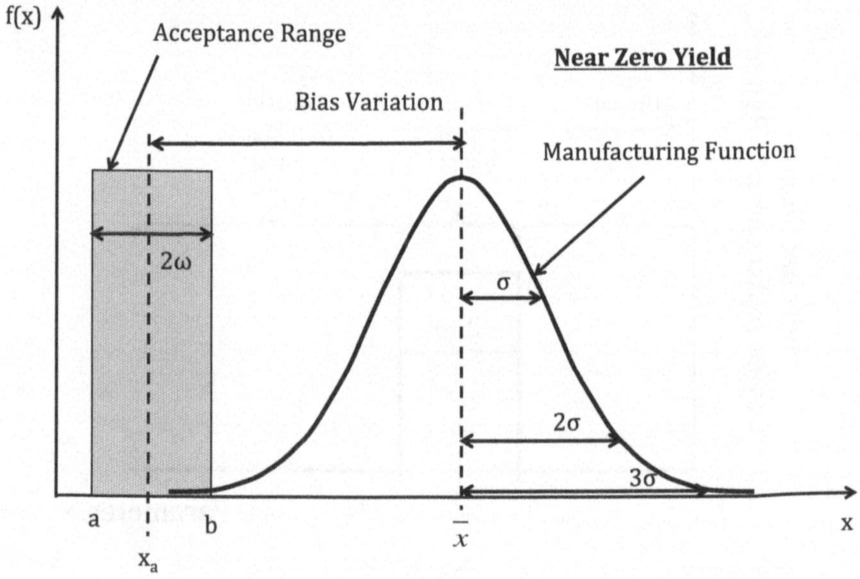

Fig. 9.4 An illustration of the manufacturing yields for a microsystems manufacturing process wherein the resultant yield is given by the overlap of the manufacturing function, which is a normal distribution, and an acceptance range shown as a shaded rectangle. In this scenario, the yield would be very low since the overlap between the manufacturing function and the acceptance range results in very few of the manufactured devices being within the acceptance range

acceptance function gives the parametric yield, which would be a very small resultant yield in this scenario. It can be observed that one reason for the low yield is that the middle of the acceptance range is displaced from the mean of the manufacturing function. The amount of displacement is equal to the bias variation present in the manufacturing process. Methods to move the manufacturing function so that it better aligns with the acceptance range are called design centering and will be discussed in the next chapter.

As noted above, the manufacturing functions as well as acceptance functions can be used on any of the parameters of the microsystems devices, including both the device input parameters such as the dimensions and material properties and the output performance parameters. It should also be obvious that the same sort of determinations about yield of individual processing steps during manufacturing can be made regarding both input and output parameters. For example, a manufacturing function of the radii of microchannels after etching could be an input parameter that is routinely measured during manufacturing. Such measurements of the device input parameters during manufacturing are in fact one of the important means of quality control during production. Nevertheless, depending of the type of MEMS device and the parameters involved, it may not be possible or desirable to measure each and every of the input device parameters during manufacturing. For example, the lateral dimensions that may be an important input device parameter may not be directly measurable using metrology methods. In those cases, it may be possible to measure the parameter indirectly or even occasionally perform destructive analysis on some small number of the devices in order to gather some statistical information about such a parameter. In short, manufacturing functions and acceptance ranges can be used for yield estimation both on the devices during manufacturing as well as the devices after manufacturing.

The device parameter space was portrayed in Figs. 9.2, 9.3 and 9.4 (and related text) as having only one dimension. In most practical situations, the parameter space will have a number of different parameters whether the device input parameters or the device output performance parameters are being considered. In such cases, the parameter space must be represented by a dimensional space equivalent to the number of parameters. If there are a number of n input parameters associated with a microsystems device, then that parameter space will have n input dimensions. A similar statement can be made for the output parameter space.

In considering an input parameter space of two dimensions, a single point in this parameter space represents a single MEMS device. The variations in the device input parameters are represented by a region of parameter space that can be called the variation (or tolerance) region and labeled as R_V. Every device manufactured is represented by a single point that lies within R_v. In general, if there are two input parameters associated with a microsystems device given by x_1 and x_2, they can be illustrated on the two-dimensional plot of Fig. 9.5 by a single point.

Importantly, the actual values of the input parameters x_1 and x_2 for any microsystems device may not be known, but the nominal or design parameter values will be known from the engineering design. That is, only the design value of a parameter, $x_1{}^o$ and $x_2{}^o$, may be known as well as their respective variations, given by $\pm\Delta x_1$ and $\pm\Delta x_2$. The variation region for the input parameters x_1 and x_2 can be represented by Fig. 9.5. Therefore, while a specific point represents an individual

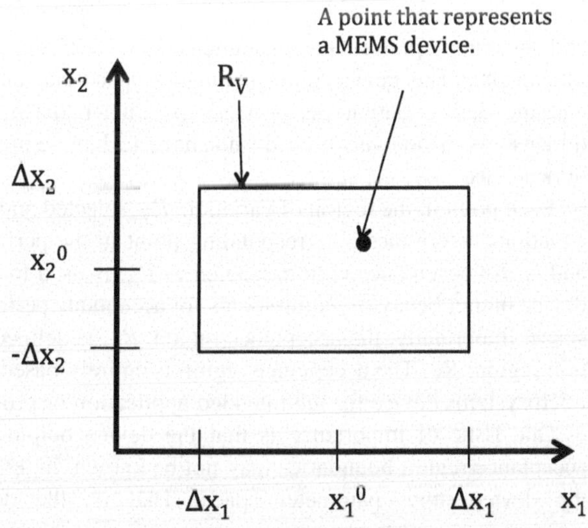

Fig. 9.5 An illustration of a device with two input parameters, showing the desire values, the ranges of each of the parameters, and the variation region

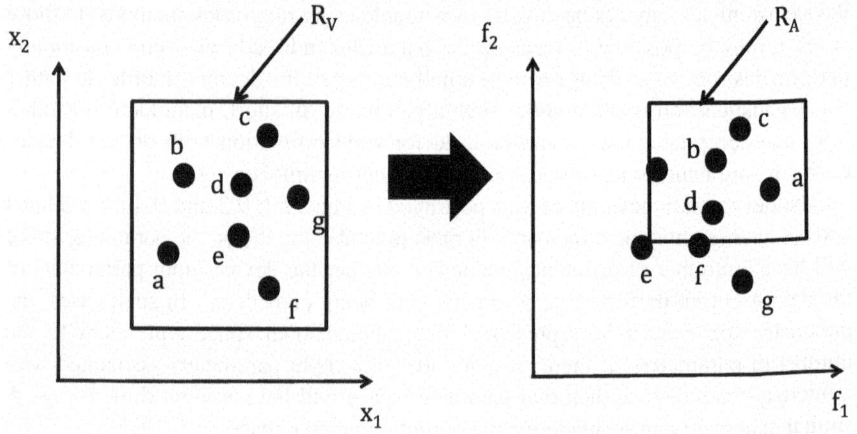

Fig. 9.6 A two-dimensional parameter space (left) and performance space (right) describing a microsystems device and its output behavior. Within the parameter space, each point represents a microsystems device with associated parameter values. The points in parameter space must reside within the region of variation, R_V. If points are selected in the parameter space and an analysis of the device output is performed, the result is a point within the performance space. Points in performance space falling within the region of acceptance represent microsystems devices that would be deemed to be passing the performance requirements

microsystems device, the variation region represents region of all possible values of the device input parameters x_1 and x_2 for a microsystems device.

It is important to point out the differences between points in the input parameter space from the points in the device output performance space. Points within the two-dimensional input parameter space as shown in Fig. 9.6 (labeled a through g) are used in the device analysis equations to determine points in the output parameter performance space (also labeled a through g) representing the output behavior of the microsystems device. In Fig. 9.6 are shown an input parameter space (shown on left) and an output performance parameter space (shown on right) both having two dimensions. The device input parameter space can be one or more dimensions, and the device output performance parameter space can also be one or more dimensions. Moreover, both do not have to have equal number of parameters or dimensions.

Each point in the region of variation, R_V, selected and used in the device analysis equations determines a corresponding point in the performance space with axes f_1 and f_2. An acceptance region, labeled as R_A, is used to represent the microsystems device output behavior requirements for acceptable performance in the performance space. Importantly, the acceptance region, R_A, is defined separately from the variation region, R_V. The acceptance region is entirely based on the requirements of the microsystems device for the intended application or product.

One issue of importance is that the device output requirements, namely, the acceptance region boundaries may not be known, in terms of their locations within the device input parameter space. That is, the device output performance requirements must be transformed into limits within the input parameter space in order to determine the relationship between the acceptance region, R_A, and the variation region, R_V. This can be a difficult task, particularly when most microsystems devices are dependent on a number of different input parameters. As a matter of practice when there are a number of input parameters, computer algorithms will be required in order to determine the boundaries of R_A in the device input parameter space, and in many cases, the problem is so complex that ingenious methods may be required in order to determine R_A efficiently. The acceptance region in the device input parameter space will generally be of irregular and perhaps even discontinuous shape(s). This is due to fact that the output behavior of the device may be a nonlinear function of the input device parameters. Some of the possible shapes that R_A can have in the input parameter space include multiple separate R_A's, regions of unacceptability located entirely within R_A, and others. Methods for determining the region of acceptance in the device input parameter space will be discussed in the next section.

Once the acceptance region is mapped onto the device input parameter space, the manufacturing yield for the production of a microsystems device can be found by the overlap of variation and acceptance regions, R_V and R_A as shown in Fig. 9.7, for a device input parameter space of two dimensions. As seen in this case, the manufacturing yield would be less than 100%. If on the other hand, R_A was located entirely within R_V, then the yield would be 100%. If R_A and R_V do not intersect at all, the yield would be 0%. It is worth noting that Fig. 9.7 only provides an *estimate* of the actual manufacturing yield.

If the performance parameter space is multidimensional, the yield can be written as follows:

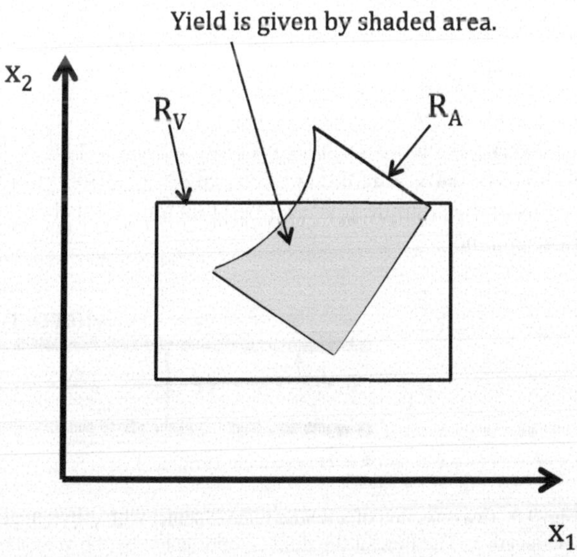

Fig. 9.7 A plot of the intersection of the variation and acceptance regions for a microsystems device having two parameters. The yield is given by the shaded region

$$Y = \int_{L_1}^{U_1} \int_{L_2}^{U_2} \cdots \int_{L_n}^{U_n} \Omega(x_1, x_2, \ldots, x_n) dx_1\, dx_2, \ldots .dx_n, \tag{9.28}$$

where $\Omega(x_1, x_2, \ldots, x_n)$ is the n-dimensional Gaussian probability distribution manufacturing function. It should be noted that performing the integration of Eq. (9.28) can be very challenging.

For non-Gaussian probability distribution manufacturing functions given by $f(x)$, based on use of the Chebyshev's inequality, the probability of a random variable occurring less than a distance of $\alpha\, \sigma_{f(x)}$ from the mean, $\mu_{f(x)}$, is greater or equal to $(1 - 1/\alpha^2)$ regardless of the form of the PDF, where α is a constant. The yield can be determined for this arbitrary PDF using the known value of σ_f, as follows [9]:

$$Y \geq (1 - \frac{1}{\alpha^2}), \tag{9.29}$$

where it is assumed that the upper and lower device specifications are symmetrical about the mean. This can be expanded to situations with multiple dimensions.

9.4.2.2 An Example of a Parametric Yield Model

In Chap. 8, we examined the impact of the physics on the standard deviation of device input parameters raised to higher powers and estimated the variation in the flow resistance output behavior of a microchannel device. In this section, we estimate the manufacturing yields for the microchannel. It will be recalled that the volumetric flow rate in the microchannel device is given by the Hagen–Poiseuille equation that is repeated here for convenience as follows:

$$Q = \frac{\pi r^4 \Delta P}{8 \mu L}, \tag{9.30}$$

where ΔP is the differential pressure across (equal to $P_1 - P_2$ in Fig. 9.8), μ is the fluidic viscosity, L is the conduit length, Q is the volumetric flow rate, and r is the radius of the conduit. It is assumed that the flow is laminar and well-developed in the flow channel.

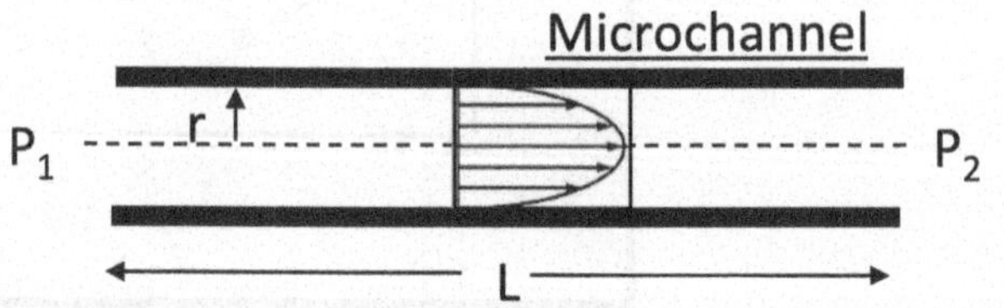

Fig. 9.8 Cross section of a fluidic microchannel with differential pressure $\Delta P = P1 - P2$, radius r, and length L. The flow of the fluid is represented by the velocity profile in the microchannel

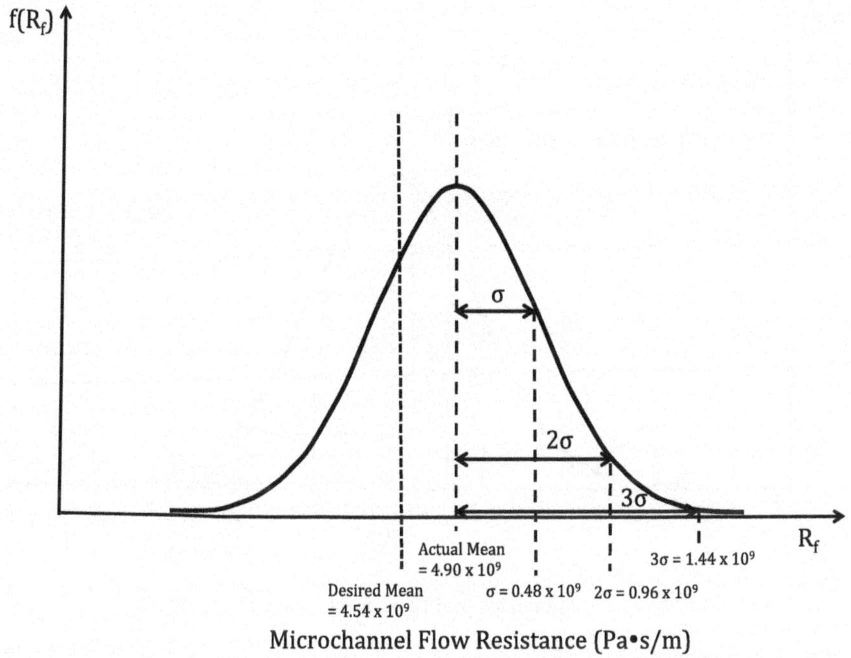

Fig. 9.9 The calculated probability distribution manufacturing function of the flow resistance of the microchannel device

It will also be recalled that the output of the microchannel device was expressed in terms of the flow resistance given by the ratio of the differential pressure across the device, ΔP, and the volumetric flow rate, Q. The desired (i.e., nominal) design parameter value of the radius of the microchannel was 100.00 microns, and the desired (i.e., nominal) parameter value of the microchannel length was 2.00 mm. This would result in a desired flow resistance value of 4.54×10^9 Pa•s/m, which is derived from the application specification.

It is assumed that there exist probability distribution functions for the device input parameters, that is, the microchannel radius and length, there are a sufficiently large number of samples, and a sufficiently extensive Monte Carlo analysis was performed to generate the probability distribution of the device output behavior shown in Fig. 9.9, which is the microchannel device's manufacturing function.

It is found in the microchannel device variation analysis that the mean value of the flow resistance is 4.9×10^9 Pa•s/m with a tolerance interval of 0.48×10^9 Pa•s/m to contain 95% of the population with a level of confidence of 99%. It is noted that there appears to be a bias variation since the mean value of the flow resistance is shifted to the right (i.e., to larger values of flow resistance) from the desired flow resistance. The shift appears to be of an amount of 0.36×10^9 Pa•s/m.

To estimate the yield of the manufacturing process sequence for the microchannel, the acceptance range must be known. This will depend on the

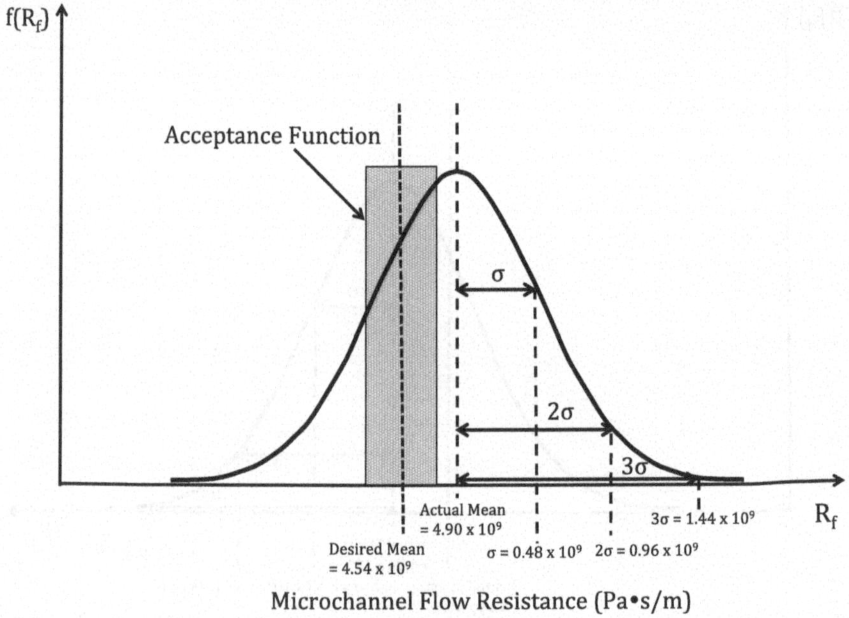

Fig. 9.10 The calculated yield is found by superimposing the acceptance range over the probability distribution of the flow resistance manufacturing function of the microchannel device

application specifications. If the application requires that the flow resistance for a given application have a nominal value of flow resistance of 4.54×10^9 Pa-s/m with a permissible spread in the flow resistance of $\pm 5\%$, or an upper and lower flow resistance of 4.77×10^9 Pa-s/m and 4.31×10^9 Pa-s/m, respectively, then the predicted yield would be less than 40% as shown in Fig. 9.10. If the yield calculation is performed on a computer, the yield would be given integral of the manufacturing function with limits of the lower and upper limits of the acceptance region.

9.5 Yield Estimations Using Sampling Methods

Regionalization and simplicial approximation are both forms of deterministic sampling methods (see Fig. 8.26) wherein the device input parameter sample points are selected using deterministic methods. Monte Carlo analysis is a pseudorandom sampling process. All of these methods can be used to determine the manufacturing yields of microsystems.

9.5.1 *Yield Estimation Using Regionalization*

Regionalization is an analysis method whereby a grid is overlaid onto the variation region of the device input parameter space and an analysis is performed at a representative point within each of the grid elements to determine if the device output meets or exceeds the device output performance specifications at that point in the device input parameter space [10, 11]. Typically, each grid element is given a "passing" or "failing" score depending on whether the analysis of the device's output performance passes or not. An example of this technique of analysis is shown in Fig. 9.11 for a situation in which there are only two input device parameters, x_1 and x_2, and Fig. 9.12 shows the result of the analysis wherein the acceptance region has been estimated.

If the device parameter probability distribution functions are uniform and uncorrelated, then the yield is simply estimated by dividing the number of passing elements in the grid by the total number of elements in the region of variation. However, in most situations the device parameters will have a non-uniform probability distribution functions and can also be correlated. This slightly complicates the analysis.

If the joint probability distribution function $\Phi(x_1, x_2)$ of the parameters x_1 and x_2 is known and correlated, then using the joint PDF along with a random number generation algorithm without performing analysis on the device output allows a reasonable estimate of the variation region according to the joint PDF as shown in Fig. 9.13. Importantly, since no device analysis was performed to generate these points, there is little expenditure of analysis effort. The variation region is then separated into a grid pattern as shown in Fig. 9.14 with the corresponding number of points determined within each element of the grid as shown in Fig. 9.15. These numbers of points within each grid element can be divided by the total number of

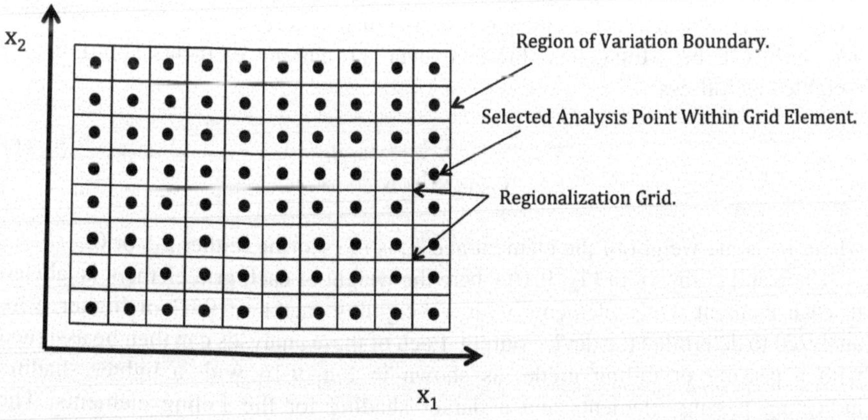

Fig. 9.11 A two-dimensional parameter space that has been divided into elements of equal size over a region of variation of the device parameters. Points within each grid element at the center of the element have been selected as the location points for the device output analysis

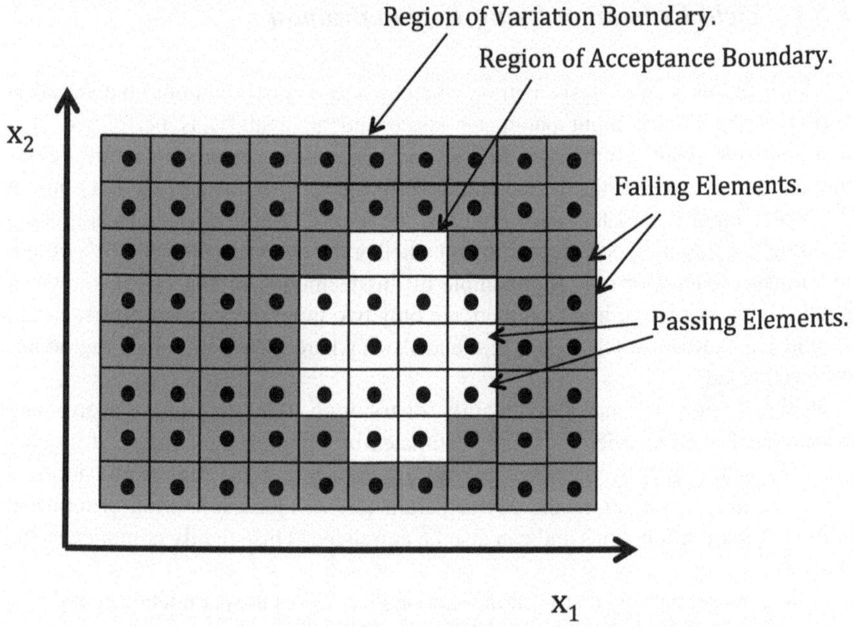

Fig. 9.12 The same grid pattern shown in Fig. 9.11 after an analysis has been performed at each center point within each of the elements. A passing or failing score is assigned to each of the elements depending on whether the device analysis met or exceeded the output specifications. The failing elements are shaded gray, and the passing elements are un-shaded

points created, in this case 75, to calculate a weighting factor for each element in the grid as shown in Fig. 9.16.

Subsequently, an analysis of the device output using the central parameter values in the element with a weighting factor above a certain threshold is performed, and based on this analysis, each element can be assigned a score of either "pass" or "fail." The yield can be estimated as the sum over the passing elements appropriately weighted as follows:

$$Y \approx \sum_{\text{All selected elements}} w_e h_e, \tag{9.31}$$

where w_e is the weight of the element and h_e is the scoring, either a 1 or 0.

The result is shown in Fig. 9.16 where the weight of each grid element is labeled in each element. Only elements with a weighting factor of 0.03 or higher were analyzed to determine the device output. Each of these analyses can then be assigned with a passing or failing grade, as shown in Fig. 9.16 with a lighter shading indicating passing elements and a darker shading for the failing elements. The yield in this instance would be the sum of the weights of the lighter shaded elements, which is 0.41 or 41%.

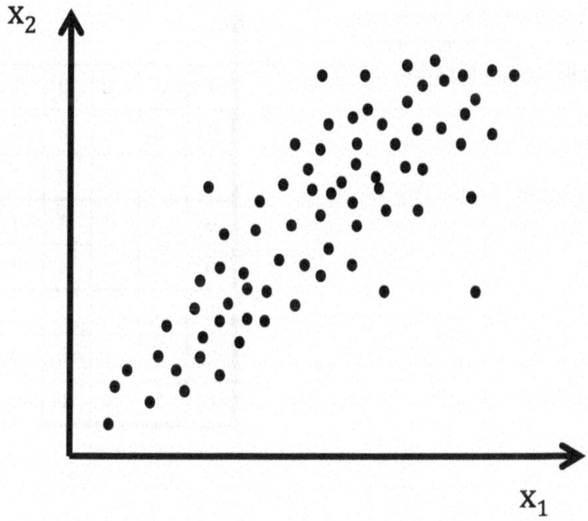

Fig. 9.13 A plot of the variation region of two parameters, x_1 and x_2, that has been generated using their respective probability distribution functions

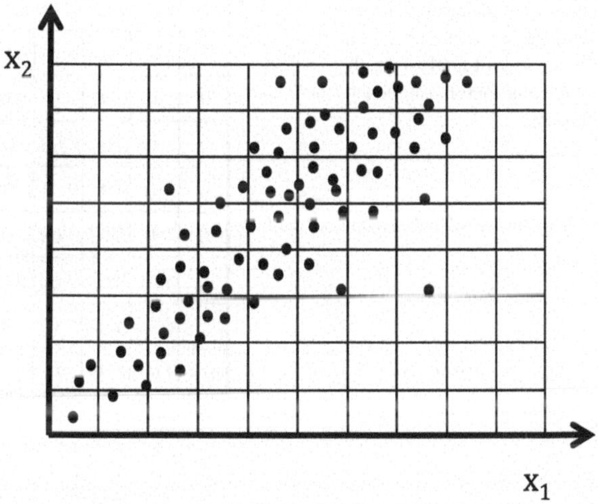

Fig. 9.14 The plot of points generated as shown in Fig. 9.13 overlaid with a grid to define elements

Higher accuracy can be obtained by performing device analysis on every element as opposed only to those that meet or exceed some weighting factor or making the grid elements smaller.

The computational cost of the regionalization method is related to the number of parameters involved in the microsystems device. If a device is described by n parameters, and each parameter has its respective variation range separated into m regions, defined by m, then the number of elements would be m^n, assuming that all of the elements are analyzed. This shows that the number of calculations can grow to a very large number if there are a large number of parameters. Therefore, this method can only realistically be used when the number of device input parameters is below about 5.

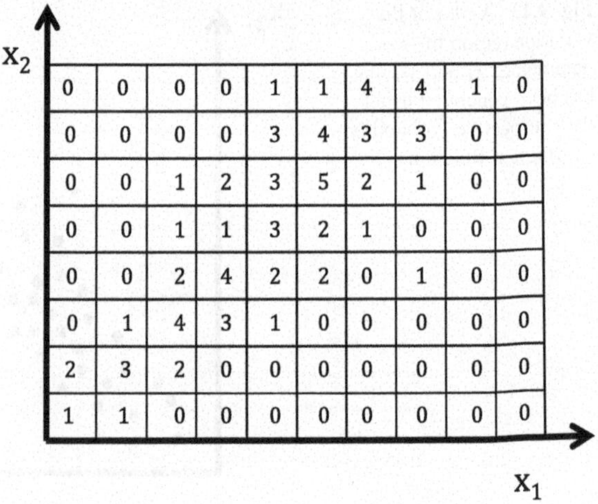

Fig. 9.15 The number of points with each of the elements of the grid

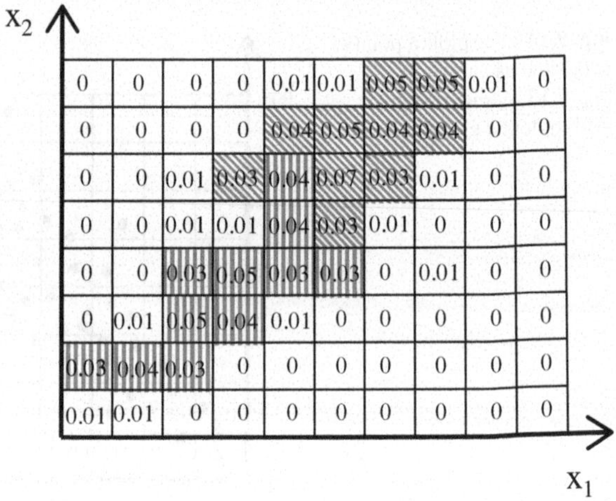

Fig. 9.16 Passing (lighter shading) and failing (darker shading) elements based on device output analysis with weighting factors of each element

9.5.2 *Yield Estimation Using Simplicial Approximation*

Simplicial approximation uses a point-by-point routine to approximately outline the boundary of an unknown acceptance region in the input parameter space to estimate the yield [12, 13]. Using the device input parameter values of selected points, a calculation of the device output is performed to determine if each selected point results in an acceptable output or not (i.e., the point was either a "pass" or a "fail"). If the point results in a pass, the analysis process is repeated by incrementing in one device input parameter direction while keeping all of the other parameters constant. If a point is found to be a fail, then a new direction is selected to choose new points

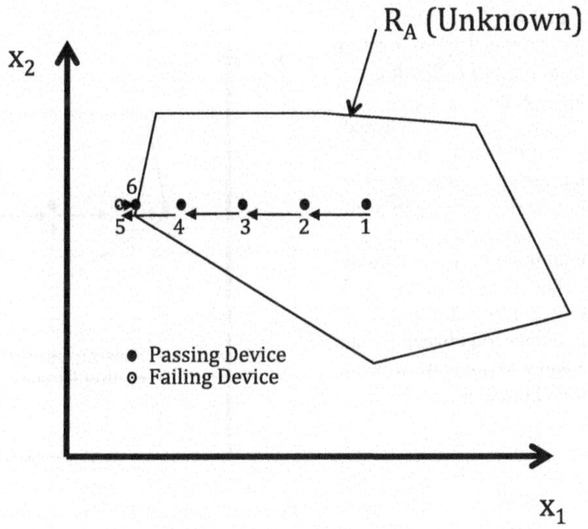

Fig. 9.17 An illustration of simplicial approximation. An acceptance region, labeled R_A is unknown before starting the analysis. The method begins with first finding a point, labeled 1 that has a device output that is passing. A distance is stepped along one parameter axis to test a new point, which is labeled point 2 and is also found to be passing. This process continues to point 5 that is found to fail. A smaller distance is then stepped in the reverse direction from point 5 is then performed until point 6 that is passing is found

for analysis. This searching process continues using a detailed methodology until the boundary of the acceptance region has been approximately outlined in the device input parameter space.

The method of simplicial approximation is illustrated in Fig. 9.17, which shows an unknown acceptance region, given by R_A that is dependent on only two parameters, x_1 and x_2. Only two parameters are shown in this illustration for simplicity, but the technique can be used for cases where there is any number of device parameters.

An initial point labeled 1 is selected for the first device output analysis. It is found that the output at point 1 is passing. Parameter x_1 is then changed by a certain amount while keeping x_2 constant, thereby resulting in point 2. A device output analysis is performed at point 2 and also found to be passing. This process continues using equal distance displacements along the same parameter direction to eventually point 5, which is found to fail. A distance in the reverse direction is determined using interpolation between points 4 and 5 to locate point 6 where another analysis is performed and found to be passing. It is assumed that point 6 approximates the location of the boundary of R_A at one x_1 and x_2 location.

Next, starting again at point 1, equal distant points are selected at increasing values of x_1 as shown in Fig. 9.18. This step-and-repeat process is performed to locate the point 9, which is found to fail. A distance in the reverse direction is determined using interpolation between points 8 and 9 to locate point 10 where another analysis is performed and found to be passing. Point 10 is approximately at one location on the right side of the area R_A at a new value of x_1 and x_2.

Then, starting at point 1, points equal distance along a line orthogonal to the line between points 6 and 10 are selected as shown in Fig. 9.19. This results in the points 11 through 14, with the resultant approximation of the boundary of R_A at point 14.

A triangle can be overlaid of the corner points 6, 10, and 14 as shown in Fig. 9.20. Then the largest circle is drawn that stays within the bounds of the triangle. The

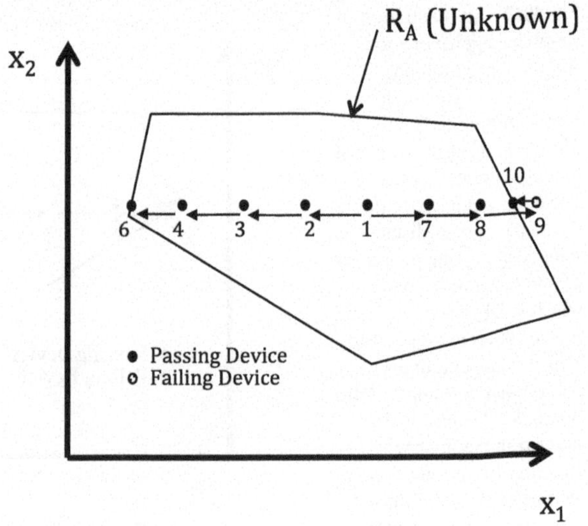

Fig. 9.18 Continuation of the simplicial approximation from Fig. 9.17. Starting again at point 1, equal distance points are selected and tested along the increasing x_1 direction. The testing of points continues along that parameter axis until point 9, which is found to fail. Then the direction is reversed by a distance interpolated between points 8 and 9 to point 10, which is found to pass

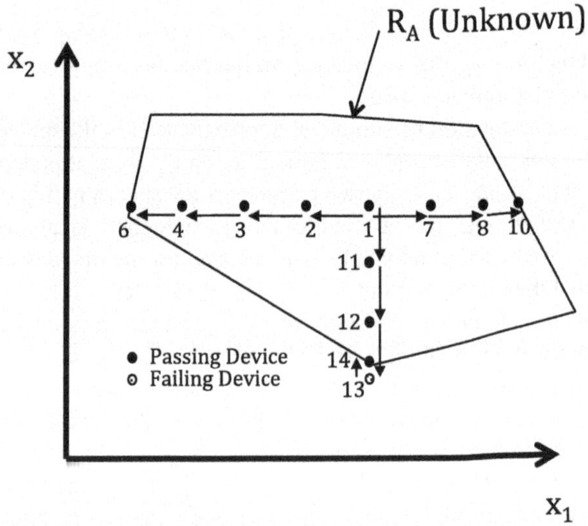

Fig. 9.19 Starting again at point 1, points equal distance along a line orthogonal to the line between points 6 and 10 are selected and tested. This results in the points 11 through 14, with the resultant approximation of the boundary of R_A at point 14

longest side of the triangle is assumed to represent the worst estimate of the boundary of the acceptance region. Therefore, from a midpoint of this longest side of the triangle, new points are tested along points 15, 16, 17, and 18 in the perpendicular direction away from the triangle as shown in Fig. 9.20 to determine another boundary of the acceptance region at point 18. This process is repeated a number of times in order to find the boundaries of R_A with increasing accuracy.

The example of the methodology of simplicial approximation illustrated above was only for a case of two parameters. In general, there will be a number n

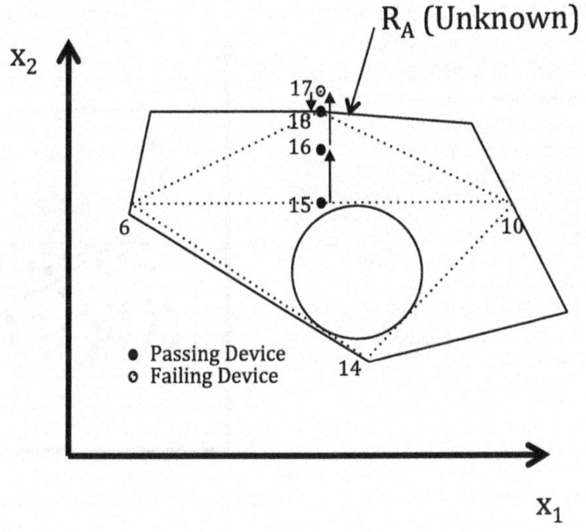

Fig. 9.20 A triangle is drawn between points 6, 10, and 14, and then the largest circle possible is drawn that stays within the bounds of the triangle. The longest side of the triangle represents the worst estimate of the boundary of R_A, so a midpoint is selected along the longest side, and new points are selected and tested along pathway from points 15 to 18 to determine the boundary location of RA at point 18

parameters, and therefore the parameter space will be n-dimensional. This process is performed using an algorithm on computers.

Simplicial approximation requires multiple calculations to determine the boundary points of R_A. Moreover, as the number of parameters increases to n, R_A is represented by a multiple dimensioned hyper-plane, which becomes computationally intensive. Additionally, simplicial approximation requires the acceptance region to be convex and simply connected. Simplicial approximation also does not provide a mechanism for estimating the amount of error involved in its use.

Once the boundaries of the acceptance region, R_A, have been approximated to the designed level of accuracy, the variability region, given by R_V, can be overlaid with the acceptance region R_A as shown in Fig. 9.21.

Importantly, the probability distribution functions for the parameters must be known in order to plot sample points within the variation region, R_v. If the number of passing devices in the overlap of the R_A and R_V regions is N_P, and the total number of points generated is equal to N, then the yield is given by: $Y = N_p/N$.

9.5.3 Monte Carlo Yield Estimation

As discussed in Chap. 8, Monte Carlo analysis is the most powerful and commonly used method to predict the variations in the output performance behavior of a microsystems device based on the random variations in the device input parameters. Monte Carlo can also be used as an efficient estimator of the manufacturing yield as will be shown below.

Consider the two-dimensional parameter space for a microsystems device shown in Fig. 9.22. Both of the parameters, x_1 and x_2, are described by Gaussian probability

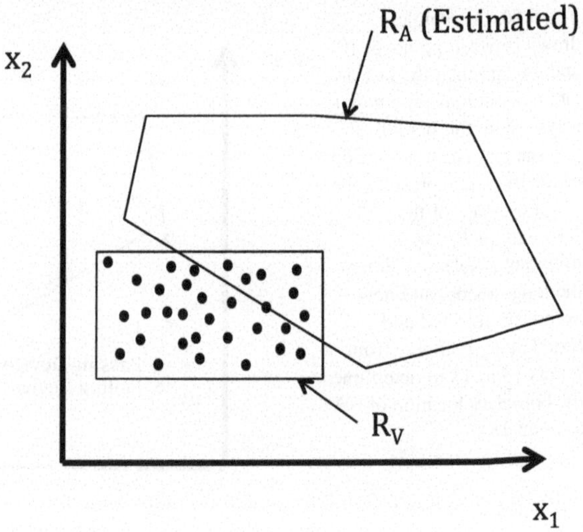

Fig. 9.21 Showing how the yield is estimated using simplicial approximation wherein the variation region is superimposed onto the region of acceptance in parameter space for a microsystems device having two parameters. The yield is given by the sample points in the overlapping R_V and R_A regions divided by the total number of sample points

distribution functions as shown, although it should be noted that this method can be used on any type of distribution. The probability distribution functions are used to appropriately bias the pseudorandomly selected samples of the parameters x_1 and x_2 used in the analysis of the output behavior of the microsystems device. With each of the pseudorandomly selected samples, x_{1n} and x_{2m}, the output behavior of the microsystems device can be determined using the appropriate physical equations describing the device. Depending on the values of these calculated device outputs, the device at that sample point can be declared as either a "pass" or "fail" depending on whether it meets or does not meet the device output performance requirements as shown in Fig. 9.22. The yield is estimated by counting the number of passing points, given by N_p, divided by the total number of samples, N, or:

$$\hat{Y} = \frac{N_p}{N}. \tag{9.32}$$

An important item to note with using Monte Carlo methods to estimate the yield is that the location and boundaries of the acceptance region is not required. This is unlike the other methods described above that involve complicated analysis methods to attempt to approximate the boundaries of R_A. This is another advantage of Monte Carlo analysis.

9.5.3.1 Confidence Intervals for Monte Carlo Yield Analysis

An important and unique characteristic of Monte Carlo analysis is that the accuracy of an estimated yield can be calculated. Other methods of variation analysis do not have this property. Specifically, if a large number of Monte Carlo analyses are

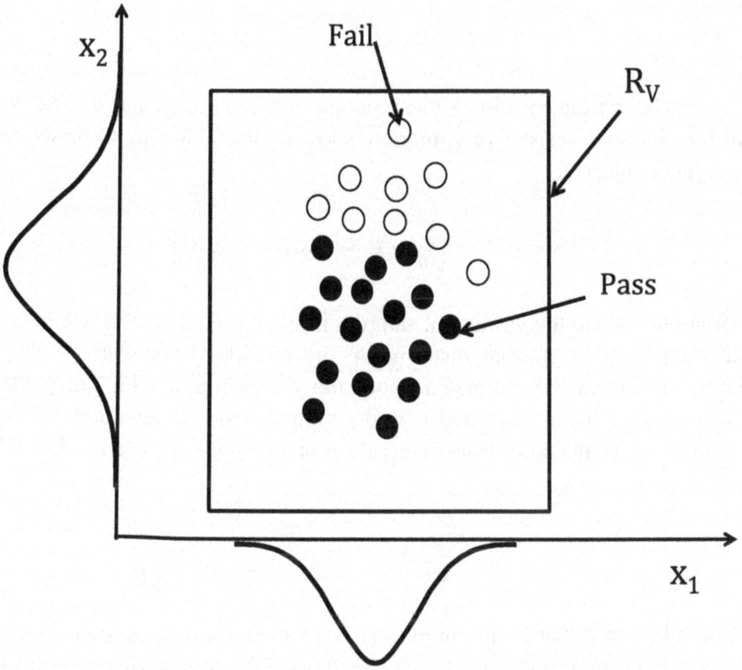

Fig. 9.22 The use of Monte Carlo methods to estimate the manufacturing yield. The device has two input parameters, $x1$ and $x2$, that are represented by Gaussian probability distribution functions. Sample points are pseudorandomly selected for analysis of the device output behavior at each selected point in the parameter space. Depending on the calculated output and whether it meets the performance requirements or not, the device can be declared as either "passing" or "failing." The yield is the number of passing points divided by the total number of points

performed on a microsystems device (i.e., more than 30), the central limit theorem states that probability distribution function of the yield will have a normal distribution, and the following can be written:

$$P(-z_{\frac{\alpha}{2}} < \frac{\bar{Y} - \mu}{\frac{\sigma}{\sqrt{n}}} < z_{\frac{\alpha}{2}}) \approx 1 - \alpha, \tag{9.33}$$

where $z_{\frac{\alpha}{2}}$ is the z-score, $\bar{Y}$ is the distribution of the sample mean of yields derived from averaging random samples of size n drawn (with replacement) from measurements of the yield, μ is the mean, σ is the standard deviation, and n is the number of samples.

It can be recalled from the central limit theorem that if there are multiple random samples of the yield, given by Y_1, Y_2,. . . .Y_n that have a mean μ and standard deviation σ, then if n is sufficiently large, $\bar{Y}$ has a normal distribution with a mean given by μ, and a standard deviation given by $\frac{\sigma}{\sqrt{n}}$. Rearranging this relationship for m results in a confidence interval for μ with a confidence level of $100(1 - \alpha)\%$ or:

$$\bar{Y} - z_{\frac{\alpha}{2}}\frac{\sigma}{\sqrt{n}} < \mu < \bar{Y} - z_{\frac{\alpha}{2}}\frac{\sigma}{\sqrt{n}}. \tag{9.34}$$

$\overline{Y}$ and $\boldsymbol{\sigma}$ can be estimated with a yield sample mean Y_{Average} and a yield sample standard deviation s, respectively, thereby allowing the following to be written for the confidence interval:

$$Y_{\text{Average}} - z_{\frac{\alpha}{2}}\frac{s}{\sqrt{n}} < \mu < Y_{\text{Average}} - z_{\frac{\alpha}{2}}\frac{s}{\sqrt{n}}. \tag{9.35}$$

In situations where the number of samples is small (i.e., less than 30), but can be determined to have a normal distribution, the confidence interval for the yield population mean can be estimated using the t-distribution with the population standard deviation being estimated with the sample standard deviation, s, resulting in the following for the confidence interval for m with a confidence level of $100(1 - \alpha)\%$:

$$Y_{\text{Average}} - t_{\frac{\alpha}{2},n-1}\frac{s}{\sqrt{n}} < \mu < Y_{\text{Average}} - t_{\frac{l\alpha}{2},n-1}\frac{s}{\sqrt{n}}. \tag{9.36}$$

It should be noted that if the sample size is small and it is not known whether the distribution is normal, then the distribution should be tested for normalcy before using Eq. (9.36).

It is useful to estimate the number of iterations, n, of the Monte Carlo method that has to be performed in order to obtain a certain confidence interval. Knowing that the confidence interval of $100(1 - \alpha)$ is given as:

$$Y_{\text{Average}} \pm z_{\frac{\alpha}{2}}\frac{s}{\sqrt{n}}, \tag{9.37}$$

allows this to be solved for n. However, s may not be known until some number of iterations has been performed.

9.6 Statistical Process Control (SPC)

Statistical process control (SPC) methods are principally used for quality control and assurance purposes to help ensure higher manufacturing yields but can also be used to improve device performance as well. A key element of SPC is the measurement of various characteristics of the microsystems manufacturing process that are indicative of whether a quality standard has been achieved or not. These measurements are normally performed during manufacturing to monitor individual processing steps, or groups of processing steps, but can also be conducted at the completion of the manufacturing. Importantly, SPC is directly related ensuring that the device input parameters are within their acceptance ranges, thereby resulting in device output performance parameters also being within their acceptance ranges.

As an example, consider the deposition of a thin-film layer using PECVD. After performing this processing step, thickness measurements using a metrology tool, such as an interferometer or ellipsometer, would likely be performed. One or more measurements of the thickness of the deposited thin-film layer would be taken on each wafer, and this data would then be analyzed to calculate representative statistics for this processing step such as the sample mean, range, and/or the standard deviation. This data can then be examined and compared to expected results to determine if the processing step was run correctly or not as well as track the behavior over time.

A key premise of SPC is the recognition that there will always be some amount of random variability in any manufacturing process. A manufacturing process that is operating within some pre-defined limits of random variability is said to be in statistical control. These pre-defined limits will be determined individually for each processing step and manufacturing process based on the data from past performance history, an understanding of what measurements to take and how to group them, and the application of statistical methods.

In addition to the expected randomness of the manufacturing process, there may also be other kinds of variability that can be exhibited. These kinds of variations could be caused by a number of reasons, but some of the more common include a process tool operating out of specification; operator equipment error; or defective materials. As a general rule, these kinds of variations are non-random and are often larger than the random variations. Since these variations are not random in nature, they are said to be due to "assignable causes," and when exhibited, the processing step is considered to be not in control.

The variations in a manufacturing process are depicted in Fig. 9.23 in the form of the manufacturing functions of a process parameter over time. The centerline corresponding to the average value of the quality manufacturing process characteristic is plotted along the *x*-axis, with the calculated lower and upper control limits, LCL, and UCL, respectively, also shown on the *x*-axis. These process control limits are determined by collecting statistics to understand the characteristics of the random variances that occur in the manufacturing process. Time is plotted along the *y*-axis, and there are measurements taken on sample production units from the manufacturing process at times, t_0, t_1, t_2, and t_3. The magnitudes of the process parameter manufacturing functions are plotted along the *z*-axis. At time, t_0, the measurements taken on sample units from the manufacturing process have a mean and a standard deviation given by $\overline{x_o}$ and s_0, respectively. The sample mean at t_0 equals the calculated expected mean $\overline{x_o}$, and the sample standard deviation is equal to the expected standard deviation, s_0. Since both $\overline{x_o}$ and s_0 are within the limits of LCL and UCL, the manufacturing process is considered to be in control at time t_0.

At time t_1 it can be seen that the sample process mean has shifted to a new value given by $\overline{x_1}$, where $\overline{x_1} > \overline{x_o}$. Moreover, the value of the sample mean at t_1 is outside of the upper control limit, UCL. The sample standard deviation at time t_1 has not changed and is $s_1 = s_0$. Nevertheless, the manufacturing process at time t_1 would be

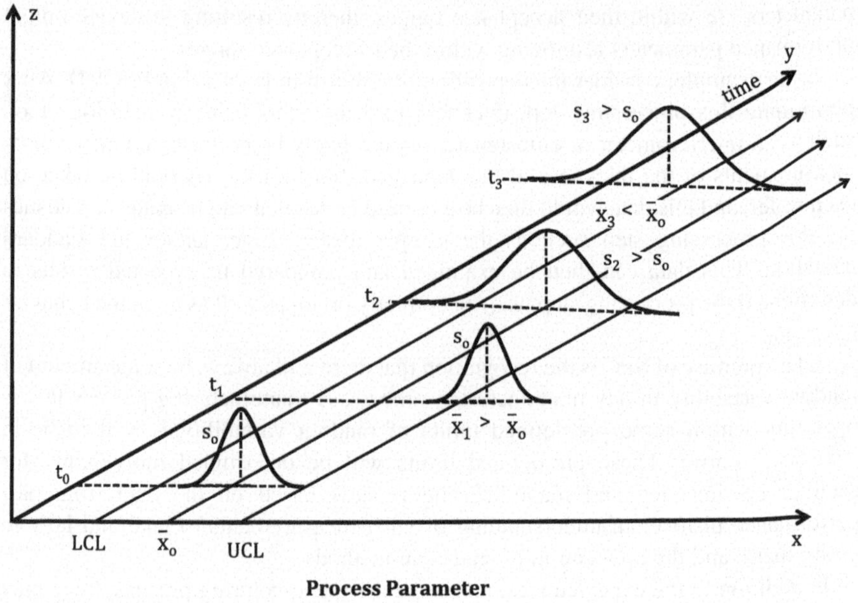

Fig. 9.23 A depiction of random and assignable causes of variations as a function of time and sample of a manufacturing process [9]

considered as being not in control any longer. The amount of shift in the sample mean would be beyond the variations established in the control limits, and a cause for the shift of the manufacturing process mean would be worthy of investigation.

At time, t_2, the sample mean has not changed from the value at t_0 and is equal to $\overline{x_o}$, but the sample standard deviation at this point in time has significantly broadened to a new value of $s_2 > s_0$. Therefore, at time t_2 the process would also be considered as being out of control since the tail of the standard deviation extends beyond one of the control limits and an investigation as to the cause of the broadening of the process standard deviation would be initiated.

At time, t_3, it can be seen that the sample mean has now shifted wherein $\overline{x_3} < \overline{x_o}$ and the sample standard deviation have also broadened to $s_3 > s_0$. In this case, the mean is outside of one of the control limit boundaries, and the tail of the distribution also extends beyond the control limit boundary. Therefore, at time t_3 the process would be considered as being out of control, and the cause(s) of the shift in the mean and of the broadening of the process standard deviation would be initiated.

The important element conveyed in Fig. 9.23 is that the control chart plots the expected random behavior of a quality characteristic of a sample set from a manufacturing process[3] and if either the sample mean, the sample standard deviation,

[3]Control charts can be used on the outcome of processing steps, groups of processing steps, or the entire process sequence.

or both vary beyond the predetermined upper and lower control limits, then the process is considered to be operating outside of the boundaries of expected random variation and would be considered out of control. Once the process is out of control, then action needs to be taken by the manufacturing team to determine and correct the cause of the non-random variations and thereby place the manufacturing process back into a state of being in control.

As this example illustrates, control charts allow the manufacturer to monitor and detect problems in the device production process that would result from non-random occurrences. Moreover, since the control chart is plotted as a function of time and production samples, it is possible to identify trends in the data over time possibly indicating that the process is drifting toward an out-of-control state, and this can provide the manufacturer with the ability to make corrective actions even before the process has crossed the boundary into being classified as out of control.

While the graph in Fig. 9.23 is useful for understanding the purposes of control charts, this specific type of depiction is not commonly used in practice. Instead, the more common type of control chart is illustrated in Fig. 9.24. The data points of the manufacturing characteristic are plotted as a function of time or sample number. The manufacturing characteristic plotted on the control chart depends on what measurements are deemed to be needed to monitor a process. A control chart could be used to plot a dimensional parameter of the manufacturing process, such as the radius of a microchannel, the thickness of a deposited thin-film layer, the depth of an etch, as well as any metrology or testing measurement involved in microsystems

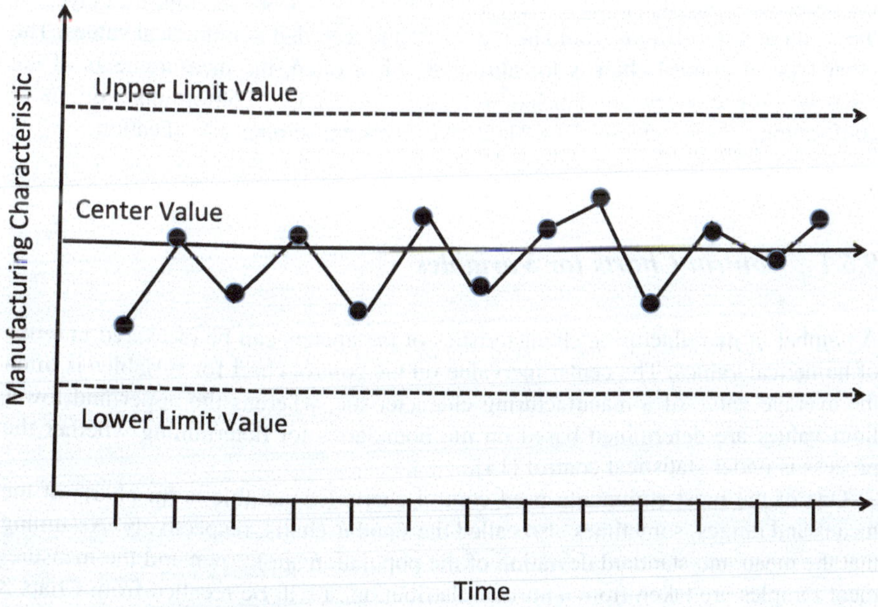

Fig. 9.24 An illustrative control chart

manufacturing. The manufacturing characteristic could also be the output performance parameter response of a microsystems device during testing, such as the fluidic resistance of a microchannel device.

The manufacturing process engineers will be responsible for determining what manufacturing characteristics are to be measured, how often they are to be measured, what number of measurements will be needed, and where the measurements should be taken for each sample in each particular microsystems manufacturing circumstance. The amount of SPC performed in each situation is usually a balance between the need to ensure maintenance of good quality control weighed against the costs associated with performing measurements and analysis. How many quality control measurements are needed is dependent on many complex factors including the degree of difficulty of the individual processing steps in the process sequence; how often the process trends to an out-of-control state; whether the cost of producing devices that fail to meet the application requirements is higher than the cost of performing measurements; and others. It is also very common that the amount of measurements taken on a new manufacturing process is often much higher than one that is relatively mature. As a process sequence matures, the process engineers will often be able to identify which processing steps require more rigorous monitoring as well as those that do not.

In general, there are two different types of control charts used in microsystems manufacturing. The first is the control chart for variables, which translates into control charts for numerical values of parameters. Many of the measureable quality characteristics of microsystems can be expressed as a numerical value. For example, the radius of a microchannel device, the thickness of a deposited thin-film layer, or the width of micro-beams would be measured and recorded as numerical values. The other type of control chart is for attributes. Most often, the measurements of the microsystems devices are labeled as "pass" or "fail," "conforming" or "non-conforming," "non-defective" or "defective," or some similar classification.

9.6.1 Control Charts for Variables

A number of manufacturing characteristics or parameters can be measured in terms of numerical values. The centerline value on the control chart for variables is often the average value of a manufacturing characteristic, whereas the upper and lower limit values are determined based on the boundaries for determining whether the process is under statistical control [14].

One of the most commonly used control charts for variable is for charts of the means and ranges, sometimes also called the $\overline{x}$ and R-charts, respectively. Assuming that the mean and standard deviation of the population are known and the measurement samples are taken from a normal distribution, it will be recalled from Chap. 8

that it can be asserted with a probability of $1-\alpha$ that the mean of a randomly selected sample, given by $\overline{x}$, of size n will be between the two values of:

$$\mu - z_{\frac{\alpha}{2}}\frac{\sigma}{\sqrt{n}} \text{ and } \mu + z_{\frac{\alpha}{2}}\frac{\sigma}{\sqrt{n}}. \tag{9.38}$$

These values can be used as the limits on the control chart for the mean wherein the first term is used as the lower control limit (LCL) and the second is the upper control limit (UCL) for the sample mean.

In many modern manufacturing environments, the standard of quality is often called "3 sigma," and the number 3 is used to replace the $z_{\alpha/2}$ terms. This results in the control chart limit values having a probability of 99.7% that the sample mean falls within these limits. Each manufacturing scenario is different, and therefore, the value of the $z_{\alpha/2}$ term should be determined by the manufacturing engineers based on the individual situation.

One issue with Eq. (9.38) is that the mean μ and standard deviation σ of the population are usually unknown, and therefore it is necessary to estimate their values from the sample data. In those cases, it is assumed that m samples of measurement data are taken, with each sample of size n. If $\overline{x}_1, \overline{x}_2, \ldots \overline{x}_m$ are the means of each sample, then the population mean, μ, can be estimated by taking the average of these sample means, as follows:

$$\overline{\overline{x}} = \frac{\overline{x}_1 + \overline{x}_2 + \ldots + \overline{x}_m}{m}, \tag{9.39}$$

where $\overline{\overline{x}}$ is used as an estimator of μ and is designated as the centerline of the $\overline{x}$ chart.

The standard deviation can be estimated from either the sample standard deviation or the ranges of a number of m samples. It is common in some circumstances to use the range, and therefore this will be discussed first. The range (R) is given by the difference between the maximum and minimum value of a parameter, x, as follows:

$$R = x_{\max} - x_{\min}. \tag{9.40}$$

An estimator of σ is R/d_2, where d_2 is a parameter that is tabulated as a function of the sample size n. Values of d_2 for various values of n are given in Table 9.1. If R_1, R_2,

Table 9.1 Values of the d_2 and d_3 parameters as a function of sample size n used in the equations for calculating the control limit values for the $\overline{x}$ and R-charts [14]

n	d_2	d_3	n	d_2	d_3	n	d_2	d_3
2	1.128	0.853	**8**	2.847	0.820	**14**	3.407	0.763
3	1.693	0.888	**9**	2.970	0.808	**15**	3.472	0.756
4	2.059	0.880	**10**	3.078	0.797	**16**	3.532	0.750
5	2.326	0.864	**11**	3.173	0.787	**17**	3.588	0.744
6	2.534	0.848	**12**	3.258	0.778	**18**	3.640	0.739
7	2.704	0.833	**13**	3.336	0.770			

$\ldots R_m$ are the ranges of the m samples, then the average range can be calculated using:

$$\overline{R} = \frac{R_1 + R_2 + \ldots + R_m}{m}, \tag{9.41}$$

and σ can be estimated from:

$$\widehat{\sigma} = \frac{\overline{R}}{d_2}. \tag{9.42}$$

Using $\overline{\overline{x}}$ and $\widehat{\sigma}$ as estimates for μ and σ, respectively, then the control chart for the manufacturing process characteristic mean is defined by the centerline and upper and lower control limits as follows:

$$\begin{aligned} \text{Upper Control Limit} &= \overline{\overline{x}} + \frac{3\overline{R}}{d_2\sqrt{n}} \\ \text{Centerline} &= \overline{\overline{x}} \\ \text{Lower Control Limit} &= \overline{\overline{x}} - \frac{3\overline{R}}{d_2\sqrt{n}}. \end{aligned} \tag{9.43}$$

Equation (9.43) uses a factor of 3 in the UCL and LCL terms, which would be analogous to 3σ that was discussed above. Again, the factor used in this equation should be selected based on the characteristics of any given situation.

An accompanying chart to the $\overline{x}$ chart is the R-chart, which is based on the range of measured values of a manufacturing characteristic. The centerline of the R-chart is $\overline{R}$ which is given by Eq. (9.41). The control limits for the R-chart are based on the estimate of the standard deviation of the range given by $\widehat{\sigma}_R$. The standard deviation of the range is given by [14]:

$$\sigma_R = d_3\sigma. \tag{9.44}$$

Because σ is unknown, σ_R is estimated using $\widehat{\sigma}_R$ as follows:

$$\widehat{\sigma}_R = d_3\frac{\overline{R}}{d_2}, \tag{9.45}$$

where d_3 are tabulated values that are a function of the number of samples n as seen in Table 9.1.

Using the above, the R-chart parameters can now be written as [14]:

$$\begin{aligned} &\text{Upper Control Limit} = \overline{R} + \frac{3d_3\overline{R}}{d_2} \\ &\text{Centerline} = \overline{R} \\ &\text{Lower Control Limit} = \overline{R} - \frac{3d_3\overline{R}}{d_2}. \end{aligned} \tag{9.46}$$

Again, Eq. (9.46) uses a factor of 3 in the UCL and LCL terms.

Both $\bar{x}$ and R-charts are used together in monitoring manufacturing processes [14]. The $\bar{x}$ chart monitors the average quality characteristic of a manufacturing process parameter. Sampling techniques used for the $\bar{x}$ chart are designed so as to maximize the chances for detecting non-random shifts in the average process characteristic between samples. In contrast, the R-chart is primarily used to measure the variability within the samples. Consequently, the samples used for the R-chart are selected so that the variability due to non-random events within the samples can be detected. Stated another way, the $\bar{x}$ chart is used to monitor between sample variability, while the R-chart monitors within sample variability. Since both are important for detecting whether a manufacturing process is in control or not, they are therefore used together. It should be emphasized that both the $\bar{x}$ and R-charts are designed to detect non-random events, that is, events that can be theoretically traced to an assignable cause of the problem in the manufacturing process. Random events are expected to be within the control limits of the respective chart.

The interpretation of the $\bar{x}$ and R control charts should be performed jointly. If the underlying distributions are normal, then the $\bar{x}$ and R-charts calculated from the samples are statistically independent. However, if there appears to be a correlation between the $\bar{x}$ and R-charts, such as the values on the charts appear to follow one another, then this signifies that the underlying distributions may not be independent.

With regard to a manufacturing process that is deemed to be out of control on both the $\bar{x}$ and R-charts, the assignable causes on the R-chart should be addressed first and the control charts reevaluated since often this may resolve the out-of-control issue on the $\bar{x}$ chart as well. It is strongly advised not to interpret the $\bar{x}$ chart first when the R-chart indicates that the process is not in control.

While the R-chart is often used in practice, it is sometimes more desirable to estimate and control the manufacturing process using the sample standard deviations when the sample sizes are larger (i.e., above 10 samples). However, an issue with the sample standard deviation s is that it is not an unbiased estimator of σ. Nevertheless, if the underlying distribution is normal, then it can be shown that $c_4\sigma$ can be used as an estimator of σ, where c_4 is a constant that is based on sample size n [14]. And, it can be further shown that the variation in the standard deviation can be estimated as $\sigma\sqrt{1-c_4^2}$. This allows the control chart for the standard deviation to be written as:

$$\begin{aligned} &\text{UCL} = c_4\sigma + 3\sigma\sqrt{-c_4^2} \\ &\text{Centerline} = c_4\sigma \\ &\text{LCL} = c_4\sigma - 3\sigma\sqrt{1-c_4^2}. \end{aligned} \tag{9.47}$$

Tabulated constants, B_5 and B_6, given by:

$$B_5 = c_4 - 3\sqrt{1 - c_4^2}$$
$$\text{and} \tag{9.48}$$
$$B_6 + c_4 + 3\sqrt{1 - c_4^2},$$

are commonly used in Eq. (9.47) to result in the following:

$$\begin{aligned} \text{UCL} &= B_6\sigma \\ \text{Centerline} &= c_4\sigma \\ \text{LCL} &= B_5\sigma, \end{aligned} \tag{9.49}$$

where the constants c_4, B_5 and B_6 are given in Table 9.2.

The standard deviation, σ, can be estimated by analyzing past measurement data if it is not known. If there are m samples taken of a manufacturing characteristic each of size n, the average of the m standard deviations is given by:

$$\overline{s} = \frac{1}{m}\sum_{i=1}^{m} s_i. \tag{9.50}$$

An unbiased estimator of σ is given by $\overline{s}/c_4$, thereby allowing the centerline and upper and lower control limits to the s chart to be written as [14]:

$$\begin{aligned} \text{UCL} &= \overline{s} + 3\frac{\overline{s}}{c_4}\sqrt{1 - c_4^2} \\ \text{Centerline} &= \overline{s} \\ \text{LCL} &= \overline{s} - 3\frac{\overline{s}}{c_4}\sqrt{1 - c_4^2} \end{aligned} \tag{9.51}$$

Moreover, if $\overline{s}/c_4$ is used to estimate σ, the control chart for the $\overline{x}$ chart can be written as follows:

$$\begin{aligned} \text{UCL} &= \overline{\overline{x}} + 3\frac{\overline{s}}{c_4\sqrt{n}} \\ \text{Centerline} &= \overline{\overline{x}} \\ \text{LCL} &= \overline{\overline{x}} - 3\frac{\overline{s}}{c_4\sqrt{n}}. \end{aligned} \tag{9.52}$$

Table 9.2 Values of A_3, B_5, and B_6 for different n values [14]

Samples, n	C_4	A_3	B_5	B_6
2	0.7979	2.659	0	2.606
3	0.8862	1.954	0	2.276
4	0.9213	1.628	0	2.088
5	0.9400	1.427	0	1.964
6	0.9515	1.287	0.029	1.874
7	0.9594	1.182	0.113	1.806
8	0.9650	1.099	0.179	1.751
9	0.9693	1.032	0.232	1.707
10	0.9727	0.975	0.276	1.669
11	0.9754	0.927	0.313	1.637
12	0.9776	0.886	0.346	1.610
13	0.9794	0.850	0.374	1.585
14	0.9810	0.817	0.399	1.563
15	0.9823	0.789	0.421	1.544
16	0.9835	0.763	0.440	1.526
17	0.9845	0.739	0.458	1.511
18	0.9854	0.718	0.475	1.496
19	0.9862	0.698	0.490	1.483
20	0.9869	0.680	0.504	1.470

A constant A_3, given by:

$$A_3 = \frac{3}{(c_4\sqrt{n})} \tag{9.53}$$

is used in Eq. (9.52) to result in the following:

$$\begin{aligned} \text{UCL} &= \bar{\bar{x}} + A_3\,\bar{s} \\ \text{Centerline} &= \bar{\bar{x}} \\ \text{LCL} &= \bar{\bar{x}} - A_3\bar{s}, \end{aligned} \tag{9.54}$$

where values of A_3 for various values of n are tabulated in Table 9.2.

9.6.2 *Control Charts for Attributes*

Certain manufacturing characteristics cannot be represented using numerical values but instead can be represented using attribute descriptors such as either "conforming" or "non-conforming" as well as "non-defective" and "defective." These are commonly referred to as manufacturing attributes and are handled using the discrete probabilities methods discussed in Chap. 8. There are several attribute control charts commonly used in microsystems manufacturing that will be reviewed here including the fraction that are non-conforming which is portrayed on the "*p* chart"; the number of non-conforming is portrayed in the "*c* chart"; and the non-conformities per unit is portrayed in the "*u* chart." What is described as having the attribute of being conforming or non-conforming can be applied to a device, some manufacturing process step, some dimension of a component, etc. as well as many other manufacturing characteristics. Conforming or non-conforming is usually determined on whether the unit meets some pre-defined specifications. It will be noted that the statistical techniques used for manufacturing process attributes share many of the same methods used earlier in this chapter on functional yield analysis and modeling.

The fraction of non-conforming used for generating the *p*-chart is defined as the ratio of the number of non-conforming elements in a population to the total number of items in the population and can be calculated using the binominal distribution. If the manufacturing process is stable, the probability that a unit produced will be non-conforming is given by p, and it is assumed that successively produced units are independent. Under that condition, each unit made is a Bernoulli random variable with probability p. If a random sample of n units is selected, and D units are non-conforming, then D can be expressed as a binominal distribution with probability p in n units as follows [14]:

$$P\{D = x\} = \binom{n}{x} p^x (1 - p)^{n-x} \quad x = 0, 1, 2, \ldots n. \tag{9.55}$$

Equation (9.55) has a mean and variance of np and $np(1 - p)$, respectively. The fraction non-conforming is given as the ratio of the number of non-conforming in the sample D to the number n, as:

$$\widehat{p} = \frac{D}{n}. \tag{9.56}$$

The mean and variance are given by:

$$\mu_{\widehat{p}} = p \tag{9.57}$$

and

$$\sigma^2_{\widehat{p}} = \frac{p(1 - p)}{n}. \tag{9.58}$$

Using these equations, if the actual fraction of non-conforming units given by p is known the centerline, and control limits can be defined as follows [14]:

$$\begin{aligned}
\text{UCL} &= p + 3\sqrt{\frac{p(1-p)}{n}} \\
\text{Centerline} &= p \\
\text{LCL} &= p - 3\sqrt{\frac{p(1-p)}{n}}.
\end{aligned} \tag{9.59}$$

In many situations p is not known and must be estimated from process measurement data. A quantity of m samples can be taken, each of size n, and if there are D_i non-conforming units on the i_{th} sample, the fraction non-conforming is given as:

$$\widehat{p_i} = \frac{D_i}{n} \quad i = 1, 2, 3, \ldots, m, \tag{9.60}$$

and the average probability of these fraction of non-conforming samples is given as:

$$\overline{p} = \frac{\sum_{i=1}^{m} D_i}{mn} = \frac{1}{m}\sum_{i=1}^{m} \widehat{p_i}. \tag{9.61}$$

Using these equations the control chart limits and centerline can be written as follows:

$$\begin{aligned}
\text{UCL} &= \overline{p} + 3\sqrt{\frac{\overline{p}(1-\overline{p})}{n}} \\
\text{Centerline} &= \overline{p} \\
\text{LCL} &= \overline{p} - 3\sqrt{\frac{\overline{p}(1-\overline{p})}{n}}.
\end{aligned} \tag{9.62}$$

When the fraction of non-conforming is not known, it is advisable that the number of preliminary samples taken should be at least 20–25. It is also important that the sample size, given by n, be sufficiently large to result in a high-enough probability of finding at least one non-conforming unit. Otherwise the p-chart would be essentially meaningless. It has been recommended that the sample size be sufficiently large to ensure a 50% probability of detection of a manufacturing process shift [15].

The method to determine the sample size, n, that fulfills this criterion of a 50% probability of detection of a manufacturing process shift is to note that the process shift δ should be equivalent to the control limit values, or:

$$\delta = k\sqrt{\frac{p(1-p)}{n}}, \tag{9.63}$$

where k is the control limit constant usually given by 3. For example, if it is desired to detect a process shift from $p = 0.02$ to $p = 0.06$, then δ is given by 0.04, and n can be found as follows:

$$n = p(1-p)\left(\frac{k}{\delta}\right)^2 = 0.02(0.98)\left(\frac{3}{0.04}\right)^2 = 111. \tag{9.64}$$

It should also be noted that the control limits established in Eq. (9.62) are to be treated as trial control limits and should be tested to determine if the manufacturing process is in fact in control when m samples are taken. To determine if the manufacturing process is in control for m samples, the fractional of non-conforming units in each sample are plotted. If the sample points are within the control limits and no non-random trends are identified, then it can be concluded that the process is in control, and the trial control limits can be used to monitor the process over time. However, if any of the samples plots outside of the control limits, then it is necessary to investigate the assignable cause for the point to be outside the trial control limits. If the cause is found, the out-of-bound point is disregarded, and the trial control limits are recalculated. This process is repeated until the trial control limits can be tested to show that they are valid at which point they can be adopted as the process control limits.

In some situations it is preferable to monitor the number of non-conforming units rather than the fraction of non-conforming units. The c-chart is used for this purpose. Assuming the sample has a constant size, it can be modeled as a Poisson distribution as follows [14]:

$$P(x) = \frac{e^{-c}c^x}{x!}, \tag{9.65}$$

where x is the number of non-conforming and c is a parameter of the Poisson distribution, which is given by the mean and variance of the distribution. Therefore, assuming c is known, the control chart parameters for the c-chart with 3s limits can be written as:

$$\begin{aligned} \text{UCL} &= c + 3\sqrt{c} \\ \text{Centerline} &= c \\ \text{LCL} &= c - 3\sqrt{c}. \end{aligned} \tag{9.66}$$

When c is not known, it is estimated from the average number of non-conforming units in a sample given by $\bar{c}$. Using this, Eq. (9.66) is rewritten as:

$$\begin{aligned} \text{UCL} &= \bar{c} + 3\sqrt{\bar{c}} \\ \text{Centerline} &= \bar{c} \\ \text{LCL} &= \bar{c} - 3\sqrt{\bar{c}}. \end{aligned} \tag{9.67}$$

The control chart for the average number of non-conforming units in a sample size n, or the density of non-conforming, is found by assuming that there are c non-conforming units in n samples, allowing the average number of non-conforming per sample to be written as:

$$u = \frac{c}{n}. \tag{9.68}$$

If the constants for the control limits are 3, then the centerline and control limits of the u-chart are as follows:

$$\begin{aligned} \text{UCL} &= \overline{u} + 3\sqrt{\overline{u}} \\ \text{Centerline} &= \overline{u} \\ \text{LCL} &= \overline{u} - 3\sqrt{\overline{u}}, \end{aligned} \tag{9.69}$$

where $\overline{u}$ is the average number of non-conforming units over m s groups of sample size n.

As an example of setting up a p-chart, consider a microsystems multilayer metallization process wherein a second layer metal of aluminum is deposited and patterned on the top surface of a substrate having previously fabricated vias through an insulating silicon dioxide layer that was deposited using PECVD. The vias extend through the oxide layer to make contact to a previously made first metal layer. A number of 40 samples are fabricated and measured to determine if the first and second layer metals electrically connect to one another. The testing involves passing a current through the metal 1 and metal 2 contact vias to determine if they are electrically connected or not. A non-conforming vias would register as an open circuit, while a conforming contact would have a small measurable electrical resistance. The size of each sample is 100 units from 40 wafer samples, and it is found that there are 276 non-conforming vias on this trial run. Using this information, the p-chart is constructed as follows. The average of the individual fraction of non-conforming units across the trial run is given by $\overline{p}$ where:

$$\overline{p} = \frac{\sum_{i=1}^{m} D_i}{mn} = \frac{276}{(40)(100)} = 0.07 \tag{9.70}$$

And the control limits are given as:

$$\begin{aligned} \text{UCL} &= \overline{p} + 3\sqrt{\frac{\overline{p}(1-\overline{p})}{n}} = 0.07 + 3\sqrt{\frac{(0.07)(1-0.07)}{100}} = 0.15 \\ \text{Centerline} &= \overline{p} = 0.07 \\ \text{LCL} &= \overline{p} - 3\sqrt{\frac{\overline{p}(1-\overline{p})}{n}} = 0.07 - 3\sqrt{\frac{(0.07)(1-0.07)}{100}} = 0.007. \end{aligned} \tag{9.71}$$

As noted above, the control limits calculated above are considered trial control limits until they can be tested against the m samples to determine whether any of the sample points is non-conforming.

9.6.3 *Identification of Non-random Patterns in Control Charts*

While the clearest method to determine an out-of-control manufacturing process is when one or more of the sample points fall outside of the control limits, there are also other data trends exhibited in specific patterns that are evidence of non-random behaviors and may also be indicative of a process that is not in control.

Western Electric Statistical Quality Control Handbook [16] is considered one of the best authoritative sources for establishing various rules for identifying non-random patterns possibly indicating that the manufacturing process is not in control. These rules are portrayed in Fig. 9.25. There are three regions labeled A, B, and C on either side of the centerline signifying variations of $\pm\sigma$ for region A, $\pm 2\sigma$ for region B, and $\pm 3\sigma$ for region C, respectively. The Western Electric rules for identifying non-random patterns in control charts are summarized as follows:

Rule 1: A sample point that is located beyond either of the control limits boundaries of region C at $\pm 3\sigma$

Rule 2: Two out of three consecutive sample points that are located beyond the control limits at $\pm 2\sigma$ in region C

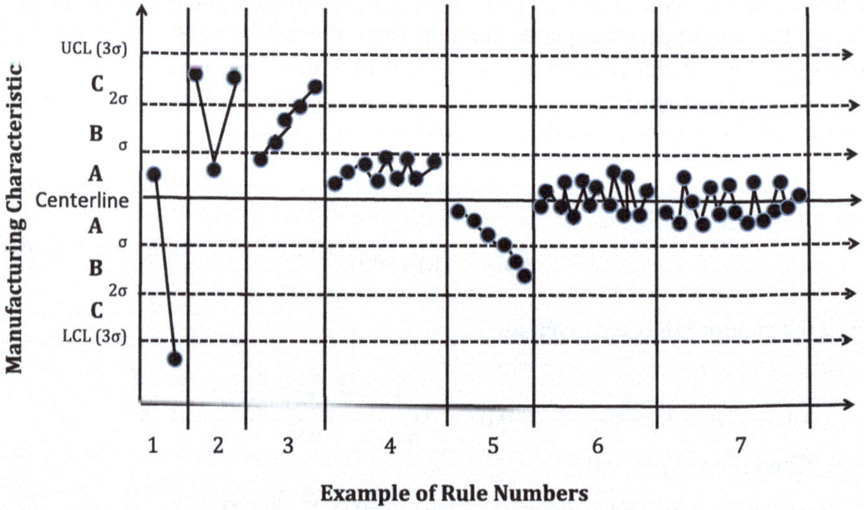

Fig. 9.25 Examples of data trends on control charts that indicate that a manufacturing process may not be in control as developed by Western Electric

Rule 3: Four out of five consecutive sample points that are located beyond the control limits at $\pm\sigma$ in regions B and C
Rule 4: Nine consecutive sample points that are located on the same side of the centerline
Rule 5: Six consecutive sample points that either increase or decrease
Rule 6: Fourteen consecutive sample points that alternate consistently up or down
Rule 7: Fifteen consecutive sample points that plot on either side of region A

The basis of these rules is that each of them has an extremely low probability of occurring, and therefore when they do occur, it could indicate a non-random event and possible evidence that the process is not in control. Among the control chart behaviors that these rules would catch include periodic behaviors of the chart points; clear patterns of points falling near or slightly outside of the control limits with relatively few points near the centerline; a discernable shift in the process level; and trends in the points wherein there is a continuous drifting of the plotted points in one direction over a number of points.

9.6.4 *Process Capability*

The process capability is a method whereby the control limits of a manufacturing process are compared to the acceptance range (also called the specification limits). It is common to express the process capability by the ratio of the difference between the upper and lower specification limits, given by USL and LSL, respectively, and 6σ, where σ is the standard deviation of a specific manufacturing characteristic for the production process. This ratio can be written as [14]:

$$C_{\mathrm{p}} = \frac{\mathrm{USL} - \mathrm{LSL}}{6\sigma}. \tag{9.72}$$

C_{p} is referred to as the process capability ratio. In Eq. (9.72), $\pm 3\sigma$ would be the control limits. Importantly, the specification limits USL and LSL are not based on the manufacturing process but instead are based on the application requirements. That is, these are externally imposed and have no direct relationship to the manufacturing process. Therefore, the process capability ratio can be seen as a measure of the ability of the manufacturing process to produce devices that meet the specifications.

Process capability is illustrated in Figs. 9.26 and 9.27. In Fig. 9.26 we have a manufacturing process with a large process capability where the acceptance range extends beyond the tails of the normal distribution of the manufacturing function. In this case, the process capability would be larger than 1. This is a desirable situation. It means that most of the manufactured product will be able to meet the application requirements since most of the devices made will fall within the acceptance range.

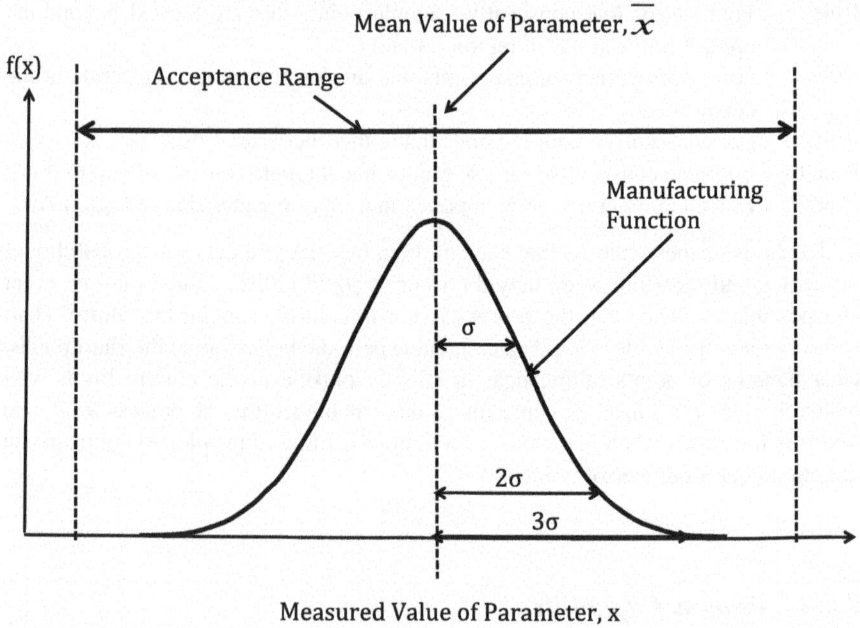

Fig. 9.26 An illustration of the manufacturing function of a certain manufacturing process and the acceptance range as defined by the device performance level. In this manufacturing process, the acceptance range extends beyond the tails of the manufacturing function and therefore the process capability would be larger than 1

In contrast, Fig. 9.27 illustrates another manufacturing process where the acceptance range does not extend beyond the tails of the manufacturing function. In this case the process capability would be less than 1. Consequently, some fraction of the devices produced would not meet the application requirements and would be scrapped. Process capability also relates to the manufacturing yield. A larger process capability is a manufacturing process that will have a larger yield. It is important to note that in both Figs. 9.26 and 9.27, it is assumed that there is no bias variation present, and the midpoint of the acceptance range coincides with the mean of the manufacturing function.

In many manufacturing situations, σ is not known. An estimate of σ can be made using the sample standard deviation, s, using the ratio of $\frac{\overline{R}}{d_2}$. This allows Eq. (9.72) to be rewritten as [14]:

$$\hat{C}_{\mathrm{p}} = \frac{\mathrm{USL} - \mathrm{LSL}}{6\hat{\sigma}}. \tag{9.73}$$

The variable d_2 is tabulated in Table 9.1 for various values of n. This takes into account the increased uncertainty of the location of the population mean when the sample size is small.

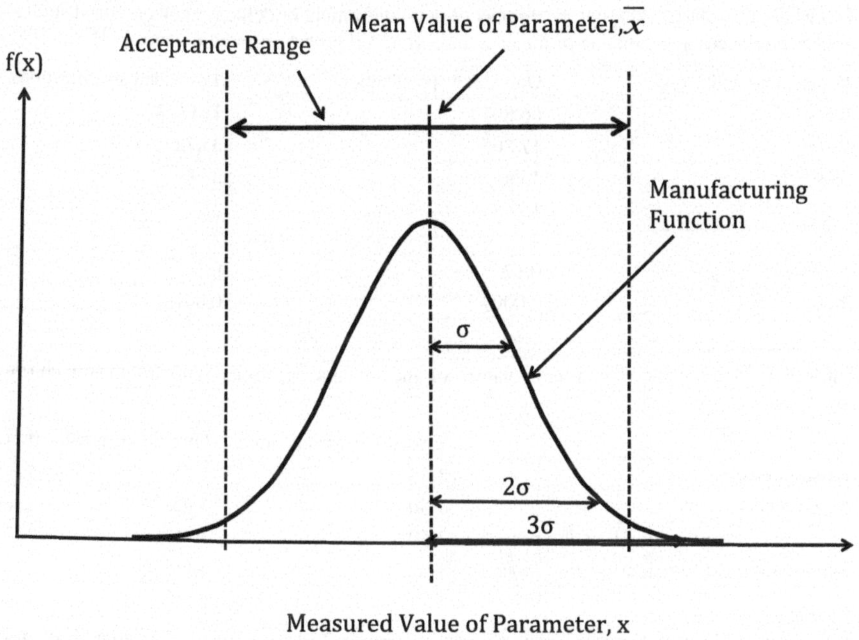

Fig. 9.27 An example of a manufacturing function of a different manufacturing process with a different acceptance range. In this manufacturing process, the acceptance range does not extend beyond the tails of the manufacturing function, and therefore the process capability would be less than 1

Equation (9.72) is used for two-sided specifications. The capacity process ratio for one-sided specifications is similarly given by [14]:

$$C_{\text{pu}} = \frac{\text{USL} - \mu}{3\sigma}, \text{ and}$$
$$C_{\text{pl}} = \frac{\mu - \text{LSL}}{3\sigma}, \qquad (9.74)$$

where C_{pu} and C_{pl} are the upper and lower one-sided capacity process ratios. Estimates of μ and σ would be $\hat{\mu}$ and $\hat{\sigma}$, respectively.

Table 9.3 shows how the values of the process capability ratio impact the number of devices that will be considered as not meeting the intended specifications (per 1 million devices (ppm)) produced [14]. This table demonstrates that higher process capability ratios result in far fewer defects. For example, a process capability ratio of 0.7 would result in over 35,000 defective devices per million devices produced, whereas a ratio of 1.5 would result in only 7 defective devices per million produced.

Industry has created recommendations for the minimum values of the process capability ratios to be used in practice, which are shown in Table 9.4 [14]. As can be seen, new processes are recommended to have higher PCRs than existing processes. Part of this is due to the imprecise statistics involved in new manufacturing

Table 9.3 Calculated values of the number of non-conforming or defective devices (per 1 million devices produced) as a function of the one- and two-sided process capability ratios [14]

Process capability ratio	One-sided specifications	Two-sided specifications
0.50	66,807	133,614
0.70	17,865	35,729
1.00	1350	2700
1.20	159	318
1.50	4	7
1.70	0.17	0.34
2.00	0.0009	0.0018

Table 9.4 Recommended minimum values of the process capability ratio for manufacturing processes [14]

	Two-sided specification	One-sided specification
Existing process	1.33	1.25
New process	1.50	1.45
Safety or critical function Existing process	1.50	1.45
Safety or critical function Existing process	1.67	1.60

processes. It is worth noting that these recommended minimums are just that, and many manufacturers use more stringent CPR values, such as a CPR of at least 2.

The discussion about process capability has assumed that the quality characteristic has a normal distribution, the manufacturing process is in control, and the mean or average value of the manufacturing characteristic is centered between the upper and lower specification limits. The first two assumptions have been noted previously. The last assumption is due to the fact that the process capability ratio does not account where the process mean is positioned relative to the specifications. Therefore, if the mean is offset, due to a bias variation, then the actual percentage of acceptable devices would be far smaller than the number calculated using the process capability. When the process has an offset mean, the process capability can be calculated using [14]:

$$C_{\mathrm{pk}} = \min\left(C_{\mathrm{pu}}, C_{\mathrm{pl}}\right). \tag{9.75}$$

C_{pk} is the one-sided process capability ratio for the specification limit that is located nearest to the process mean. If $C_{\mathrm{p}} = C_{\mathrm{pk}}$, the process is centered, while if $C_{\mathrm{pk}} < C_{\mathrm{p}}$, the process is off-center from the mean. The difference between C_{pk} and C_{p} is indicative of how off-center the process is from the mean.

Importantly, C_{pk} alone does not adequately measure how well the process is centered. The reason for this is that for a fixed value of mean in the specification interval from USL to LSL, C_{pk} depends inversely on the standard deviation, σ, thereby making C_{pk} not well suited for determining how well centered a process is. A method to address this issue is to use the following expression:

$$C_{pm} = \frac{USL - LSL}{6\tau}, \tag{9.76}$$

where τ is given by:

$$\tau = \sqrt{\sigma^2 + (\mu - T)^2}, \tag{9.77}$$

and T is given by:

$$T = \frac{1}{2}(USL + LSL). \tag{9.78}$$

In addition to the assumptions regarding process capability ratio calculations, it should also be noted that these calculations are point estimates and are prone to statistical fluctuations. Therefore, it is advisable to also determine the confidence intervals associated with any calculated process capability ratios.

By substituting s for σ, we can develop a confidence interval determine the point estimation for C_p. If the manufacturing characteristic is normally distributed, then the $100(1 - \alpha)\%$ confidence interval on C_p is given by:

$$\hat{C}_p\sqrt{\frac{\chi^2_{1-\alpha/2,n-1}}{n-1}} \le C_p \le \hat{C}_p\sqrt{\frac{\chi^2_{\alpha/2,n-1}}{n-1}}, \tag{9.79}$$

where $\chi^2_{1-\alpha/2,n-1}$ and $\chi^2_{\alpha/2,n-1}$ are the chi-square distribution with $n - 1$ degrees of freedom.

Process capability should be evaluated in terms of statistical control over time. Specifically, process capability analysis is more meaningful if it is evaluated using the variable control charts as described above. It is possible for a process to be in statistical control in the control charts but still be evaluated as being not process capable. Importantly, the determination of process capability can only be made if the process is in statistical control.

9.6.5 Rational Subgroups

The techniques used for collecting the sample data to construct control charts should be designed so as to best ensure that non-random events are quickly recognized. For this purpose, it is desired that the probability of detecting non-random shifts in the differences between subgroups is maximized, and the differences within subgroups are minimized [14]. These sampling techniques are called rational subgroups and are an important concept in the quality control methods used in microsystems manufacturing. How the sampling should be conducted for statistical analysis to ensure these goals are met will depend on the type of processing step being monitored, the specific details about the processing step, and the equipment used to implement the processing step.[4] The engineer will need to have intimate familiarity with the details of the processing step and the fabrication results on each substrate in order to devise rational subgroups for each individual situation.

There are several different types of spatial variations that can occur depending on the dimensional scale under consideration, including across die variations; across substrate variations; substrate-to-substrate variations; and lot-to-lot variations (see Fig. 9.28). These variations can be present both in the dimensions of the features and the material properties. The following can represent the commonly expected magnitude of the variations:

$$\begin{aligned}&(\text{Variations within the die}) < (\text{Variations within the wafer}) < \\ &(\text{variations within the lot}) < (\text{variations from lot-to-lot}).\end{aligned} \tag{9.80}$$

While this inequality is often true for many microsystems processing steps, there are exceptions. Therefore, the quality engineers must understand the processing step and the variation patterns before constructing sampling strategies.

As shown in Fig. 9.28b, the spatial variation across a substrate often looks like a "bull's-eye" type of pattern wherein contour lines having equal thickness can be superimposed over the wafer surface to reveal a series of concentric rings that increase or decrease from the center to the edge of the substrate in the radial direction. These ring patterns are often somewhat irregularly shaped and are commonly exhibited in the thicknesses of deposited thin-film material layers, the depths of etches, the doping concentrations, and other processing steps. These types of patterns are expected since many processing steps involve diffusion and temperature processes that exhibit temperature and process reactant density gradients in the radial directions across the circular-shaped substrates. These variation patterns are seen in batch (e.g., LPCVD, thermal oxide growth, diffusion, etc.) and single-wafer processing steps (e.g., etch, PECVD, ALD, etc.). The exact shape of the variation pattern is likely to be more closely identical for each wafer processed on single-wafer process tools, whereas for batch processes there may be similarly shaped variation patterns of each substrate of a batch and additional variations on the substrates from one end of the process tool to the other.

Since there are variations across the substrate, these will also be manifested as variations across each of the die (see Fig. 9.28a). The extent of these variations depend on the size of the die, the total amount of variation that occurs across the substrate surface (i.e., the gradient of the variation), as well as where the die is located on the substrate.

[4]This sentence implies that rational subgroups are only applied to processing steps. While it is true that rational subgroups are most often applied to processing steps, the concept can be used on process modules as well as process sequences.

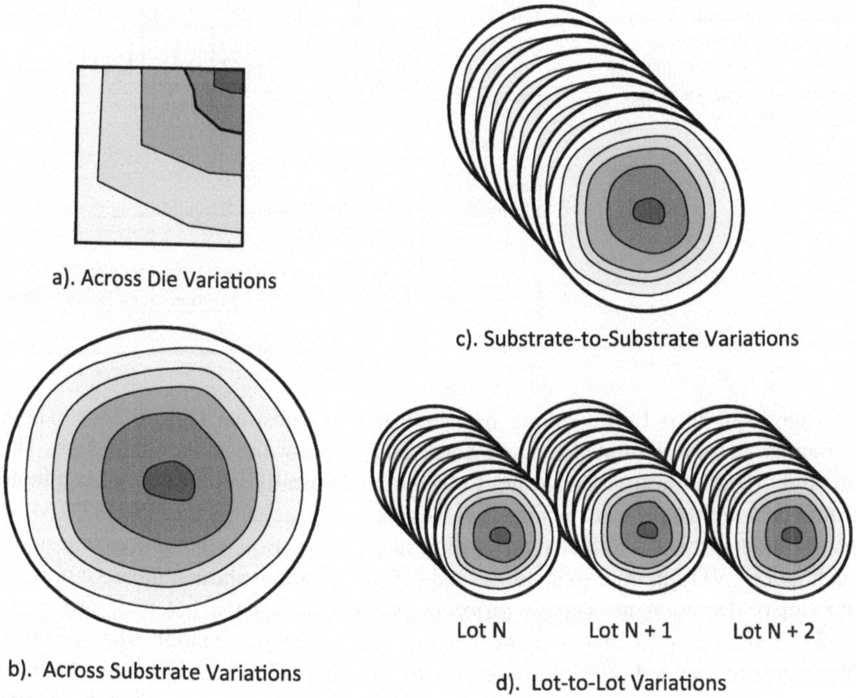

Fig. 9.28 An illustration of spatial variations on substrates at different dimensional scales, including (**a**) across the die variations; (**b**) across substrate variations; (**c**) substrate-to-substrate variations; and (**d**) lot-to-lot variations

Additionally, there may be other die-specific variations that can occur as well. One type of variation at the die level is what is termed photolithographic proximity effects [17]. These are variations in the dimensions of the features transferred onto the substrate using optical photolithography that relates to the density of small features to one another. Specifically, it is observed that a feature transferred onto a substrate using photolithography will exhibit differing dimensions in areas where these features are isolated compared to areas where the feature density is higher. This happens even when the features have the same "as-drawn" dimensions. This is called an "iso-dense print bias." It typically begins to appear as the features approach the resolution limit of the photolithographic tool used and when the features are relatively close to one another. This print bias increases in magnitude as the feature sizes decrease. For example, for a 365 nm wavelength exposure and a NA of about 0.5, the print bias starts to be seen at 0.7 um and increases below this wavelength. The amount of print bias effect seen in a particular situation will depend on the exact dimensions of the features and gaps between the features, the photolithographic tool and processes used, and other effects and therefore usually needs to be determined for each particular situation.

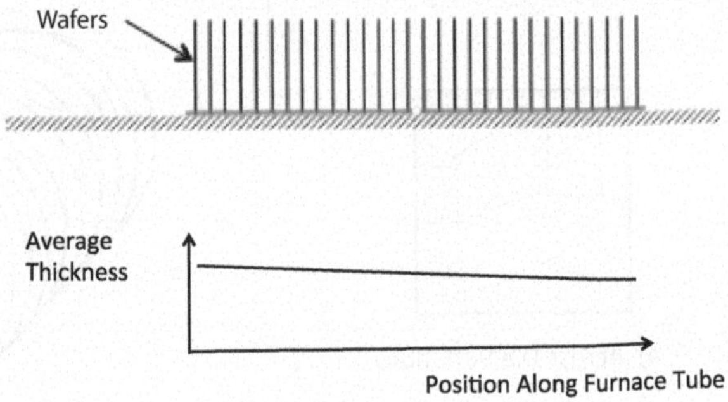

Fig. 9.29 Illustration of a common variation seen across a lot of wafers in thin-film thickness deposition such as LPCVD

There will also be variations from substrate-to-substrate (Fig. 9.28c). These variations can be exhibited in both batch and single-wafer processing steps. The amount of variation from one wafer to the next is typically very small in magnitude from one wafer to the next in single-wafer processing steps, such as PECVD, ALD, photolithography, and RIE etch processing steps. In contrast, batch processing steps, such as LPCVD, thermal oxidations, and diffusions, often show a more significant amount of variation as well as differing patterns across the batch of processed wafers. Consequently, there may be a significant amount of variation when comparing wafers at one end of the furnace tube to the other (Fig. 9.29). This results from issues such as gas depletion and temperature gradient effects across the tube during depositions and thermal processes.

Lot-to-lot variations are also exhibited in microsystems manufacturing processing steps (see Fig. 9.28d). In most instances, the amount of variation from one lot to the next should be fairly small, with any variations becoming more perceptible when comparing a number of processed lots over time. These kinds of variations can be seen in deposition and etch steps. For example, in a LPCVD deposition, there is material that is deposited on the inside of the furnace tube walls, and over time this can affect the temperature profile of the tube, thereby resulting in a variation from one lot to the next. However, it usually takes several lots before the temperature profile is altered sufficiently to have a noticeable effect on the uniformity of the deposition. Lot-to-lot variations are sometimes also seen in etching processes. For example, when performing deep etching into fused silica using an ICP process, a metal hard mask, such as nickel, is used [18]. The etch results in deposited nickel on the etch chamber walls that can affect the etch results from one lot to the next.

It is common to see dimensional and material property variations that occur in processing steps over periods of time (i.e., temporal variations). These types of variations are seen in both single-wafer and batch processing steps. Temporal variations and wafer-to-wafer (on single-wafer tools) or lot-to-lot variations are essentially the same since both are based on a processing step being performed at different times.

9.6.5.1 Sampling Methods for Rational Subgroups

An important point from the preceding discussion is that many microsystems manufacturing processing steps exhibit variations of different types and magnitudes. Depending on how the rational subgroups are designed and constructed has a significant effect on the quality of the data used for statistical analysis and quality control purposes. Therefore, it is important to develop suitable rational subgroups for each processing step for use in quality control charts for monitoring microsystems processing steps. For example, taking measurements on wafers across the furnace tube for a LPCVD deposition process to develop a $\bar{x}$ control chart would likely result in upper and lower control limits that were very far apart. This is due to the large amount of variation in the mean from one wafer to the next across the furnace tube. As a result, variations with assignable causes that occur during the depositions can be missed or only detected far later than otherwise.

The question is how to construct sampling strategies that ensure that non-random changes in a processing step can be quickly detected? Attempting to develop rules for sampling methods that can be applied to every situation is not possible. Instead, the microsystems manufacturing quality engineers need to carefully analyze the variations for each processing step in order to develop methods of sampling that provide data that is useful and can be statistically analyzed. This requires not only an understanding of the statistical methods discussed above but also a deep understanding of the causes and effects of the variations in each processing step used and at each of the dimensional and time scales. Unfortunately, each process step and tool is different, and therefore there is no defined method that works for each and every situation. This is a complex problem that requires process data, analysis, significant experience, and considerable thought about the subtle nuances of the physics of each processing step and how these can affect the process outcomes.

For example, consider the within-die variations due to the iso-dense print bias discussed above. Rather than treat all of the critical dimension (i.e., CD) measurements equally, a better strategy would be to know and understand how the iso-sense print bias effect is manifested across the die for the specific patterns being transferred in the photolithography processes. Obviously the amount of print bias will depend on the specific layout design, specifically when there are regions on the die layout where small dimensional features are closely spaced to one another. Once the print bias distribution effect across the substrate is known, this information will enable the identification of locations within each die that would be expected to have nearly identical CD measurement outcomes, and therefore any large variations present in the CD should be only due to non-random effects. The strategy is to take sampling measurement data at the same positions on different die across the wafers for statistical analysis. Same die position refers to the same feature at the same exact position on each die. In some cases, it may be useful to take measurements on some number of these positions on each die and then group the measurements from these same die positions together into appropriate subgroups for statistical analysis. This

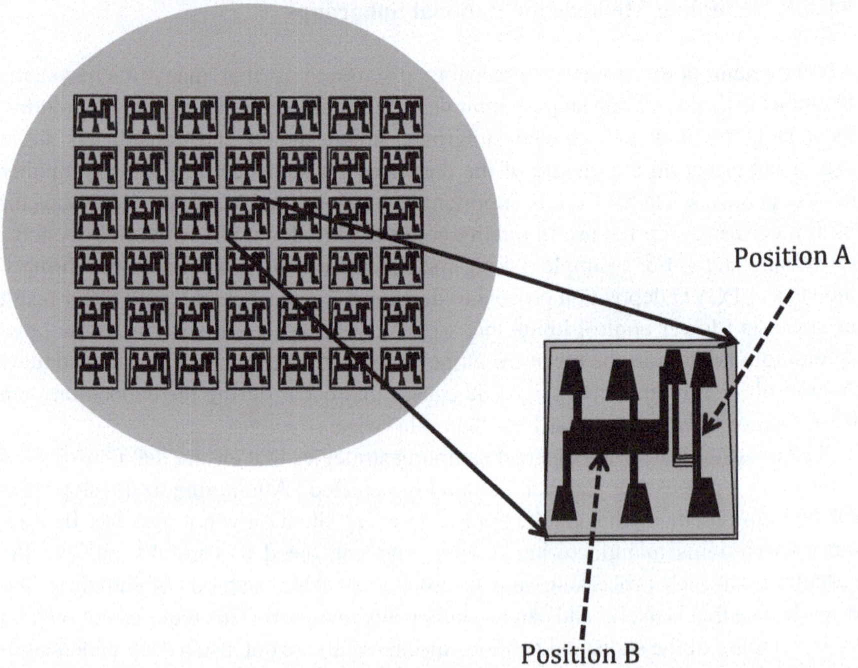

Fig. 9.30 Illustration of selecting locations on die that will exhibit approximately equal iso-dense print bias effects in order to develop suitable rational groupings of the sampling data for quality control charts

method is illustrated in Fig. 9.30 wherein a wafer with multiple die is shown and one die near the center has been selected and enlarged to show the different sizes feature.

Two positions on the enlarged die are noted: position A and position B. Measurements taken on the die indicate that there are print bias effects on this die. Location A represents a die location where the dimensions of the printed features are small and closely packed. Therefore, some amount of print bias is expected to occur at Location A. Location B is where the printed dimensions are larger, and there are no other features in close proximity. Therefore, the expected print bias at Location B is less or nonexistent compared to Location A. A possible sampling method would take CD measurements at Location A on multiple die across the wafer and group them together for statistical analysis and also take CD measurements at Location B on multiple die and group these measurements in a separate group from the measurements at Location A, for statistical analysis. This sampling strategy would be preferred over sampling at locations A and B and averaging these together. Indeed, if sampling at locations A and B was combined, it would be very difficult to tease out changes occurring in the photolithography process at locations A since they would be obscured by the samples taken at the B locations.

As should be obvious to the reader, methods to ensure rational subgroups of the sampling measurement data require expert knowledge of the fabrication processing step and careful thought. The manufacturing engineer will have to determine the specific methods that provide accurate data for statistical analysis in quality control while still keeping the costs and time of processing monitoring within acceptable bounds.

9.7 Summary

The importance of ensuring quality and performing yield analysis in microsystems manufacturing was discussed. Methods for determining device functional yield as well as whether it meets pre-defined specifications derived from the application requirements for determination of the parametric yield were covered. The types of defects that result in nonfunctional devices and various analytical techniques used for yield modeling based on point defects were explained. Parametric yield was then reviewed wherein a manufacturing function that incorporates all of the statistical distributions of the parameters describing the microsystems device's output response is overlaid by an acceptance range (or region), which are taken from the application requirements, to determine the yield. Methods to ensure quality during manufacturing using statistical process control (SPC) were covered. The concept of control charts showed how processing steps can be monitored for whether they are in control or not. Various types of control charts were discussed for both variables and attributes. Process capability, which is the ratio of the range of the specifications to the spread in the manufacturing function, was shown to be an excellent gauge for whether a manufacturing process will be able to produce microsystems devices having acceptable yields. Lastly, techniques for sampling of the measurements on wafers during and after manufacturing so as to obtain data that is rationally subgrouped so as to identify non-random changes were discussed.

References

1. R. Cappe, C. Perry-Sullivan, Yield and cost challenges at 16 nm and beyond, Solid State Technology, See: http://electroiq.com/blog/2016/02/yield-and-cost-challenges-at-16nm-and-beyond/
2. G.S. May, C.J. Spanos, *Fundamentals of Semiconductor Manufacturing and Process Control* (Wiley, New Jersey, 2006)
3. B. Murphy, Cost-size optima of monolithic integrated circuits. Proc. IEEE **52**(12), 1537 (1964)
4. J. Pineda de Gyvez, D. Pradhan, *Integrated Circuit Manufacturability* (IEEE Press, New York, 1999)
5. T. Okabe, M. Nagata, S. Shimada, Analysis of yield of integrated circuits and a new expression for the yield, in *Defect and Fault Tolerance in VLSI Systems*, ed. by C. Stapper, vol. 2, (Plenum Press, New York, 1990)
6. C. Stapper, Fact and fiction in yield modeling. Microelectron. J. **210**(1-2), 129 (1989)

7. J. Cunningham, The use and evaluation of yield models in integrated circuit manufacturing. IEEE Trans. Semicond. Manuf. **3**(2), 60 (1900)
8. C. Stapper, R. Rossner, A simplified method for modeling VLSI yields. Solid State Electron. **25** (6), 655 (1973)
9. H.J. Larson, *Introduction to Probability and Statistical Inference* (Wiley, New York, 1969)
10. T.R. Scott, T.P. Walker, Regionalization: A method for generating joint probability distributions. IEEE Trans. Circuits Sys **CAS-23**(4), 229 (1976)
11. K.H. Lueng, R. Spruce, Efficient statistical circuit analysis. Electron. Lett. **10**, 377 (1974)
12. S. Director, G.D. Hachtel, The simplicial approximation approach to design centering. IEEE Trans. Circuits Sys **CAS-24**(7), 363 (1977)
13. S. Director, G.D. Hachtel, L.M. Vidigal, Computationally efficient yield estimation procedures based on simplicial approximation. IEEE Trans. Circuits Sys **CAS-25**(3), 121 (1978)
14. D.C. Montgomery, *Introduction to Statistical Process Control*, 5th edn. (Wiley, New Jersey, 2005)
15. A. Duncan, *Quality Control and Industrial Statistics* (Irwin, Chicago, 1974)
16. Western Electric, *Statistical Quality Control Handbook* (Western Electric Corporation, Indianapolis, 1956)
17. C.A. Mack, Optical proximity effects, See: http://www.lithoguru.com/scientist/litho_tutor/TUTOR13%20(Spring%2096).pdf, Accessed Jan 2018
18. M. Pedersen, M. Huff, Plasma etching of deep high-aspect ratio features into fused silica. IEEE J. Microelectromech. Syst **28**(2), 448 (2017)

Chapter 10
Managing Parameter Variations in Microsystems Device Design

10.1 Introduction

The present chapter is focused on how to manage device parameter variations when performing the engineering design so that the devices can be manufactured with sufficient yield[1] and satisfy the application requirements. Specifically, this chapter is directed at methods of developing a good quality microsystems device design. What constitutes a good quality microsystems design? First, the device design must include all of the functional aspects required for the application with an acceptable level of performance. Second, the design should be as simple as possible while still incorporating the functionalities and meeting the performance requirements. Third, the microsystems design must be manufacturable; specifically the device design and associated manufacturing process for the device must have a sufficiently high yield. Fourth, the cost of the manufacturing at the production level to satisfy the market(s) must result in a profit. Note that this criterion qualified the statement as "a production level to satisfy the market." This is important since almost anything can be made less expensively at higher production levels. However, it is only the production level that will satisfy market demand that is relevant. Fifth, the devices must be suitably reliable for the market and application intended.

Developing a microsystems design that meets all of these requirements is extremely challenging. Proof of the level of challenge involved is exhibited when comparing the number of MEMS microsystems devices reported in the literature to the number of MEMS products in the commercial market. Indeed, there have been thousands of new MEMS microsystems devices reported, but the number in the commercial market is at most a few dozen devices.

[1]The reader may wonder why the objective is not to maximize the yield. Sometimes obtaining maximum yield is the objective, but not always. In some situations, it may be beneficial to have a reduced, but satisfactory yield in order to lower the manufacturing costs.

M. Huff, *Process Variations in Microsystems Manufacturing*, Microsystems and Nanosystems, https://doi.org/10.1007/978-3-030-40560-1_10

The roadmap of this chapter will be as follows. First, a high-level description of MEMS design and development will be reviewed. This will discuss how the elements of the system-level design are broken down into manageable pieces as well as the tools generally employed for these activities. Then the topic of *design for manufacturability* as it relates to MEMS microsystems design will be discussed. Design for manufacturability is a commonly used term in industry to the point that it has devolved into a marketing phrase. Nevertheless, the overall concept is very important, and there are certain general guiding principles as to what this means as it relates to MEMS device design. Subsequently, the topic of manufacturing variation (tolerance) design is discussed. Variation design is the primary focus of this chapter and will be discussed in detail using examples of MEMS devices for reinforcement of the concepts. The approach will be to use simple examples of devices dependent on only one parameter and then expand into a discussion of tolerance design for multiple parameters. Lastly, yield sensitivity analysis will be discussed.

It is also important to disclose what this chapter will not cover. MEMS design and development is usually based on a number of concurrent activities including system-level design; device functional design; process sequence design; electronics design (if needed for application); and package design. Moreover, all of these activities may involve participation and input from the engineering staff as well as from the business and marketing people. This chapter will not discuss the business and marketing activities, as well as the electronic and package development. While these topics are extremely important, these activities usually do not involve or are directly impacted by MEMS device parameter variations due to manufacturing.

This chapter will also not discuss how to design a process sequence for manufacturing a MEMS microsystems device. Additionally, this chapter will not review all of the physical equations that are commonly used in modeling MEMS microsystems devices. Instead we will show how device parameter variations are dealt with during device design using the device examples that have been revisited throughout this volume. Once the method of effectively managing device parameter variations is understood, the reader should be able to use these methods for any microsystems device design activity no matter the transduction mechanisms or physical describing equations that are involved.

Given the importance of managing device parameter variations in MEMS manufacturing, it is surprising how little this subject is covered in the existing literature. There have been a few papers that discuss various methods for dealing with MEMS device parameter variations, but all of these are directed on the design of a specific type of device and not on general techniques that can be used on any MEMS device [1–4].

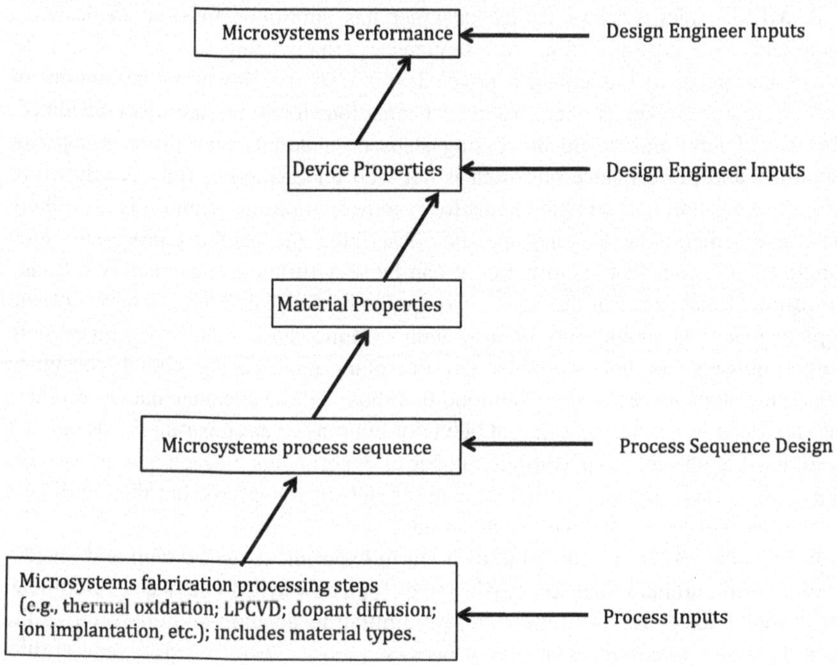

Fig. 10.1 The technology levels involved in a MEMS device

10.2 Relationships Between Process Sequence and Parameter Variations

Figure 10.1 is an illustration of different levels involved in microsystems technology. At the bottom level are the individual microsystems fabrication processing steps to be used in the implementation of a MEMS device. Most fabrication steps will result in some amount of variation in the device parameters. These processing step variations propagate into the next level of technology, namely, the process sequence, whereby the parameter variations in each of the multiple processing steps result in additional parameter variations when combined with the other processing steps also having parameter variations. Included in the parameter variations due to the processing steps and process sequence are the variations in pattern transfer activities and lithographic misalignments. These were discussed in the previous chapter on the topic of process integration. All of these device parameter variations result in output variations of the microsystems. The amount and type of device parameter variations at each level have a complicated resultant impact on the output behavior of the device.

A MEMS microsystems device designer has enormous freedom to design a completely new customized process sequence, perhaps using new and customized processing steps, to implement a novel device design.[2] However, the amount of freedom that the designer exercises in each situation should be carefully considered. The use of new and exotic processing steps or materials may promise superior capability and performance but often is not well understood or fully developed to provide consistent and reliable results for commercial products, and it is also likely that there will not be existing statistics describing the resultant processing step variations. Without that information, it can be very difficult to accurately estimate the resultant variations in the device's output behavior. Therefore, new processing steps or materials should only be used with extreme caution and when other more mature options are not available. On the other hand, using already designed processing steps increases the likelihood that there will be adequate data to estimate the variations in the device's output behavior but may over-constrain the designer's freedom. It is possible to perform a number of experiments on each new processing step or material to develop sufficient data for statistical analysis, but there will be a substantial cost associated with such a study.

In practical terms, it is possible to perform experiments to develop and characterize a small number of new processing steps for a new MEMS device development, but it would likely be cost (and time) prohibitive to develop and characterize the majority of the processing steps in a process sequence. Therefore, as a general rule, most MEMS designers would develop and characterize a very small number of new processing steps and only if this was absolutely essential to the device development activity.

Similarly, the MEMS designer has the design freedom to create a customized process sequence. A customized process sequence allows a higher level of design freedom to implement a device design, yet it also makes it all but certain that there will not be statistics on parameter variations. Without that information, it will likely be very difficult to accurately estimate the variations in the device's output behavior. On the other hand, using an already designed process sequence increases the likelihood that there will be adequate data to estimate variations in the device's output behavior but may over-constrain the designer's freedom.

It will always be less costly in both time and money and lower in risk to develop a MEMS process sequence leveraging from existing developed processing steps, materials, and process sequences to the extent feasible. Having said that, nearly all MEMS designers develop process sequences for new device designs that are to some extent are customized. The same is true with regard to new microsystems device designs; most all MEMS development activities develop a new design to some extent.

The reasons why there is customization of process sequences and device designs in MEMS are due to the couple of factors. First, the MEMS device applications are

[2]In contrast an IC designer often does not have any freedom on the selection of the processing steps or process sequence.

tightly tied to the device design, and the diversity of MEMS applications is truly enormous. Each type of application usually requires something that is different and unique in the device design. Since there are a very large number of applications that MEMS can be used in, there are also a large number of device design variants. Additionally, the device design is also tightly coupled to the process sequence used to implement the device. This means that there is also a wide diversity of process sequences used to implement different MEMS devices. Consequently, it has proven to be extremely difficult to take an existing MEMS device designed for a specific application and use the device design or the process sequence to implement a device for a different application without at minimum making some significant changes. Additionally, most MEMS developers desire to protect both the device and process sequence design intellectual property (IP) as a competitive advantage. Therefore, there is an incentive for designers to create something new in their designs. Lastly, most MEMS device and process sequence designs are often highly coupled to a specific foundry.

While creating a good MEMS device design is a difficult endeavor, it also provides some notable opportunities. In comparison to microelectronics design, MEMS design encompasses many physical domains (e.g., electrical, mechanical, thermal, fluidic, chemical, etc.) and therefore has a much larger breadth of applications that arguably allows considerably more creativity. While IC design only involves transistors, resistors, and capacitors, MEMS design on the other hand involves just about any type of device or system known. In short, designing MEMS is both challenging and interesting and fun.

10.3 Overview of MEMS Device Design and Modeling

The topic of MEMS device design was well explained by Senturia in the book entitled *Microsystem Design* [5]. Many of the concepts reviewed in this section are covered in more detail in that resource, and the reader is encouraged to reference it for a more complete discussion of MEMS device design and modeling.

The design and modeling of MEMS microsystems devices has many similarities to that of IC devices and involves multiple hierarchical layers that can be represented as illustrated in Fig. 10.2. The four levels[3] include the (1) system design and modeling level; (2) device design and modeling level; (3) physical design and modeling level; and (4) process design and modeling level.

The lowest bottom level is the process design and modeling level. This level involves the design of the process sequence and the design of the associated mask layout design that is used to pattern the material layers and/or substrates used in the

[3]Conventional semiconductor devices are more commonly represented by three layers including process; device; and circuit layers [6]. We show four layers, which is how the models are described for MEMS by Senturia [5].

device fabrication. The design and modeling at this level is fundamental since it determines the geometrical shape and dimensions of the device and selects and describes the materials and material properties and the processing steps (and possibly the process equipment) used in the device implementation. The mask layout specifies the shapes and dimensions at a two-dimensional level that will be used for photolithographic patterning. Mask design is highly automated using layout generation software, and advanced forms of mask layout software developed by the IC industry incorporate a number of design verification capabilities including design rule checking (DRC); layout versus schematic (LVS); electrical rule checking (ERC); and others. Rule checking for MEMS is only possible for an already developed process technology, and since MEMS development efforts often involve a customized new-to-be-developed process sequence, rule-checking capability may not be available. In those cases, the designer of the MEMS device must have sufficient experience and knowledge to be able to develop reasonable *estimated* design rules for the layout artwork and the associated process sequence. Moreover, when using estimated design rules, the rule checking is usually manually performed. Once the mask layout design has been completed, the mask layout software is translated into an industry-standardized format called Graphic Database System II, or GDSII. GDSII is a binary file format that represents the photolithography layers, the shapes of the devices and alignment marks, as well as other important information about the layout. Additionally, technology CAD, or TCAD for short, composed of two different types of software, process TCAD and device TCAD, are used to model the processing steps such as diffusion of dopants, oxidation, ion implantation, etc., model the process sequence, and predict the device geometry and dimensions based on the process sequence and mask set.

The next level up in Fig. 10.2 is the physical design and modeling level. This level uses the 3-D representations of the device developed at the process level to

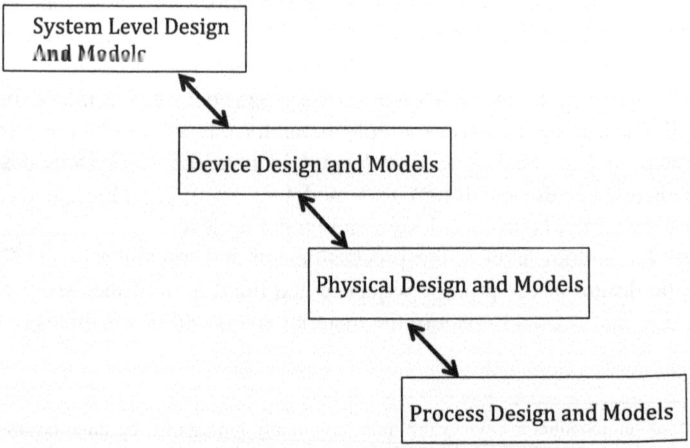

Fig. 10.2 Four levels shown for the hierarchy of models used in MEMS development [5]

model the devices often using partial differential equations. Analytical techniques can be used for finding closed-form solutions if the geometries are either ideal or approximated as being ideal. Alternatively, numerical techniques, such as finite element method (FEM), boundary element method (BEM), etc., applied to meshed representations of the device can be used to find solutions for non-ideal geometries as well as more accurate solutions for the actual 3-D device structures. Multiple and coupled fields can simultaneously be handled in these simulations, which is difficult to do with analytical representations. It should be noted that the use of meshed representations and numerical techniques such as FEM methods can often be unwieldy for use in the initial design phase where multiple device concepts are being evaluated. Therefore, these techniques may be employed after the design concept is more concrete.

The device level in Fig. 10.2 usually consists of macro-models or reduced-order models that incorporate the crucial physical behavior of the device that can be effectively used at the next higher level, the system level. One important caveat of the device-level model is that it usually does not contain the complete physics of the device but instead uses a partial model that is manageable for higher-level simulations of more elements, and this partial model may only be valid over a limited range of possible parameters. Therefore, the designer must be cognizant of the limits of the device-level model used and careful not to exceed its capability for modeling the device behavior.

The highest level is the system design and modeling level and uses either block diagrams or lump-element circuit representations. The system model for a typical MEMS device is composed of an equivalent electrical circuit representation of the lump-element components wherein the lump-element components are derived from the device level. A set of coupled ordinary differential equations can be used to mathematically describe the behavior of the system.

The MEMS designer(s) will provide inputs on each level shown in Fig. 10.2, and the work at each level is iterative and interactive with the other levels, as shown by the two-sided arrows. The design process can begin at any level desired but often starts at the system level and works downward.

Another part of the design activity is simulation, verification, and validation. Simulation is the process of using a model of a device to approximate the operation of the device. The model is a representation of the device, and the simulation is a representation of the device behavior for a given set of inputs. The model of the device includes the device's physical characteristics (i.e., shape, geometry, dimensions, materials, etc.). Verification of a device model is the process of confirming that the model correctly matches the specifications and assumptions for it to be considered acceptable for the given application. Verification usually involves iterative device testing and correcting of the model so that it more accurately represents the behavior of the real device. Validation checks the accuracy of the model compared to the real device that the model is intended to imitate. Obviously, validation requires that actual devices have been fabricated and are sufficiently workable that they can be tested (even if their behavior and performance are not as desired).

Software tools for modeling of IC microsystems are extremely advanced and accurate for most applications. Modeling tools for MEMS devices are a more recent development and have become available in the commercial market in the 1990s and are steadily improving as the field matures and grows in economic importance.

10.4 Example of the Design Levels for a MEMS Device

As an example of the design levels discussed in the previous section, the fluidic microchannel device that was originally discussed in Chap. 8 will be reconsidered. The microchannel device at the process level is represented by the process sequence as illustrated in Fig. 10.3.

The process design and associated models would be represented in the case of the microchannel by a mask layout and an associated process sequence. The mask layout in this case would be composed of two mask layers, first for patterning the thin-film material masking layer and silicon substrate and second for patterning the glass cover substrate. The process sequence and mask set could be inputted to TCAD software to develop a 3-D representation of the device structure. The lateral dimensions of the microchannel would be set by the dimensions of the features on the mask for the patterning of the silicon substrate, and the depth of the microchannel would be established by the processing step(s) for the etching. If statistical information on these processing steps is available, then the models developed at this level may incorporate estimates of the expected variations in the microchannel dimensions.

The physical level of the microchannel device is developed using Navier-Stokes equations in 3-D cylindrical coordinates, which can be expressed as follows [7]:

$$\rho \frac{\delta \overline{v}}{Dt} = -\nabla p + \mu \nabla^2 \overline{v} + \rho \overline{g}, \tag{10.1}$$

where ρ is the fluid density, v is the flow velocity, p is the fluid pressure, and g represents body accelerations acting on the continuum. Based on the situation of a well-developed flow of an incompressible fluid, a set of assumptions for the microchannel can be written as:

$$\begin{gathered}\text{Fluid flow is steady}\,(\delta(\ldots)/\delta t = 0);\\ \text{Radial and Angular components of the fluid velocity are zero}\,(u_r = u_\theta = 0);\\ \text{Fluid flow is axisymmetric}\,(\delta(\ldots)/\delta\theta = 0);\text{ and}\\ \text{Fluid flow is fully developed}\,(\delta u_z/\delta z = 0),\end{gathered} \tag{10.2}$$

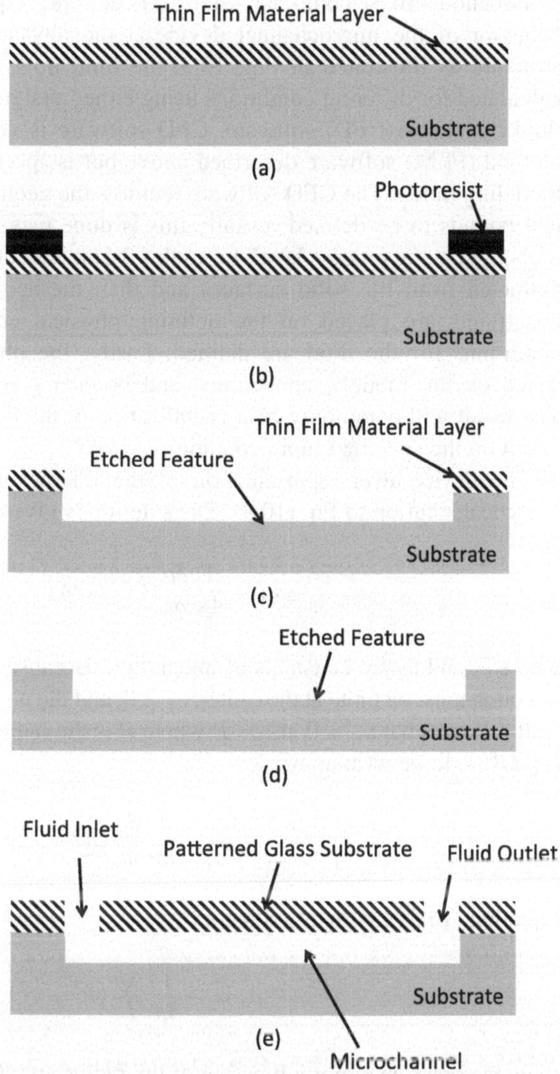

Fig. 10.3 Cross section of substrates illustrating the manufacturing process to implement the microchannel device

thereby resulting in the following reduced form of Navier-Stokes applicable to the microchannel device as follows:

$$\frac{1}{r}\frac{\delta}{\delta r}\left(r\frac{\delta u_z}{\delta r}\right) = \frac{1}{\mu}\frac{\delta p}{\delta z}, \tag{10.3}$$

where μ is the viscosity of the fluid.

Equation (10.3) could be used as a defining equation for representing the behavior of the microchannel device at the physical level.[4] Using the device structure as illustrated in Fig. 10.3, the fluid flow in the microchannel can be calculated for different conditions using either analytical methods or computation fluid dynamics (CFD) software. CFD software is very similar to finite element method (FEM) software described above but is specifically developed for use in modeling fluids. The CFD software requires the geometry, dimensions, and physical bounds to be defined; usually this is done using the computer-aided design tools used at the process level. From this 3-D model of the fluid system, the fluid is extracted from the solid surfaces and then meshed into cells. The appropriate constraints are placed on the defining physical equations, and the boundary conditions for the fluid are defined. Lastly, the physical equations are solved based on the models, constraints, and boundary conditions using simulations. The result will commonly be a visualization of the fluid flow in the microchannel based on the specified input conditions.

The device-level representation of the microchannel could be based on an analytical solution to Eq. (10.3). The solution is given by:

$$u_z = \frac{1}{4\mu}\frac{\delta p}{\delta z}r^2 + c_1 \ln r + c_2, \tag{10.4}$$

where c_1 and c_2 are constants of integration. Because the fluid velocity in the axial direction must be finite at the center, $c_1 = 0$, and the no slip boundary condition at the wall requires that $u_z = 0$ at $r = R$ where R is the radius of the channel. This allows Eq. (10.4) to be written as:

$$u_z = -\frac{1}{4\mu}\frac{\delta p}{\delta z}\left(R^2 - r^2\right). \tag{10.5}$$

Equation (10.5) can be rewritten as:

$$u_z = -\frac{1}{4\mu}\frac{\Delta p}{L}\left(R^2 - r^2\right), \tag{10.6}$$

since the pressure is not a function of the radius and the radius of the microchannel does not vary over its length. Using Eq. (10.6), the total volumetric fluid flow, Q, can be summed by integrating the velocity in the axial direction of the microchannel over the radius as follows:

[4]Equation (10.3) is arguably too simplified as the defining equation for incompressible fluid flow in the microchannel since it ignores the regions of the flow path at the entrance and exits wherein the fluid flow is not well developed and also ignores the 90-degree turns that the fluid must make as it enters the inlet port and exits the outlet port. These effects would normally be included at the physical level since they will represent additional flow resistances. However, the inclusion of these effects greatly complicates the equations and, therefore, is not included in the example above.

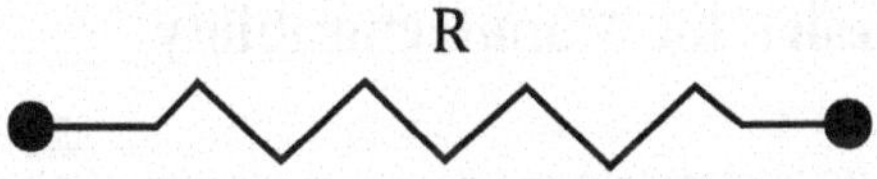

Fig. 10.4 An electrical resistor representation of the microchannel flow resistance

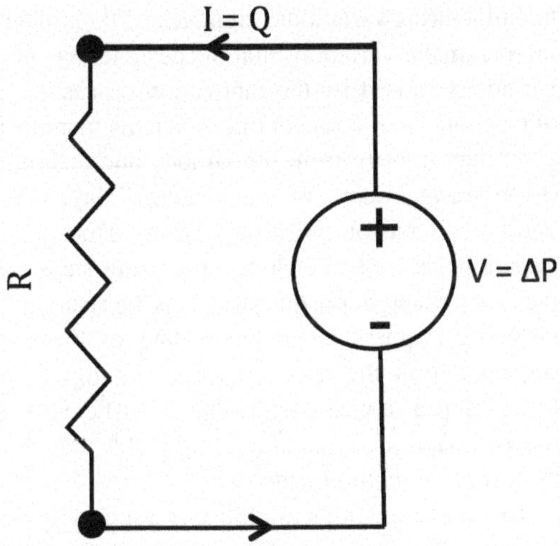

Fig. 10.5 Equivalent electrical circuit of the microchannel with a differential pressure across the device and a volumetric flow rate

$$Q = -\frac{\pi}{2\mu}\frac{\Delta P}{L}\int_0^R \left(rR^2 - r^3\right)\, dr = \frac{\pi R^4 \Delta P}{8\mu L}. \tag{10.7}$$

This is the Hagen-Poiseuille relationship that was first seen in Chap. 8 and was used to represent the device behavior. Equation (10.7) can be placed into a form that expresses the fluid flow resistance as:

$$R = \frac{\Delta P}{Q} = \frac{8\mu L}{\pi r^4}. \tag{10.8}$$

The device-level depiction of the microchannel can be represented as a lump-parameter element as shown in an electrical resistor in Fig. 10.4 wherein the value of the resistor element is given by Eq. (10.8).

The system level for the microchannel is illustrated in Fig. 10.5 and consists of an equivalent circuit of the flow resistance of the microchannel, represented by a resistor, and the pressure source, represented by a voltage potential, V. Additionally, the volumetric flow rate of the fluid, Q, is represented by the current, I, in the circuit.

10.5 MEMS Design for Manufacturability

The performance of any microsystems device is strongly linked to the design. An important goal in designing for manufacturability is to *desensitize the device's behavior from the device parameter variations that occur in manufacturing*. The first requirement in desensitizing a device design is to understand where the manufacturing variations originate and their magnitude and impact. The manufacturing parameter variations that the designer has most control over are the dimensional variations caused by the fabrication processes.[5] There are three major sources for dimensional variations in microsystems manufacturing processes: planar dimension variations (mostly from lithography and etching processes); planar location offsets (from misalignment of one patterned layer to another during lithography); and vertical dimension variations (from deposited thin films or substrate thickness variations as well as etching processing steps). Each of these was discussed in the previous chapters on microsystems fabrication technologies. Design for manufacturability requires an understanding of these variations for a particular process sequence, how the resultant variations impact the device behavior, and the ability to develop a device design that is sufficiently insensitive to these variations and meets the required manufacturing yields, which in turn is dependent on the product performance requirements.

Design for manufacturability is not easy to describe or teach. Understanding the source and magnitude of manufacturing variations associated with microsystems process sequences employing a multitude of physical processes is challenging in its own right, but when coupled to the need to be able to conceive of device designs that are insensitive to these variations increases the challenge immensely. Some of the fundamental methods employed to understand these variations are a good understanding of processing technologies, a good quality statistical data gathering mechanism, and appropriate modeling to connect design expectations to physical behavior. However, it also requires a deep understanding of the device subsystems and a thorough multi-domain thinking process. In short, it requires considerable experience to do well. Nevertheless, there are some recommended practices that can be employed that can assist the microsystems developer to achieve a basic level of design for manufacturability. This section provides a high-level review of the major areas of microsystems design wherein device parameter variations are important. Some general recommendations regarding design for manufacturability are also given.

[5]The other sources of variations are the variations in material properties. However, as a general rule, the MEMS designer does not often have the ability to control the material properties, and even when control is possible, it usually will be only over a limited range of values as discussed in Chap. 6.

10.5.1 MEMS Device Design for Manufacturability

The photolithographic and etching processes define the lateral dimensions in most microsystems devices. This process typically has three pattern transfer steps: (1) from layout database to mask, (2) from mask to photoresist, and (3) from photoresist to the desired layer pattern, usually through an etching process. At each lithographic step, there can also be both registration and dimensional variations. Additionally, there are variations in the thicknesses of deposited material layers. The result of all these variations is that the microsystems device as fabricated will not be a perfectly accurate dimensional representation of the device design as specified in the mask layout.

One mechanism for achieving design for manufacturability in microsystems devices is to experimentally understand the variations observed in the fabrication processes and feed them back into the device design and models so as to scale the dimensions accordingly. For example, if one is designing a comb-finger-based variable capacitance structure, the width of the gaps between the fingers is directly proportional to the capacitance. A typical photolithography-defined pattern followed by a chemically based etch process will result in comb-fingers that are slightly narrower than as drawn in the mask layout. This is a form of bias variation. Moreover, the amount that the comb-fingers are narrower will vary from device to device, from wafer to wafer, and from lot to lot. This is a form of random variation. Understanding the offset, as well as the mean and standard deviation of the "as-etched" comb-finger widths, allows the designer to translate these dimensional variations to the device behavior and performance variations. This understanding of performance in light of manufacturing variations gives the designer the information on which to develop methods to desensitize the design to these variations. In the current example, imagine that the comb-finger variable capacitor is part of an accelerometer whose mechanical stiffness is formed from the same material layer as the comb-fingers. In that case the same manufacturing variation that causes a reduced sensitivity due to an increase in the comb-finger gap may also result in an increased sensitivity due to reduced suspension stiffness. Careful design can sometimes partially balance these effects and desensitize the design to this and other manufacturing variations.

Another way to desensitize the design from fabrication variations is to avoid using the undesirable aspects of a fabrication process to define a device's critical dimensions. For example, the dry etching sidewall is typically somewhat vertical, but the smoothness and sidewall angles are usually not well controlled. In the case of a device design requiring a perpendicular sidewall with 0.5 degree variation or mirror-like sidewall to achieve the critical performance, it will take an enormous amount of process development effort to obtain these process requirements and therefore would not be considered a robust design. Similarly, if the thickness variation in a sacrificial silicon oxide layer is critical to the performance, a robust design at the device level and system level should accommodate the resulting thickness variation expected from the process technology used (a few percent thickness variation is common for LPCVD). When considering process design for manufacturability, a thermal oxidation process will typically exhibit less thickness variation than either LPCVD or PECVD processes. Thus, we can consider a thermal oxide process that minimizes the thickness variation as a more robust design choice since it would more desensitize the device's behavior from the oxide thickness variations.

10.5.2 MEMS Process Sequence Design for Manufacturability

There is a very strong coupling between device design for manufacturability and process sequence design for manufacturability. Therefore, it should not come as a surprise that the methods for understanding and desensitizing the device performance to variations resulting from processing methods overlap the issues discussed in the previous section. Building on the example discussed above of making comb-fingers in a material layer, the manufacturing variations in the fingers will also be strongly dependent on the process methods used to implement the fingers. For example, the amount of lateral etch observed in a material layer (i.e., the amount of undercut of the masking layer) will vary depending on the material used in the layer, the type of etch process used (both equipment and process recipe), the depth of etch, the amount of over-etch required, and loading effects due to the amount of substrate area being etched. Most wet chemical etchants are isotropic and will undercut the masking layer very significantly during an etch process. In comparison, RIE, particularly DRIE, are highly anisotropic, and therefore the amount of undercutting of the mask will be greatly reduced. Therefore, use of DRIE to etch the comb-fingers will result in less variation in their widths.

In addition, since the dry etch rate varies within the wafer and from wafer to wafer, it may be difficult to control etch depth using a timed etch as a robust design. Instead, an etch stop mechanism may offer better control. Consequently, etching a silicon layer on top of an oxide layer (that acts as an etch stop) would be a better strategy than using a timed etch process.

Etches of all types have a characteristic called "loading." This refers to the effect on the material etch front due to the presence or lack of adjacent material also being etched. The etch of a finely spaced pattern of lines and spaces will be different (i.e., slower) than the etch than larger features. One way to minimize the loading effect is to design a layout with the same sized features so that the etching rate is approximately the same everywhere. If this is not possible, then a suitable over-etch will likely be required to clear the smaller features, but this will also result in some additional lateral etch on the larger features as well. The amount of over-etching required for each situation will be different and may require some development.

Like the planar dimensions described above, variations in material thicknesses can have a strong effect on device performance. In many MEMS processes, the thicknesses of the material layers are highly dependent on the type of process employed to deposit the layer. For example, many thin-film deposition methods, such as LPCVD and PECVD, have an across-wafer thickness non-uniformity of at least a few percent or more. Additionally, there will be some variation from the nominal film thickness desired and the actual thickness after deposition. Similarly, methods of wafer bonding and thinning-back to create single-crystal silicon layers can translate into variations in thinned device layers.

The manufacturing variations in the critical dimensions of MEMS devices will also be highly dependent on the process sequence that is used to implement the MEMS device. That is, all processing steps before and after and the processing step used to define a critical layer's dimensions may also have an effect on the dimensions of that layer and must be considered. Therefore, as can be appreciated, there is truly a myriad of complicated physical mechanisms that can give rise to geometric manufacturing variations in MEMS. The key to good design for manufacturability is to fully understand these variations in all of the key dimensions with the potential to affect the device performance. Developing this understanding requires considerable data gathering and statistical analysis.

In addition to geometric variations, MEMS devices are subject to significant variations in material properties.[6] Typical material properties with significant variations are mechanical properties, such as the residual stress, stress gradient, and the elastic modulus. Again, the key to good design is to fully understand the material property variations and how they affect device performance. Good process design can improve manufacturing variations caused by material variations. For example, a material's elastic modulus is intimately linked to its crystal structure. To obtain repeatable material properties, the process must be designed to produce repeatable material morphology. In addition to repeatability, the material properties can be engineered to be more or less isotropic, based on the material deposition process settings.

Thin-film material's stress state is very important in released MEMS structures, where the stress state is strongly dependent on the film's crystal structure. In particular, variation of the crystal structure through the thickness of the film can cause stress gradients that can result in bending moments, which cause the released structure to curl. This curvature can vary widely with small changes in the material's morphology, thereby causing large manufacturing variations in released structures. Design for manufacturability requires careful process design to deposit a uniform film and anneal the film to reduce its residual stress, as well as device design to reduce the sensitivity of the device's performance to any variations in curvature. As this discussion demonstrates, the design of a material deposition process is much more complicated than simply getting the desired thickness of a given material.

---

[6]Material properties were discussed in Chap. 6.

10.5.3 MEMS Microsystems Partitioning

MEMS developers are often confronted with important questions regarding how to partition the device design, as well as how to partition the fabrication. The first and often the most important partitioning question is whether the MEMS are to be integrated onto the same die with microelectronics, and if so, what kind of microelectronics and how much?

Many MEMS processing steps are not considered to be compatible with microelectronics processes. MEMS uses a much broader and diverse spectrum of processes and materials than microelectronics fabrication, and some of the materials often used in MEMS are known to pose contamination problems for integrated circuits (e.g., gold). Additionally, during processing MEMS wafers can generate particulates that can degrade the yield of microelectronics processes. For example, a MEMS process technology may involve the etching of silicon or some other material to create very thin membranes or other fragile structures across the surface of the wafer. These structures may be susceptible to breakage during subsequent handling and processing. The breakage of these structures would create a large amount of particulate contamination that would be very problematic for wafers undergoing integrated circuit fabrication. A large amount of particulate matter on the surface of the wafer would degrade the yield of any subsequent processing steps, and this degradation would increase as the line width resolution is reduced. As a general rule, the process restrictions for integrated circuits will always be more restrictive than for MEMS. Consequently, the processing equipment used for MEMS fabrication is frequently physically separated from the equipment used to make integrated circuits for most integrated MEMS process sequences.

Many microelectronics foundries are operated as continuous flow manufacturing environments wherein every wafer is produced according to exactly the same ordered sequence of exactly the same processing steps from start to finish. This coupled with the highly strict controls on contamination and particulates dictates that the wafers cannot easily be removed from a manufacturing line, have one or more MEMS processing steps performed, and then reinserted into the integrated circuit processing line. In contrast, the customized nature of MEMS processing sequences often requires that the foundries are operated more akin to a job shop where some wafers are processed according to one process sequence, another group of wafers are processed according to a different process, and so on. This customization usually requires that the wafers jump from one type of process and the associated equipment tool to another. This has a large impact on the partitioning of the process and equipment manufacturing of MEMS devices.

The decision on whether to integrate microelectronics with MEMS will also greatly increase the cost of development of a manufacturing process with the devices and significantly lengthen the development time. Therefore, in general, it is usually prudent to have the MEMS and microelectronics on separate die, unless there is an absolute requirement that they be integrated together. Such a requirement may be generated if the MEMS are composed of a very large array of devices that need to be individually addressed on a relatively short time scale. An example like this would be the Texas Instruments' Digital Light Processor (DLP)® technology discussed in Chap. 7 wherein there are millions of mirrors that need be individually addressed with actuation voltages to create the display image. It would not be possible to use off-chip electronics for this addressing given the complexity of the addressing over such a large array. If the application requires MEMS to be integrated with microelectronics, then it is usually advisable to limit the amount of electronics to only that is absolutely required on-chip.

When fabricating integrated MEMS, based on the concerns of contamination and particulates from MEMS processing steps as well as the restrictive nature of integrated circuit fabrication, often the most practicable approach for integrated MEMS manufacturing is to perform the microelectronics fabrication first (i.e., start fabrication within facilities having the strictest restrictions) and then transfer the wafers to a less restricted foundry (i.e., a MEMS foundry) wherein the remaining portion of the fabrication is completed. This is the approach of the Texas Instruments' Digital Light Processor (DLP)® technology discussed in Chap. 7.

Another partitioning question is whether two (or more) different MEMS device technologies should be integrated together onto the same die. If the devices use the same process sequence, then integration of them together on the same die will likely be a good approach. However, if the two devices have different processing steps and materials, integration will be difficult, and strong consideration to a two (or more)-die solution would be warranted.

Partitioning will also have a large impact on packaging since if they are multiple dice in the system, the packages and external connections (possibly including non-electrical connections) must be considered and addressed.

Another issue of concern is the partitioning of the fabrication. Partitioning of MEMS fabrication typically involves decisions at multiple levels. This issue is made more difficult in MEMS compared to IC fabrication since many MEMS foundries do not have everything needed for complete fabrication and the volumes often involved are relatively low and require foundries to manufacture several device process technologies in order to maintain higher capacity utilization levels. Therefore, unlike IC fabrication, wafers in MEMS production sometimes have processing work performed at different foundries. Depending on the restrictions of each individual item of equipment, location within a foundry, or from foundry to foundry, there are often issues such as wafer cleanliness and cross contamination when transferring wafers among tools and foundries that must be considered and addressed.

10.6 Overview of MEMS Development

The IC industry has a mature array of advanced modeling software capabilities combined with sets of developed and standardized process technologies (including design rules, standard cell libraries, etc.) available to designers for new circuit development activities. This allows designers to undertake a new design activity with a variety of automated and standardized software tools that are accurate predictors of the actual circuit behavior and performance. Equally important, the IC designers will almost always be able to leverage an existing manufacturing process technology for producing the circuits. In short, the IC industry design and development process is very automated, standardized, and efficient.

The same is not true of MEMS microsystems design and development. In most instances, the developer starts with either a MEMS idea that (hopefully) has a market opportunity or has identified a market need that a MEMS may be suited to address. However, since the MEMS device is usually new and entails some novel and never before demonstrated functionality, the process sequence and mask design that are used for the implementation are almost always new and customized. In this situation, there are unlikely to be design rules or other formalized existing technology available.

As a starting place, MEMS designers often first develop a system-level model of what is to be implemented in either a block-level diagram or a lump-element equivalent circuit and then perform a series of analyses (perhaps with the help of software tools) on a number of possible design concepts. Once a design at the system level appears to be producing fruitful results, the designer then moves to the device, physical, and process levels wherein the design will progressively become more detailed and tangible. The models at these levels are commonly analyzed using a range of different device, dimensional, process, and material parameter values to determine how these parameters impact the device performance and behavior. It is not unusual to have what appears to be a satisfactory design at any level only to find out at another level that either the process must be revised or even the entire design process must be restarted from scratch. The creative space is so large when designing MEMS microsystems that the analytical modeling approaches can often be more efficient early in the design activity given the prohibitive effort and time expense to attempt to conduct detailed numerical design activities without a reasonably concrete device design. Once a MEMS device initial design is available, then either the device is fabricated and tested or detailed numerical simulations can be performed in order to more accurately develop the device design.

A major issue with performing numerical simulation at an early point in a MEMS development activity is that it is not known if the process sequence will produce any working devices and the material properties, being process sequence dependent, may not be known. Consequently, it may be prudent to attempt the fabrication of the device when using a new process sequence before performing detailed numerical simulations since this can result in the discovery of various processing issues that may require device or process sequence design changes. The running of a newly designed process sequence can also allow test structures for material property determination to be implemented that can be used to feed more accurate material property information back into the models.

When the device design and fabrication process sequence are developed sufficiently to obtain (probably small number of) working and testable devices, then it is common to undertake detailed numerical analysis on the device design in order to

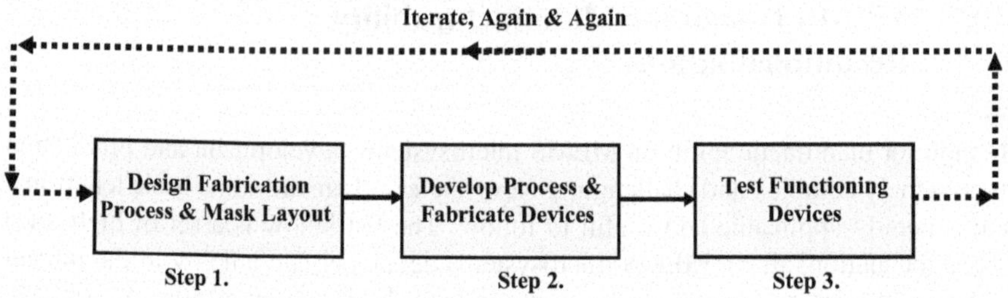

Fig. 10.6 Open-loop manufacturing process typically employed in MEMS production

optimize its performance. These models can use the statistical dimensional data and material property information derived from previously performed process development runs of the process sequence and suitable test structures and devices. The results from the numerical analysis may be used to again revise the device design and possibly the process sequence. Even after the process sequence is yielding, there are often additional revisions that are needed in both the process sequence and the device design. This iterative process of design, fabricate, test, and repeat may be performed several times before the designers are confident that the design is satisfactory for the intended application or product.

The presence of device parameter variations has significant implications for microsystems device development. As shown in Fig. 10.6, the development of a manufacturing process for the production of microsystems is essentially an open-loop iterative process. That is, the fabrication process sequence and the parameter settings of the process sequence (i.e., tool settings, process recipes, etc.) are first defined and set. However, it is not known beforehand if this sequence and parameter settings will result in working devices. A major effort in any new process sequence is the process development that is required. This is a major undertaking in its own right. Combined with process development activities are the measurements of the parameters on multiple samples in order to perform meaningful statistics. This allows the designers to know the magnitude and spread of the parameter variations as well as possibly develop schemes to lessen these variations or their impact on the device performance. The parameter variations are tightly tied to the processing steps, and therefore any changes in the processing steps and process sequence can have large implications on the parameter values and their variations. Since the development of a working process sequence is highly iterative and can have a large impact on the parameter variations, the development of a microsystems device is commonly a lengthy and expensive endeavor.

The important point of this discussion is that the device design and the process sequence design have an enormous impact on how easy it is to produce a device that meets its required performance specifications. Although very difficult to do in practice, it is extremely important to ensure that the process sequence, processing steps in the sequence, materials, equipment, and design suitably handle the variations in the fabrication process and their impact on the resultant device behavior as needed for the intended application.

10.7 MEMS Design for Manufacturability Recommendations

Design for manufacturability in MEMS microsystems development and production is extremely complex and challenging. Nevertheless, there are some basic tenets that are generally applicable and useful to follow. The list below is a set of high-level recommendations that a MEMS microsystems developer can employ in the pursuit of achieving design for manufacturability in their development efforts.

1. Keep the device design and process sequence as simple as possible – more processing steps and a more complicated design will make for lower yields, longer development times, and higher costs and risks.
2. Leverage from existing process modules and technologies to the extent possible – this will lower the development (nonrecurring engineering or NRE) cost and time as well as reduce risk.
3. Know the physics of your device and how it relates to the dimensions and process sequence. Avoid fabrication processes that result in large relative tolerances of the dimensions for critical elements, particularly those having higher-order dependencies (i.e., to the square, cube, or fourth power). Use bistable mechanical elements if possible since these types of elements are less sensitive to device parameter variations. The large relative tolerances of microsystems devices can make continuous-movement device types hard to implement and costly to calibrate.
4. Make critical mechanical elements from materials having known and reproducible properties if possible – making devices from materials such as single-crystal silicon can be far easier than making them from thin-film materials.
5. Avoid designs that critically depend on the hard-to-develop or hard-to-control fabrication processes such as perpendicular etched sidewalls, timed etching depth, and sidewall smoothness.
6. Use processing recipes, tools, and materials that are well proven to the extent possible – avoid new equipment, new process capabilities, and new materials until they are well proven and shown to be reliable and able to deliver reproducible results.
7. Attempt to use appropriate etch stop layers and end point detection to accommodate etch rate variations.
8. Minimize the number of materials and layers that comprise the critical mechanical elements of the microsystems device – managing the behavior of device elements made from stacks of multiple material layers of thin films is difficult.
9. Understand and account for the respective relative variations in the device parameters for the fabrication methods employed. Use processing steps that have existing statistics on dimensional and material property parameters to the extent possible.
10. Expect to make several design and process changes as well as multiple iterations in the development of a viable process sequence – multiple iterations are common in MEMS development.

11. Avoid merging MEMS with microelectronics on the same substrate unless the application absolutely requires it – integrated MEMS is a costly and risky strategy.
12. Engage packaging, assembly, and test engineers into the device development effort at the beginning of the design process – all these parties must work closely together, listen, and be ready to compromise to achieve a successful outcome. The fabrication–validation cycle takes a long time. Perform multi-physics simulation to optimize the design critical elements and reduce the development cycle time.
13. Embed test structures and devices in the layout to determine material properties and performance. These test devices should also be designed to collect measurement data on the device parameters for statistical analysis. Test structures and devices are critical and useful when diagnosing the failure mode and understanding the fabrication problems.
14. Early in process development efforts, it is advisable to use layout designs of the devices that have a range of dimensional values rather than specific individual values derived from the engineering design. There are two direct benefits of this approach: first, it allows the collection of better parameter variations statistical data; and second, it is more likely to result in some number of working and testable devices.

10.8 Managing Device Parameter Variations in MEMS Design

Most often in microsystems design activities, the developers are aware of the device performance requirements before the development activities start since the MEMS device is being developed for a particular product and market. Therefore, it is the task of the device designers to develop a MEMS design that sufficiently manages the device parameter variations inherent in microsystems fabrication to enable the manufacturing yield to be high and at an acceptable cost. Estimation of the variations in the device output is called variation (or tolerance) analysis and includes methods whereby variation valuations are assigned to the device parameters to determine the variations of the device performance and how well the device behavior meets the output requirements. These topics were covered in Chaps. 8 and 9. While these methods are useful for analyzing the impact on manufacturing parameter variations in the device output and yield, a method of designing MEMS microsystems that manages the parameter variations is also needed. The method of designing microsystems taking into account anticipated manufacturing variations of the fabrication is commonly called variation (or tolerance) design and is the topic of this chapter.

The approach taken here is to first review the primary methods of variation design and show how these are used in some very simple examples having only one-dimensional parameter spaces. This will be followed by a discussion of how these methods are used in more general cases where there are multiple device parameters and device performance functions.

The methods used in microsystems design to manage parameter variations and increase yield include the following:

1. Design centering – This process involves adjusting the nominal values of one or more device parameters so that the manufacturing function of the device output more closely aligns and overlaps with the acceptance range so as to increase the yield. Design centering often consists of adjusting the nominal values by removing bias variations in the parameter values. Design centering is usually the most simple and cost-effective means to increase the yield and is usually performed before any other methods are attempted (with the exception of the best practices on mask layout design discussed below).
2. Parameter variation reduction – This is the process of reducing the amount of random variation in one or more impactful device parameters. This method essentially is reducing the standard deviations of the device parameters. This often involves using different processing technologies to reduce the parameter variations and frequently results in an additional manufacturing cost.
3. Device size scaling – This method has a similarity to the parameter variation reduction technique described above since it will also result in a reduction in the relative amount of variation of one or more parameters. This method is based on the fact that many of the random variations in lateral dimensions of microsystems devices are not percentages of the dimensional size of an element but instead have a fixed or constant value. For example, the random dimensional variations of features using contact photolithography are commonly expressed as ± 0.5 microns. Therefore, if the dimension of a device is increased, the corresponding relative variation will decrease. For instance, suppose the initial design specified a 2 micron feature made with contact photolithography, which would result in a relative variation of $\pm 25\%$. However, if this feature were increased to 10 microns, the relative variation would decrease to $\pm 5\%$. Note that since the device features are made larger and therefore consumes more substrate area, this also comes at an increased cost.
4. Acceptance range increase – This process increases the acceptance range boundaries of the device output behavior. This can be done without making any changes in the device design with the result of increasing the yield. In many situations, this method is not an attractive option since it probably results in a device having reduced performance levels and/or not meeting the required application specifications.
5. Best practices on mask layout design – The mask layout of microsystems devices can have an impact on how sensitive the design is to manufacturing variations, and therefore there are several practices that can be used in reducing this sensitivity. The best practices for mask layout of microsystems include matching; consider nearest neighbor(s); symmetry; inter-digitation; common centroid; and others.

10.8.1 Design Centering

The example of the microchannel flow device was covered in Chaps. 8 and 9 wherein the variations in the output behavior, specifically the flow resistance, and the resultant yields were examined. We now re-examine this example in order to show how the yield can be increased using the method of design centering.

In reviewing the results of the variation and yield analysis performed on the microchannel device, it will be recalled that the application requirement specified a value of the microchannel flow resistance to be 4.54×10^9 Pa•s/m and the designed device dimensional values were selected as a device length of 2000.00 microns and a radius of 100.00 microns. Additionally, the desired dynamic viscosity of the fluid (water at room temperature) used in the microchannel was given as 8.90×10^{-4} Pa s. The output of the microchannel device was expressed in terms of the flow resistance given by the ratio of the device output, Q (i.e., volumetric flow rate), and the device input, ΔP (i.e., differential pressure across device).

It is assumed that a large production lot of microchannel devices is manufactured with the radius and length dimensions measured, and it was found that the mean value of the radius was 102.08 microns with a standard deviation of 2.44 microns. The mean value length was found to be 2002.08 microns with a standard deviation of 2.52 microns. The probability distribution functions for both the radius and the length are given in Figs. 10.7 and 10.8.

It was further found from the Monte Carlo analysis (or measurements of fabricated devices) that the output behavior of a large number of microchannels had a mean value of flow resistance of 4.90×10^9 Pa•s/m with a standard deviation of 0.48×10^9 Pa•s/m. It was also noted that there appeared to be a bias variation in the output since the mean value of the flow resistance was shifted to the right (i.e., to larger values of flow resistance) from the desired flow resistance. The shift appeared to be of an amount of 0.36×10^9 Pa•s/m as seen in Fig. 10.9. The cause of this bias variation in the flow resistance can be traced back to the probability distribution functions for the microchannel radii and lengths, which also appear to be shifted slightly from the desired design values due to bias variations.

As explained in Chap. 9, the yield of the manufacturing process for the microchannel is estimated by overlaying the acceptance range onto the probability distribution function of the manufacturing function, which in this case is the microchannel flow resistance. The acceptance range is based on the product or application specifications. The application requires that the flow resistance for a given application has a nominal value of flow resistance of 4.54×10^9 Pa-s/m with a permissible spread in the flow resistance of $\pm 5\%$. The predicted yield was found to be less than 40% as shown in Fig. 10.10. In most circumstances, a yield of 40% would be considered too low, and therefore methods that can be used to increase the yield would be pursued.

Usually the easiest and often first pursued method to improve the yield is to perform the method of design centering. Design centering is achieved by adjusting the nominal values of one or more parameters in order to move the manufacturing

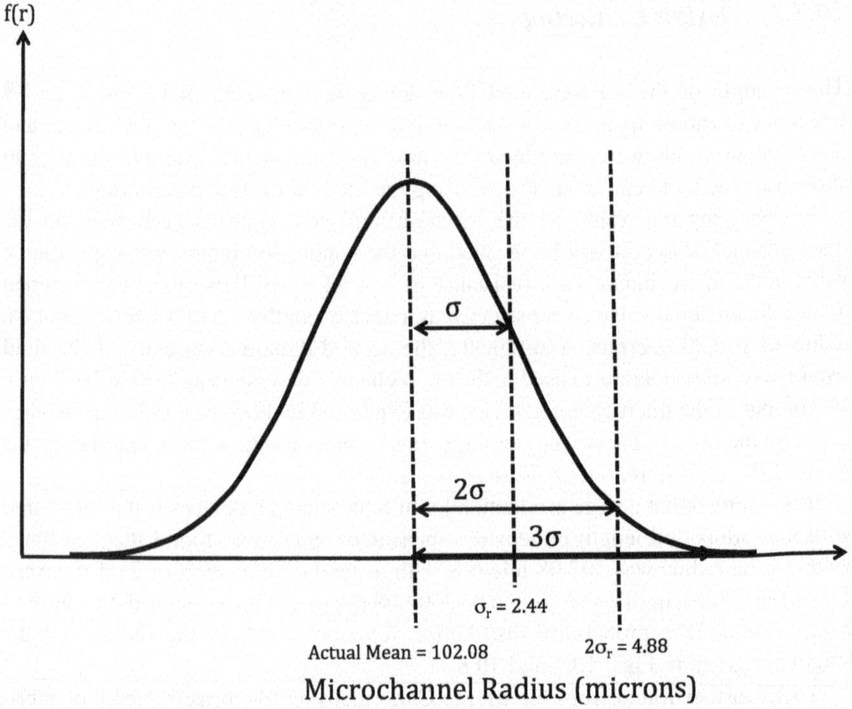

Fig. 10.7 The probability distribution function of the microchannel radii

function so that it better aligns with the acceptance region. The most obvious method is to remove the bias variations in the parameter(s), such as the radius and length of the microchannel. These bias variations are the result of the processing steps used to implement the microchannel into the silicon and will most likely be due to the small amount of deviation from the dimensions on the mask to the printed features in the photoresist and the small amount of lateral etch under the mask since no etch process is perfectly anisotropic. Removal of these bias variations is relatively simple and inexpensive since all is needed is to slightly bias the photolithography mask used to pattern the silicon microchannel so as to account for the dimensional inadequacies of the processes used. Specifically, if the mean value of the microchannel radius is 102.08 microns, then an adjustment of the mask dimension would be made to reduce the as-drawn radius from 100.00 to 97.92 microns. Similarly, the mean of the length is 2002.08 microns, and therefore the new dimensions for the mask would have a dimension of 1997.92.

The result of these dimensional adjustments to the microchannel radius and length would cause the manufacturing function, which is the probability distribution function of the microchannel flow resistance, to shift to the left, thereby resulting in a mean value of flow resistance equal to the desired value of 4.54×10^9 Pa•s/m as shown in Fig. 10.11. It is noted that while the manufacturing yield has improved, it is still below 50%.

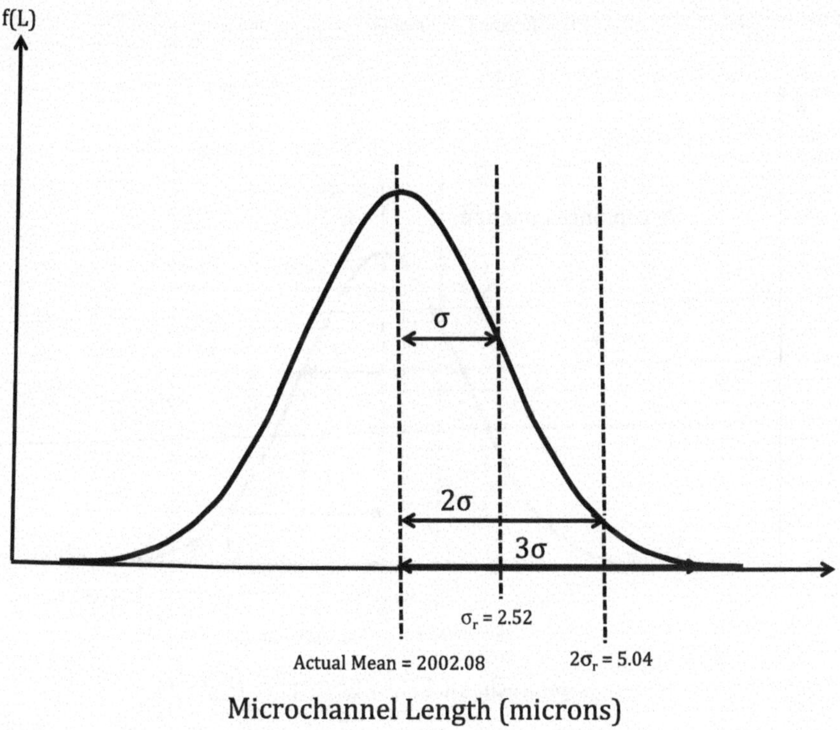

Fig. 10.8 The probability distribution function of the microchannel length

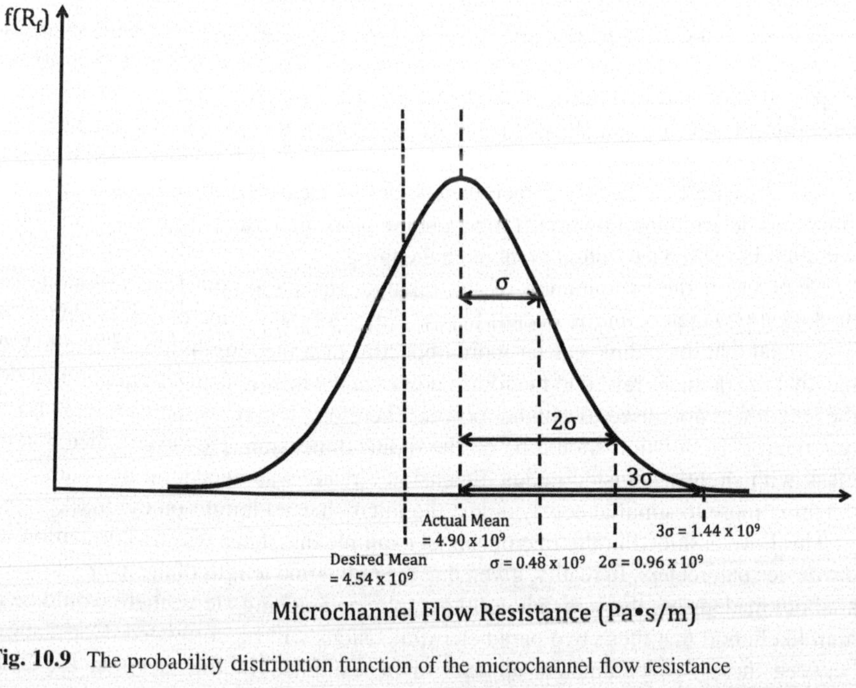

Fig. 10.9 The probability distribution function of the microchannel flow resistance

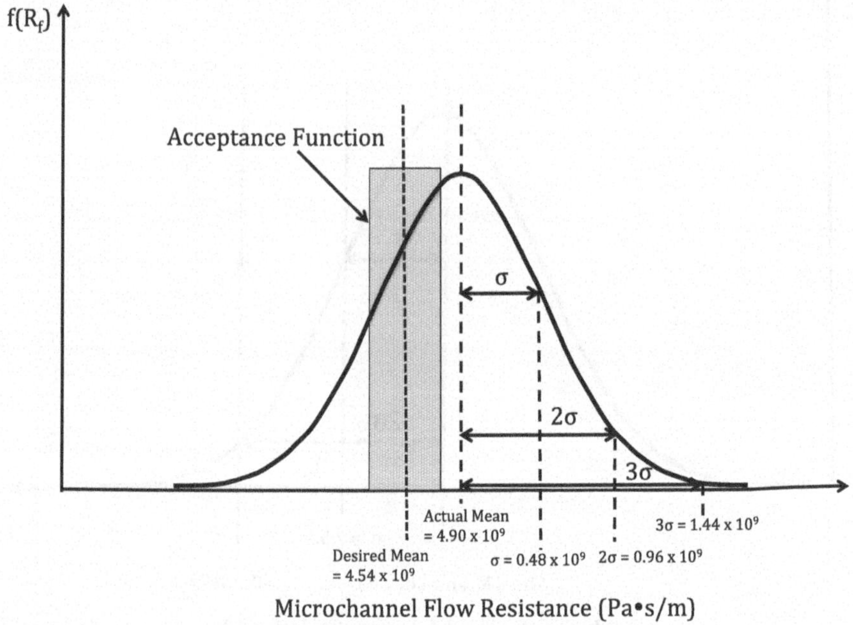

Fig. 10.10 The probability distribution function of the microchannel flow resistance with the acceptance range overlaid. The manufacturing yield is the shaded region under the PDF, which is below 40% and likely would be considered as being unacceptable

A few comments about design centering. First, the example of the microchannel is one that allows the dimensions of the radius and length of the flow path to be inspected and measured during manufacturing. Therefore, this not only enables the detection of any bias variations in the device parameters, but it also enables the device designer to know how to adjust these parameters with relative ease. It is not always the case that the device parameters can be measured directly, and in some situations determining how much to adjust the parameters so as to remove any bias variation in the device output is more challenging.

Second, in the examination of the relative impacts of the bias variations of thedevice parameters on the overall bias variation of the device output behavior, it was found that the radius was far more impactful than the length, mostly due to the fact that the radius is raised to the fourth power in the flow resistance equation while the length was not raised to a higher power. Therefore, it may seem that the focus on removing bias variation should be on the radius dimension. However, since a new mask with slightly adjusted radius dimension already was needed, it does not cost anything more to simultaneously adjust the microchannel length on the mask.

Third, we assumed in the microchannel example that there was no correlation in the device parameters. In reality, given that the radius and length of the microchannel are both made using the same photolithography and etching steps, there would be a high likelihood that these two parameter dimensions will be correlated. Correlation between these two parameters appears to be exhibited in the fact that the bias variation magnitudes of these two dimensions have identical values (i.e., 2.08

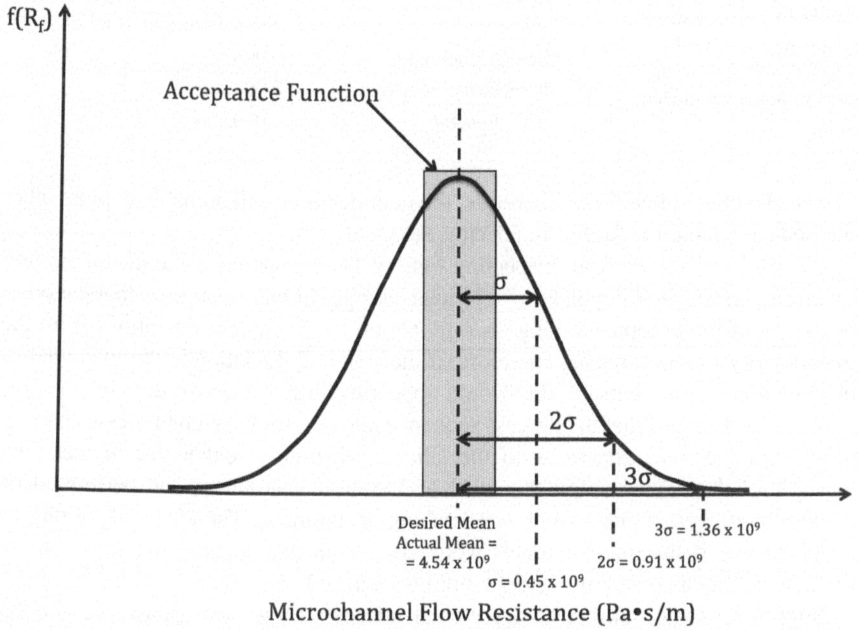

Fig. 10.11 The probability distribution function of the microchannel flow resistance after the manufacturing function has been design centered by removing the bias variation. As before, the acceptance range is overlaid onto the manufacturing function, and the yield is the shaded region under the PDF, which is still below 50%

microns). Regardless of the correlation between these two parameters, the bias variations can be removed using the technique described above. Nevertheless, in other situations any correlations between parameters should be examined, and how the adjustment of one parameter may impact other parameters needs to be carefully taken into account.

Fourth, it is worth pointing out that when the nominal values of the parameters are adjusted, this action may also impact other calculations such as the estimated standard deviations of the parameters and the manufacturing function. Therefore, in real situations these effects should be taken into account.

Lastly, it is important to note that removal of bias variations is not the only technique available for design centering. As can be appreciated by the above example, various changes in the nominal values of the parameters can be used to shift the manufacturing function so that it better aligns with the acceptance region.

10.8.2 Parameter Variation Reduction

The method of design centering improved the yield in the example of the microchannel device, but even with this improvement, the yield is still relatively

Table 10.1 Normalized standard deviations for variations in dry etched microchannel parameters

Parameter	Normalized standard deviations
Microchannel radius	0.0062
Microchannel length	0.00031
Fluid viscosity	0.0005

low at a level of below 50%. Therefore, the manufacturer will probably explore other methods available for further improving the yield.

As can be observed from inspecting Fig. 10.11, the spread in the manufacturing function extends well beyond the acceptance range. In fact, it appears that the outer boundary of the acceptance range is only about ±1/2 standard deviations from the manufactory function mean. Therefore, if the standard deviation in the manufacturing function can be reduced, this would appear to further improve the yield.

It will be recalled that the process sequence allowed for the etch processes used to implement the microchannel into the silicon substrate is either dry or wet. Our example of the device output variation and yield analysis has been performed on devices that were made using wet etching techniques. Therefore, it would be expected that if the microchannels were made with dry etching, the variations of the device dimensional parameters would be reduced.

Suppose a large number of dry etched microchannels were fabricated and the resultant mean and standard deviation of the dry etched microchannel radii were found to be 100.01 microns and 0.62 microns, respectively. The microchannel length of the dry etched microchannels had a mean value of 2000.01 microns and 0.62 microns, respectively.

The means of the radii and length do exhibit some small amount of bias variation, but this can be easily removed using the technique described above, and these adjustments do not significantly affect the standard deviations. Additionally, in our previous variation analysis, the fluid viscosity had a ±1.1% standard deviation. This could be improved by performing more accurate and precise measurements of this material property and/or having the device operated in a more temperature-controlled manner. Therefore, it is assumed that the fluid viscosity standard deviation could be improved to a ±0.05% or better.[7]

Table 10.1 lists the normalized values of the standard deviations of the microchannel parameters fabricated using dry etching techniques. As can be seen, the normalized standard deviations are significantly smaller than those obtained with wet etching in the example of Chap. 8.[8]

Using Monte Carlos analysis (or measurements of fabricated devices), it was determined that the flow resistance has a value of $4.54 \times 10^9 \pm 0.22 \times 10^9$ Pa-s/m or a percentage variation in the flow resistance of ±4.85% for 95% of the population at a 99% confidence interval. The revised manufacturing function of the output flow

[7]The fluid viscosity is not like some material properties such as the Young's modulus, due to the fact that the value of the viscosity is not process dependent.

[8]It will be noticed that the normalized standard deviations in the table use the adjusted values of the device dimensional parameters to remove the bias variations.

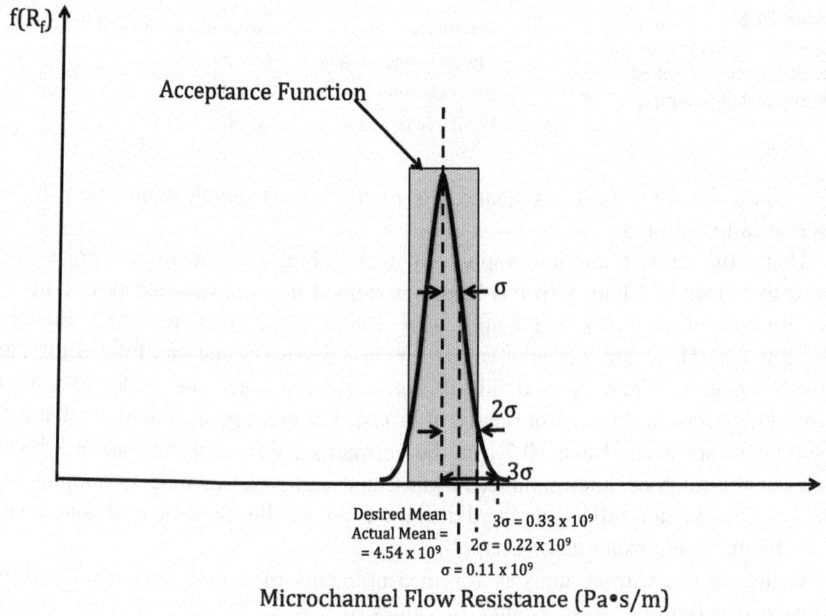

Fig. 10.12 The probability distribution function of the microchannel flow resistance after the manufacturing function has been revised by using dry etching instead of wet etching to make the microchannels. As before, the acceptance range is overlaid onto the manufacturing function, and the yield is the shaded region under the PDF, which is now above 95%

resistance of the microchannel devices made with dry etching as opposed to wet etching is shown in Fig. 10.12. As can be seen, the yield is now above 95%. This is a much more desirable situation and fortunately was available due to the existence of an etching technology that has very little lateral undercut of the mask.

Reducing the variations of one or more parameters will often result in improved yield. However, it also often increases the manufacturing cost. Therefore, the microsystems device designer will need to determine if the increased manufacturing cost is offset by the yield improvements that can be obtained using this method.

10.8.3 Device Size Scaling

Device size scaling can also be used to reduce parameter variations. Using the example of the microchannels, it is assumed that the photolithogrpahy used to pattern the masking layer for the substrates had a variation of ±0.5 microns that is independent of the dimensions of the features. Therefore, suppose that the dimensions of the microchannel were significantly increased. For example, assume that the microchannel nominal width is increased from 100.00 to 200.00 microns.

Table 10.2 Normalized standard deviations for variations in wet etched microchannel parameters

Parameter	Normalized standard deviations
Microchannel radius	0.012
Microchannel length	0.00079
Fluid viscosity	0.0005

To obtain the same flow resistance, the microchannel length would need to be increased to 3.20 mm.

Using the same photolithography and wet etching equipment and processing steps as before in Chap. 8, it has been determined that the standard deviations of the microchannel width and length were found to be 2.44 and 2.52 microns, respectively. These are the same values for the standard deviation of the width and length found in Chap. 8, and this is not surprising since the same fabrication processing steps were used presently; it is just that the dimensions of the features have been increased. Table 10.2 lists the normalized values of the standard deviations of the microchannel parameters fabricated using wet etching techniques. As can be seen, the normalized standard deviations are smaller than those obtained with wet etching in the example of Chap. 8.[9]

Using Monte Carlos analysis (or measurements of actual devices), the flow resistance is found to have a range of values of $4.54 \times 10^9 \pm 0.43 \times 10^9$ Pa-s/m or a percentage variation in the flow resistance of ±9.47% for 95% of the population at a 99% confidence interval. The revised probability distribution function (also the manufacturing function) of the output flow resistance of the microchannel devices is shown in Fig. 10.13. As can be seen, the yield is about 68% since the acceptance range coincides with about the one standard deviation points on the manufacturing function. Therefore, the yield is higher than was achieved with design centering but is less than the yield obtained using parameter variation reduction. Obviously, if the microchannel was fabricated using dry instead of wet etching, the variations would be reduced more significantly. Alternatively, the dimensions of the microchannel could be further increased since this would also further reduce the manufacturing function of the flow resistance variations.

While the device size scaling did increase the yield of the manufacturing process for the microchannels, it should also be recognized that this came at the cost of the consumption of additional substrate real estate. Specifically, the area of the microchannel increased by a factor of over three times. This means that the cost of the devices made may also increase by a similar factor. Therefore, a major disadvantage of this technique is that it can significantly increase the manufacturing cost to the point where the yield improvement obtained is not justified by the increased cost.

An important issue regarding device size scaling is that it is effective only on parameter variations that are absolute in value. As seen in the example above where the lateral dimensions made in the fabrication of the microchannel had a variation of

[9]It will be noticed that the normalized standard deviations in the table use the adjusted values of the device dimensional parameters to remove the bias variations.

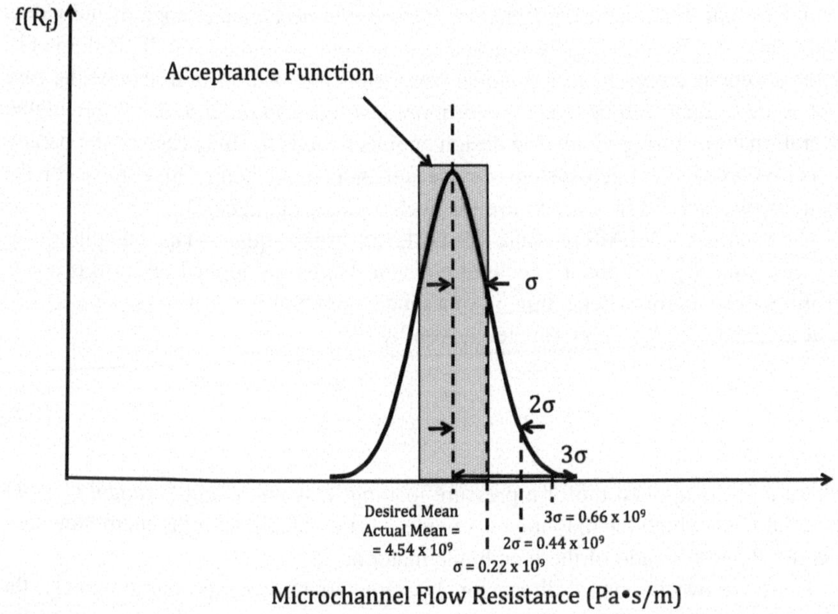

Fig. 10.13 The probability distribution function of the microchannel flow resistance after the manufacturing function has been revised using the device size scaling method whereby the dimensions of the microchannels are increased. As before, the acceptance range is overlaid onto the manufacturing function, and the yield is the shaded region under the PDF, which is now around 68%

about ±0.5 microns, making the lateral dimensions larger resulted in a decrease in the variations of width and length. However, this method often will not usually work for the thicknesses of deposited layers, such as thin films deposited using LPCVD or CVD or thin films grown using thermal oxidation. The reason is that the variations in these layers are usually a percentage of the thin-film thickness, and therefore increasing the thickness does not result in a reduction in the amount of variation. This is important since for many MEMS devices, the output behavior strongly depends on the thickness of the mechanical structure, and moreover, the thicknesses of these mechanical layers often represent the largest relative parameter variation in the physical equations describing the device output behavior.

10.8.4 Acceptance Range Increase

Since the acceptance range is most often set by the requirements of the customer or product, increasing the acceptance range is usually not an appealing or acceptable method to improve the manufacturing yield. However, it is important to note that increasing the acceptance range does not necessilary have to be only applied to the device output manufacturing function. Instead, the acceptance range of individual parameters can be individually adjusted so as to obtain a higher overall yield. That is, if the acceptance regions of individual parameters are examined, it may be the case that some of them can be made less or more restrictive in order to result in a higher overall manufacturing yield. The design engineer needs to study each of the parameters involved in the microsystems device and understand their relative impact on the manufacturing yield in order to use this technique advantageously.

For example, a MEMS pressure sensor device uses a square-shaped diaphragm as the mechanically compliant structural element. Under an applied uniform pressure loading, the maximum deflection, w, of a square membrane with an edge length of $2a$ is at the center and is approximately given by [8]:

$$w \approx \frac{12Pa^4(1 - \upsilon^2)}{47Et^3}, \tag{10.8}$$

where P is the applied uniform pressure loading, E is the Young's modulus of the material from which the diaphragm is made, t is the thickness of the membrane, and ν is the Poisson's ratio of the membrane material.

There are two important dimensional parameters in this expression, namely, the edge length of the membrane, a, which is raised to the fourth power, and the thickness, t, which is raised to the third power. Most likely the thickness of the membrane will have the largest contribution to the variation of the output of the pressure sensor, and the designer may not be able to reduce its contribution. Therefore, in order to obtain a certain specified variation in the device output, the designer's focus should be on reduction of the variation of the edge length so as to attempt to compensate for the evitable variation of the membrane thickness that cannot be reduced.

10.8.5 Best Practices in Layout Design

The techniques used in the development of device designs, specifically the layouts that are used in the patterning of microsystems, are also important to desensitizing the devices to the effects of parameter variations.

The matching of components is a commonly used practice in microsystems design. The term matching refers to components that are specified to have identical values or values that are multiples of one another. One of the most common examples of matched components is the switched capacitor circuit, which is often used as an electronic filter [9]. The success of the switched capacitor circuit is because its response depends on the ratio of certain capacitors in the circuit and not the absolute values of the capacitors. Ratios of components, such as capacitors, are much more accurately defined using microfabrication techniques compared to the absolute values due to the presence of correlated parameter variations in microfabrication. There are similar circuit configurations using resistors, active

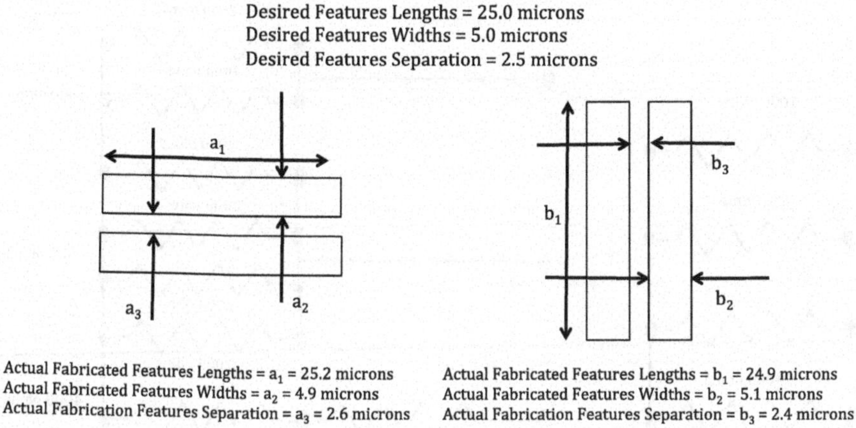

Fig. 10.14 Illustration of the effect of feature orientation on the resultant variations of the dimensions of the features

electronic devices, and MEMS. It is a good design practice to place matching elements in as close proximity to one another as possible. The reason is that if they are close to one another, they will exhibit stronger dimensional correlations than if they are further apart.

The orientation of the components and devices, whether matched or not, also is important. This is due to the fact that it is common in microfabrication for the resultant dimensions to be dependent on the orientation of the substrate. An example is given in Fig. 10.14, which illustrates the dependence of orientation on the resultant dimensions of etched features made in a material layer. As can be seen, the etching process used to make the feature results in the feature sizes not only differing from the desired design values of the features but also differing depending on the orientation of the features. Therefore, for components and devices that are to have consistent parameter values, it is prudent to orient the features the same on the substrate if possible.

It is also useful to attempt to layout the devices in the design so that they have matched geometrical shapes. This helps in ensuring that the device design has matched variations. Additionally, when there are large devices involved in a design, it can be advantageous to segment these devices into parts and interweave these parts in order to better cancel out any device parameter variations.

When dealing with a number of components or devices that are to be matched in their parameter values, it is wise to select a middle value as the root value rather than the lowest common value. For example, suppose that a design requires a resistor network as shown in Fig. 10.15 (on left side of figure). There are four resistors, and it is desired that the ratios of these resistors are as nearly equal to one another as possible. The selection of a base value of resistor to implement the network is needed, but the question is what is the best base value? It might be thought that using the lowest common denominator would be the best choice, which would be a base value of 500 ohms. However, this would require that the 4000 ohms resistor be

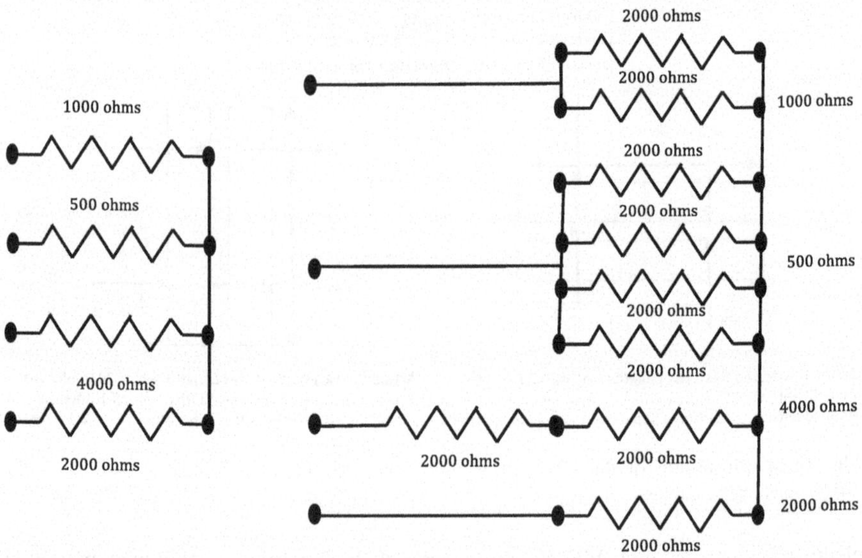

Fig. 10.15 The desired resistor network configuration on the left. On the right is the network implemented using one value of resistor wherein the 2000 ohms resistor is the base value

implemented with eight resistors in a series configuration. Since the contact resistances of such a long series connection will likely be relatively large, this would result in the ratios of the resistors in each of the parallel resistors not being equal as desired. A better configuration is to select a midpoint value from the resistor values, such as 2000 ohms from which to implement the network as shown in Fig. 10.15 on the right side.

In addition to reducing the overall contribution to the contact resistance of some of the elements, another benefit is that by using larger values of resistors (that will have larger dimensions), this also results in lower variations in the resistor values compared to smaller-sized elements. While this concept works when using a resistor network, it is also applicable to other components and devices.

Inter-digitation of the components or devices is a layout method of placing the elements in a configuration wherein a base value element is at the center and the other elements are kept as close as possible. Inter-digitation is sometimes advisable in layout design since it can keep the components and devices in closer proximity to one another and therefore results in their values being less sensitive to parameter variations. An illustration of inter-digitation is shown in Fig. 10.16 where in Fig. 10.16a a circuit diagram to be implemented is at the top and is composed of two parallel connections each having four resistors in series. If the resistor elements were laid out as shown in Fig. 10.16a, resistor R1 and R4 would be far apart, and resistor R5 and R8 would also be far apart, thereby making their values more prone to parameter variations such as the thickness of the layer used to make the resistors.

A better configuration is shown in Fig. 10.16b where the elements can be located closer to one another. One caveat of Fig. 10.16b is that a two-layer metal scheme is needed for the electrical crossovers, whereas in Fig. 10.16a a single-layer metal

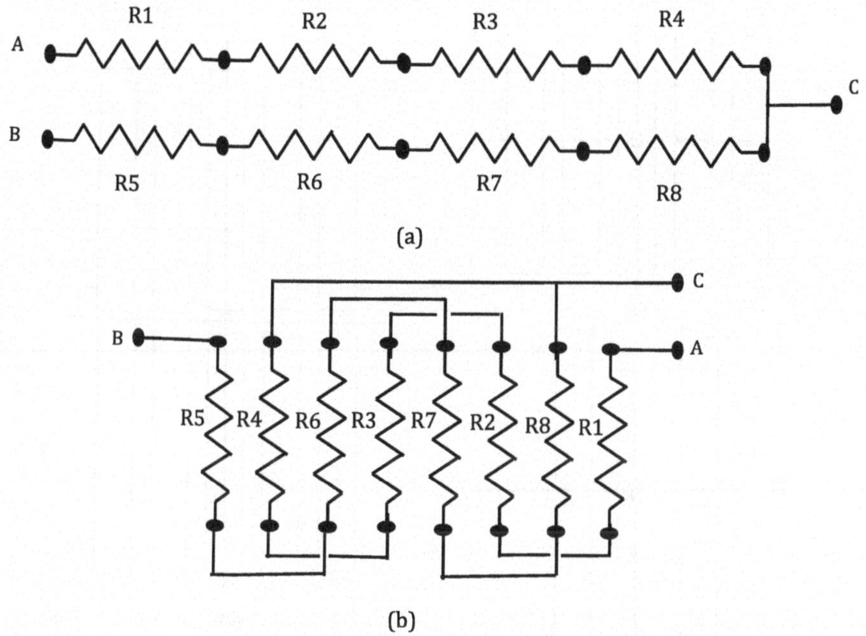

Fig. 10.16 The use of inter-digitation of a network of resistors in order to place them closer to one another

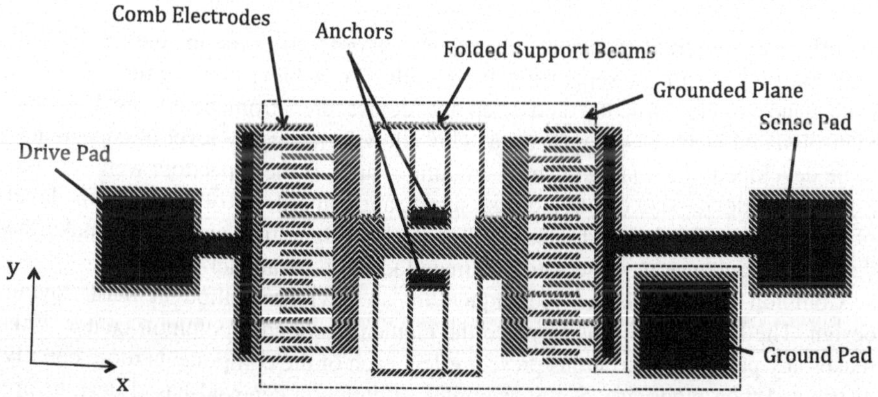

Fig. 10.17 An example of the use of inter-digitated electrodes on a MEMS resonator device

would suffice. However, most microsystems process technologies have multiple metal layers so this should not be an obstacle.

Inter-digitation is also used commonly in MEMS designs for electrode configurations as shown in the electrostatically actuated resonator in Fig. 10.17. The device has periodic voltages applied to the pairs of inter-digitated electrodes on either side of the shuttle that are 180 degrees out of phase to move the shutter back and forth.

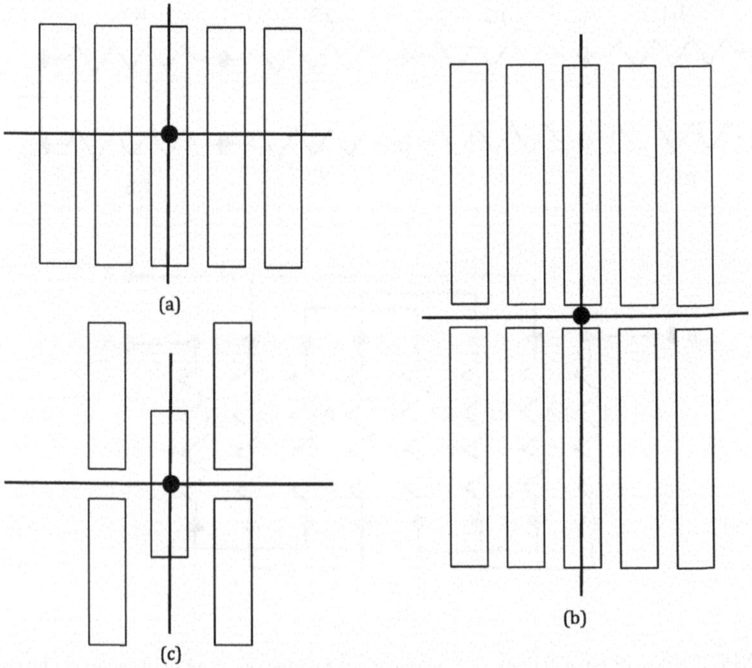

Fig. 10.18 Some examples of the layout of components or device that use a common centroid

A sufficient amount of capacitance is required in order to overcome the mechanical stiffness of the springs supporting the shuttle. To achieve the required level of capacitance in the structure and keep the device area from being much larger, inter-digitated electrodes can be used. This allows a necessary level of capacitance to be developed in a relatively small amount of area. If the capacitors were laid out linearly, the device size would increase significantly in order to obtain the same level of capacitance. The larger dimensionally sized configuration would also make the device's capacitance more sensitive to manufacturing parameter variations.

Common centroid design techniques are an advised method in mask layout design. The basic concept is to place the elements around a common center. This makes any parameter variations present effect each of the components more equally distributed than otherwise. Some examples of common centroid-based designs are shown in Fig. 10.18a–c.

The use of symmetry in the design of components and devices is also a good practice in device design. If the design is asymmetric, there may be imbalances in the device that can make it more sensitive to parameter variations and affect its behavior. For example, in Fig. 10.19 is an illustration of the electrical interconnections between three components or devices where the layout is asymmetrical in Fig. 10.19a and the same three components when using a symmetrical design configuration in Fig. 10.19b. Not only is the configuration in Fig. 10.19b less

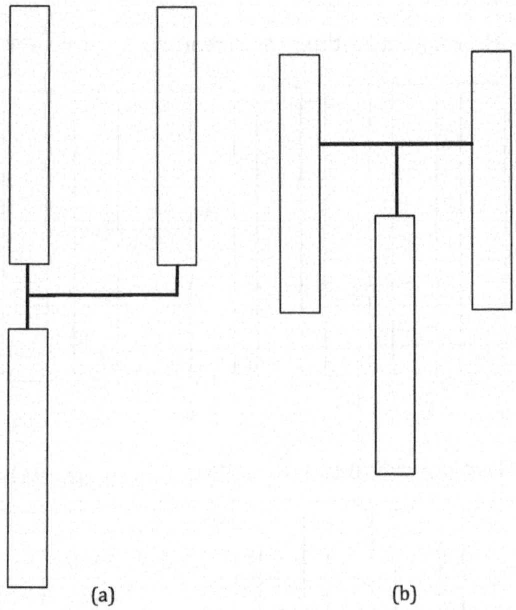

Fig. 10.19 An example of an asymmetric design layout in (**a**) and the same three elements in a symmetrical layout in (**b**)

sensitive to parameter variations in fabrication, but the electrical interconnections between the components will have more equal resistances and capacitances.

Dummy devices can be used in microsystems in order to cancel out the end effects in the manufacturing of component or device arrays. This is illustrated in Fig. 10.20 where an array of devices is made. In Fig. 10.20a is shown a mask layout of some features to be implemented and in Fig. 10.20b what these features look like after the etching has been performed. Due to end effects of the etching process, the features on both ends of the array have a width that is less than the features in the middle resulting in parameter variations. This is a common effect observed in microfabrication and is mostly due to the loading effect differences in the etch process between the features on the ends compared to those in the middle. A solution to this problem is to place "dummy" features on each end of the array as shown in Fig. 10.20c. The dummy features also exhibit the same end effects of the etching as shown in Fig. 10.20d, but since they have no contribution to the microsystem (they are not connected to the actual device electrically, mechanically, etc.), their reduced widths have no consequence on the device performance.

When performing etches it is advisable to ensure that the etched features have the same dimensions. The reason is that dry etched features that are larger in size tend to etch faster than the smaller ones due to process gas diffusion effects. If the layout has features with both large and small widths, the features with the larger widths will complete etching sooner than those with small widths. Therefore, in order to ensure all of the features have completed etching, an over-etch is usually performed until the smaller features have cleared. However, this means that the larger features will be exposed to the etchant for longer than it takes to clear these features resulting in an

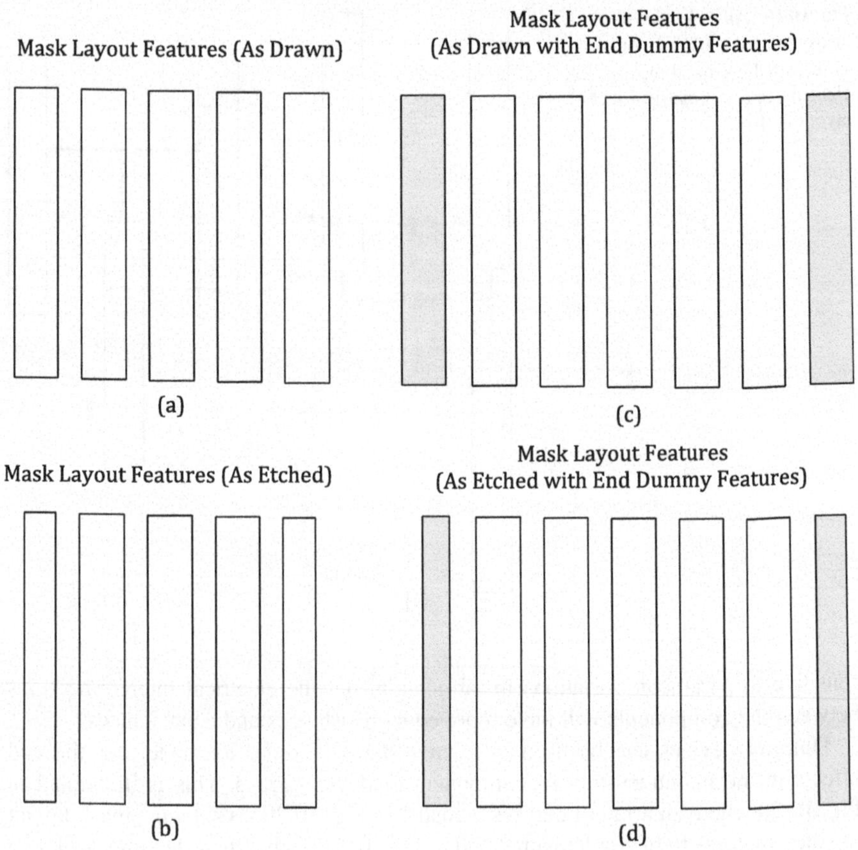

Fig. 10.20 The use of dummy features on the ends of an array of features as shown on the right to prevent the desired device features on the ends to have reduced widths as shown on the left

increase in the amount of the lateral etch of the larger features compared to the smaller ones. This will result in undesired parameter variations.

A method to prevent this problem is to make all of the etched feature dimensions the same. This is illustrated in Fig. 10.21, where in Fig. 10.21a is shown a mask that has openings (the clear regions) of different sizes. In Fig. 10.21b are shown the features after etching has taken place in the clear regions, and the larger features have enlarged since they have been etched for a longer time than it would take to clear these features. An improved design solution is shown in Fig. 10.21c wherein the larger features are drawn so that their widths are the same as the smaller features. The result after the etch has been completed as shown in Fig. 10.21d, and it can be seen that this allows the etching to proceed at equal rates at all locations on the features, thereby reducing the amount of parameter variations in the etched dimensions.

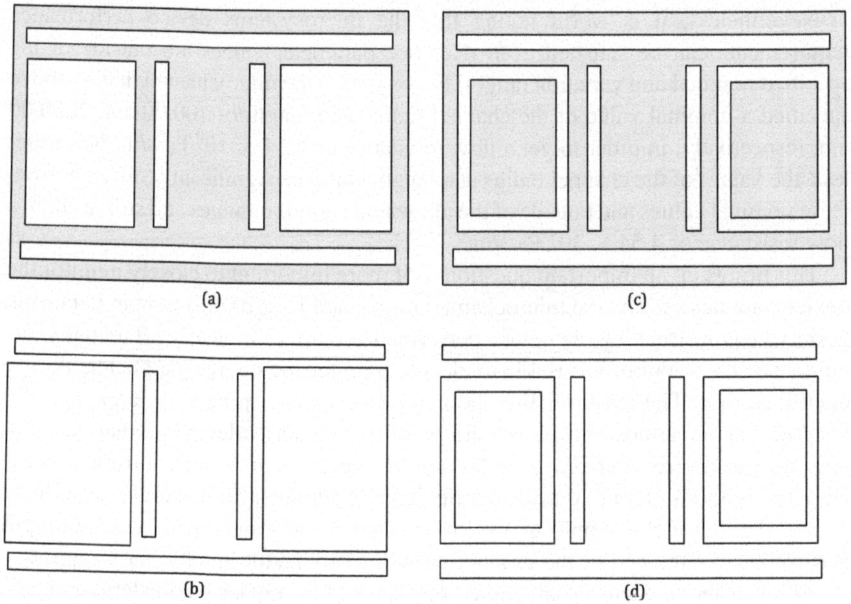

Fig. 10.21 An illustration of a method where large and small features can be etched to maintain the desired dimensions

It is also important to carefully consider the signal conduction paths in the layout design. It is desirable to keep signal paths as short as possible in order to avoid distortion or loss of the signals. In electrical circuits, longer electrical interconnects increase both the resistance and capacitance of the connections between components or devices resulting in a loss and timing delay of the signal. The same applies to other mechanisms of physical conduction including mechanical, thermal, etc. Additionally, if there are to be parallel conduction paths, for example, a differential signal to be conducted, it is important to keep these parallel signal paths of the same length, geometry, and orientation.

10.8.6 Further Comments About MEMS Design Methods

A few further comments about the above described design methods are in order. First, it should be noted that it is often possible to determine the device design parameter values during manufacturing. Of course, this will depend on whether these device parameters can be measured using a suitable means of metrology and are observable. In some situations, the device parameters may be covered by other material layers and therefore not observable. Additionally, there are cases in MEMS where there are deep, high-aspect ratio features where it can be difficult to perform accurate metrology.

Nevertheless, it is worth noting that the microsystems device performance requirements can be satisfied even if device parameter values are outside of the specified nominal and variation ranges. For example, the microchannel device above specified a nominal value of the channel radius and length of 100.00 and 2000.00 um, respectively, in order to get a flow resistance of 4.54×10^9 Pa-s/m. Nevertheless, the values of the channel radius and length could be significantly different from these nominal values and outside of the allowable variation ranges and still result in a flow resistance of 4.54×10^9 Pa-s/m.

This brings up an important question: is it more important to closely monitor the device parameters (i.e., the microchannel radius and length) during manufacturing or only be concerned by the device performance output? In nearly all manufacturing instances, it will be advisable to closely monitor the device parameters during manufacturing. The reason is that the parameter space can be very large, thereby resulting in an almost limitless variety of parameter value combinations that provide the desired performance output. However, if a manufacturing process were run without proper controls on the device parameter dimensions, it is almost certain that the yields would be far lower. That is, only measuring device output performance at the end of the production will likely result in a higher scrap rate.

As noted above, there are situations were some of the device parameters cannot be measured during production. In these cases, the manufacturing engineers will perform quality control measurements during production including using indirect means to monitor any parameters that cannot be measured directly, within the constraints of allowable metrology costs.

Second, it will be the designer's responsibility to develop device designs that allow suitable quality control measurements to be performed during manufacturing. An excellent device design that cannot be properly monitored for quality control during production is unlikely to be a successful device in production. Therefore, the engineers need to consider this aspect of the device when developing the design.

Third, as can be appreciated by the review of the design methods discussed above, managing all of the complexities of microsystems design combined with coming up with a design that satisfies the application and product requirements, including manufacturing cost goals for the given size market, is a very challenging engineering task.

10.9 MEMS Design in Multidimensional Spaces

The previous section demonstrated how the yields of microsystems manufacturing processes can be increased using various design techniques to manage the device parameter variations. The examples shown were intentionally simplified by having only two important design parameters (i.e., the microchannel radius and length considered with the fluid viscosity not being considered since it is non-varying) exhibiting manufacturing variations and one performance output function (i.e., the

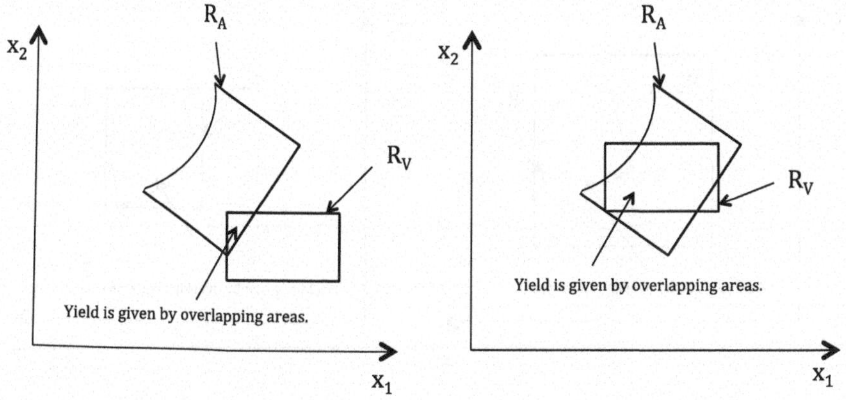

Fig. 10.22 Two device parameters being displayed along with the regions of variation and acceptance. The left shows the starting place wherein the regions do not overlap significantly resulting in a low yield, whereas on the right the nominal values of the parameters have been altered to result in a better alignment and higher overlap of these regions and a higher yield

flow resistance of a microchannel). In most microsystems design circumstances, there will be a larger number of device parameters and performance output functions.

Shown in Fig. 10.22 is a situation where there are two device parameters, x_1 and x_2. The variation region is shown as R_V and the acceptance region as R_A. As shown on the left, the overlap of the variation and acceptance regions is not substantial, and therefore the yield would be low. With an adjustment of the nominal parameters of the device, shown on the right of Fig. 10.22, the variation region is shifted so as to increase the overlap, thereby significantly increasing the yield. This would be an example of design centering.[10]

Figure 10.23 shows an example of parameter reduction when there are two device parameters. On the right of this figure, the yield is relatively low due to the fact that the region of variation is much larger than the acceptance region. In this case, if the manufacturing variations for the relevant device parameters can be reduced, resulting in the region of variation shrinking, the yield can be improved as shown on the right side of Fig. 10.23. Such a reduction in the device parameters could be enabled by using a dry as opposed to a wet etch process as described in the case of the microchannel, or using a lithography capability with lower variations for the given nominal dimensions, or other means.

While reduction in the device parameter variation can result in a higher yield in some situations, there are other situations where that is not the most desirable course of action. Such a situation is illustrated in the left side of Fig. 10.24 where the regions of acceptability and variation indicate that the yield could be potentially improved.

[10]The microchannel example of the previous section could have been presented this way since there are two important parameters as well, but to simplify the discussion, the parameters were examined separately. It is also important to note that altering the nominal values can result in a change in the tolerances and the boundaries of the region of variation.

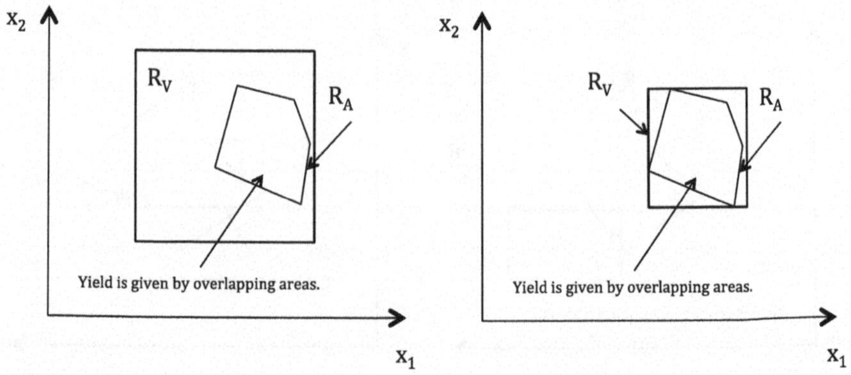

Fig. 10.23 Two device parameters being displayed along with the regions of variation and acceptance. The left side displays a situation where the region of variation is much larger than the region of acceptance. This results in a relatively low manufacturing yield. On the right side of this figure is shown where the variations of the parameters have been decreased. This results in a higher manufacturing yield

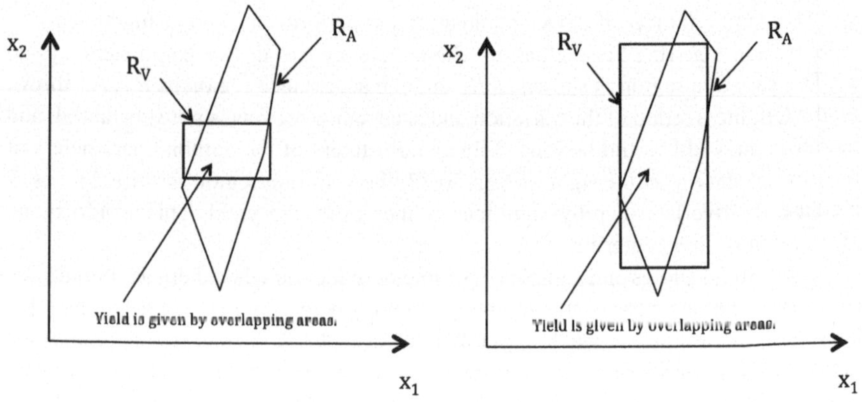

Fig. 10.24 Two device parameters being displayed along with the regions of variation and acceptance. The left side displays a situation where the region of acceptance is larger than the region of variation. This results in a relatively low manufacturing yield. However, the manufacturing cost is too high. The right shows a relaxation of the variation region which can result in a lower manufacturing cost. The yield may also be slightly lower as well

The yield could be increased by reducing the variations in the parameters but will likely come at the expense of an increased manufacturing cost.

In some circumstances, the manufacturing cost is not acceptable and must be further reduced even if this was to result in a reduction of the yield. Since the manufacturing cost is often related to the specific fabrication methods used for implementation, with lower amounts of variation in the parameters corresponding to increased costs, the opposite is also often the case wherein the manufacturing cost

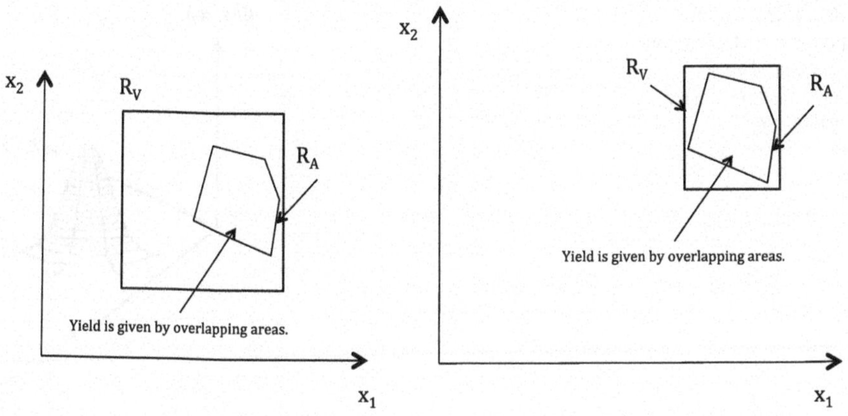

Fig. 10.25 Two device parameters being displayed along with the regions of variation and acceptance. The left side displays a situation where the region of variation is larger than the region of acceptance. This results in a relatively low manufacturing yield. On the right side of this figure is shown where the parameter nominal values are scaled to larger values and consequently the region of variation shrink relative to the nominal parameter values. This has the effect of increasing the overlap in the variation and acceptance regions and thereby the yield

can be reduced by using fabrication methods with higher levels of parameter variation. An increase in the device parameter variations is shown in the right side of Fig. 10.24 where the region of variation is shown having boundaries larger than on the left. As can be observed, this has the effect of decreasing the yield as expected but may be compensated for by the lower manufacturing cost.

An example of parameter device scaling for a device having two parameters is illustrated in Fig. 10.25. As shown on the left side, the overlap of the region of variation and acceptance results in a relatively low yield. However, if the two device parameters are scaled to larger values, that is, the nominal values are made larger, then the relative variations in these parameters may decrease in magnitude, as shown on the right of Fig. 10.25. As a result of this parameter scaling, the overlap of the acceptance and variation regions increases, as does the yield.

While the above representations are useful for visualizing the methods of design for devices having two device parameters, they leave out an important piece of information, namely, the probability distribution functions for each of the parameters. As noted previously, in most situations, the device parameters will be represented by Gaussian probability distribution functions as illustrated in Fig. 10.26 for a case where there are two parameters, and the joint probability distribution function is given by $\Phi(x_1, x_2)$. When viewed in two-dimensional space, the joint probability distribution function can be described by concentric rings, since there is no correlation in these parameters. If on the other hand there is a correlation between the two parameters, then the rings would have an elliptical shape.

The example of the microchannel in the previous section also only had one performance function, namely, the flow resistance of the device. In most cases,

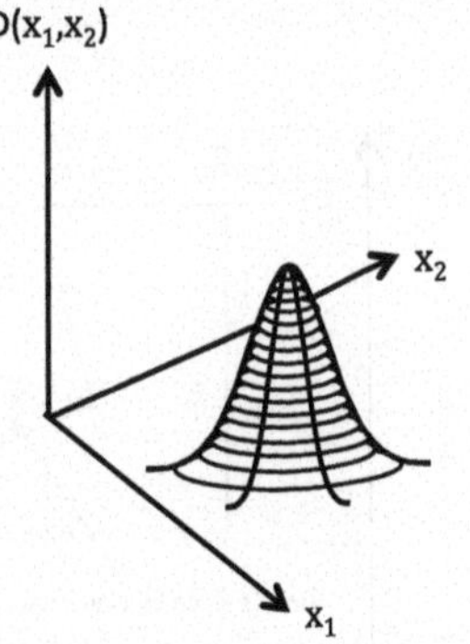

Fig. 10.26 A joint probability distribution function of two microsystems device parameters

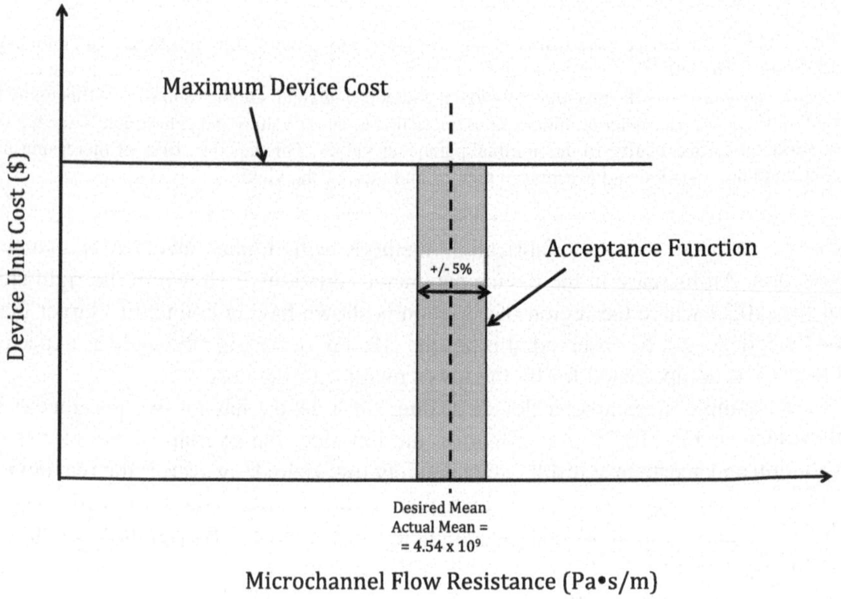

Fig. 10.27 A representation of the performance functions for the microchannel device including the flow resistance on one axis and the maximum device unit cost on the other axis

there will be more than one performance function. Even in the case of the microchannel, another performance criteria for the device would almost certainly be the maximum manufacturing cost and perhaps an upper and lower bound of the allowable device areas. While the device area and cost are directly related, the device area may have a constraint placed on it based on how the device fits into a larger system.

Therefore, another representation of the performance space could be illustrated in Fig. 10.27 wherein the flow resistance is on one axis and the maximum device cost is on the other axis. In this case, the device requirement as defined by an

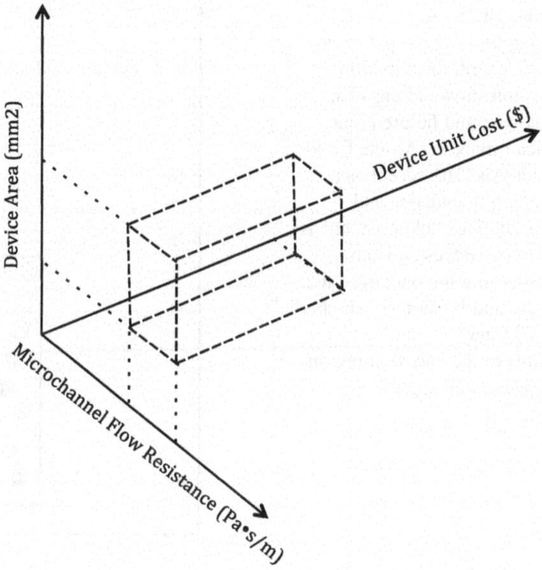

Fig. 10.28 A representation of the performance functions for the microchannel device including three performance criteria: flow resistance; maximum device unit cost; and device area

acceptance region would be that the flow resistance value should reside within the boundaries on the x-axis and the device unit cost is below the maximum value on the y-axis.

An even more inclusive performance function representation is shown in Fig. 10.28 wherein there are three performance criteria including flow resistance; maximum cost; and device area to define the acceptance region in three-dimensional performance space.

As can be appreciated by observing the above illustrations, it becomes impossible to illustrate device parameters or performance functions when there are more than three parameters (two if presented with the probability distribution function). Nevertheless, these situations can be easily handled using computers.

10.10 MEMS Design Methods Using Monte Carlo Techniques

Various methods of microsystems design such as design centering, device size scaling, etc. were discussed above and illustrated using the example of the microchannel device. While these discussions were useful to help understand the important concepts for how to design microsystems devices so as to improve yield, it was assumed that the location, size, and boundaries of the acceptance region, R_A, in the parameter space were known. However, as a practical matter, this information may not be known. It will be recalled that the mapping from parameter space to performance space is performed using the microsystems analysis equations and tools

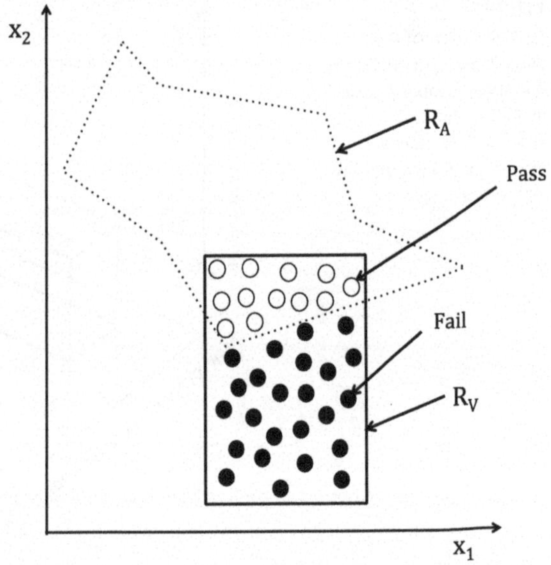

Fig. 10.29 A two-dimensional parameter space with the variation region shown along with passing and failure points determined by Monte Carlo analysis. The acceptance region boundaries and location are unknown, but as can be suggested from observing the pattern of the pass and fail points, a higher yield may be obtained by a shift of the variation region upward and slightly to the left

with the acceptance region being defined in the performance space. In contrast, the mapping of the acceptance region from performance space back into parameter space was performed using search techniques, such as regionalization or simplicial approximation covered in Chap. 9. While these searching techniques can be effective, they are computationally expensive. Monte Carlo analysis techniques can also be used for this operation and are more efficient and provide other significant benefits.

This section outlines systematic methods using the results from Monte Carlo analysis for locating the acceptance region in parameter space as well as methods for increasing the manufacturing yields. These methods are performed on computers and are essential when there are a relatively large number of device parameters and performance functions, that is, when the parameter and performance spaces are multidimensional.

The basic concept is illustrated in Fig. 10.29 where a two-dimensional parameter space is shown along with the points that have been evaluated for either pass or failure using Monte Carlo analysis. The variation region, R_V, and the acceptance region, R_A, are also shown. However, the location and boundaries of the acceptance region in parameter space are usually not known beforehand and therefore are portrayed using a dotted boundary line. The yield is given by the ratio of the number of passing points to the total number of points.

An important item observed in Fig. 10.29 related with the passing (signified by circles with unfilled centers) and failing (signified by black circles with black centers) points is that the pass points are grouped at the top third of the variation region, while the failing points are grouped in the bottom two-thirds of the variation region. This suggests that if the variation region was moved upward with respect to

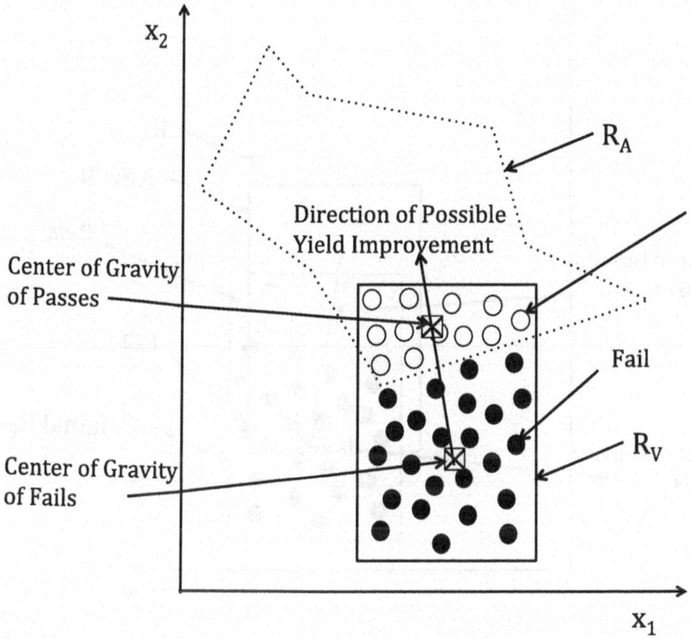

Fig. 10.30 The same two-dimensional parameter space with the variation region shown along with passing and failure points determined by Monte Carlo analysis as seen in Fig. 10.29. The center of gravity of both the passing and failing points has been calculated and plotted above. It is reasonable to conclude that the yield can be improved by shifting the region of variation along the line joining these centers of gravity

the acceptance region (i.e., higher values of parameter x_2), an improvement in yield may be obtained. Additionally, since the pass points are also weighted toward lower values of parameter x_1, an improvement in yield may also be gained by a shift of the variation region to the left. If the centers of gravity of both the passing and failing points were calculated, as shown in Fig. 10.30, then it would appear that the yield could be improved by moving along the line joining these centers of gravity.

If a sufficiently large number of samples (i.e., microsystems simulations) are performed, the centers of gravities of the passing and failing points will become equal to the actual centers of gravities of these two regions weighted by their respective probability distribution functions.

Figures 10.29 and 10.30 provide a high-level illustration of how the manufacturing yield can be improved using the results of Monte Carlo analysis, but more methodical techniques are needed for practical use, which are described in the following sections. The overarching goal is usually to maximize the manufacturing yield.[11]

[11]The goal may be more complicated, such as maximizing the yield within certain cost constraints. Nevertheless, these considerations can be taken into account during the design process.

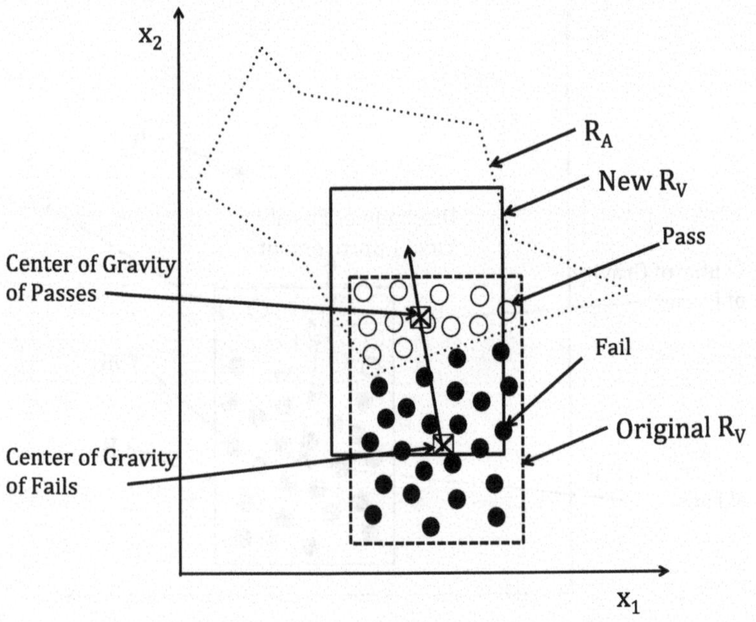

Fig. 10.31 An illustration of the use of centers of gravity algorithm for design centering. The center of the variation region is shifted to coincide with the center of gravity of the passes as shown

10.10.1 Design Centering Using Monte Carlo

As demonstrated in the section above, the results of Monte Carlo analysis provide insightful information about where to move the variation region (and thereby shift the device nominal parameter values) to increase the manufacturing yield. This section gives an in-depth explanation as how to perform design centering based on information derived from performing Monte Carlo analysis.

A centers of gravity algorithm is used on the Monte Carlo-generated data for design centering. This was graphically illustrated in Figs. 10.29 and 10.30. The focus now is one of the mechanics of how to employ this technique. As seen in Fig. 10.31, the centers of gravity of both the Monte Carlo analysis passes and fails have been calculated, and a line is drawn between these two centers. As noted previously, it appears that if the variation region is moved along this line toward the center of gravity of passes, this would result in a yield improvement. The movement of the variation region in Fig. 10.31 is to reposition the center of the variation region at the location of the center of gravity of the passes. After this movement, another Monte Carlo analysis is re-performed using the new nominal values of the device parameters. This process is repeated until there is confidence that the yield has been maximized. This is either when the movement results in a yield decrease or after repeating the process of back and forth along the centers of gravity joining line with increasingly smaller step lengths until some comfortable lower limit on the step lengths is reached.

Of course it is not known if the amount of shift portrayed in Fig. 10.31 is the correct amount or not since there is no information about the points in the new variation region from the original location. Therefore, an important question is what is the correct amount of the step length along the line between the centers of gravity? As it turns out, there is not a consensus answer to this question, and in most circumstances the shift amount is either to the center of gravity of the passes in the original variation region or some fraction of the distance from the two centers of gravity toward the center of gravity of the passes in the original variation region [10, 11].

It is also important to ensure that the yield estimates for the successive yield improvement activities performed by shifting the variation region actually result in an improvement in the yield with some level of confidence. This is important since there is always some amount of error involved in the yield estimates.

If it is assumed that the variation region is shifted in location in the parameter space from R_{V1} to R_{V2} and the yield estimates have values of $\widehat{Y}_1$ and $\widehat{Y}_2$ in the first and second variation regions, respectively, then the estimated change in yield can be written as [12]:

$$\widehat{\Delta Y} = \widehat{Y}_2 - \widehat{Y}_1. \tag{10.9}$$

Since the yield estimates $\widehat{Y}_1$ and $\widehat{Y}_2$ are random variables, their difference, $\widehat{\Delta Y}$, is also a random variable. Moreover, since the yield estimates are Gaussian distributions, their difference is also a Gaussian distribution.

The mean of the yield estimate difference with a sufficiently large number of estimates will converge to the actual yield difference, and therefore the mean of $\widehat{\Delta Y}$ is given by ΔY. The variance of the probability distribution function of $\widehat{\Delta Y}$ can be expressed as:

$$\mathrm{var}\left(\widehat{\Delta Y}\right) = \mathrm{var}\left(\widehat{Y}_1\right) + \mathrm{var}\left(\widehat{Y}_2\right) - 2\,\mathrm{cov}\left(\widehat{Y}_1, \widehat{Y}_2\right). \tag{10.10}$$

Figure 10.32 illustrates the probability density function of $\widehat{\Delta Y} = \widehat{Y}_2 - \widehat{Y}_1$. The confidence of $\widehat{Y}_2$ being greater than $\widehat{Y}_1$ is equal to the shaded area under the curve, that is, ΔY has a positive value. This area is given by:

$$\mathrm{Area} = \frac{1}{2} + \mathrm{erf}\left[\frac{\Delta Y}{\widehat{\sigma_{\Delta Y}}}\right], \tag{10.11}$$

where erf is the error function and $\widehat{\sigma_{\Delta Y}}$ is the square root of $\mathrm{var}\left(\widehat{\Delta Y}\right)$.

Since both ΔY and $\widehat{\sigma_{\Delta Y}}$ are unknown, it is necessary to use estimates of these values; specifically $\widehat{\Delta Y}$ is used to estimate ΔY and $\widehat{\sigma_{\Delta Y}}$ is estimated using:

$$\widehat{\sigma_{\Delta Y}} = \sqrt{\mathrm{var}\left(\widehat{\Delta Y}\right)} = \sqrt{\mathrm{var}\left(\widehat{Y}_1\right) + \mathrm{var}\left(\widehat{Y}_2\right)}, \tag{10.12}$$

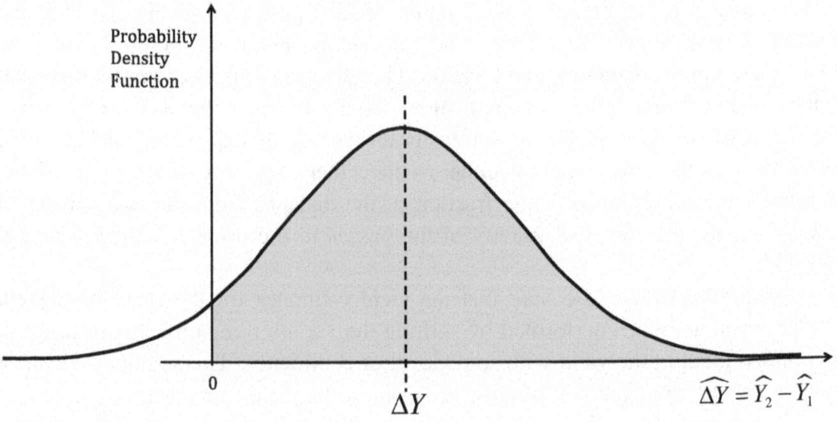

Fig. 10.32 The probability density function of $\widehat{\Delta Y} = \widehat{Y}_2 - \widehat{Y}_1$

where:

$$\mathrm{var}\left(\widehat{Y}_1\right) = \frac{\widehat{Y}_1\left(1 - \widehat{Y}_1\right)}{N} \text{ and } \mathrm{var}\left(\widehat{Y}_2\right) = \frac{\widehat{Y}_2\left(1 - \widehat{Y}_2\right)}{N}, \quad (10.13)$$

with N as the number of Monte Carlo samples or runs performed.

The above describes how to estimate the confidence that the yield is actually improving with the centers of gravity method, but there are also methods of improving the center of gravity method as well.

One method is to use correlated sampling [12]. This method is illustrated in Fig. 10.33 where a variation region, R_{V1}, is defined in parameter space having a center value of x_1^{o} and a Monte Carlo analysis is performed using pseudorandom points from the probability distribution function representing the region to develop the yield estimate Y_1. Instead of simply shifting the original variation region, R_{V1}, to a new region, R_{V2}, and using a completely new set of pseudorandomly selected points in a Monte Carlo analysis to determine the estimated yield in the new region, the approach of correlated sampling is to use the same pattern of points in region R_{V2} as was used in region R_{V1}. Specifically, each point in R_{V1} is translated by an amount equal to Δx to create the set of points used in the yield estimate of R_{V2} and having a center x_2^{o}. The technique for performing this function is to use the same seed for the pseudorandom number generation for points in R_{V2} as was used in R_{V1}. The estimated yield is then calculated for R_{V2}. It should be noted that it is assumed that the variation region boundaries are fixed.

The benefit derived from using correlated sampling is that it purposely introduces a positive correlation between the sample points in the two variation regions assuming Δx is small, and therefore there will also be a positive correlation in the two yield estimates. If a point in R_{V1} is passing, it is likely that the translated point in

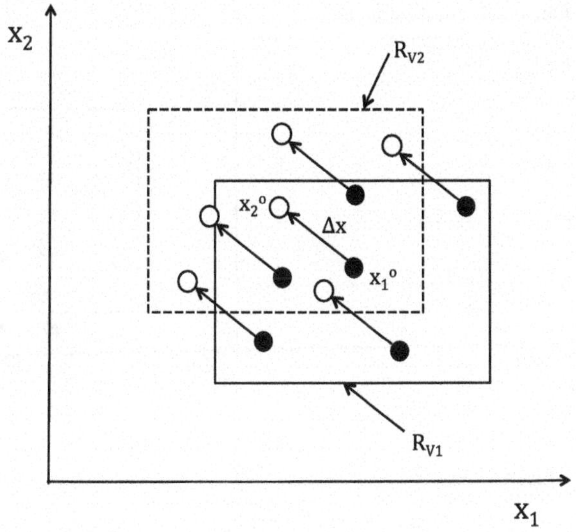

Fig. 10.33 An illustration of correlated sampling wherein the original variation region with pseudorandomly selected points. The points in R_{V1} are each translated by a small distance Δx to create a new set of points in variation region R_{V2}. This means that the points in R_{V2} are positively correlated with the points in R_{V1}

R_{V2} will also be passing. The outcome of this is that even though each of the yield estimates for R_{V1} and R_{V2} may have some error, they will likely have the same amount of error. This means that their difference will be relatively accurate. That is, using this form of design centering using the centers of gravity algorithm is likely to provide a higher level of confidence in yield improvement than if the yield estimates were calculated using two independent Monte Carlo analyses.

The net effect of selected positively correlated points in the variation regions is to reduce the value of $\widehat{\sigma_{\Delta Y}}$. Since $\widehat{\sigma_{\Delta Y}}$ is the square root of $\text{var}\left(\widehat{\Delta Y}\right)$, which is the denominator in Eq. (10.11), if $\widehat{\sigma_{\Delta Y}}$ is reduced, the area under the curve shown in Fig. 10.32 is increased. The reduction in $\text{var}\left(\widehat{\Delta Y}\right)$ is due to the increase in the value of $2\text{cov}\left(\widehat{Y_1}, \widehat{Y_2}\right)$ in Eq. (10.10).

The confidence in correct ranking of the estimated yields in the two correlated variation regions can be calculated using [12]:

$$\text{Area} = \frac{1}{2} + \text{erf}\left[\frac{\sqrt{\widehat{\text{var}}\left(\widehat{\Delta Y}\right)}}{\widehat{\Delta Y}}\right]^{-1}, \tag{10.14}$$

where: $\widehat{\text{var}}\left(\widehat{\Delta Y}\right) = \text{var}\left(\widehat{Y_1}\right) + \text{var}\left(\widehat{Y_2}\right) - 2\widehat{\text{cov}}\left(\widehat{Y_1}, \widehat{Y_2}\right)$.

The other technique to increase the confidence of the yield estimates is called the common points method. This concept is illustrated in Fig. 10.34 where the first variation region, given by R_{V1}, has Monte Carlo analysis performed on the pseudorandomly selected sample points to calculate an estimate yield, Y_1. A new

Fig. 10.34 An illustration of the common points method

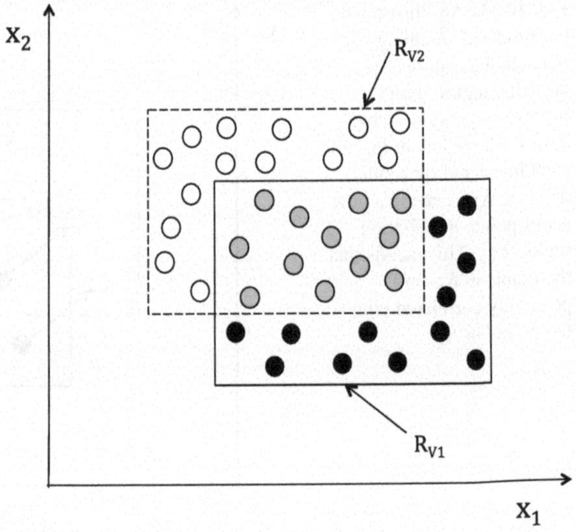

variation region is defined using the centers of gravity algorithm resulting in R_{V2}. There is a significant overlap between R_{V1} and R_{V2}. Importantly, there are a number of points in this overlap region (shown as gray points) that are shared between the two regions. In the common points method, since the points in the overlap region have already been determined to be passing or failing based on the results of the Monte Carlo yield analysis of variation region R_{V1}, these points do not need to be reevaluated in the Monte Carlo yield analysis of variation region R_{V2}. The confidence in the ranking of the estimated yields of R_{V1} and R_{V2} can be calculated using Eq. (10.11). The net effect of selected positively correlated points in the variation regions is to increase the value of ΔY in Eq. (10.11). A significant advantage of the common points method is that it reduces the computational effort. This is due to the fact that the points in the overlap region do not have to be reevaluated using Monte Carlo analysis.

10.11 Sensitivity Analysis

Sensitivity analysis is a useful and important microsystems device design method for a number of reasons including improvement of the manufacturing yield and reduction of the manufacturing cost. A simple example of the use of sensitivity analysis will be reviewed followed by a more general discussion.

Assume for simplicity that a microsystems device having only two device parameters has been designed and the nominal parameter values have been selected. It is also assumed that a Monte Carlo analysis has been performed at the nominal parameter values and the manufacturing or performance function for this point in

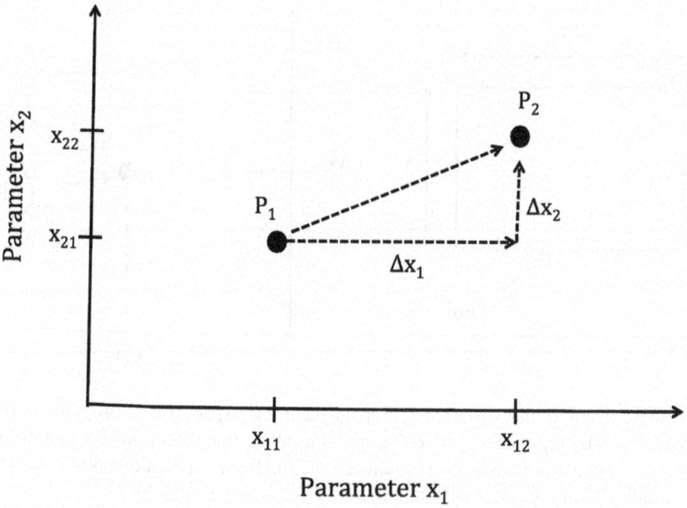

Fig. 10.35 An illustration of evaluation of the performance or manufacturing function at a nearby point, P_2, based on the results of the analysis at a near proximity original point, P_1, in parameter space

parameter space has been estimated, hereby labeled as P_1. The performance function could be the yield or some other type of function of importance in the manufacturing of the microsystems device, such as the device unit cost. It is desired to estimate the microsystems device performance function value at a nearby point in parameter space, namely, P_2. This involves a translation in the device parameter space from the original point at x_{11} and x_{21} to a new position point at x_{12} and x_{22} as illustrated in Fig. 10.35. The movement from the original point to the nearby point involves a translation in the two device parameters given by Δx_1 and Δx_2, where $\Delta x_1 = x_{12} - x_{11}$ and $\Delta x_2 = x_{22} - x_{21}$. A Taylor's series can be used to approximate the performance function at the new point based on the value at the original point in parameter space as given by:

$$P_2 \cong P_1 + \left.\frac{\delta P}{\delta x}\right|_1 \Delta x, \qquad (10.15)$$

where the Taylor's series has been truncated after the first partial derivative term for simplicity, $\frac{\delta P}{\delta x}$ is the differential sensitivity of the performance function evaluated at the original point x_{11} and x_{21}, and Δx is a matrix given by $[\Delta x_1 \; \Delta x_2]$.

Therefore, if the differential sensitivity is given as:

$$\frac{\delta P}{\delta x} = [3 \; 1], \qquad (10.16)$$

then the differential performance function estimate can be written as:

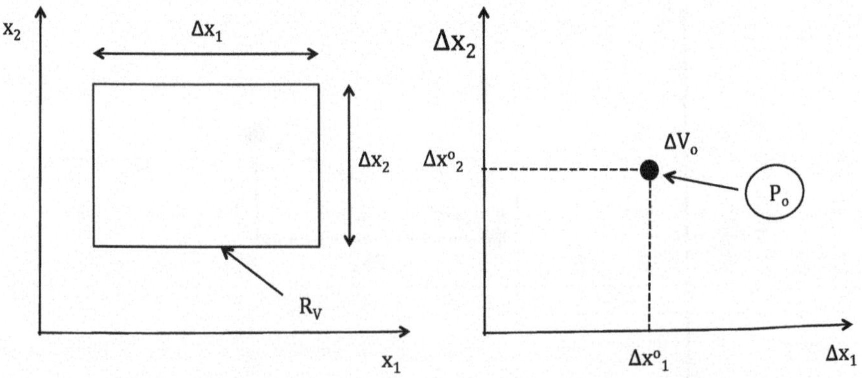

Fig. 10.36 On the left is the variation region in parameter space for microsystems having two device parameters. The right shows the point, given by ΔV_o, that represents the variation region in variation space. An evaluation of the performance function for this variation point in variation space is performed using Monte Carlo analysis, and the value is represented by P_o

$$\Delta P = P_2 - P_1 \cong \left.\frac{\delta P}{\delta x}\right|_1 \Delta x = (3 \;\; 1)\begin{pmatrix} \Delta x_1 \\ \Delta x_2 \end{pmatrix} = 3\Delta x_1 + \Delta x_2. \tag{10.17}$$

As noted above, sensitivity analysis can be used in the device parameter space to estimate the performance functions, such as the yield, at nearby locations in that space. It can be used to analyze the effects of parameter variations on performance functions as well.

Consider the two-dimensional device parameter space shown in Fig. 10.36 having the variations in both of the parameters, x_1 and x_2, plotted on the left and the variation region, labeled as R_v, also shown. These variations can also be plotted in variation space as shown in the right side of Fig. 10.36 wherein the parameter variations can be represented by a single point given by ΔV_o. The performance function, whether it be the yield, device unit manufacturing cost, or some other metric, can be estimated at the point in variation space labeled ΔV_o with a value given as P_o as shown in Fig. 10.36 using Monte Carlo analysis.

Using the truncated Taylor's series from above, it is a relatively simple matter to find the relative values of the performance function at points in near proximity of P_o and thereby determine the direction where the performance function may be increasing (or decreasing) most rapidly. Such a scenario is shown in Fig. 10.37 wherein a number of point estimates of the performance function have been plotted around the central point P_o. Again, each of these point estimates can be easily evaluated using the differential sensitivity of the performance function (see Eq. (10.15)) relative to the original point P_o. Moreover, the differences in the values of these point estimates from that the original point can also be easily calculated. These differences can be improvements (or worsening) of performance function from the original point in variation space. If the performance function is the parametric yield, then the preferred direction from the original point would be based on yield improvement, whereas if the performance function was the device manufacturing cost, the favored

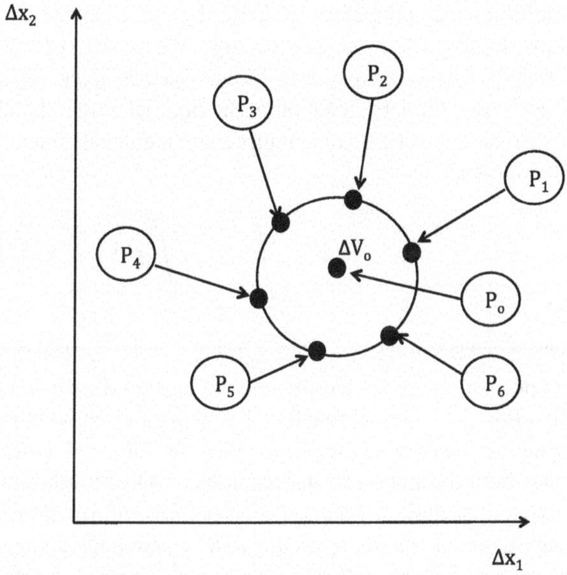

Fig. 10.37 An illustration of the determination of the direction for optimizing the performance function by evaluating the performance function at nearby locations to the original point in variation space using differential sensitivity analysis. Each of these points is then combined using a weighted average to determine the optimal direction. Points along that direction are then evaluated using the same techniques until a local minimum or maximum is located. Then a new Monte Carlo analysis is performed and the process repeated

direction would be based on cost reductions. Depending on the performance function in question, improvements or reductions may be desirable. Since the performance function estimates along the perimeter of the circle shown in Fig. 10.37 are made using a truncated Taylor's series, the radius is kept relatively small, such as some small fraction (e.g., 1/10th) of the value of ΔV_o. A weighted average of those points P_1 through P_6 relative to the original point can be evaluated to estimate the optimal search direction. Subsequently, points along the line in the search direction are evaluated until the performance function is maximized (or minimized), and then a new Monte Carlo analysis is performed at this new location in variation space, and the process is repeated until the design criteria have been satisfied.

While the examples shown in Figs. 10.35, 10.36, and 10.37 involve only two parameters, these same techniques can be used for microsystems devices having any number of parameters. In those cases, the parameter space would be multidimensional, and the performance function would be evaluated along the surface of a hypersphere in variation space.

10.11.1 Generalized Sensitivity Analysis Methods

This section reviews more generalized sensitivity analysis methods, particularly in use for yield improvement activities. The yield gradient (i.e., first derivative of the yield function) can be used to provide useful information about what direction in multidimensional parameter space the yield function is increasing or decreasing (i.e., the slope of the yield function), and the second-order yield gradient (i.e., second derivative of the yield function) provides information about what direction in multidimensional parameter space the yield function is increasing or decreasing the most rapidly (i.e., rate of change of the slope) [12]. Yield gradients can be easily computed from the results of Monte Carlo analysis and can be used to maximize the yield based on determination of the optimal nominal and variation parameter values.

In the general case of a multivariant parameter space, the yield can then be written as follows:

$$Y = \int_{R_V} \cdots\cdots \int [g(X)]\,\phi(X, X^{o})\,dX, \tag{10.18}$$

where X^o is defined as $X^o = x_1{}^o, x_2{}^o, \ldots x_N{}^o$ and represents the nominal parameter values for the N device parameters in multidimensional parameter space, $g(x)$ is a function that is equal to either 1 or 0 depending on whether each point in parameter space is determined to either pass or fail, and $\phi(X, X^o)$ is the joint probability distribution function of the parameter values in multidimensional space. The probability distribution function incorporates information about the variations of each parameter in the form of the PDF's standard deviation (assuming the PDF is a normal distribution). Taking the partial derivative of this yield function with respect to a parameter of interest gives the following:

$$\frac{\delta Y}{\delta x_i^o} = \int_{R_V} \cdots\cdots \int [g(X)]\,\frac{\delta\phi(X, X^o)}{\delta x_i^o}\,dX. \tag{10.19}$$

It is observed that Eq. (10.19) is taking the partial derivative of the yield function with respect to a specific parameter. Equation (10.19) can be rewritten as follows:

$$\frac{\delta Y}{\delta x_i^o} = \int_{R_V} \cdots\cdots \int \left[\frac{g(X)}{\phi(X, X^o)}\,\frac{\delta\phi(X, X^o)}{\delta x_i^o}\right]\phi(X, X^o)\,dX, \tag{10.20}$$

by multiplying the numerator and denominator by $\phi(X, X^o)$. It is assumed that the function $\phi(X, X^o)$ is not zero and can be differentiated. The integrals in these equations are difficult to develop analytical solutions, and therefore other techniques are needed.

It will be recalled from Monte Carlo analysis that the estimated yield was given as:

$$\widehat{Y} = \frac{1}{N}\sum_{i=1}^{N} g(x_i), \tag{10.21}$$

where $g(x_i)$ is either 1 or 0 depending on whether the parameter sample point is a pass or fail. Comparing this expression to that of Eq. (10.20) shows that the yield gradient can be estimated as follows:

$$\frac{\delta \widehat{Y}}{\delta x_i^o} = \frac{1}{N} \sum_{i=1}^{N} \left[\frac{g(x_i)}{\phi(x_i, x^o)} \frac{\delta \phi(x_i, x^o)}{\delta x_i^o} \right] \tag{10.22}$$

Using this equation and the results from the Monte Carlo analysis allows the estimated yield to be easily computed.

A similar mathematical manipulation can be performed to develop the second-order partial derivative of the yield estimate as follows:

$$\frac{\delta^2 \widehat{Y}}{\delta x_j^o \delta x_i^o} = \frac{1}{N} \sum_{i=1}^{N} \left[\frac{g(x_i)}{\phi(x_i, x^o)} \frac{\delta^2 \phi(x_i, x^o)}{\delta x_j^o \delta x_i^o} \right]. \tag{10.23}$$

In the more general case of a multivariant parameter space where there may be correlations between device parameters, the yield function can be written as follows:

$$Y = \int_{R_V} \ldots\ldots \int [g(x)] \Phi(x, \mathbf{X}) dx, \tag{10.24}$$

where $\mathbf{X}$ is a vector representing the nominal parameter values and device parameter variations, as well as any correlations that may exist between each of the N device parameters in multidimensional parameter space, $g(x)$ is a function that is equal to either 1 or 0 depending on whether each point in parameter space is determined to either pass or fail, and $\Phi(x, \mathbf{X})$ is the joint probability distribution function of the parameter values in multidimensional space with a point in parameter space given by x, and $\mathbf{X}$ represents the dependency of the joint probability distribution function on the parameter nominal and variation values and any correlations.

Following a procedure as performed above allows the estimated yield gradient to be expressed in more general terms as follows:

$$\frac{\delta \widehat{Y}}{\delta \mathbf{X}_j} = \frac{1}{N} \sum_{i=1}^{N} \frac{g(x_i)}{\phi(x_i, \mathbf{X}_j)} \frac{\delta \phi(x_i, \mathbf{X}_j)}{\delta \mathbf{X}_j}. \tag{10.25}$$

Additionally, without much more computational effort, the second derivative of the yield function for the more general care can be evaluated as follows:

$$\widehat{\frac{\delta^2 Y}{\delta \mathbf{X}_i \delta \mathbf{X}_j}} = \frac{1}{N} \sum_{i=1}^{N} \frac{g(x_i)}{\phi(x_i, \mathbf{X}_j)} \frac{\delta^2 \phi(x_i, \mathbf{X}_j)}{\delta \mathbf{X}_i \delta \mathbf{X}_j}. \tag{10.26}$$

The method of yield maximization using yield gradients has similarity to the other methods described above. Specifically, an initial trial design with the associated nominal and variation values of each parameter is first selected based on an initial

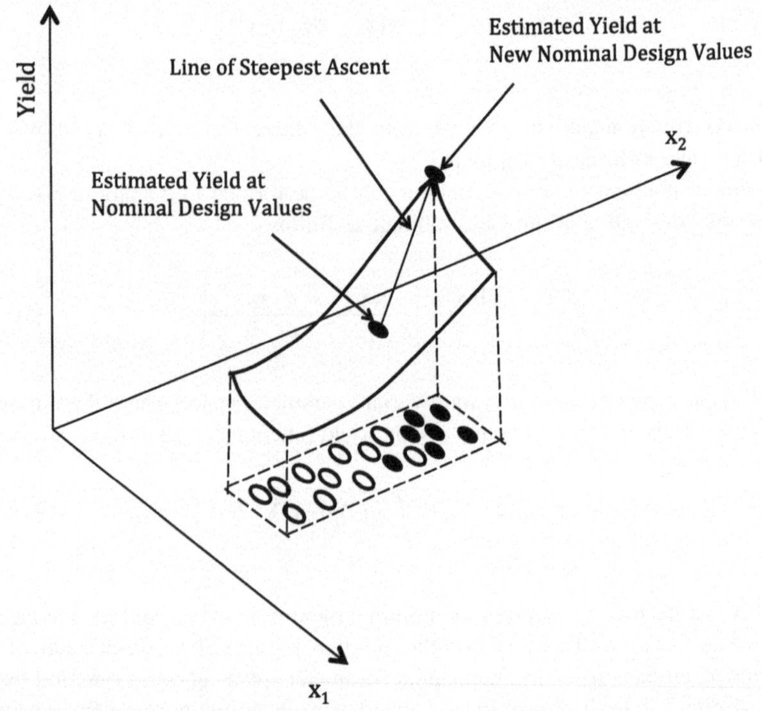

Fig. 10.38 An illustration of the use of yield gradients for determining the direction for improvement of the yield

design exercise. The yield is then estimated using Monte Carlo analysis, and the yield gradients are calculated. The yield gradients indicate in which direction of the parameter space the yield is increasing as well as how quickly it is increasing. This allows new parameter nominal values to be selected. That is, the results of the Monte Carlo and yield gradient calculations are used to select a new design that may have a higher yield.

This process is illustrated in Fig. 10.38 where there are two device parameters shown as a function of the yield, and a Monte Carlo analysis was performed on the variation region around the nominal, with passing and failing points from the analysis shown as either black or white circles, respectively. The yield gradients provide information about where and what direction on the yield function that the yield is maximized. A new point is then chosen as the new nominal value for the parameters, and the Monte Carlo analysis is performed again. This iterative process of design, evaluation using Monte Carlo analysis to estimate yield and yield gradients, followed by selecting a new design, and repeating, proceeds until the designer is satisfied with the resultant design. While the techniques covered in this section appear daunting, they are easily performed on computers.

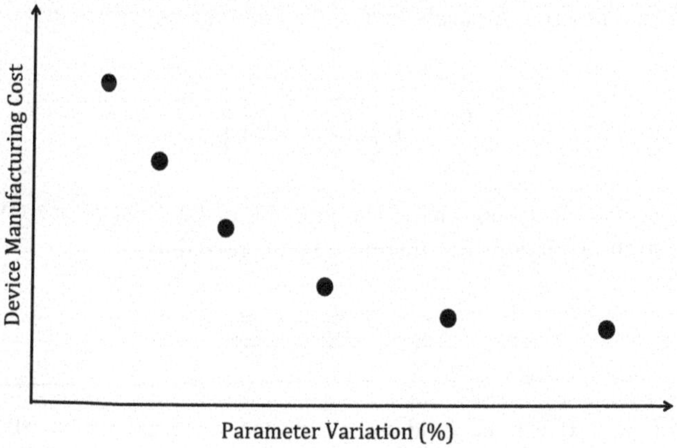

Fig. 10.39 An illustration of the relationship between the variation of a parameter and the manufacturing cost

10.11.2 Optimizing Manufacturing Cost Function

In some microsystems situations, it may be the device cost that is to be minimized, and making parameter variation assignments is a common approach for obtaining this goal. It is assumed that the nominal parameter values are fixed, and the selection of device parameter variation values has an impact on the average cost of acceptable manufactured microsystems. The latter assumption is usually true in nearly all types of manufacturing environments including MEMS microsystems. In most manufacturing environments, the cost of each device made can be assumed to vary somewhat inversely with the manufacturing variations as illustrated in Fig. 10.39.[12] Therefore, the total cost, C, of the parameters in the microsystems device can be written approximately as:

$$C = \sum_{i=1}^{K} \alpha_i + \frac{\beta_i}{3\sigma_i}, \tag{10.27}$$

where α_i and β_i are manufacturing fixed cost constants that depend on the fabrication process(es) used for a specific device parameter, $3\sigma_i$ is the variation in a specific device parameter (assumed to be 3σ but could be a different multiple of σ depending on circumstance), and K is the number of parameters in the device.

[12] In MEMS microsystems manufacturing, the relationship between cost and parameter variations is usually more complicated than shown in Fig. 10.39. The actual relationship often more resembles a descending staircase wherein the variations using a specific set of fabrication equipment have a relatively fixed cost over a range of parameter variations values that increase or decrease with the use of other sets of fabrication equipment having higher or lower resolution.

The total cost function for each device, C_{d}, can then be written as:

$$C_{\mathrm{d}} = \frac{1}{Y}\left[\sum_{i=1}^{K} \alpha_i + \frac{\beta_i}{3\sigma_i}\right], \tag{10.28}$$

where Y is the manufacturing yield. The gradient of the total device cost function with respect to the variations can therefore be expressed as:

$$\frac{\delta C_{\mathrm{d}}}{\delta 3\sigma_i} = \frac{Y\frac{\delta C}{\delta 3\sigma_i} - C\frac{\delta Y}{\delta 3\sigma_i}}{Y^2}. \tag{10.29}$$

In solving for Eq. (10.29), the yield and yield gradient terms can be estimated from the results of the Monte Carlo analysis, and the terms dealing with the cost function and its partial derivative with respect to the variation can be easily determined from Eq. (10.27). It should be noted that the cost model will notably have a constraint on the yield that can be expressed as $Y \geq Y_{\mathrm{min}}$ wherein Y_{min} is the minimum allowable yield.

10.12 Summary

This chapter has reviewed the techniques that can be used to design MEMS microsystems devices so as to manage the device parameter variations that are inherent in micro- and nano-fabrication. Design for manufacturability was first discussed as it relates to device and process sequence design. Some general recommendations for developing device designs that adhere to design for manufacturability principles were then covered. A review of the design techniques used in MEMS microsystems included design centering; device parameter variation reduction; device size scaling; acceptance region increase; and best practices for layout. Each of these was substantiated with simple examples in a one-dimensional parameter space, followed by a more generalized discussion for multidimensional spaces. The use of Monte Carlo analysis techniques for these design methods was then reviewed. Lastly, sensitivity analysis for design manufacturing or performance function improvement was outlined.

References

1. S.P. Vudathu, K.K. Duganapalli, R. Laur, D. Kubalinska, A.B. Gerstner, Parametric Yield Analysis of MEMS VIA Statistical Methods. DTIP of MEMS and MOEMS, Stresa, IT, Apr. 26–28, (2006)
2. L. Gao, Q.-A. Huang, Modeling of the effect of process variations on a micromachined doubly-clamped beam. Micromachines J. **8**, 81 (2017). https://doi.org/10.3390/m30081

3. L. Rong, B. Paden, K. Turner, MEMS resonators that are robust to process-induced variations. J. Microelectromech. Syst. **11**, 505–511 (2002)
4. Z. Luo, X. Wang, M. Jin, S. Liu, MEMS gyroscope yield simulations based on Monte Carlo method. Proceeding of 2012 IEEE 62nd Electronic Components and Technology Conference, May 29 – June 1, (2012), pp. 1636–1639
5. S.D. Senturia, *Microsystems Design* (Springer, New York, 2001)
6. R. Dutton, "CRC Electronic Design Automation for IC Handbook. Vol II, Chapter 25, 2016
7. R.B. Bird, W.E. Stewart, E.N. Lightfoot, *Transport Phenomena* (Wiley, New York, 1960)
8. A. Ttimoshenko, S. Woinowky-Krieger, *Theory of Plates and Shells*, 2nd edn. (McGraw-Hill, New York, 1959)
9. P.R. Gray, P.J. Hurst, S.H. Lewis, R.G. Meyer, *Analysis and Design of Analog Integrated Circuits*, 5th edn. (Wiley, New York, 2009)
10. R.S. Soin, R. Spence, Statistical exploration approach to design centering. Proc. IEEE, Pt. G **127**(6), 260–269 (1980)
11. I.R. Ibbotson, E. Compton, D. Boardman, Improved statistical design centering for electrical networks. Electron. Lett. **20**(19), 757–758 (1984)
12. R. Spence, R.S. Soin, *Tolerance Design of Electronic Circuits* (Imperial College Press, London, 1997)

Index

M. Huff, *Process Variations in Microsystems Manufacturing*, Microsystems and Nanosystems, https://doi.org/10.1007/978-3-030-40560-1

The manufacturer's authorised representative in the EU is Springer Nature Customer Service Centre GmbH, Europaplatz 3, 69115 Heidelberg, Germany. If you have any concerns regarding our products, please contact ProductSafety@springernature.com

Printed and bound by CPI Group (UK) Ltd, Croydon, CR0 4YY
05/10/2026
02228120-0018